FIRE IN ECOSYSTEMS OF BOREAL EURASIA

FORESTRY SCIENCES

Volume 48

The titles published in this series are listed at the end of this volume.

Fire in Ecosystems of Boreal Eurasia

Edited by

JOHANN GEORG GOLDAMMER

Max Planck Institute for Chemistry-Freiburg University, Mainz-Freiburg, Germany

and

VALENTIN V. FURYAEV

V.N. Sukachev Institute of Forestry and Timber, Russian Academy of Sciences, Krasnoyarsk, Russia

KLUWER ACADEMIC PUBLISHERS

DORDRECHT / BOSTON / LONDON

A C.I.P. Catalogue record for this book is available from the Library of Congress.

ISBN 978-90-481-4725-0

Published by Kluwer Academic Publishers,
P.O. Box 17, 3300 AA Dordrecht, The Netherlands.

Kluwer Academic Publishers incorporates
the publishing programmes of
D. Reidel, Martinus Nijhoff, Dr W. Junk and MTP Press.

Sold and distributed in the U.S.A. and Canada
by Kluwer Academic Publishers,
101 Philip Drive, Norwell, MA 02061, U.S.A.

In all other countries, sold and distributed
by Kluwer Academic Publishers Group,
P.O. Box 322, 3300 AH Dordrecht, The Netherlands.

Printed on acid-free paper

CONTENTS

IV. Geographical Analysis

V. Pyrological Classification of Landscapes, Sites and Fuel Types

VI. Fire Characteristics: Behavior and Modelling

Preface

One of the first priority areas among joint East/West research programs is the rational use of natural resources and sustainable development of regions. In the boreal zone of North America and Eurasia forests are economically very important and, at the same time highly vulnerable to disturbances. Because of its size and ecological functions the boreal forest zone and its most dynamic disturbance factor - fire - play an important role in ecosystem processes on global scale.

Interest within the global change research community in Northern Eurasia (Fennoscandia, European Russia, Siberia, and the Far East of Russia) has grown dramatically in the last few years. It is a vast area about which very little is known. It is a region where temperature rise due to anthropogenic climate forcing is predicted to be the greatest, and where the consequent feedbacks to the atmosphere are potentially large. In addition, it is poised to undergo rapid economic development, which may lead to large and significant changes to its land cover. Much of this interest in Northern Eurasia, as in the high latitude regions in general, is centerd on its role in the global carbon cycle, which is likely to be significantly affected under global change.

New research initiatives between Western and Eastern countries have been designed to address a series of phenomena, problems and management solutions. Cooperative research agreements under the International Geosphere-Biosphere Programme (IGBP) and the International Global Atmospheric Chemistry (IGAC) project, in conjunction with the International Boreal Forest Research Association (IBFRA), provide instruments to initiate joint research.

In June/July 1993 two scientific events established a new platform of joint East-West fire research, the international scientific conference "Fire in Ecosystems of Boreal Eurasia" and the start of the Fire Research Campaign Asia-North (FIRESCAN).

This volume presents the results of the conference which aimed to compile, interpret, and discuss the state of knowledge of the role and impacts of fire in boreal ecosystems, with special emphasis on Eurasia. For the first time this publication makes available the in-depth knowledge in fire science achieved in the former Soviet Union and in the Russian Federation. Together with the contributions from Fennoscandia and boreal North America this volume aims to stimulate a new era of pan-boreal fire research, especially considering the need to put basic and specific local aspects of fire ecology into the broader context of the newly emerging global fire science.

The book is particularly aimed at supporting the upcoming IGBP Northern Eurasia Study, a joint effort of several IGBP Core Projects, the Biospheric Aspects of the Hydrological Cycle (BAHC), International Global Atmospheric Chemistry (IGAC) and Global Change and Terrestrial Ecosystems (GCTE) Projects.

The conveners of the conference and editors of this volume are indebted to the sponsors and organizers of the meeting as well as to the inputs by the fire science community. Particular acknowledgements are given to the VOLKSWAGEN Foundation (Germany) which supported the Max Planck Institute for Chemistry, Biogeochemistry Department, and the V.N. Sukachev Institute for Forestry and Timber, Siberian Branch, Russian Academy of Sciences, Krasnoyarsk, Forest Fire Laboratory, for hosting the conference, Dimitri Odintsov, Deputy Chief of the Federal Forest Service of Russia, and Nikolay A. Andreev, Eduard P. Davidenko, and Nikolay A. Kovalev, all from the Aerial Forest Fire Protection Service

Avialesookhrana, actively supported the organization and logistics of the conference and the international Bor Forest Island Fire Experiment in the frame of the Fire Research Campaign Asia--North (FIRESCAN) which took place immediately after the conference (cf Annex III).

The International Boreal Forest Research Association (IBFRA) provided the frame for countries and scientists cooperating in the Stand Replacement Fire Working Group to actively participate in the conference (cf Annex II). The opening remarks by Eldon W. Ross, former president of IBFRA, underscored the importance of this very first active program of IBFRA (cf Annex I).

The conveners gratefully acknowledge the tremendous efforts by colleagues who helped to translate and interpret the Russian manuscripts into English. While Ms. Irina Savkina, interpreter at the V.N. Sukachev Institute for Forest, provided the base translations, much work had to be done to identify appropriate English terminology without violating the Russian style of scientific writing.

First, the editors are intebted to Susan Conard (Riverside Fire Laboratory, USDA Forest Service) for helping edit many more manuscripts than previously envisioned. A tremendous amount of work was done by Frank Albini (University of Montana) in reviewing all manuscripts dealing with the mathematical modeling of forests, insects and fire. Anders Granström (University of Umeå), Günter Helas (Max Planck Institute for Chemistry), Eino Mälkönen (The Finnish Forest Research Institute), Brian J. Stocks and Bruce Lawson (both Canadian Forest Service), and Ross W. Wein (Canadian Circumpolar Institute) also devoted considerable time to help review and edit Russian manuscripts.

The Max Planck Institute for Chemistry, the Canadian Forest Service (Forest Fire Research, Ontario, B.J.Stocks) and the US Forest Service, Washington (W. Sommers and M. Fosberg), provided financial support for producing high-quality figures and color plates. Ms. Soo Ing, Georg Buchholz, and Hans Page, all students at Freiburg University, helped type and copy-edit the manuscripts.

Freiburg (Germany) and Krasnoyarsk (Russian Federation)

May 1996

Johann Georg Goldammer
Valentin V. Furyaev

J.G.Goldammer and V.V.Furyaev (eds.), Fire in Ecosystems of Boreal Eurasia, 1-20.

Fire in Ecosystems of Boreal Eurasia: Ecological Impacts and Links to the Global System

J.G. Goldammer [1] and V.V. Furyaev [2]

1. Introduction

The circumpolar belt of the boreal zone stretches in two broad trans-continental bands across North America and Eurasia. The northern boundary of the zone corresponds to the July 13°C isotherm, while the southern boundary is limited by the July 18°C isotherm (Kuusela 1990). The boreal zone has been classified into three sub-zones, the maritime, continental and high-continental sub-zones. The maritime sub-zone has mean summer temperatures of 10-15°C, winter temperatures of 2-3°C, and annual precipitation of 400 to 800 mm. The continental sub-zone has long, cold winters with mean temperatures from -20 to -40°C, and summer mean temperatures from 10 to 20°C. The growing season is between 100 and 150 days, and annual precipitation ranges between 400 and 600 mm. The high continental sub-zone covers the largest portion of the boreal zone and is characterized by more extreme winters and milder summers. In Europe, the influence of maritime airmasses decreases from west to east, reaching West Siberia as far as the Yenisei river. East Siberia and the Far East are characterized by high-continental climate.

The definition of the boundaries of the boreal zone is often considered synonymous with the occurrence of northern coniferous forests. However, the northern forest limit which exceeds 70°N latitude only in Eurasia, in fact, is a broad forest-tundra ecotone characterized by the transition between tundra associations and discontinuous forest cover (Treter 1993). The southern limit of the boreal forest zone is at ca. 45°N. As a consequence of geography (size of the continent, orography, oceanic and atmospheric circulation) ca. 40% of the North American boreal forest is between 45 and 55°N, much further south than in Eurasia's boreal forests which are mainly north of 55°N latitude (cf map on p. 496).

The distinct climatic seasonality with a short vegetation period and low average temperatures facilitates the accumulation of organic layers and widespread permafrost soils in the boreal zone. Together with topography both features critically determine species composition and dynamics of the forest landscapes in which bogs and grasslands are intermixed (Shugart et al. 1992; Treter 1993).

[1] Max Planck Institute for Chemistry, Biogeochemistry Department, Fire Ecology Research Group, c/o Freiburg University, D-79100 Freiburg, Germany

[2] V.N. Sukachev Institute for Forestry and Timber, Siberian Branch, Russian Academy of Sciences, 660036 Krasnoyarsk, Russian Federation

The ecologically and economically most important coniferous tree species are pine (*Pinus* spp.), larch (*Larix* spp.), spruce (*Picea* spp.), and fir (*Abies* spp.); the main broadleaf tree species are birch (*Betula* spp.), poplar (*Populus* spp.), and alder (*Alnus* spp.) (Nikolov and Helmisaari 1992).

Boreal forests cover ca. $1.2x10^9$ ha, of which $920x10^6$ ha are closed forest. The latter number corresponds to ca. 29% of the world's total forest area and to 73% of its coniferous forest area (ECE/FAO 1985). About $800x10^6$ ha of boreal forests with a total growing stock (over bark) of ca. 95 billion m^3 are exploitable (41% and 45% respectively of the world total). The export value of forest products from boreal forests is ca. 47% of the world total (Kuusela 1990).

The carbon stored in boreal ecosystems corresponds to ca. 37% of the total terrestrial global carbon pool (plant biomass and soil carbon) (Apps et al. 1993). Thus, the magnitude of the boreal forest area suggests that it may play a critical role in the global carbon budget and its influence on the climate system of the earth (potential sink or source of atmospheric carbon).

More than seventy percent of the global boreal forest cover is in Eurasia, mainly in the Russian Federation, and represent the largest unbroken forested area of the globe; the remainder is in Canada and Alaska, and relatively small areas of boreal forests are found in the North East of China and in the Nordic countries (Fennoscandia).

The total area of the Russian Forest Fund comprises ca. $1{,}181x10^6$ ha, of which $886x10^6$ ha (= 75,0%) are forested and $763x10^6$ ha (= 64%) are stocked (Federal Forest Service of Russia 1994). The Federal Forest Service of Russia exercises control over 94% of the total Forest Fund area and 91% of the total growing stock of Russia. Depending on the criteria used to define "boreal forest", the area of closed boreal forest in the Russian Federation varies from 400 to $600x10^6$ ha (Pisarenko and Strakhov 1993). These numbers correspond to a 43-65% share of the world's closed boreal forest.

2. Disturbances in Boreal Ecosystems

Over evolutionary time periods boreal ecosystems have been subjected to climate changes, and species were forced to migrate in accordance with advancing and retreating glacial land ice cover. At the end of the last glacial (Weichselian) major parts of Eurasia's present forests and wetlands were still covered by inland freshwater lakes (Grosswald 1980). During the present interglacial - starting ca. 10,000 years ago - the boreal forest biome has been subjected to inter- and intra-annual climate variability associated with multi-year drought periods and extreme dry years (cf Schweingruber, this volume), which in turn are associated with insect outbreaks (cf Holling 1992) and lightning fires (FIRESCAN Science Team 1996; Clark and Richard, this volume).

Among these natural disturbances, lightning-ignited fire is the most important factor controlling forest age structure, species composition and physiognomy, shaping landscape diversity, and influencing energy flows and biogeochemical cycles.

Small and large fires of varying intensity have different effects on the ecosystem. High-intensity fires lead to the replacement of forest stands by new successional sequences, offering a rich variety of floristic and faunistic habitats. Low-intensity surface fires favor the selection of fire-tolerant trees such as pines (*Pinus* spp.) and larches (*Larix* spp.) and may repeatedly occur within the lifespan of a forest stand.

In Eurasia fire has for a long time been an important tool for land clearing (conversion of boreal forest), silviculture (site preparation and improvement, species selection) and in maintaining agricultural systems, e.g. swidden agriculture, pastoralism, and hunting societies (Viro 1969; Pyne 1995 and this volume). In addition to natural fires, these old cultural practices brought a tremendous amount of fire into the boreal landscapes of Eurasia. In the early 20th century, the intensity of fire use in the agricultural sector began to decrease since most of the deforestation had already been accomplished for agriculture, and traditional small-sized fire systems (treatment of vegetation by free burning) became replaced by mechanized systems (use of fossil-fuel driven mechanic equipment). Despite the loss of traditional burning practices, however, humans are still the major source of wildland fires; only 15% of the recorded fires in the Russian Federation are caused by lightning (cf Korovin, this volume).

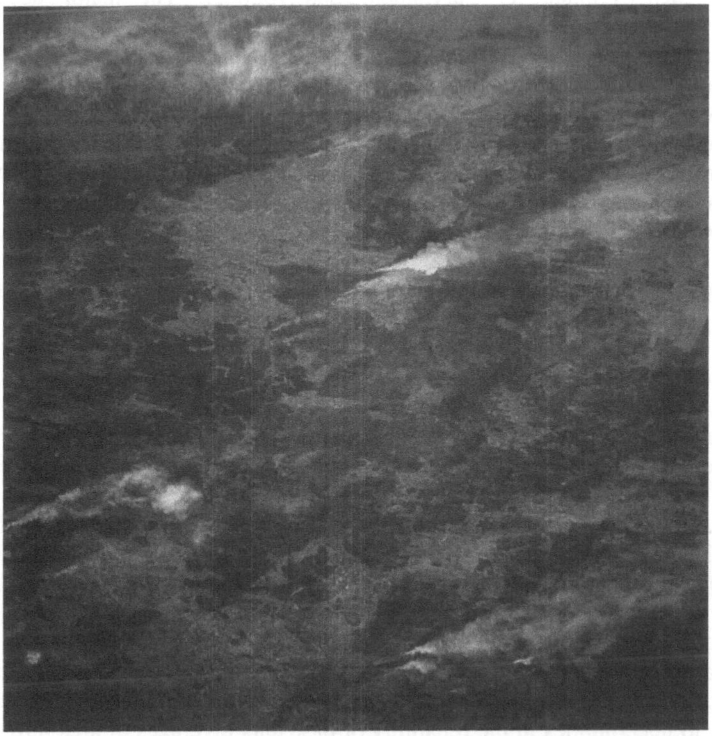

Fig.1. Forest and swamp fires and agricultural burning in Western Siberian lowlands documented by the Space Shuttle STS-39 mission. This photograph was taken at altitude 145 nm a.s.l. in the morning of 29 April 1991 (nadir at 57.3°N - 70.0°E). Photo: courtesy NASA.

Large natural and human-caused fires have been reported in this century, e.g. as a consequence of the Tunguska meteorite impact (ca. 60°54'N-101°57'E) on 30 June 1908. This cometary nucleus explosion at ca. 5 km altitude was one of the more exceptional events which caused large-scale forest fires in the region of impact (cf Grishin, this volume). Several years later, from June to August 1915, the largest fires ever recorded, occurred as a consequence of an extended drought in Central and East Siberia (Tobolsk, Tomsk, Yeniseisk, NE Irkutsk, S Yakutsk regions). Shostakovich (1925) estimated that the fires were burning ca. 50 days in the region between 52-70°N and 69-112°E. The main center of the fires was between the Angara River and Nijnya Tunguska, and the total area burned was estimated at 14.2×10^6 ha. However, the smoke of these fires covered the region between 64-72°N and 61-133°E, corresponding to ca. 680×10^6 ha. Shostakovich estimated continuous smoke (visibility ca. 100 m) on 284×10^6 ha, heavy smoke (visibility 25-100 m) on 215×10^6 ha and thick smoke (visibility 5-20 m) on ca. 181×10^6 ha. It is not clear, however, whether lightning, humans or a combination of the two were the primary cause of the extended fires of 1915.

In recent years wildfires have been more or less eliminated in the Nordic countries (Norway, Sweden, Finland; cf Parviainen, this volume). The major occurrence of contemporary Eurasian fires is on the territory of the Russian Federation. Statistics compiled by the Russian Aerial Fire Protection Service *Avialesookhrana* show that between 10,000 and more than 30,000 forest fires occur each year, affecting up to $2\text{-}3 \times 10^6$ ha of forest and other land. Since fires are monitored (and controlled) only on protected forest and pasture lands, it is estimated that the real figures on areas affected by fire in Eurasia's boreal vegetation are much higher. For instance, satellite-derived observations by Cahoon et al. (1994) indicate that during the 1987 fire season approximately 14.5×10^6 ha were burned (cf also Stocks et al., this volume). In the same fire season ca. 1.3×10^6 ha of forests were affected by fire in the montane-boreal forests of Northeast China, south of the Amur (Heilongjiang) River (Goldammer and Di 1990; Ende and Di 1990; Cahoon et al. 1994).

3. Forest Fires in Boreal Eurasia as a Cyclic Source and Permanent Sink of Carbon

The contributions in this volume reveal that natural fire-return intervals in boreal forests range between several years, decades, and centuries, maintaining a dynamic equilibrium between site potential, climate and vegetation. Theoretically there are no fire-induced net C fluxes to the atmosphere in the present interglacial because carbon released by fire will be sequestered by new growth at varying time scales. However, past climate fluctuations, as established by dendrochronological and densitrometric analyses in the boreal zone, show that decadal and centennial periods warmer or cooler than the long-term average must have changed carbon fluxes periodically.

Modern carbon emissions from forest fires have been estimated on the basis of official fire statistics and average fuel consumption data. These estimates indicate a broad range of numbers due to uncertainties in the reliability of base data (averages of area burned and fuel consumed). Shvidenko et al. (1995a) estimated a net release of C by forest fires in the Russian Federation in 1990 at 58.1 Tg. Taking into account the post-fire C emissions, the authors conclude that a total of 150 Tg C yr was released by forest fires in 1990. Kolchugina and Vinson (1995) estimated the total C emissions associated with forest fires at 137 Tg yr^{-1} (all figures include surface and crown fires in forests, fires in non-forested land and fires in organic layers, e.g. peatland fires). Higher estimates were given by Dixon and Krankina (1993) who estimated an average fire-related C emissions of 199 Tg yr^{-1} for the period 1971-1991. Our own estimates, based on experimental data (e.g. FIRESCAN Science Team 1996), reveal the annual area burned in Eurasia (mainly on the Russian territory) in protected and non-protected lands may well reach $10x10^6$ ha yr^{-1} and consume between 20 and 40 t of ground, surface and aerial fuels per hectare. This would be equivalent to 200-400 Tg plant biomass combusted, equalling ca. 100-200 Tg C released as direct effect of fire.

Tab.1. Selected data on global boreal forest fires

Boreal North America (Annual Average)	1-5 x 10^6 ha
Extreme Years, e.g. Canada 1989/1995	7.4/7.1 x 10^6 ha
Boreal Fennoscandia (Annual Average)	< 4,000 ha
Extreme Years, e.g. Sweden 1933	30,000 ha
Boreal China (Annual Average)	< 55,000 ha
Extreme years, e.g. NE China 1987	1.3 x 10^6 ha
Boreal Eurasia, Recorded Fires (Annual Average)	2-3 x 10^6 ha
Extreme Years, e.g. 1987	16 x 10^6 ha
Boreal Eurasia, Unrecorded Fires (on Non-Protected Lands)	10 x 10^6 ha ?

Specific characteristics of boreal fire emissions

Certain radiatively active trace gases, e.g. incompletely oxidized reactive combustion products such as CO and CH_4, are emitted from boreal fires in a larger proportion (CO_2-normalized emission ratio [ER]) compared to other ecosystems, e.g. the tropical and subtropical savannas. This was confirmed by the Bor Forest Island Fire Experiment, conducted in a boreal coniferous forest in the Krasnoyarsk Region in 1993. This large fire experiment revealed that the ERs of CO and CH_4 are consistently higher than those resulting from fires in other major global ecosystems (FIRESCAN Science Team 1996; Cofer et al. 1996 and this volume).

Some boreal fires are characterized by specific behaviour, e.g. high-intensity stand replacement fires producing strong convective activity and injecting smoke emissions into the high altitudes of the free troposphere. One of the major objectives of the Bor Forest Island Fire Experiment in 1993 was the analysis of methyl bromide (CH_3Br) and methyl chloride (CH_3Cl). Decay products of these compounds are, like the longer-lived chloro-fluorocarbons (CFC), known to induce depletion of stratospheric ozone. (It should be noted here that bromine is much more efficient on a per atom basis than chlorine in breaking down ozone by a factor of about 40; cf Manö and Andreae [1994]; WMO [1992].) The emission ratios of CH_3Br and CH_3Cl measured in the Bor Forest Island Fire were in the range of $0.11\text{-}3.1 \times 10^{-6}$ and $8\text{-}140 \times 10^{-6}$ respectively. This was considerably higher than those found in savanna and chaparral fires or in laboratory experiments (cf. Manö and Andreae 1994). Highest values were found over smouldering surface fuels. This can be explained by the lower combustion efficiency of the smoldering process associated with boreal fires as compared to the prevailing flaming combustion of grass-type fuels.

Estimates of global pyrogenic emissions of CH_3Br from all vegetation fires and other plant biomass burning falls in the range of 10-50 Gg yr^{-1}, or 10-50% of the total source strength (Manö and Andreae 1994). The share of boreal fires still needs to be defined through improved estimates of boreal vegetation area and biomass affected by fire. Special attention must be given to the specific behavior of boreal fires, including the injection of ozone-destroying gases into the high troposphere and possible further transport into the stratosphere.

Fire effects: a carbon sink ?

A close look at soils and organic layers in boreal forests and other organic terrain (raw humus, peat) reveal the abundance of charcoal. Most of the charcoal basically consists of black carbon (BC, also called elemental carbon), which is formed during pyrolysis. Basically BC is biologically non-degradable, chemically inert and not available for uptake by plants.[2] In addition to the deposition of larger BC-containing charcoal particles, BC is also emitted in small fractions and transported as aerosol; quantitative data are not yet available on this feature.

In general, most boreal forest fires do not destabilize the ecosystem (e.g. towards a lower phytomass productivity or lower biomass carrying capacity) in the long term, regardless of the sustainable fire return interval. Thus, the formation of BC represents a net atmospheric carbon sink because the cycling uptake of atmospheric carbon (through photosynthesis) remains constant, and the deposition of ground and soil charcoal as well as the aerosol BC deposited in sites distant from the fire are not available for plant life and are not subjected to degradation.

The fire science community is just beginning to quantify the potential role of BC formation and deposition as a global atmospheric carbon sink (Kuhlbusch and Crutzen 1995; Kuhlbusch et al. 1996). This will require data collection on charcoal storage in soils, peatlands, and sediments at pan-boreal scale. The present state of knowledge allows the

---

[2] The state of international discussion on a standard definition of black carbon and charcoal will be published in the pages of the proceedings of the NATO Advanced Research Workshop "Global Biomass Burning and Climate Change" (Clark et al., 1996).

conclusion, however, that fires in sustainable fire ecosystems, particularly in the boreal zone, may help to explain at least a part of the global "missing sink" of carbon.

4. Global Climate Change: Boreal Forest in Possible Transition from Carbon Sink to Carbon Source

Estimates of carbon stored at present in living and dead plant biomass (without soil organic matter) above- and below-ground in the global boreal forest area range between 66 and 98 Pg (e.g., Apps et al. 1993). Additional large amounts of carbon are stored in boreal forest soils (ca. 200 Pg) and in boreal peatlands (ca. 420 Pg) (Apps et al. 1993).

Expected global warming over the next 30-50 years, as projected by GCMs for a doubled carbon dioxide equivalent greenhouse gas forcing scenario ("2xCO_2 climate") will be most evident in the northern circumpolar regions (Bolin et al. 1986; Maxwell 1992; Shugart and Smith et al. 1992; Shugart et al. 1992). According to these models, zonal warming may lead to the shift of vegetation belts, e.g. causing the boreal forest to shift north (e.g. Kauppi and Posch 1988). The shift of ecosystems is projected to have a considerable impact on the distribution of phytomass. Zonal warming will also affect the balance of the pan-boreal carbon pool. The processes involved, however, are rather complex and should not be generalized.

A model developed by Smith et al. (1992) shows that in a 2xCO_2 climate 72% of today's boreal forest area would be covered by temperate mixed-deciduous forests. Based on the assessment that mixed-deciduous forests in the region will have ca. 43% more carbon in living biomass than do *taiga* forests (Kolchugina and Vinson 1993), the total carbon in living biomass in the region occupied by today's boreal forests would increase by approximately one-third over the long term (Kasischke et al. 1995).

Climate-induced changes in carbon stored in the ground layer, however, are different. The predicted increase of average temperatures in the boreal zone will increase the decomposition rate of dead and dissolved organic matter in the ground and mineral soil layers, thus reducing the amounts of carbon stored (Bonan et al. 1990; Kasischke et al. 1995). A model developed by Bonan and Van Cleve (1992) predicts a net 6-20% decrease in total ground layer carbon in response to a 5°C increase over a 25-yr period.

Climate change and fire regimes

As Flannigan and van Wagner (1991), Stocks (1993), Wein and de Groot (this volume), and Stocks et al. (this volume) underscore, fire may be a driving force in changing the taiga under zonally warming conditions. The prediction of an increasing occurrence of extreme droughts in a 2xCO_2 climate indicates that fire regimes will undergo considerable changes. An increase in the length of the fire season would lead to a higher occurrence of large, high-intensity wildfires. Wotton and Flannigan (1993) predicted an increase in the length of the fire season in Canada by an average of ca. 30 days in a 2xCO_2 climate, resulting in an additional 20% increase in the annual area burned in Canadian boreal forests.

Fosberg et al. (this volume) used the Canadian Climate Center GCM (McFarlane et al. 1992) to predict forest fire danger and severity in Canada and Siberia. The climate model projects a global mean temperature increase of 3.5°C for a doubled carbon dioxide equivalent greenhouse gas forcing scenario ("2xCO_2 climate"). Regional warming during winter would be as follows: continental regions of Siberia and Canada: + 6-8°C, Alaska + 2-4°C, and Scandinavia with little change. Spring temperatures are projected to be uniformly 2-6°C warmer and spring precipitation 8-30% greater than at present. Early fire season temperature changes show up to +6°C in western Siberia, with precipitation greater than at present. While mid and late fire season temperatures will be nearly the same as present, the precipitation is projected to decrease.

In their analysis Fosberg et al. (this volume) used two measures of fire danger and severity (the Russian *Nesterov* Index of ignition [Nesterov 1949] for calculating the ignition potential, and the Canadian Fire Weather Index System [Van Wagner 1987] for assessing fire severity). They used the 90th percentile level of the indices at each of the 224 climate stations in Russia and 191 stations in Canada.[3] The results show that the present worst 10 percent which are currently classed as moderate or high, in future are classed as having extreme fire ignition and severity potential. Extreme monthly severity will be close to double the area for boreal North America, with an extreme ignition index virtually blanketing Eurasia in future.

Combined effects of zonal warming and fire

Kasischke et al. (1995) concluded that changes of above- and below-ground biomass characteristics due to zonal warming would also affect the flammability of vegetation. Over the longer term they expect flammability to decrease for the above-ground biomass because of the long-term shift towards less flammable deciduous trees. In the near term the surface fuels (ground layer) would become drier and more flammable, thus increasing the overall fire risk of forests in transition to the new equilibrium. Consequently, over the shorter term (the next 50-100 yr) there would be an overall increase between 20 and 50% in the annual area burned, resulting in a decrease in the fire return interval from the present average of 150 yr to 125-100 yr.

Considering the fact that it is still largely unknown where exactly the ground layer carbon is stored, Kasischke et al. (1995) developed a carbon flux model in which two baseline carbon levels were established ([1] all ground layer carbon is stored in litter, humus and peat; [2] half of the ground layer carbon is stored in the mineral soil). The model predicts that the net loss of carbon in the ground layer due to zonal warming only (no change of annual area burned) in the short term (the next 50-100 yr) would range between 2.8 and 3.9 kg m^{-2}, or 33.0 to 46.0 Pg on a global basis.

An increase in the annual area burned of 20% would lead to a net loss of ground layer carbon ranging between 3.1 and 4.7 kg m^{-2}, or 36.6 to 55.5 Pg on a global basis; an increase of annually burned area by 50% would result in the global decrease in ground layer carbon between 41 and 66 Pg. Considering the net gain of carbon by increasing aboveground biomass, there would still be a net carbon loss between 46.0 and 53.7 Pg from the global boreal forest to the atmosphere.

---

[3] This corresponds to the worst 10 percent of the weather-related fires which will result in 90% of environmental and social impacts. The evaluation of the change in the highest 10 percent of the indices give a more accurate depiction of the change in risk, since this is the range in which fire control becomes extremely difficult (Fosberg et al., this volume).

5. Interactions Between Modern Anthropogenic Disturbances and Wildland Fire

Traditional forestry practices and low-impact and sustainable use of non-wood forest products in boreal Eurasia are being subjected to dramatic changes which are stimulated by increasing national and international demands for boreal forest products. This has resulted in the widespread use of heavy machinery, large-scale clearcuts, and thereby in the alteration of the fuel complexes. Many clearcut areas are reportedly not regenerating into forest but are rather degrading into grass steppes which may become subjected to short-return interval fires. The opening of formerly closed remote forests by roads and subsequent human interferences bring new ignition risks. These direct effects on the ecosystem are, in addition to the indirect effects, induced by climate change, and both together will certainly contribute to an unprecedented change in fire regimes.

Additional fire hazards and environmental consequences, which are still mainly unpredictable, are created on forest lands affected by industrial emissions. Krankina et al. (1994) report that in the Russian Federation ca. 26.5×10^6 ha of forest lands are severely damaged by industrial pollution (cf. also Kharuk, 1993; Shvidenko et al. 1995b). While it is known in general that the availability of inflammable fuels makes dying and dead forest stands more susceptible to fire than living stands (Alexeyev 1995), other mechanisms are still unknown. For instance, what will be the effects of combusting those chemical depositions which have caused the die-back of forests? How will these agents be converted and re-distributed? Research on fire effects in vegetation polluted by industrial nitrogen deposition in the U.S.A. revealed that post-fire wathershed NO_3 yield contributes to regional groundwater pollution (Riggan et al. 1985).

Radionuclear contamination on ca. 7×10^6 ha land on the territory of the Russian Federation may create considerable problems in the redistribution of radionuclides through forest fires. The mechanisms have been investigated by research on the contaminated Savanna River Site in the U.S.A. (Murphy 1991) and in forests in the impact region of the 1986 Chernobyl nuclear power plant accident (Dusha-Gudym 1992 and this volume). The latter accident caused massive deposition of radionuclides on the surrounding land, e.g. plutonium (^{239}Pu) within the 30 km zone around the plant. Cesium (^{137}Cs) and strontium (^{90}Sr) radionuclides contaminated a number of districts in more distant sites. These contaminated sites were abandoned by humans up to a distance of 100-120 km from the accident site. In the Russian Federation a total of ca 4.5×10^6 ha of forest land is considered contaminated from nuclear fallout. Contamination is also observed in the successional vegetation which developed on large areas of abandoned agricultural lands.

In the years after the 1986 accident wildfires occurred repeatedly in the Chernobyl region, predominantly in successional vegetation and forests. In May 1992 a 500 ha wildfire occurred within the 30-km zone around the power plant. With the smoke plumes of the wildfires, radionuclides were lifted from the contaminated litter layers. Within the 30-km zone, the level of radioactive cesium in the aerosols increased about ten times. The contaminated aerosols were injected into the atmosphere and caused nuclear fallout in distant places (Dusha-Gudym, this volume); the most recent fire event in contaminated terrain was in April 1996.

This unprecedented example of interaction between anthropogenic environmental pollution and wildland fire shows a new dimension of fire problems that may become of increasing importance in the technologically altered global environment - a challenge to modern fire research and environmental management.

6. Fire Research in Eurasia in Transition: Linking with the Global System

With the increasing awareness of the multi-functional values of forests, the need for fire protection became a major focus in Russia's forestry literature of the last century (e.g., Semyonov 1845; Shafranov 1872; Baranetsky 1880). It was soon followed by the exploration of basic fire-vegetation relationships, e.g. the description of fire adaptations by Tkachenko (1911). It was also Tkachenko who introduced the term "forest pyrology" (Tkachenko 1931). In the same year fundamental work on the impact of fire on the biogeocenosis was published, e.g. by Gulisashvili (1931), Kazanski (1931) and Sushkina (1931). In his fundamental work on "Forest Pyrology" Melekhov (1944) based applied fire protection concepts on the study of the "nature of fire". In 1949 N.P.Kurbatski established a Forest Fire Protection Department affiliated with the St.Petersburg Institute of Forestry, and ten years later he founded the Forest Fire Research Laboratory in the V.N.Sukachev Institute of Forest, Siberian Branch, Russian Academy of Sciences. His fundamental work on "Fires in the Taiga Forests. Patterns of Their Occurrence and Growth" (Kurbatski 1964) and numerous other publications on fire basics and management technologies made him widely recognized as the leading fire scientist in Russia (cf Pyne, this volume).

Early concepts for forest fire prevention as an integral part of the forestry system were suggested by Nat (1902) and Yatsenko (1917), but implemented only several decades later. Development efforts during the post-revolutionary decade (1918-27) concentrated on organizational and methodological questions of fire protection. In the 1930s fire suppresssion technology development dominated the field of fire, especially in aerial fire fighting and advanced fire suppression technologies, e.g the use of chemicals and explosives (for more details on Russia's history in fire science and development cf Kurbatski [1967] and Pyne [this volume]). The use of prescribed fire was first applied in West Siberian forests which had been infested and destroyed by the Siberian silkworm (*Dendrolimus sibiricus*) on a total area of 3×10^6 ha between 1952 and 1957. The large-scale broadcast burning program was planned on the scientific base of fire behavior and fire effects. It proved to be successful for inducing the regeneration of a new forest (Furyaev 1966).

In Fennoscandia the use of fire for improving grazing in forests, for slash and burn agriculture, and for forest regeneration was discussed throughout the 18th and 19th century, but hardly anything was done in terms of research.[4] When modern science developed in this field towards the end of the 19th century, wildfires were really not a major threat anymore and the research relating to fire was driven by forest regeneration problems. Soil nutrient dynamics (Hesselman 1917) and plant colonization (Kujala 1926; Sarvas 1937) was analysed in burnt forests with the aim of understanding natural processes. Researchers were aware that the forests had developed earlier under the repeated influence of fire, and the underlying idea was that practical forestry would benefit from a better understanding of the role of fire. Extensive documentation of the influence of wildfire and fire practices were published by Heikinheimo (1915), Wretlind (1934) and Tirén (1937). Prescribed burning had been used locally for a long time, but was fine-tuned by Wretlind in the 1920s and 1930s. After the first three decades of this century, there seems to have been a slack in the interest of fire research. Nevertheless, prescribed fire was used extensively and in the 1950s and 1960s some research was done around the effects of prescribed fire on e.g. vegetation and soil (Uggla 1957), seedling establishment (Yli-Vakkuri 1961) and nutrient dynamics (Viro 1970).

[4] The short outline on Fennoscandia's fire research history is based on Granström (1995)

From 1965-70 prescribed fire was no longer applied, and virtually no research was done on fire. In the 1970s, Zackrisson (1977) made the first quantitative study of fire history. In recent years there has been a renewed interest in the role of fire in natural forest ecosystems and the use of prescribed fire to maintain biodiversity in the managed boreal forests of the future.

As in many other disciplines of science, the historical development of boreal forest fire research clearly suffered from being separated by political boundaries between East and West for a critical phase of at least four to five decades. Except for some courtesy contacts between government institutions during the Cold War and some occasional joint participation at international meetings (e.g., Wein and MacLean 1983), no exchange or cooperation in fire science took place until very recently. Under the given and predicted developments of the global boreal forest, it is evident that new initiatives in cooperative fire research are needed. The differences in research structures and economic situations in the East and the West, however, will require specific approaches to perform joint research.

The research facilities of the former USSR, capable of supporting regional fire research programs, are now mainly part of the Russian Academy of Sciences. These rather large research complexes are suffering heavy financial problems due to Russia's current economic transition. At present, the total funding made available to the Academy of Sciences is so insufficient that adequate salaries cannot be paid. "Brain drain" and the release of personnel has weakened the scientific potential dramatically.

In western countries, the present economic depression coupled with high unemployment rates has also had severe repercussions on research. Increasing costs for advanced research hardware coincide with decreasing research budgets. Fire research in Canada and the U.S.A. has been affected dramatically. While the US Forest Service reduced funding and personnel for fire research successively during the past years, the reduction in Canada was sudden and occured mainly in 1995. The shrinking research resources in the East and in the West unfortunately take place in a period in which interdisciplinary research programs are becoming mandatory and require increasing research investments (cf Weber 1995).

However, the first phase of East-West cooperation in forest fire research is now underway. Its outline, objectives and implications for further activities are described in the following pages. First steps towards cooperative fire research were initiated in 1990-91. At the invitation of the Aerial Forest Fire Protection Service (*Avialesookhrana*) and the Academy of Sciences of the USSR, two fire scientists from the U.S.A. and Germany explored the most important fire research and management facilities and programs of the Soviet Union in 1991 (Pyne 1991, 1992). In 1992 a meeting of the **International Boreal Forest Research Association (IBFRA)** took place in Krasnoyarsk and resulted in signing a formal agreement for cooperation in fire research (Fosberg 1992). In the same year an exchange program between U.S. and Russian fire scientists and managers was established (Davidenko and Goldammer 1994). One of the outcomes of the 1991-92 activities was the decision to jointly prepare and conduct a conference and a field campaign in the summer of 1993, both devoted to the exploration of fire in ecosystems of boreal Eurasia. The objective of the conference, which was held in Krasnoyarsk, 29 June to 3 July 1993, was to compile, interpret, and discuss the current state of knowledge on the role and impacts of fire in ecosystems of boreal Eurasia.

The volume presented here is the result of this first major East-West encounter in fire science with the participation of prominent fire scientists of all boreal countries. The publication of the re-examined knowledge aims to stimulate a new era of pan-boreal fire research, especially considering the need to put basic and specific local aspects of fire ecology into the broader context of the newly emerging global fire science (Crutzen and Goldammer 1993).

The re-examination of the state of knowledge in boreal fire science was followed by the first joint fire experiment. The **Fire Research Campaign Asia-North (FIRESCAN)** was prepared and implemented in its first phase under the co-sponsorship of the **International Geosphere-Biosphere Programme (IGBP)** and IBFRA. The research campaign was supported by the **Volkswagen Foundation** and various national sources of IBFRA member countries.

The IGBP provides a platform for international and interdisciplinary research programs. One of the operational IGBP core projects is the **International Global Atmospheric Chemistry Project (IGAC)**. One of the activities of its foci (Focus 2: Natural Variability and Anthropogenic Perturbations of the Tropical Atmospheric Chemistry) is oriented towards investigating the impact of biomass burning on the atmosphere and biosphere ("Biomass Burning Experiment" [BIBEX]) (IGAC 1992; Annex II).

The boreal fire research program is a joint activity with the IBFRA. Following the recommendations of the **White Sea Declaration** of 1990, IBFRA was founded in 1991 in Mezhgorje, Ukraine (Fosberg, 1992). The two priority research areas are (1) Inventory, Monitoring, and Classification of Boreal Forests and (2) Global Climate Change and Ecosystem Function of Boreal Forests. Under the second area the first IBFRA Working Group established was on **Stand Replacement Fire** (Annex III).

Several international meetings of fire scientists, managers and policy makers provided a platform to develop new visions and concepts in research and policies. The international conference "Fire in Ecosystems of Boreal Eurasia" (Krasnoyarsk, 1993) was the starting point, accompanied by the first phase of the **Fire Research Campaign Asia-North (FIRESCAN)**. Additional joint activities are manifested in three conferences, the NATO ASI "New Mechanisms for Scientific Collaboration Between East and West" in Novosibirsk (1993) (Goldammer and Furyaev 1995); the International Conference on Mathematical and Physical Modelling Forest Fire and Ecology Problems at the University of Tomsk (1995); and the World Congress of the International Union of Forestry Organizations (IUFRO), in Tampere, Finland (1995), during which the Forest Fire Research Group (S.1-09) presented the results of FIRESCAN phase 1993-95 (IUFRO 1995).

In November 1993, the "Taiga Aerospace Investigations using GIS Applications (TAIGA) Workshop" was held in Moscow (Stocks 1994). The workshop, sponsored by NASA's Office of Mission to Planet Earth and the International Forestry Institute (IFI) of the Russian Academy of Sciences, was aimed at developing a multi-national science plan for studying the boreal forests of Russia. The impetus for this workshop was the decision to place new satellite receiving stations in Siberia, a region of the world where previous satellite coverage was less than adequate, and to use this increased capability to intensively study the vast boreal forests of Russia to determine their role in global biogeochemical cycles. The first NASA HRPT (High Resolution Picture Transmission) station was installed in Krasnoyarsk in 1994. Cooperative projects developed at the workshop will make use of both AVHRR (Advanced Very High Resolution Radiometer) data from NOAA polar-orbiting satellites and existing ground and aircraft based datasets.

The United Nations FAO/ECE/ILO Seminar "Forest, Fire and Global Change", which will be held shortly after the publication of this volume in Shushenskoye, Krasnoyarsk Region (August 1996), will be based on the achievements of the joint research and exchange programs between 1993 and 1995. The UN activity, under the auspices of the FAO/ECE Team of Specialists on Forest Fire, aims to communicate with policy makers in order to build a bridge linking fire science with applied systems and international agreements for regional and global fire management.

6.1 IBFRA/FIRESCAN: The First Operational Fire Research Campaigns

The various regional fire research campaigns organized under the umbrella of IGBP/IGAC/BIBEX, IBFRA and other research agreements are operational or in the planning stage (Goldammer 1994; Annex II). These campaigns are designed to establish a comprehensive analysis of environmental impacts of fire at a global scale. The East European/North Asian part of this global research approach, as represented by the regional FIRESCAN program, is a contribution in which East-West collaboration is mandatory.

The first FIRESCAN activity mentioned above was a large experimental forest fire in Central Siberia, the **Bor Forest Island Fire Experiment**. In this fire experiment the ecological and atmospheric chemical impacts of a high-intensity fire were investigated by a mixed group of scientists from all boreal countries, representing a broad range of research disciplines (FIRESCAN Science Team 1994, 1996). A long-term follow-up research area has now been established, the first one in Eurasia's taiga forests. This site will serve as a research focus for future decades and hopefully for the next two centuries.

Fig.5. Aerial view of the Bor Forest Island Fire Experiment of July 1993. The results of this first East-West fire experiment have given important insights into the ecology and atmospheric chemical impacts of boreal forest fires (Source: FIRESCAN Science Team 1994).

The research will be complemented by various experiments, either directly linked to FIRESCAN or independent. In the late 1990s two major experiments are planned in Canada and the U.S.A.. The International Crown Fire Modelling Experiment will take place in the Northwest Territories, Canada (ca. June 1997). This is a cooperative international experiment, bringing together fire modelling experts from Canada, the U.S., and Russia, to address the prediction of high-intensity fire behavior. The NWT Fire Experiment will serve this purpose, with the goal of conducting a replicated series of highly-instrumented crown fires in order to quantify parameters essential to modelling the initiation and spread of crowning fires.

FROSTFIRE is a fire research program being proposed for the Caribou-Poker Creeks Research Watershed, near Fairbanks, Alaska. Under this proposal, a prescribed fire would be conducted over a 550 ha second-order basin within the watershed, permitting a variety of research activities, including catchment hydrologic response, permafrost thermal regimes, landscape stability, water quality, stream biota, regeneration and succession of terrestrial and riparian vegetation, and fire dynamics, emissions, and effects. FROSTFIRE is proposed for 1998 or 1999.

6.2 The IGBP Northern Eurasia Study: Integrated Global Change Research and its Fire Component

Interest within the global change research community in Northern Eurasia has grown dramatically in the last few years. Much of this interest in Northern Eurasia, as in the high latitude regions in general, is centred on its role in the global carbon cycle, which is likely to change significantly under global change.

During two planning meetings in Sweden (Stockholm, 1994) and Japan (Tsukuba, 1995) a prospectus was developed for an integrated hydrological, atmospheric chemical, biogeochemical, and ecological global change study in the tundra/boreal region of Northern Eurasia, the **IGBP Northern Eurasia Study**. The preparation of the research outline and justification of the study is a joint effort of scientists representing several IGBP Core Projects, the Biospheric Aspects of the Hydrological Cycle (BAHC), International Global Atmospheric Chemistry (IGAC) and Global Change and Terrestrial Ecosystems (GCTE) projects.[5] The unifying theme of the IGBP Northern Eurasia Study is the terrestrial carbon cycle and its controlling factors, and the most important overall objective of the study is to determine how these will change under the rapidly changing environmental conditions projected under global change. Two aspects are particularly important. First, understanding the interaction between ecosystem composition and ecosystem process is critical. Second, the overall objective demands a projection not just of net primary productivity or net ecosystem productivity but of "biome productivity", that is, net ecosystem productivity integrated over large spatial and temporal scales to account for such processes as successional dynamics and major disturbances such as fire, timber harvesting and insect outbreaks.

The IGBP Northern Eurasia Study will consist of an integrated set of experimental and observational studies at a number of scales, modelling and aggregation activities, and supporting databases and GIS systems. The major elements are:

---

[5] The materials presented here are taken from the prospectus of the study and will be available soon as a IGBP publication (Steffen and Shvidenko 1996).

- North-south transect(s) in eastern Siberia (East Siberia/Far East) including sites in tundra, tundra-forest transition and in larch forest. This is a critical region for study; it has a high degree of continentality, it contains a vast expanse of larch forest, about which more needs to be known, and the potential for land-use change over the coming decades is high. The transect(s) will join others in Alaska, Canada and Scandinavia/northern Europe as part of the high latitude set of IGBP terrestrial transects.

- A network of sites throughout Northern Eurasia focussing on trace gas emissions and their controlling factors. The network will be concentrated on the extensive wetlands in western Siberia but should include at least two or three other sites in wetlands in the northern areas of European Russia and in eastern Siberia (including the area of continuous permafrost), one site in the coniferous evergreen forests, and one site in the tundra. Of particular importance is an apparent significant zone for gas emissions in the wetlands at the southern edge of western Siberia.

- A water, energy and carbon flux study, integrated with the transect and network components. The design of this component requires further background research and modelling sensitivity studies before it is finalized, but its linkage to the trace gas and ecological components as much as possible will provide a powerful mechanism for integrating much of the work in the IGBP Northern Eurasia Study.

- Studies of disturbance regimes. The studies should include work on all major disturbances of Northern Eurasian ecosystems, such as timber harvesting and insect infestations, but particular emphasis will be placed on fire. They should include work on both gas emissions from fires, and on the ecological role of fire in influencing the successional dynamics of boreal forests. An essential component is the development of a fire database which includes extent, frequency and controlling factors so that fire regimes under a changing environment can be predicted.

- Scaling up, modelling, integration. A wide variety of models will be used to help design the experimental and observational studies, interpret the data, and project the impacts of global change on high latitude systems into the future. These include ecosystem physiology and dynamics models at various scales, trace gas flux models, models on soil-vegetation-atmosphere transfer, landscape hydrological models, atmospheric transport models, biome-scale simulations of the carbon and hydrological cycles, and comparisons to paleo-climate and paleo-vegetation studies. Scaling methodologies such as nested watershed studies and the boundary-layer averaging technique will be used, as appropriate, to scale up results of the process studies carried out at individual sites. Remote sensing, data assimilation and management, and GIS technology are important methods that will be used in the study.

- Associated studies. The IGBP Northern Eurasia Study provides a framework for related global change research. Examples include global change impacts on managed forests, observational and modelling studies of land-use/cover change in the region, and research on global change and ecological complexity (biodiversity). The land-use cover change study, in particular, is crucial to long-term projections of change in the carbon cycle in the Northern Eurasian region. Taken together, the IGBP Northern Eurasia Study and associated studies will provide valuable information to assist the development of sustainable management strategies for Northern Eurasia, in addition to elucidating its role in the global system.

The IGBP Northern Eurasia Study will be implemented in a phased approach. Where there is already general agreement that research is required in a particular area, such as a north-south transect(s) in eastern Siberia, appropriate sites can be selected and long-term observational studies can be initiated as soon as practicable. Care should be taken, however, that sites selected are compatible with the requirements of more intensive studies that will be phased in later. This is a critical consideration, as the "whole system" approach recommended for this study requires that the various individual studies be carried out at the same sites, if at all possible.

Other components of the study, such as the intensive measurements of water, energy, carbon and trace gas fluxes by eddy correlation or similar techniques, require preparatory (sensitivity) analyses before an experimental design can be finalized and work begun. Given the overall requirement for a general understanding of the composition/structure and physiology/productivity of Northern Eurasian ecosystems at a large scale, an early task is to survey the existing research that is being carried out in the region, analyse past work, and, where appropriate, retrieve and standardize data from existing and past work.

Design of the fire component

Fire-induced successional dynamics are a critical element in determining the productivity and carrying capacity of phytomass. Succession studies and measurements of plant biomass, carbon storage and fluxes made at point locations during relatively short time steps, must be scaled up over decades and over large regions containing ecosystems in various stages of succession. It is essential that the extent and frequency of fires under changing environmental conditions and management scenarios be predicted, based on data of present-day fire regimes and their relationship to climate and biomass.

The fire component of the IGBP Northern Eurasia Study will have four parts: (i) fire manipulations at individual forest sites; (ii) a series of campaigns based on air- and spaceborne research platforms; (iii) the construction of a fire database, relating the extent, frequency and intensity of fires to vegetation and climatic conditions for present-day and historic conditions; and (iv) development of aggregated models of forest fire frequency and extent, responsive to global change variables.

First, fire will be used as a treatment in the manipulative experiments at some of the intensive study sites in the transect and in the network. The experiments will measure the effects of fire on the successional dynamics of the vegetation, particularly on the composition of the regrowth forest, and will determine the effects on the nutrient dynamics of the soil and the vegetation. Special attention will be given to the consequences of fire on permafrost sites and to emission characteristics and black carbon formation of peat fires and other less explored fuel types.

Second, more research campaigns similar to FIRESCAN (or a continuation of FIRESCAN) are required to couple air- and spaceborne measurements of biogenic trace gas and aerosol emissions with pyrogenic sources. The newly established NOAA AVHRR receiving stations in Krasnoyarsk, Yakutsk and Khabarovsk will play an important role in identifying fires in real-time or in post-campaign reconstruction.

Third, given the importance of fire in terms of long-term and large scale ecosystem dynamics, a comprehensive ecologically-oriented fire database for Northern Eurasia is essential. The database will build on the National Resources Geographic Information System developed by the Russian International Forestry Institute and should include, in a geographically explicit format, the timing, areal extent, frequency, and intensity (in terms of effects on vegetation as well as gas emissions) of fires as well as the state of the vegetation and the climatic conditions at the place and time of the fires. A strong component of paleo-fire and historic fire research will link information from tree ring fire chronologies and densitrometric analyses with palynological and charcoal data in peat layers and lake sediment cores. This component will be a major inter-core project link and provide the base for the reconstruction of past fire regimes.

Fourth, the ultimate aim is to develop a model of change in fire frequencies and patterns under global change, based on an environmental response surface for fire. Such a model must also be responsive to natural disturbances such as lightning and to fire patterns caused by socio-economically driven land-use changes.

7. Conclusions and Outlook

In the second half of the 1990s the international fire research community has recognized the importance of a cross-disciplinary global fire science. The boreal zone of Eurasia is a vast, heterogeneous set of fire ecosystems which may become dramatically altered by anthropogenic climate forcing and the overall impacts of global change. The consequent feedbacks to the atmosphere are potentially large.

The conveners of the conference **Fire In Ecosystems of Boreal Eurasia** and editors of this volume are indebted to all of those scientists who contributed in preparing this comprehensive analysis of the role of fire on biosphere, atmosphere and climate. It is possible now for the first time to integrate the accumulated fundamental knowledge in fire science accomplished in the former USSR and in the Russian Federation into the context of global change research. At the same time it is hoped that the beginning scientific cooperation between fire scientists in the East and West will be deepened and new initiatives stimulated.

References

Alexeyev, V.A. 1995. Impacts of air pollution on far north forest vegetation. Sci. Total Environ. 160/161, 605-617.

Apps, M.J., W.A. Kurz, R.J. Luxmoore, L.O. Nilsson, R.A. Sedjo, R. Schmidt, L.G. Simpson, and T.S. Vinson. 1993. The changing role of circumpolar boreal forests and tundra in the global carbon cycle. Water, Air, and Soil Pollution 70, 39-54.

Baranetsky, P.V. 1880. Forest protection. St. Petersburg <in Russian>.

Bolin, G., B.R. Döös, J. Jäger, and R.A. Warrick. 1986. The greenhouse effect, climate change and ecosystems. SCOPE 29. J. Wiley and Sons, Chichester, 539 pp.

Bonan, G.B., H.H. Shugart, and D.L. Urban. 1990. The sensitivity of some high-latitude forest to climate parameters. Climatic Change 16, 9-29.

Bonan, G.B., and K. van Cleve. 1992. Soil temperature, nitrogen mineralization, and carbon-sink relationships in boreal forests, Can. J. For. Res. 22, 629-639.

Cahoon, D.R., B.J. Stocks, J.S. Levine, W.R. Cofer, and J.M. Pierson. 1994. Satellite analysis of the severe 1987 forest fires in northern China and southeastern Siberia. J. Geophys. Res. 99(D9), 18627-18638.

Clark, J.S., H. Cachier, J.G. Goldammer, and B.J. Stocks (eds.). 1996. Sediment records of biomass burning and global change. Springer-Verlag, Berlin-Heidelberg-New York (in press)

Cofer, W.R., E.L. Winstead, B.J. Stocks, L.W. Overbay, J.G. Goldammer, D.R. Cahoon, and J.S. Levine. 1996. Emissions from boreal forest fires: Are the atmospheric impacts underestimated? In: Biomass burning and global change (J.S. Levine, ed.), The MIT Press, Cambridge, MA (in press).

Crutzen, P.J., and J.G. Goldammer (eds.). 1993. Fire in the environment. The ecological, atmospheric chemical, and climatic importance of vegetation fires. Environmental Sciences Report 13. John Wiley and Sons, Chichester, England, 400 pp.

Davidenko, E.P., and J.G. Goldammer. 1994. Exchange of fire research and fire management personnel with the U.S.A. Int. Forest Fire News (ECE/FAO) No.10 (January 1994), 17.

Dixon, R.K., and O.N. Krankina. 1993. Forest fires in Russia: Carbon dioxide emission to the atmosphere. Can. J. For. Res. 23, 700-705.

Dusha-Gudym, S. I. 1992. Forest fires on areas contaminated by radionuclides from the Chernobyl power plant accident. Int. Forest Fire News (ECE/FAO) No.7 (August 1992), 4-6.

ECE/FAO (Economy Commission for Europe/Food and Agricultural Organization of the United Nations) 1985. The forest resources of the ECE region (Europe, the USSR, North America). ECE/FAO/27, Geneva.

Ende, J., and Xueying Di. 1990. The forest conflagration of May 1987 in Northeastern China. In: Fire in ecosystem dynamics. Mediterranean and northern perspectives (J.G. Goldammer and M.J. Jenkins, eds.), 169-174. SPB Academic Publ., The Hague.

Federal Forest Service of Russia (ed.) 1994. Russian forests. All-Russian Research and Information Centre for Forest resources (ARICFR), Moscow.

FIRESCAN Science Team. 1994. Fire in boreal ecosystems of Eurasia. First results of the Bor Forest Island Fire Experiment, Fire Research Campaign Asia-North (FIRESCAN). World Resource Review 6, 499-523.

FIRESCAN Science Team. 1996. Fire in ecosystems of boreal Eurasia: The Bor Forest Island Fire Experiment, Fire Research Campaign Asia-North (FIRESCAN). In: Biomass burning and global change (J.S. Levine, ed.), The MIT Press, Cambridge, MA (in press).

Flannigan, M.D., and C.E. van Wagner. 1991. Climate change and wildfire in Canada. Can. J. For. Res. 21, 61-72.

Fosberg, M. 1992. International Boreal Forest Research Association, Stand Replacement Fire Working Group. Int. Forest Fire News (ECE/FAO) No.7 (August 1992), 6-8.

Furyaev, V.V. 1966. The taiga silkworm moth and its control by fire. Moscow, Nauka <in Russian>.

Goldammer, J.G. 1994. Interdisciplinary research projects for developing a global fire science. A paper presented at the 12th Confer. Forest Meteorology, October 26-28, 1993, Jekyll Island, Georgia. Society of American Foresters Publ. 94-02, 6-22.

Goldammer, J.G., and Xueying Di. 1990. The role of fire in the montane-boreal coniferous forest of Daxinganling, Northeast China: A preliminary model. In: Fire in ecosystem dynamics. Mediterranean and northern perspectives (J.G. Goldammer and M.J. Jenkins, eds.), 175-184. SPB Academic Publ., The Hague.

Goldammer, J.G., and V.V. Furyaev. 1995. Global change, the boreal forest, and fire: Search for new strategies in science policies and research mechanisms. Science Policy: New Mechanisms for Scientific Collaboration between East and West (V.A.Koptyug and J.Klerkx, eds.), 45-61. NATO ASI Series 4, Science and Technology Policy Vol.1. Kluwer Academic Publishers, Dordrecht-Boston-London, 256 p.

Granström, A. 1995. Boreal fire research in Sweden - Past and present. Paper presented at IUFRO XX World Congress, 6-12 August 195, Tampere, Finland. S1.09-00 Forest Fire Research. Abstract volume, p.93. IUFRO, Vienna.

Grosswald, M.G. 1980. Late Weichselian ice sheet of Northern Eurasia. Quat. Res. 13, 1-32.

Gulisashvili, V.Z. 1931. The experimental studies of fire influence on forest reproduction. In: Studies in forestry (M.E.Tkatchenko, ed.). Moscow-Leningrad, Selkolkhozgiz <in Russian>.

Hesselman, H. 1917 On the effect of our regeneration measures on the formation of salpeter in the ground and its importance in the regeneration of coniferous forests. Meddelanden Från Statens Skogsförsöksanstalt 13-14.

Heikinheimo, O. 1915. Der Einfluss der Brandwirtschaft auf die Wälder Finnlands. Acta forestalia Fenn. 4, 1-59.

Holling, C.S. 1992. The role of forest insects in structuring the boreal landscape. In: A systems analysis of the global boreal forest (H.H. Shugart, R.Leemans, and G.B.Bonan, eds.), pp.170-191. Cambridge University Press.

IGAC (International Global Atmospheric Chemistry Project). 1992. Biomass Burning Experiment: Impact on the Atmosphere and Biosphere. An Activity of the International Global Atmospheric Chemistry (IGAC) Project, IGAC Core Project Office, The MIT, Cambridge, MA, 19 pp. + App.

IUFRO (International Union of Forestry Research Organizations). 1995. Caring for the forest: Research in a changing world. Abstracts of invited papers, IUFRO XX World Congress, 6-12 August 195, Tampere, Finland. S1.09-00 Forest Fire Research, pp.86-94. IUFRO, Vienna.

Kasischke, E.S., N.L. Christensen, and B.J. Stocks. 1995 Fire, global warming, and the carbon balance of boreal forests. Ecol. Applications 5, 437-451.

Kauppi, P., and M. Posch. 1988. A case study of the effects of CO_2-induced climatic warming on forest growth and the forest sector: A. Productivity reactions of northern boreal forests. In: The impact of climatic variations on agriculture, Vol. 1. Assessments in cool temperature and cold regions (M.L.Parry et al., eds.), pp.183-195. Reidel, Dordrecht.

Kazanski, N.A. 1931. The experiment on studying the influence of fire on pine reproduction. In: Studies in forestry (M.E.Tkatchenko, ed.). Moscow-Leningrad, Selkolkhozgiz <in Russian>.

Kharuk, V.I. 1993. Forest decline in the Siberian North. World Resource Rev. 5, 72-76.

Kolchugina, T.P., and T.S. Vinson. 1993. Equilibrium analysis of carbon pools and fluxes of forest biomes in the former Soviet Union. Can. J. For. Res. 23, 81-88.

Kolchugina, T.P., and T.S. Vinson. 1995. Role of Russian forests in the global carbon balance. Ambio 24, 258-264.

Krankina, O.N., R.K. Dixon, A.Z. Shvidenko, and A.V Selikhovkin. 1994. Forest dieback in Russia: Causes, distribution and implications for the future. World Resource Review 6, 524-534.

Kuhlbusch, T.A.J., and P.J. Crutzen. 1995. Toward a global estimate of black carbon in residues of vegetation fires representing a sink of atmospheric CO_2 and a source of O_2. Global Biogeochemical Cycles 9, 491-501.

Kuhlbusch, T.A.J., M.O. Andreae, H. Cachier, J.G. Goldammer, J.-P. Lacaux, R. Shea, D.E. Ward, and P.J. Crutzen. 1996. Black carbon formation by savanna fires: measurements and implication for the global carbon cycle. Special Issue SAFARI-92: J. Geophys. Res. (in press.).

Kujala, V. 1926. Untersuchungen über den Einfluss von Waldbränden auf die Waldvegetation in Nord-Finnland. Comm. Inst. Quaest. Forest. Finl. 10, 1-37.

Kurbatski, N.P. 1964. The fires in the taiga. Patterns of occurrence and growth. Doctoral thesis. Synopsis. Krasnoyarsk <in Russian>.

Kurbatski, N.P. 1967. Forest pyrology and the most important stages of its development. In: Forestscience. The achievements of 50 years, pp. 122-137. Krasnoyarsk <in Russian>.

Kuusela, K. 1990. The dynamics of boreal coniferous forests. The Finnish National Fund for Research and Development (SITRA) Publ. 112, Helsinki, Finland, 172 pp.

Kuusela, K. 1992. Boreal forestry in Finland: a fire ecology without fire, Unasylva 43 (170), 22-25.

Manö, S., and M.O. Andreae. 1994. Emission of methyl bromide from biomass burning. Science 263, 1255-1257.

Maxwell, B. 1992. Arctic climate: potential for change under global warming. In: Arctic ecosystems in a changing climate (F.S.Chapin, R.L.Jefferies, J.F.Reynolds, G.R.Shaver, and J.Svoboda, eds.), pp.11. Academic Press, New York.

McFarlane, N.A., G.J. Boer, J.-P. Blancet, and M. Lazare. 1992. The Canadian Climate Centre second-generation general circulation model and its equilibrium climate. J. Climate 5, 1013-1044.

Melekhov, I.S. 1944. On theoretical fundamentals of forest pyrology. Arkhangelsk, ALTI <in Russian>.

Murphy, C.E. 1991. A method to estimate the concentration of elements in smoke from burning vegetation growing in contaminated soil (U). Westinghouse Savanna River Company, WSRC-TR-91-109, Aiken, SC, 20 pp.

Nat, S.G. 1902. Forest fires and the means of fire fighting. Moscow <in Russian>.

Nesterov, V. G. 1949. Combustibility of the forest and methods for its determination, USSR State Industry Press <in Russian>.

Nikolov, N., and H. Helmisaari. 1992. Silvics of the circumpolar forest tree species. In: A systems analysis of the global boreal forest (H.H. Shugart, R.Leemans, and G.B.Bonan, eds.), pp.13-84. Cambridge University Press.

Pisarenko, A.I., and V. V. Strakhov. 1993. Global role of the Russian boreal forests: A viable assessment, Paper presented at the Ann. Meeting International Boreal Forest research Association, Biri, Norway.

Pyne, S.J. 1991. Wildland fire protection in the USSR. Report of a study tour, 15 July to 10 August 1991, at the invitation of the Air Forest Fire Protection Service of the USSR, conducted by S.J.Pyne and J.G.Goldammer (mimeo, available through the author).

Pyne, S.J. 1992. The Russian fire establishment. Impressions from a study tour. Int. Forest Fire News (ECE/FAO) No.6 (Jan. 1992), 3-4.

Pyne, S.J. 1995. World fire. Henry Holt, New York, 379 pp.

Riggan, P.J., R.N. Lockwood, and E.N. Lopez. 1985. Deposition and processing of airborne nitrogen pollutants in mediterranean-type ecosystems of Southern California. Environ. Sci. Technol. 19, 781-789.

Sarvas, R. 1937. Beobachtungen über die Entwicklung der Vegetation auf den Waldbrandflächen Nord-Finnlands. Silva Fennica 44, 32-64.

Semyonov, V. 1845. Forest protection. St. Petersburg <in Russian>.

Shafranov, N.S. 1872. Forest protection. St. Petersburg <in Russian>.

Shostakovitch, V.B. 1925. Forest conflagration in Siberia. With special reference to the fire of 1915. J. For. 23, 365-371.

Shugart, H.H., R. Leemans, and G.B. Bonan (eds.). 1992. Boreal forest modeling. A systems analysis of the global boreal forest. Cambridge University Press, Cambridge, England, 565 pp.

Shugart, H.H., and T.M. Smith. 1992. Modeling boreal forest dynamics in response to environmental change. Unasylva 43 (170), 30-38.

Shvidenko, A., S. Nilsson, R.E. Dixon, and V. Rojkov. 1995a. Burning biomass in the territories of the former Soviet Eurasia: Impact on the carbon budget. Quat. J. Hungarian Meteorological Serv. 99, 235-255.

Shvidenko, A., S. Nilsson, and V. Rojkov. 1995b. Possibilities for increased carbon sequestration through improved protection of Russian forests. IAASA Working Paper WP-95-86. International Institute for Applied Systems Analysis, Laxenburg, 27 pp.

Smith, T.M., R. Leemans, and H.H. Shugart. 1992. Sensitivity of terrestrial carbon storage to CO_2-induced climate change: comparison of four scenarios based on general circulation models. Climatic Change 21, 367-384.

Steffen, W.L., and A.Z. Shvidenko (eds.). 1996. The IGBP Northern Eurasia Study. Prospectus for integrated global change research. The International Geosphere-Biosphere Programme: A Study of Global Change (IGBP) of the International Council of Scientific Unions (ICSU), IGBP Stockholm (in press).

Stocks, B.J. 1993. Global warming and forest fires in Canada. The Forestry Chronicle 69, 290-293.

Stocks, B.J. 1994. Taiga Aerospace Investigations using GIS Applications (TAIGA) Workshop. Int. Forest Fire News (ECE/FAO) No.10 (January 1994), 30-31.

Sushkina, N.N. 1931. On microbiology of forest soils in view of fire influence. In: Studies in forestry (M.E.Tkatchenko, ed.). Moscow-Leningrad, Selkolkhozgiz <in Russian>.

Tirén, L. 1937. Forestry historical studies in the Degerfors district of the province of Västerbotten. Meddelanden Från Statens Skogsförsöksanstalt 30.

Tkatchenko, M.E. 1911. Northern forests. Transact. Forestry in Russia, Vol. XXV <in Russian>.

Tkatchenko, M.E. 1931. Introduction to the book by P.I.Chudnikov "The influence of fires on the reproduction of forests in the Urals". Moscow-Leningrad, Selkolkhozgiz <in Russian>.

Treter, U. 1993. Die borealen Waldländer. Westermann, Braunschweig, 210 pp.

Uggla, E. 1957. Mark och lufttemperatur vid hyggesbränning samt eldens inverkan på vegetation och humus. Norrlands Skogsvårdsförbunds Tidskrift 74, 443-474.

van Wagner, C. E. 1987. Development and structure of the Canadian Forest Fire Weather Index System. Canadian Forestry Service Tech. Rep. 35. Ottawa.

Viro, P.J. 1969. Prescribed burning in forestry. Comm. Inst. For. Fenn. 67 (7), 1-49.

Weber, M.G. 1995. Fire research and the global village. The Forestry Chronicle 71, 584-588.

Wein, R.W., and D.A. MacLean (eds.). 1983. The role of fire in northern circumpolar ecosystems. SCOPE 18, John Wiley & Sons, Chichester, UK.

WMO (World Meteorlogical Organization). 1992. Scientific Assessment of Ozone Depletion. WMO Rep.25. Geneva.

Wotton, B.M., and M.D. Flannigan. 1993. Length of fire season in a changing climate. The Forestry Chronicle 69, 187-192.

Wretlind, J. 1934. Naturbetingelserna för de nordsvenska järnpodsolerade moränmarkernas tallhedar och mossrika skogssamhällen. Sveriges Skogsvårdsförbunds Tidskrift 32.

Yatsenko, I. 1917. Forest fire and forest management. Forest Journal Vol. 9-10 <in Russian>.

Yli-Vakkuri, P. 1962. Emergence and initial development of tree seedlings on burnt-over forest land. Acta Forestalia Fennica 74.

Zackrisson, O. 1977. Influence of forest fires on the north Swedish boreal forest. Oikos 29, 22-32.

J.G.Goldammer and V.V.Furyaev (eds.), Fire in Ecosystems of Boreal Eurasia, 21-44.

Wild Hearth
A Prolegomenon to the Cultural Fire History of Northern Eurasia

S.J. Pyne [1]

The last and wildest stretch of earth
Where Europe's genius built a hearth
- Zachris Topelius

"But perhaps this was all possible because every farm had its comforter. One that came to rich and poor alike, that never failed and never tired...it was nothing but the fire, burning on the hearth on a winter's night."
- Selma Lagerlöf, The Löwensköld Ring

1. Introduction

Since the time of *Homo erectus* hominids have, through fire, interacted with the Earth. Anthropogenic fire has been present throughout northern Eurasia whenever ice sheets, sea levels, and climate have permitted; throughout the Holocene, occupation by *Homo sapiens* has been continuous. Through their fire practices humans have sought, from the earliest times, to reshape and render habitable their surroundings. The ways by which humans have protected themselves from wildfire and the means by which they have projected their own domesticated fires onto the land have varied considerably - they have changed over historic time as well as across geographic space. But their fires mark human presence as surely as flint arrowheads or burial mounds. Everywhere fire has mediated between humans and the land. The biota of northern Eurasia has thus coexisted, if not coevolved, under the pressures of anthropogenic fire.

The details of this environmental dialectic are not well understood, however. For some regions or subregions fire history is known with confidence. For others it has been barely researched, or if studied, its records remain in relatively inaccessible forms. For some portions or times, particularly in the distant past, reconstruction must rely on analogy. The fire histories of Europe and North America can thus help illuminate the fire history of Russia, especially Siberia, through a process of analogical triangulation.

Such a fire history is an immense task, for which this paper can serve, at most, as a prolegomenon. What is important even in this preliminary formulation is to include culture with nature. It is not sufficient to recognize the ecological role of fire in boreal regions; it

[1] American Studies Program, Arizona State University West, Phoenix, Arizona 85069-7100, U.S.A.

is equally necessary to appreciate the role of anthropogenic fire. The human presence is everywhere predicated on fire. Their fire practices - the ways in which humans withhold and apply fire - explain much about why northern Eurasia looks the way it does.

2. Wild Fire

Flame

Nature of course is rife with fire, and would be so quite independently of humanity. Most human fire practices in fact mimic natural events - slash-and-burn agriculture, for example, simulates burning from windfall and lightning. Fire rites and ceremonies have evolved out of the practical observation that fires and smoke do purge sites temporarily of weeds and pests, do fertilize, and do transform whatever they touch. It is no accident that humans tend to prefer those landscapes and species that are most amenable to manipulation by fire. Where fire is difficult, settlement tends to be dispersed, nomadic, or marginal.

The boreal environment is thus a paradox. Certainly its great forests and hardy grasslands (less so its mires and tundra) are fire-sculpted, almost certainly fire-adapted. But the growing season is short, the opportunity for burning brief, and the capacity of fire *per se* to reshape the landscape limited by the dynamics of weather and fuel. Fires tend to be either small or very large. The ability to produce a range of fires more or less on demand, to underburn frequently, to reshape the landscape with routine fire, and to extract vital resources from *in situ* burning is less pronounced in taiga, tundra, mire, and forest steppe than elsewhere.

Accordingly settlement came slowly, and then when fire could be allied with other technologies, including biotic, by which to transform the land - when, for example, farming could supplement gathering; when herding could augment hunting; when fire could interact with steel axe or railroad. In brief, the boreal environment was impossible to inhabit without fire, but fire practices allied with simple technologies could not alone sustain a large society. Instead it was as common to transport the forest to the hearth fire for burning as to project the cultural hearth over the forest.

Torch

People carry fire with them, a perpetual flame. Anthropogenic fire competes, both directly and indirectly, with lightning to restructure the fire regime of a given landscape. In some environments the fire regime is exclusively that of humans. In others, fire and lightning interact to shape a mobile matrix. Through applying and withholding ignition, by modifying the fuels on which all fires depend, humans alter the timing, frequency, intensity, and extent of burning.

This competition occurs over historic time as well as geographic space. As fire and fire-catalyzed technologies allow for greater human settlement, lightning recedes. When for whatever reason humans withdraw, lightning fire returns. Fire remains throughout, although its regime varies. This pattern has characterized northern Eurasia as elsewhere, but with a difference. In the boreal landscape the possibility for large lightning-kindled wildfires to rage can persist unless and until settlement strips the land of its natural fuels and replaces them with domesticated surrogates. A robust agriculture, however, came slowly - the climate was too harsh, the soils too hostile. It came spasmodically, unevenly, different from the paradigmatic model of mixed agriculture that characterized the European hearth. Until it arrived in force the biota reflected a complex fugue between lightning and torch.

The aboriginal fire history is everywhere uncertain in its details. The most basic fire uses were domestic, fire for warmth, for cooking, for light, for saunas. Shelter - for fire as well as for humans (the needs of the two were inextricable) - was the first order of survival. Smoke partly shielded humans from the worst plagues of insects. If fire were removed, the boreal environment became uninhabitable. Without domestic fire humans could not exert their other influences on the landscape.

But humans also applied the domesticated fire to the land around them. Probably peoples migrated toward the margins of burned areas, or to those steppe-like environs that could be easily burned. Human were fire foragers, the patchiness of fire helping to shape the texture of human geography and human seasons. Fire aided hunting, if not through fire drives, then with fire- or smoke-baited traps, rings of bonfires, evening torches. Evidence supports such fire hunting for bear in Scandinavia, boar in Transbaikalia, and reindeer throughout the taiga. Probably the earliest paleohunters pursued mammoth and mastodon with the aid of fire, as archaeological evidence suggests they did in Torralba, Spain and as explorers have recorded for modern Africa. Equally, they used fire to shape favored grazing grounds, thus exploiting fire-catalyzed grass and browse to attract deer, moose, muskrats, and so on to particular sites. (Fires in the steppes or forest-steppe were frequent, even annual. Explorers like Peter Pallas and John Bell routinely reported them. Bell explained that the "Mongalls, in the spring, during high winds" often set such fies "to procure early pasture for their cattle" (Bell 1966). In Manchuria Arthur Adams remarked on the "prodigious number of those charred and blackened trees that strew the ground in every direction," the outcome of "wandering shooting and fishing parties of Manchu tartars, who always fire the scrub and burn down the trees, to clear the land and make it yield good pasturage..." (Adams 1870). Hunters likewise burned along traplines to stimulate forage for the small mammals on which fur-bearers like fox and sable fed. They fired marshes to encourage habitat favorable to ducks and muskrats. They carried torches to enhance fishing at night.

Much of the edible harvest of the forest derived from fire-cycled plants. Bees thrived on fireweed and other fire-flushed forbs, and inhabited trees often hollowed by fire; certain mushrooms sprouted best on burned lands; berries blossomed profusely under the proper regimen of burning. It is estimated that an apiary of 100 hives requires 5000 ha of suitable habitat, a condition almost certainly created by fire. So, too, proper burning promoted grasses or shrubs useful for baskets. Where nut-bearing trees flourished, such as oak, fire assisted their harvesting. Like much of the game they hunted, humans sought out that rough edge between ecotones or helped create it, a biotic flint chipped by fire.

The aboriginal fire practices that Henry Lewis has documented for boreal Canada are very likely applicable to aboriginal Eurasia (Lewis and Ferguson 1988). Frequently travelled corridors, regularly burned, joined patches of more extensively burned lands, a sinewy geography shaped by fire. Wherever fire could be used, it almost certainly was used. "Carelessness" no doubt also accounted for many burns, the flaming refuse of abandoned campfires, smoking fires, and trees torched for amusement.

Even so, the land was formidable, and the resources extracted were not large. Populations were sparse. Small groups could scavenge among the residue of lightning-kindled burns. Yet it is all but certain that aboriginal populations did more. After all they carried fire with them wherever they went; and it was a simple matter to progress from an attraction to burned borders to the creation of those favored edges. These aboriginal practices did not cease with the advent of other economies. Rather they continued, sometimes pushed to the margins, sometimes integrated with agricultural activity. They continue today.

Hearth

So anthropogenic fire was a tool, one derived from natural precedents. But it was also something more. The possession of fire has been a unique hominid attribute, a species monopoly, the one ecological niche humans occupy that no other organism does. Anthropogenic fire defined the human identity as a unique fire creature on a unique fire planet. Inevitably it entered into the cultural, and ultimately moral, universe created by human societies. Fire came to symbolize family, tribe, polis, nation, civilization. When anthropologists speak about the homeland or core of distinctive peoples, they call that region a cultural hearth.

While the homeland for many peoples, some still extent, some extinct or absorbed, northern Eurasia is not the hearth for any of the world's great civilizations. That designation has been reserved traditionally for large syncretic cultures, all of which have been based on agriculture and all of which are far removed from the boreal environment. Boreal Eurasia was an agricultural sink, not a source, and it received the Neolithic revolution only at considerable labor. Some horticulture penetrated northward from China and Korea. Pastoralism flourished on the central Asian steppes, and in various forms seeped northward through the mountain walls that segregate north from south Eurasia. The dominant infusion of agriculture, however, has come from Europe.

The neolithic revolution entered Europe from Asia Minor during the height of the Hypsithermal. It appears that elements of agriculture - domesticated animals and possibly cereals - followed the final recession of the ice sheets from Scandinavia and probably other parts of the Eurasian plain. Since then, the colonization of northern Eurasia has, to varying degrees, been one of agricultural settlement, with the dominant cultures representing one form or another of farming and herding. Agriculture branded new fire regimes across the region. Its fire history tells the story of how the expansion of European peoples extended the European hearth, in both its literal and symbolic senses, across the boreal biota of northern Eurasia.

3. Vestal Fire

Agricultural Exemplar

European agriculture resembled a kind of biotic ark, a rich mix of flora and fauna, and one potentially self-contained. Among its domesticated fauna were grazers, browsers, foragers, scavengers, and carnivores; among its flora were fruit and nut trees, cereal grains, tubers, and vines. Wherever it settled it sought to remake the environment to suit its requirements, even as it was compelled to adjust its elements to suit that environment and the indigenous organisms. Sometimes elements were added, sometimes dropped.

The image of an ark is appropriate, however, in that it suggests the sense in which a whole biota, not simply individual bands of humans, reclaimed the land (Crosby 1987). Europe's expansion into boreal Eurasia was as much biological as political or economic; the encounter remade the landscape as fully as it did the society of indigenous peoples and the ancient trade in spice and silk. As this domesticated biota moved, fire was everywhere present as cause, catalyst, and consequence. It made possible other technologies, not only mechanical but biological, and these reciprocated by influencing the subsequent character of fire. Spring burning, for example, helped shock dormant grasses to life, which allowed pastoralism to flourish; herds complemented fields by helping fertilize arable lands; grazing, in turn, helped restructure the fuels upon which fires fed. Similarly, fire practices interacted with other land usage such as logging, draining, and harvesting; the one redefining the dominion of the other.

The European ideal remained a mixed economy of livestock and farming, supplemented by specialty gardens and orchards. It was, when completed, a closed system in which a portion of the fodder and pasture went to domestic animals who not only supplied meat, milk, wool, or hides but manure. Aggregated manure was recycled onto fields to sustain the productivity of the soil. It is clear, too, that this prescription requires a stable, fairly dense population that must function on a fixed land base. It is an interesting question whether this style of agriculture encouraged these social patterns, or whether these social expectations promoted this version of agriculture.

In fact the European exemplar found its full expression only in certain places and at certain times. Elsewhere environmental or social conditions biased one or another part of the formula. Thus pastoralism tended to prevail in the Mediterranean region, particularly following the disintegration of the Roman Empire. Slash-and-burn farming persisted in much of the boreal region where long winters and sparse fodder made it difficult to build up herds (and thereby manure). And the paradigm itself evolved. Land use intensified; animal husbandry replaced herding; crop rotation substituted for field rotation. The amount of land left to long fallow shrank until, by the 15th century, the three-field system began to prevail. Still the paradigm characterized the heartland of Europe - the agronomic ideal celebrated by the major scientific, colonial, and (later) industrial powers; by England, Holland, France, and the German states. With adaptations it would characterize the agricultural regime of northern Eurasia.

There was plenty of fire in European agriculture. Fire cleared and fertilized land, recharged pastures, purged fields of pests and disease, vaccinated vulnerable lands against later wildfire. But increasingly, as agriculture intensified, the use of fire became more circumscribed - banished to remote summer grazing sites, left to stubborn lands that resisted sedentary farming and demanded some kind of swidden, nurtured in ceremonies for which fire practices became symbols (Fraser 1922). Europe itself came to resemble a fire, burned out at the core, flaming along its perimeter.

By the time of the Enlightenment and the agricultural revolution that, in Europe, preceded industrialization, intellectuals almost universally condemned agricultural fire as primitive, slovenly, and wasteful. It destroyed humus, the universal index of environmental health. Any agriculture that relied on fire was by definition "primitive"; any that found surrogates for burning was "rational."

In effect Europe's demands for a sedentary agriculture required that the "landscape" be cycled through a farm rather than cycling the farm through the landscape. But regardless of whether crops rotated through a plot of land or the plot rotated among a wider landscape, that movement demanded fire. For boreal Europe, in particular, agriculture could not proceed without it. Those flames burned well into the 20th century. The colonization of Russia resembled a flaming front, driven by the winds of European expansionism.

Boreal Europe: The Scandinavian Experience

The mixed geography of Swedish fire coevolved with humans who, in different places at various rates and with assorted means, profoundly restructured its ensemble of wet and dry sites. From the beginning there were humans present, outfitted with the initial technologies of the Neolithic revolution, tramping along the fast-marching frontiers of retreating ice and falling sea, as opportunistic as the birch, alder, and pine that seized the disturbed landscape. Unlike them humans could, through fire, prolong the disturbances. Over the centuries their successors converted forests to fields, drained mires, harvested peat, grazed meadow and marsh, built reservoirs, introduced exotic flora and fauna, constructed roads, logged woodlands, transformed forests to heath and then replanted heath to forests - every niche experienced some kind of change. Within limits anthropogenic fire could halt spruce invasion, reverse mire-formation, inhibit paludification, counteract lake acidification, and stimulate a host of other challenges to what humans considered to be a degrading climate. Only by such means could they, in the end, colonize the new lands.

Fire Practices

Without fire that land was virtually uninhabitable. The torch shielded humans from the ice, the cold, the wet, and the endless winter darkness. People lived in tents or houses to preserve the nurturing fire. Fire heated house and sauna; smoke drove away pestilential insects; flame warded off wolf and witch. Torches helped hunt for bear and assisted fishing in the evening. Fire promoted foraging for berries and herbs, and (for Finns) the harvesting of mushrooms. Pastoral burning penetrated heath, forest, mire, and meadow; through their livestock, humans projected their influence and their fires far beyond the limited range of settlements; and the institution of summer grazing at remote cabins, a kind of transhumance, extended their range of influence even further. Slash-and-burn agriculture reworked forest and peatland. Through anthropogenic fires were made possible the essential technologies that integrated the region into national and international markets - fire mining; charcoal for smelting and ironworks; tar for naval stores; potash and saltpeter for a primitive chemical industry.

Their fires defined relationships not only between humans and the land but between humans. Saame, Finns, and Swedes, all at one time or another used a scorched earth to define the limits of their claims, to project or prevent encroachment by miners, cattle, or reindeer. Armies fought with fire sleds, fire ships, fire arrows, and fire missles flung from catapults. Blackmail often took the form of a "fire-tax," with money extorted to forestall arson. The response to plague was prophylactic fire, not only to infected houses but to the countryside. Fire towers, bonfires, or special trees torched on cue, announced vital news or warned of new Russian invasions along the Gulf of Bothnia. The fiery cross called villagers to meeting. Ritual fire attended the major ceremonies of Swedish life, from the candle crowns of St. Lucia Day to the great bonfires of Beltane and **Midsommarsdagen**.

People penetrated everywhere, and they forced their fires to compete with those of lightning. The evolved landscapes of Sweden were those that fire, or those that fire-assisted agriculture and industry, helped bring into being. In the south fire ranged routinely in extensive heaths, the stubble of cultivated fields, forage-producing bogs, and **svedjed** forests; in the north, among grazed forests and mires; and almost everywhere in the many manifestations of fire-based cultivation. New fire regimes proliferated not solely through new ignition patterns but by the interaction of fire with other processes like grazing and draining. Some provinces of Norrland, for example, previously had a third or more of their landscape under mires, sites that once drained became exposed to new kinds of burning.

The range of burning was vast in sum, but intricate in practice. A new mosaic replaced the old one, partly because the emergent anthropogenic landscapes built upon and retained some elements of the old, partly because land ownership and use by farmer and village favored innumerable subdivisions. Still, some burning could be extensive, or become large through fires that escaped into surrounding heath or forest. Most were more delicate, like etchings on Orrefors glass. A slope was burned here, a berry patch there; here a farmer fired the slash from a new swidden, there a fallow field, still elsewhere his pastures - an expanse of *Calluna* heath, the grassy understories beneath birch and alder forests, the grass-moss-**ljung** cover beneath pine. Farmers even burned along the slopes of small lakes so that a slight nutrient flush would descend into the slack waters and stimulate sedges useful as fodder or thatch. Almost every component of the biota could be refashioned with some kind of fire, or became exposed to new fire sources that resulted in novel fire regimes. Anthropogenic fire worked and reworked the landscape like lumps of bog iron forged and hammered and fired again into swords of Swedish steel.

Fire Farming

But perhaps the most vital, and most characteristic, manifestation of folk fire culture was the practice of swidden or slash-and-burn cultivation. Settlers adapted swidden, however, to a great panorama of landscapes, each of which had peculiar features and for each of which they gave particular names. But however called, fire farming was the most fundamental expression of environmental manipulation because it first broke the land to agriculture. Sites once felled and fired supplied the tiles from which a radically different landscape mosaic could evolve, the pieces with which Scandinavian agriculture could reshape the environment to arable field, meadow, fallow, forest pasture, and shifting cultivation. It made possible the rapid recycling of landscapes.

Fire farming assumed many forms, and was given many names. In Sweden it became known as **svedjebruk**. In Finland the vocabulary was richer, but two forms assumed dominance - **huuta**, which described the slashing and burning of primary forest, and **kaski**, which referred to the same practices in secondary forest. So prominent was the practice, in fact, that a number of European observers believed that **sveden** was the origin of "Sweden," that **svedjebruk**, which had so shaped the countryside, had given one of its variant names to the country as a whole. Other variants described adaptations; **kärrbruk**, for example, characterized its adaptation to mires, peat, and other organic soils. Another version, "swidden," applied to periodic heath burning, entered English with Nordic invaders, and later became the generic term adopted by anthropologists to describe slash-and-burn cultivation everywhere.

If swidden remade the land, it is also true that swiddening assumed many of its characteristics from the lands it touched. Swedish **svedjebruk** served grazing as much as cereal cultivation; Finnish swidden, primarily the cultivation of rye and turnips. Both obeyed a similar rhythm of cutting, firing, planting, refiring, grazing, and fallow, followed by the repetition of the sequence that gives both expressions their close resemblance. Each, in turn, adapted its specific expressions to the landscapes, fuels, and peoples with which it interacted. One form characterized old-growth forest; others, various varieties of forest fallow; an important variant, organic soils otherwise impervious to farming; still others, mires, peatlands, and stony hills, any site with agricultural potential. The varieties were legion.

A swidden landscape was complex and intricate, rife with biotic niches. In any one year settlers had an abundance of sites in various stages of development - some felled, some freshly burned, some put to rye, oats, barley, or turnips, some grazed, some reburned, some abandoned to fallow. There was room for most of the species that had existed before **svedjefolk** appeared, and room for the new species that followed and served them. But swidden also preserved habitats, a diversity of landscapes and at least some of the processes, fire included, that had made those environments. If anything the diversity increased at any one site because swiddening speeded up the rates of change. What had burned in a century now burned in decades. Scandinavia became an environmental kaleidoscope that humans constantly turned with ax and fire.

The torch was in fact indispensable. Only by burning could the land be refashioned into usable forms; only with fire could vital secondary processes like grazing by goats, cattle, and sheep or the chemical distillation of wood into charcoal, tar, potash, and pitch be effective. In the absence of adequate animal manure, the mixed agriculture of central Europe was impossible, a condition not overcome in marginal regions until the industrial revolution made chemical fertilizers widely available. Until then burning assumed that function - and others. And although anthropogenic fire could only modulate, not overcome, climate, it helped define climatically "good" years from "bad" ones. Bad years were those in which drought replaced **svedje** with wildfire, and those wet years that made vigorous burning impossible. Without fire swidden failed. Without a good burn swiddeners faced famine.

Swidden was, moreover, the critical technology for pioneering, for converting wildland to cultivation. It had a dynamic - a capacity to expand - that the sedentary cultivation of central Europe lacked. Here the link between social order and fire practices becomes again apparent. The prevailing political structures distrusted broadcast burning, even as swidden, in favor of sedentary farming. Instead swidden was relegated to the frontiers, the perimeters of expansion by monastic orders like the Cistercians or island-colonizing Norse or pioneering peoples like the Finns.

The great internal colonization of Finland proceeded through swidden. Apparently some kind of fusion occurred in the 10th century between Finns and Slavs around Novgorod such that Finns, both pushed and pulled - crowded by other peoples and attracted by lightly-settled lands - moved across Lake Ladoga and into the northern interior. Century by century the line of settlement advanced northward. Not surprisingly, when the Vasas wanted to expedite the colonization of Sweden's interior and northlands, they brought in the more dynamic Finns who promptly slashed-and-burned beyond the grasp of bailiff and tax collector. The advent of new Nordic settlers in fact was everywhere marked by a horizon of charcoal - on Iceland's steppes, among Greenland's distant fjords, at Sweden's abortive colony along the Delaware River, and later, the American North Woods to which Scandinavian immigrants flocked.

Boreal Eurasia: The Russian Experience

From their first appearance in the 6th century the Slavs, already both a forest and a farming people, practiced a full-service European agriculture. From their cultural hearth in the Vistula, they migrated northward, then eastward. The process accelerated with the growth of the early Russian state, and concentrated first in the mixed forests that characterized much of Europe, then along the forest-steppe border. From the perspective of fire practices, the Slavic expansion across the boreal landscapes of the Eurasian plain resembles the Finnish explosion across Scandinavia, although the scale was of course much vaster, the interactions with other peoples more complex, and the political context increasingly more distinctive.

Fire Farming

Swidden was the fiery wedge that helped break open the impenetrable **taiga**. As in Scandinavia the variants were many. The earliest to appear was apparently **podseka**. Practitioners distinguished between **lyada**, which targeted primary forest, and **syroseka**, which recycled sites regrown with secondary forest. Another variant, believed by some scholars to constitute a distinctive form of swidden, was **perelog**. In this widespread system cultivators worked the site for a few years, then abandoned it to 20 or more years of fallow. Adaptations of long-fallow fire farming to the more sparsely forested steppe border led to the **zalezhnaya** system. As in Scandinavia, variations were as common as the peculiarities of particular landscapes.

Fig.1. Russian monks clearing land in classic three-part rhythm: the **incisores** cut, the **incensores** fired, and the **extirpatores** removed the trunks. By reclaiming heath lands (still cursed from the Fall), the religious merged with the secular, and Europe's monastic-led expansion was an agricultural counterpart to its crusades by sword. From a 16th century tapestry.

Armed with swidden technology, Russian agriculture grew primarily by expansion, secondarily by intensification. In this too, it followed European precedents as it did the central European forest, wedging out towards the Urals. The three-field system disseminated slowly, not becoming significant in European Russia until the 16th century. Far more common were variations of two-field systems, with an infield manured for gardening and outfields subject to a regimen of swidden and grazing. Until settlement intensified, it appears that, as in Scandinavia, several systems coexisted. Clearly too, given the pull of lands eastward, intensification depended on political control, even to the point of oppression. A three-field system could conform to the needs of a stable European state more readily than could swidden, which was far better suited to pioneering, its centrifugal tendencies often at odds with the centripedal ambitions of a centralized state.

In fact fire helped define the borders of imperial Russia, both geopolitically and biogeographically. During the long contest with the pastoralists of the steppes, the Russian frontier was fortified with fire as well as logs and embankments. A scorched earth could, if properly timed, discourage invasion or raids, so the grassy borders were routinely burned. Mis-timed, such practices could backfire as they did during a 1687 campaign in which Russian and Tartar armies disintegrated in disarray before a fire 250 km long and 100 km wide, but not before 40,000 troops reportedly died. The routes of Russian travel were annually fired in the spring, as Anton Chekhov beautifully describes in his Siberian travels (Chekhov 1959). Likewise, fire - this time in the forest - often announced the Russian invasion eastward. In a famous prophecy a Buryat shaman revealed a vision in which great fires converted the dark coniferous forest to birch and light pine. The white-skinned Russians would, so the interpretation went, replace the dark-skinned Mongolians (Erman 1848). And indeed the forecast invasion came with fire as well as sword.

Fire Environments

While swidden formed the basis for permanent settlement, it complemented or coexisted with other fire-dependent economies. Lands otherwise inaccessible to farming (or abandoned swidden) could be grazed; but effective grazing typically required burning. Gathering of berries, mushrooms, and other forest edible products often depended on some regimen of fire to sustain the fields. Charcoal, potash, tar, pitch - all depended on fire to convert forest into a usable resource. Mining too often required fire. Prospectors burned over sites obscured by vegetation, used bonfires to melt through permafrost to seize gold-bearing gravels, and burned vast quantities of wood and charcoal to smelt ore. Fires also sustained the salt industry by powering artificial evaporation.

The built environment too was shaped by fire. Roadsides were burned out during the spring; peasants burned around fields, letting fire finger into the forest; extensive burning helped shield the land from wildfire. Villagers sought to protect wooden towns from lightning and conflagration. The threat of fire - cities routinely burned, much like the taiga - entered into urban design. (The forest felling, building and subsequent burning of towns was so common that it almost constitutes a macabre form of urban swidden.) Everywhere fire helped define the boundary between the domestic and the wild.

The constant recourse to the torch, however, not only restructured fire regimes but expanded the dominion of wildfire. If fire was essential, it was also dangerous. Fires often escaped, particularly during times of drought. Lightning fires found new fuels in the slash left behind from logging and landclearing. The hazards were greatest during the time of transition. If clearing continued, if agriculture intensified, if roads and fields crossed the land like so many fuelbreaks, then wildfire would recede. Ironically, the greatest fire threats then appeared in wooden villages and cities. It is no accident that the Russians have, as a patron saint, the Prophet Elijah, who disappears in a chariot of fire while tossing his mantle to earth, protection against lightning and fire.

Flaming Front

The progress of Russian colonization eastward resembled the propagation of a fire. There was a flaming front along the outward perimeter, pockets of residual burning behind, and ultimately only embers and ash. On the European model Russians confined fire to the hearth or to hearth-equivalents, often explosive, like the brevity of that season.

The effects were evident for anyone who wished to look. After the gargantuan 1915 fires the Russian scientist V.B. Shostakovitch noted the regular burning required for habitation, and observed that with "such fires being repeated from year to year, most of the new Russian settlements within the taiga are surrounded with burned out forests. This sad picture is to be observed over millions of acres" (Shostakovich 1925).

But let Fridtjof Nansen's 1914 observations, made in Manchuria, stand as an epitaph for their environmental consequences:

> *It was strange, by the way, here, as everywhere in Siberia, how seldom one saw really big trees; the forest seemed often to consist of nothing but young trees; this is not because they are felled, but rather because they are wantonly burnt; and there is no end to these fires, one sees signs of them everywhere* (Nansen, 1914).

And let S.V. Maksimov's comments describe their human consequences:

> *This same force maintains in the contemporary generation as well a constant, anxious spiritual state, and in this regard, forested and wooden Rus' is even in a special, exceptional position in comparison with other countries: she is like an inextinguishable bonfire which, never going out completely, first weakens, then flames up with such monstrous force that any idea of fighting it vanishes: an entire sea of flames is spread by a whirlwind of fire from one end of our unfortunate land to the other, and destroys without a trace forests, planted fields, villages, settlements and towns* (Frierson, 1993).

3. Industrial Fire

The industrial revolution profoundly reworked these fire regimes, and continues to do so. It influenced them in ways both direct and indirect; it expressed itself through changes in fuels, land use, and fire practices. In forms equally subtle and irresistible, industrialization redefined the geography of Eurasian fire and the nature of humans as fire creatures. The combustion chamber challenged the hearth.

As technology placed the power of fire into machines, controlled combustion began to replace controlled burning, and fossil hydrocarbons substituted for biomass. The curve of exhumed coal and oil traces the modern era of planetary fire. Fossil fuels for engines, fossil fallow for fields, fossil fodder for mechanical oxen and horses, fossil distillates for naval stores - industrial combustion substituted for agricultural burning as silica replaces lignin in petrified wood. The revolution in transportation made possible by steam power expanded the global market, allowing new economic forces to penetrate previously untouched landscapes and to remake old landscapes to new purposes. It rewired the energy circuits of the Earth's ecology. It redesigned nutrient cycles according to the trophic hierarchies of a money economy.

Industrialization redefined what in nature was a resource and what was not, what land uses were appropriate and what were not, what regions were accessible and what were not. The revolution demanded not only new combustion technologies for furnace and forge but new fire practices for field and forest. Forestry, for example, competed with agriculture; public parklands with pastoralism; active fire suppression with controlled burning.

From Vildmark to Tree Farm: Scandinavia

The revolution came to Sweden in several ways, but the most dramatic involved the development of industrial logging. The whole structure of land use changed, particularly in Norrland. The spectacular commercial success of the timber industry encouraged the belief that the boreal environment was better suited to silviculture than to agriculture. By 1903 Sweden enacted the first of its modern forest laws, one that required that harvested woods be replanted, and in this way began the process of converting farms from the production of wheat and barley to pine and spruce. For much of the century, forest products have accounted for half of Sweden's export income - ample incentive beyond a commitment to Germanic "clean floor" silviculture. By 1990 a series of ever more prescriptive forest laws had effectively transformed Sweden's forests into a vast tree farm.

The Abolition of Fire

All this influenced the country's fire environments. Land not only remained forested but former wastelands were converted and new lands created to support further planting - Sweden's heaths, for example, all but vanished, and much of it's mires were drained. In the north, domestic reindeer increased to unprecedented numbers, browsing off many surface fuels. Over time forests edged into a duo-culture of pine and spruce, the two most valuable commercial species. This highly intensive management demanded roads, which broke up fuel complexes and provided ready access to fires. Less than 2% of Sweden was committed to parks, and much of those extended in the **fjell** over the timberline. And of course forestry institutions, public and private, both sought to control wildfire. Fire no more belonged in the cultivated woods than in a wheatfield. By the mid-1980s Sweden no longer kept statistics on wildland fires.

The national fire load plummeted. Of course there were spectacular wildfires from time to time, especially where large volumes of logging slash littered the ground. There were great outbreaks in 1868, 1878, 1888, almost in lock-step with the marching timber industry. The last bust, however, came in 1933. What really diminished was the background count of burning, the quantity of land that remained too remote for active manipulation, and the regimen of agricultural fire. Gradually industrial combustion replaced the fire-assisted technologies and land usage that had defined the Swedish past. Fire loads shrank but did not vanish.

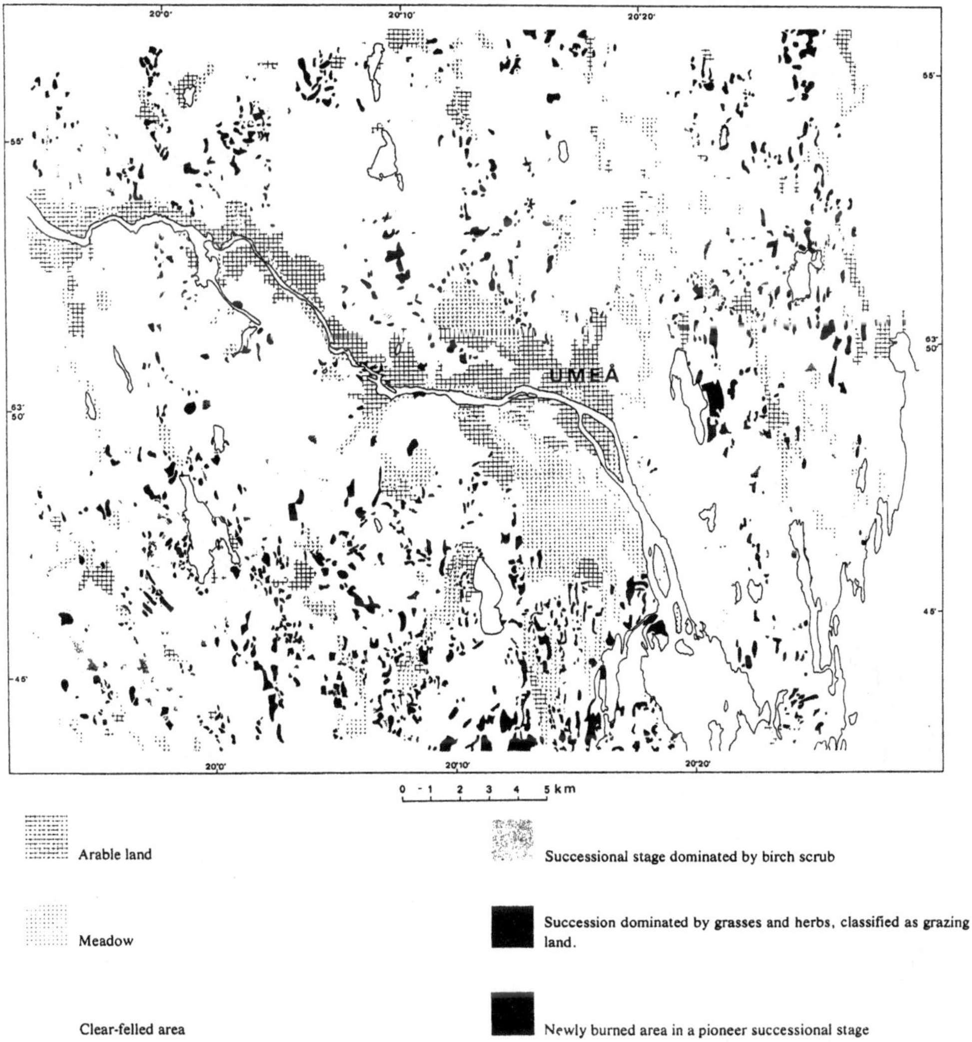

Fig.2. The swidden geography of Sweden. The vegetation mosaic around Umeå in the middle of the 19th century clearly shows the imprint of burning. Most of these fires were probably pastoral in purpose, or supported small fields soon given over to grazing, although fire for swidden and fire for pasture were rarely segregated. By this time, moreover, burning from all agricultural causes was in rapid recession. From Zackrisson (1976).

This process persisted until the 1950s when Europe's post-war reconstruction and subsequent economic boom intensified pressures on Sweden's forests. Until then, at least in the north, forestry had sought to transform rather than abolish traditional burning. To regenerate pines it adapted swidden agriculture to a swidden silviculture, modifying slash-and-burn farming to forestry, a practice known as **hyggesbränning**. But forestry, like agriculture, experienced a green revolution that condemned traditional practices and began to substitute controlled combustion and the products of fossil hydrocarbons for them. Fire became anathema.

A similar scenario applies to Finland, one not disguised even by Finland's extraordinary political odyssey during the 20th century. Conversion was delayed by the wastage of wars, and after World War II, the resettlement of Karelian refugees, much of them through a revival of traditional swidden. By the mid-1950s, however, roads, industrial policy, Swedish-model silviculture, among other factors, had more or less driven fire from the landscape. Like the Swedes, the Finns also abandoned the use of fire for forest regeneration. In 1912 the English traveller Mrs. Alec Tweedie commented on the abundance of fire everywhere in Finland, as much a plague, she believed, as locusts in Morocco (Tweedie 1913). Eighty years later those fires were gone.

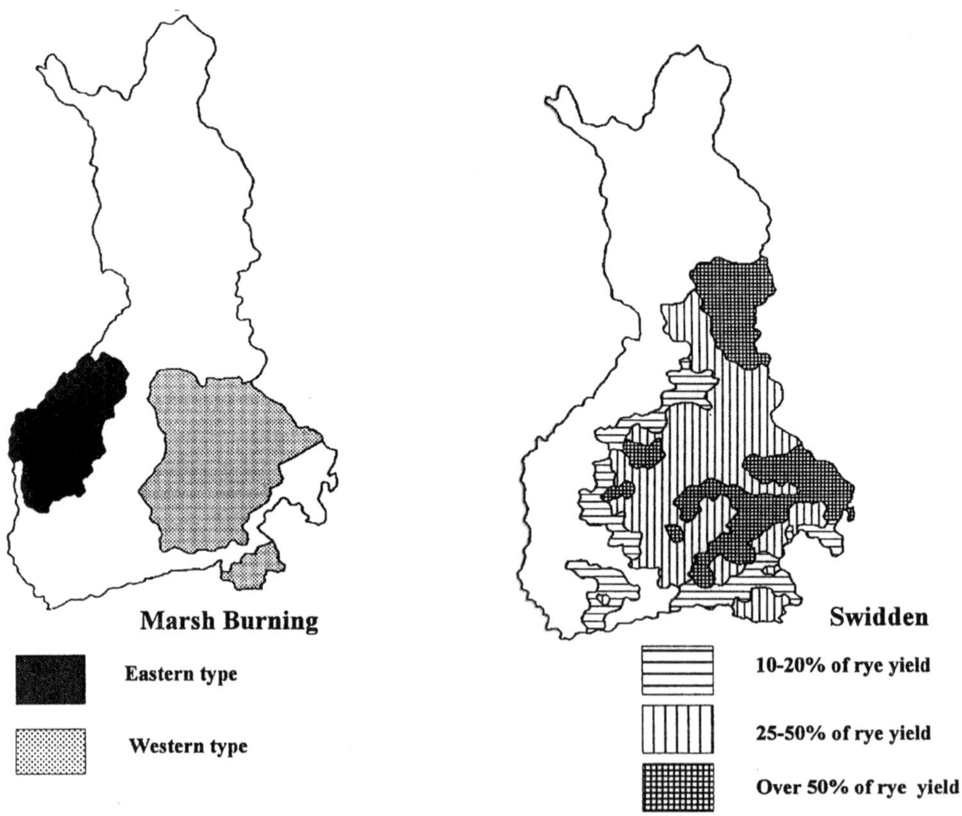

Fig.3. The swidden geography of Finland, ca. 1830s. The map on the right shows the concentration of forest-based swidden; the map on the left, the adaptation to marsh lands. In addition the use of fire to manufacture charcoal, tar, potash, and other forest products defined the accessibility of the land. Redrawn from Soininen (1974).

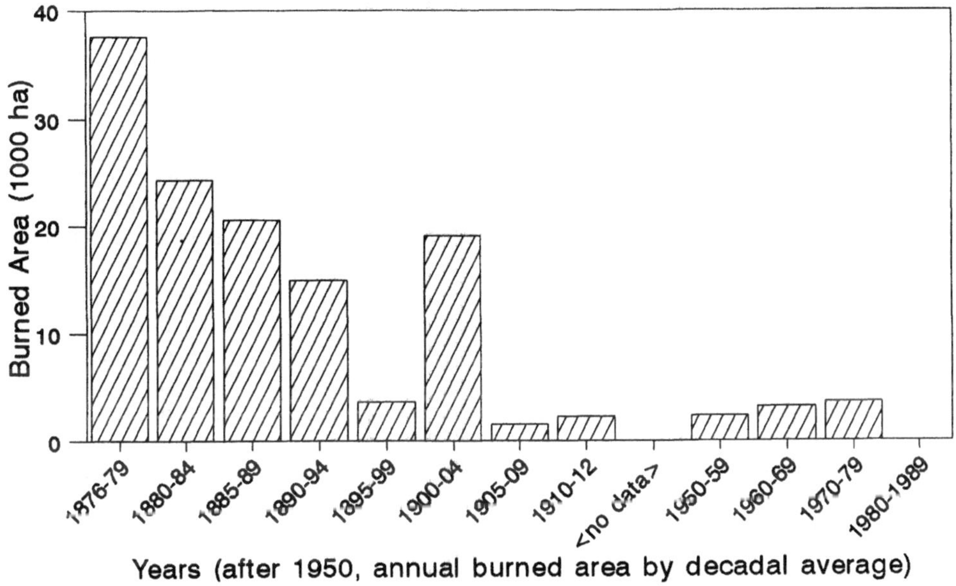

Fig.4. The suppression of fire: Sweden. Wildfire decreases rapidly and becomes virtually extinct in the post-WWII period. Not shown is the controlled use of fire for swidden and **hyggesbränning**; the first vanished in the 1930s, and the second, in the 1960s. The combined effect has been to impose a severe fire debt on the Swedish environment. Data from Lundberg, "Om skogseld," and Brian Stocks, Forestry Canada.

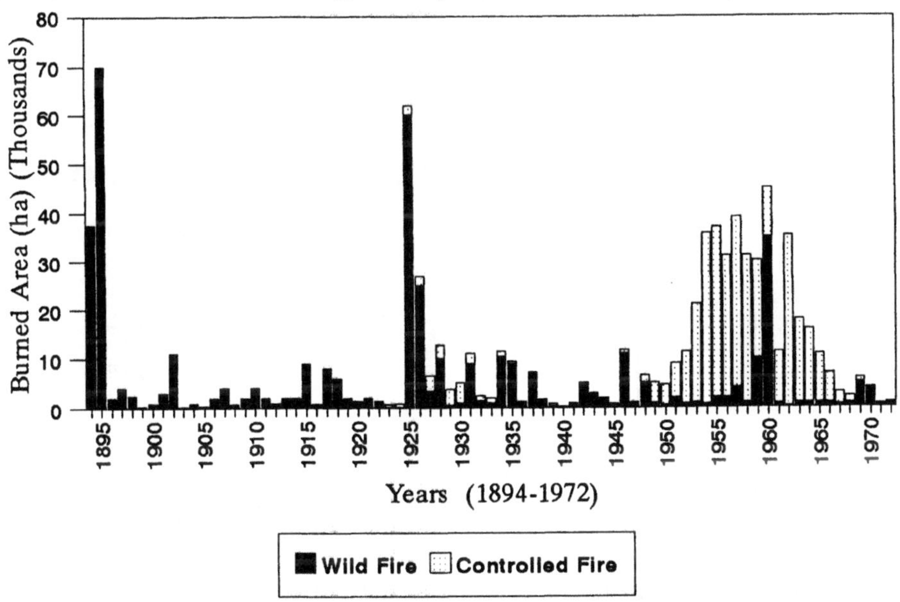

Fig.5. The suppression of fire: Finland. The figures pertain only to fires in recognized forests, and so do not include swidden which, although officially banned, continued extensively. "Controlled fires" refer to regeneration burns associated with logging, a partial surrogate for agricultural swidden. By the 1970s the final extinction was complete. Data from Sirén (1973).

The Restoration of Fire

In a sense the post-war Swedish environment had assumed the character of Swedish society. Similar principles informed both - a passion for neutrality and egalitarianism, and an abhorrence for the violent and the wild. Sweden's nature was being levelled, homogenized into a commercial plantation. But Sweden's philosophy of human relations did not readily extend to the natural world..Environmental protection meant protecting people from hazards in the environment, not protecting the environment from human activity. It was clear that many of the costs of Sweden's marvelous social structure were being downloaded onto the natural environment.

In recent years, however, interest has grown regarding the preservation of natural and traditional landscapes, and in studying environmental philosophies, even those biocentric ethics that seek to decenter humanity from a priviledged status in the cosmos. Nature reserves are growing in number, though they remain small in area. Fuels are beginning to build up, selectively but inexorably. The Swedish Riksdag is proposing to revise the forest law to accommodate matters like biodiversity. Fire history and fire ecology have found footholds among Swedish academics, and as something more than biotic archaeology. It is likely that the pressures for some kind of re-introduction of fire will increase. If so a similar movement will probably emerge in Finland.

Figure 3 A. Soot- sphere accumulation in three varved lake sediments in Sweden. For lake locations see inserted map.

B. Annual levels of coal and oil combustion in Sweden, based on data for import, production and export of coal and mineral oils.

Fig.6. The industrialization of combustion: Sweden. Fossil-fuel combustion replaces charcoal in sedimentary deposits. Not only for fuel but for other by-products, fossil hydrocarbons are replacing biomass, and their burning and distillation are substituting for other fire-catalyzed technologies. Reproduced from Wik et al. (1986).

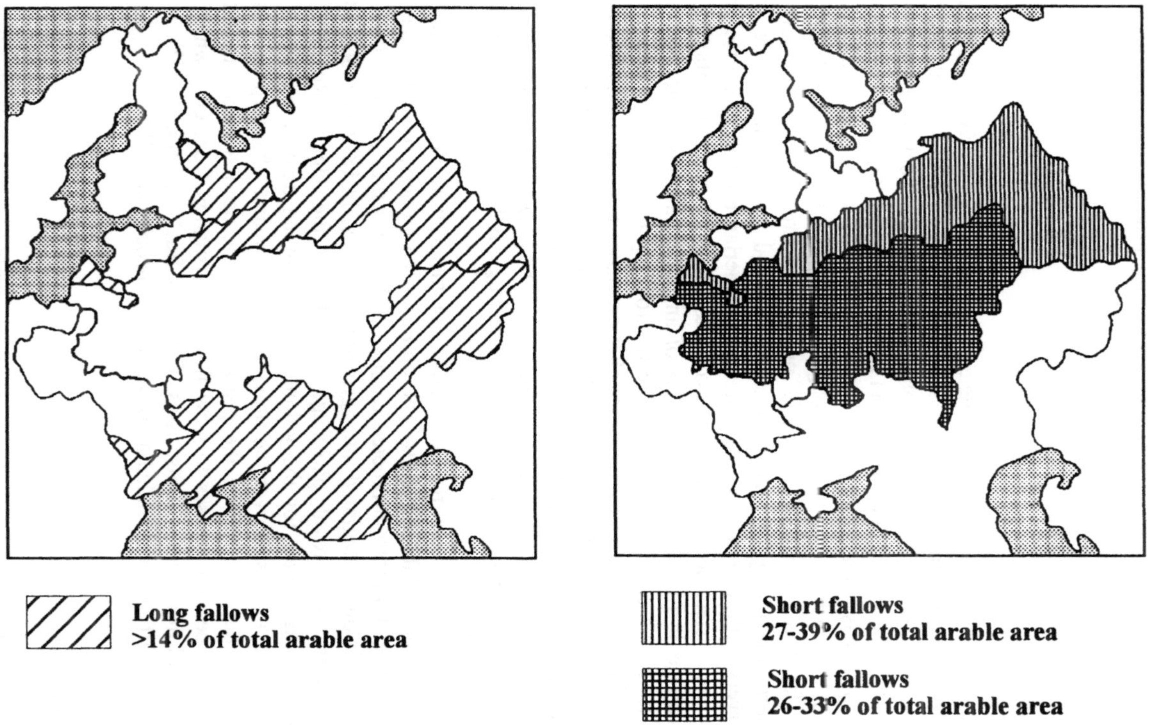

Fig.7. The fire-fallow systems of European Russia, ca. 1900. The geography reflects of long and short fallow reflects both climatic influences (the wedging out of the temperate, mixed forest) and history (the longevity of farming). Redrawn from Pallot and Shaw (1990).

The Great Transformation of Nature: the Soviet Union

Early industrialization in Tsarist Russia coincided with a host of dramatic political and economic changes - the Emancipation of the serfs (1861), imperial expansion east and south, the shift in commercial agriculture to Ukraine, a new level of integration into the global market through steam-powered transport by ship and rail. All this, and other reforms, changed how land was used, particularly forest land. Inevitably these reforms redefined what kind of fire practices were appropriate.

Pre-revolutionary Fire

In the simplest terms, forestry successfully competed with farming for private woodlands, and the need for fire control triumphed, at least in principle, over traditional fire use. Landowners sold timber in vast quantities. This, plus state needs and continued agricultural expansion, meant that by the 20th century European Russia was suffering rapid deforestation and soil deterioration. It has been estimated that between 1696 and 1914 some 67 million hectares of European Russia had been cleared, roughly 28% of its forests (French 1960).

Remoteness and far less intense demographic pressures spared Siberia this level of destruction, but there was plenty of reckless clearing at local levels and the fire scene impressed all observers with its extravagance. The Trans-Siberian railway of course drove through the **taiga** like a steel wedge, hurling sparks to all sides and redistributing the pressures for anthropogenic fire along the isograms of an industrializing geography. Thus along the Yenisey River a break in fire regimes occurs roughly in the 1870s. Before then large fires coincide with drought; afterwards, with other processes not encoded in tree rings. Almost certainly these are anthropogenic. Much of the new burning was of course concentrated into settlements and corridors, but the industrial opening of Siberia encouraged a wide scattering of fires, like a strong wind on a campfire. The great Western Siberian fires of 1915, which spewed a smoke cloud estimated to be as large as Central Europe, mocked any presumption at control.

At the same time, to protect a resource now valued for much more than fertilizing ash, the more progressive Russian establishment sought, on European models, to contain the fires that pervaded Russian life. Foresters promoted German-style fire control, at least on forest plantations and woodlands once integrated into agricultural estates. In 1893 members of the nobility and professions, with royal patronage, established the United Russian Fire Association, an "organizing center for the volunteer fire fighting association which had sprung up through European Russia over the previous twenty years." Its purpose was not only to control wildfire but to convince the peasantry "of the need to take precautions against fire..." Until the peasant "ceases to view fire as a visitation from God which one must not go against, until then fires will dispossess Orthodox Rus'" (Frierson 1993). It was clear that the conversion of land demanded a conversion of people as well.

Fire in the Minds of Men

This the Bolshevik Revolution promised. The new regime would hasten industrialization and it would modernize thinking about the world. The Soviet state forecast a Great Transformation of Nature to serve the New Soviet Man. Its commitment to full productivity meant that "waste lands" would be converted to "useful" land. Despite state ownership deforestation continued, and despite edicts, reforestation failed to keep pace. Fire control, however, did join the official duties of the sanctioned "production enterprises" that oversaw logging. The regime's commitment to a scientific ideology meant that traditional attitudes toward fire suffered official suspicion, if not outright censure. During the Great Terror, for example, forestry ceased to burn slash as part of a regeneration program; an effort at restoration did not revive until after Stalin's death.

Far from being exceptional the suppression of slash burning betrays a pattern characteristic of 20th-century Russia, a deadly dialectic between nature and culture in which thesis and antithesis were often reconciled in flames. With remarkable consistency the great fires of the century have coincided with periods of climatic drought and social unrest. The conflagrations of 1915, 1921, and 1936-1939, for example, occurred amidst war, civil war, and internal purges, and represent not only a hostile environment but a fragmented social order unable to control either its own or nature's fires. Of course Russia is vast, fires common, and political turmoil frequent; but this hall of flame registers something more than coincidence; it reveals a serious uneasiness, even a contradiction in how Russian society has related to its environment.

Initially, it appeared that the new regime would seek to control fires as it did all other aspects of Russian life. In 1921 the Executive Committee passed a special act addressing "The Struggle with Forest Fires." Research concentrated around the Leningrad Forestry Academy. In 1931 experiments adapting aircraft for fire reconnaissance proved encouraging; the next year, operational flights were organized for southern Karelia and northern Leningrad; by 1935 the program led to a new profession, the pilot-observer; and the following year it inspired a base at Irkutsk; and in 1936 smokejumpers supplemented observers. At the same time other members of the Leningrad cadre developed a surer scientific foundation for fire control.

The war devastated the Leningrad school; its leading researchers, Serebrennikov and Matreninsky, both died during the siege of Leningrad. Military demands also stripped the fledgling aerial fire protection service of planes and personnel. But in the war's aftermath, fire protection emerged stronger than ever. I.S. Melekhov, V.G. Nesterov, and N.P. Kurbatsky revived Soviet fire science; aerial fire protection (**Avialesookhrana**) began an explosive growth east of the Urals. The Soviet Academy of Science too supported research, and in 1959 established a Laboratory for Forest Fire Research at the V.N. Sukachev Institute for Forest Science in Krasnoyarsk.

Fire protection flourished in the post-war era, perpetuated by the geopolitics of the Cold War. Some of the drive to industrialize Siberia, in particular - the "transformation of nature" announced by Stalin and his immediate successors - can be viewed within a wider perspective, part of a global frontier of "development" that penetrated to such places as Amazonia, the Canadian archipelago, Alaska, Borneo, and northern Australia. But some is specific to Soviet experience, especially the distortions wrought by the Cold War, broadly conceived. Thanks to aircraft and intercontinental ballistic missiles, even remote regions of tundra and **taiga** became points of vulnerability. The next world war, both the USA and the USSR believed, would be a fire war, kindled by thermonuclear weapons. Moreover, flushed with the assertion of power made possible by the Great Patriotic War, the Soviet state sought to confirm its presence throughout its unwieldy dominion. If only symbolically, fire control did this.

Particularly after the death of Stalin, the Soviet apparatus for fire protection expanded fantastically. The Soviets established a world-class fire research program, and

Avialesookhrana fielded the largest aerial firefighting operation in the world. The fires that infested European Russia in 1972 prompted a major investment by the state in scientific and technological research, and an expansion in firefighting operations; it even inspired a national fire-prevention logo (a moose fleeing a fire). By the mid-1980s the Ministry of Forestry supported fire-related research within institutes located in St. Petersburg, Krasnoyarsk, and Khabarovsk; and the Academy of Sciences sponsored a spectrum of fire studies, some within the rubric of forest ecology, much at the dedicated Laboratory of Forest Fire Research at Krasnoyarsk.

While part of the incentive for this commitment resides within the nature of internal industrialization, part clearly derives from geopolitical considerations. Thus the Soviet repudiation of fire folkways and the intensified campaign to purge fire from the landscape correspond nicely with parallel developments in Scandinavia. But they also mimic, with eerie fidelity, moves by the Soviet Union's principal Cold War antagonist, the U.S., which dramatically expanded its own fire research facilities and its commitment to aerial fire control at exactly this same time for what appear to be many of the same reasons. Aerial fire protection came to Alaska at roughly the same time it came to much of Siberia, and for many of the same reasons.

Nor did the process end there. The Soviet Union extended its fire technology and philosophy to once and future allies. China, Mongolia, Cuba, eastern Europe - all, to varying degrees, adopted the Soviet model of fire protection, particularly the use of smokejumpers and helirappellers for initial attack. Fire protection also served the state in its relations with minority groups in Siberia, particularly those dependent on domestic reindeer. As those herds swelled (as they had also in Scandinavia), the protection of the lichens that served as winter range became ever more critical. In fact the explosion in reindeer population was impossible without such fire control. The Soviet system, and the philosophy behind it, thus extended across all of northern Eurasia, excepting only Scandinavia. The modern history of fire in the subcontinent is very largely the story of Soviet practices.

Perestroika on Fire

By the mid-1980s, however, fire control shared with many other elements of Soviet society a sense of approaching crisis. Figure 1 conveys the basic structure of change, aptly framed between the great fires of 1972 around Moscow oblast and those of 1987 in the Transbaikal region.

The initial effects of suppression were large. Creating a first-order fire control system dramatically reduced burned area. But extending those returns proved elusive. By the latter 1980s, expenditures had already completed a normal logistic curve as fire protection approached a point of diminishing returns. After the 1972 fires, authoritarian edicts banned burning, which further dropped the number of reported fires. This effect was temporary, however. It did nothing to eliminate accidental fires, arson fires, or lightning fires, and it was unenforceable, impractical, and ecologically misguided. It allowed fuels to build up in sensitive areas. It demanded that burning, essential to Russian agriculture, be conducted illegally, which is to say, without adequate supervision. Fire starts began to rise again, and, first here, then there, they encountered a favorable environment.

During the brief era of **perestroika**, these trends combined with climatic conditions to spark an explosion of fires. The 1987 fire complex burned probably 6 million hectares in Chita oblast, and perhaps over 12 million hectares throughout Transbaikalia and southern

Yakutia, a biotic meltdown that mocks the better publicized fire at Chernobyl. (These figures come from satellite imagery analyzed by Forestry Canada; the official Russian response remains one of denial, that only 94,000 ha burned, that the "incident" remained under control.) It was clear, however, that the apparatus was at odds not only with nature but with itself. The more progressive elements of the fire establishment sought reform along Western models and with modern technology. Political parallels culminated in the August revolution of 1991.

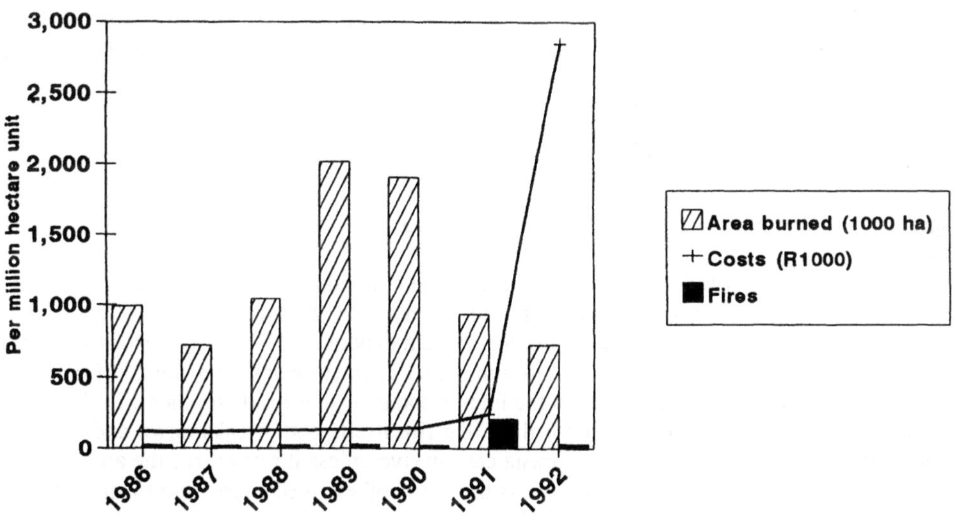

Fig.8. The suppression of fire: Russia. Officials regularly downgraded the magnitude of the burned area. Probably the numbers listed by Avialesookhrana are one and (in some instances) two orders of magnitude too small. Still trends are apparent, and are likely accurate. Source: Avialesookhrana.

The immediate consequences for fire management were the *de facto* disestablishment of **Avialesookhrana** and the fire sciences. The infrastructure for fire control began to disintegrate along with its sense of mission. Inflation, loss of personnel, escalating costs of aviation fuel, confusion over land ownership, political squabbling, the evaporation of geopolitical goals - all conspired to push costs up, depress firefighter numbers down, and send fire research scrambling for safe ground. An all-union organization shrank to fit the dimensions of the (still immense) Russian Federation. A providential drought in the summer of 1992 alarmed authorities and spared the Russian fire organization from a free-fall into disintegration.

Still, the political ferment dissolved the geopolitical mortar that had held together a more or less common fire protection system across most of northern Eurasia. Russia, Mongolia, China, Belarus, Ukraine, Kazakhstan, and others - all had to find their own logic for managing fire, and to justify appropriate institutions. Most sought contact with the West, especially with North America. **Avialesookhrana**, in particular, boldly sought exchanges with American counterparts, and the international fire science community responded with efforts to integrate Russian colleagues. By 1993 the U.S. and the Russian Federation signed a mutual assistance treaty for wildfires. Inspired by Johann Goldammer of the Max Planck Institute for Chemistry, North American and European colleagues, under the aegis of the International Boreal Forest Research Association, sponsored an international fire science conference in Krasnoyarsk and a field campaign (FIRESCAN) that helped absorb Russian fire into the global agenda of the International Geosphere-Biosphere Program. Alarm over global environmental change had made Russia as important to Western scientists as Western scientists were to a reforming Russia.

The longer term effects are unclear, as they would have been even had the August revolution not occurred. The core question remains one of mission. The Russians have to decide what kind of fire management best suits their emerging society. Almost certainly fire management will replace simple fire control, but what this means in the Russian context is not obvious. The task is daunting; but the stakes are large, for the world as well as for Russia.

Conclusion: First Nature, Second Nature, and Fire

The concept of "fire history" has two meanings. The common sense interpretation is that it describes the history of fires, fire practices, and fire regimes. The second sense refers to the study of fire as a means of better understanding human experience. Thus fire histories differ, even among regions that share similar environmental characteristics, and those differences both reflect and reveal characteristics in the human societies that reside on them. In this way fire history resembles a dialectical torch that, like fire itself, both breaks down and binds.

No system of land use in northern Eurasia can survive unless it recognizes the ancient and enduring power of fire. In some areas, fire is useful; in others, necessary; and in still others, essential and inevitable. The problem of fire management is not to abolish fire but to establish its proper proportion, to reconcile fire use with fire control. What this means specifically will vary with the specific purposes of specific lands. And those determinations will arise out of human understanding, or misunderstanding; of human will, and error; of human purpose, for good or ill. Through the dialectic of anthropogenic fire first nature will become second nature. But then fire has always been second nature to humanity.

Acknowledgements

I would like to thank Richard McNabb and Jonathan Gerber for translations from Russian to English; Hannale Mortensen, for translations from Finnish; Gunilla Oleskog, for help with Swedish; Cathy A. Frierson, for generously sharing drafts of her research regarding fire in European Russia; Henry Telitsyn, for his work with N.P. Kurbatsky in writing a historical synopsis of fire science in the Soviet Union; V.V. Furyaev, for contracted essays on aspects of Russian fire science, history, and practices; and the staff of the Russian Aerial Fire Protection Service, most especially Eduard Davydenko.

References

The following bibliography of further readings is partial - many, many more sources exist.

Adams, A. 1870. Travels of a naturalist in Japan and Manchuria. London: Hurst and Blackett.

Arrhenius, J. 1862. Handbok i Svenska Jordbruket, 3 vols. Uppsala.

Artsybachev, E.S. 1984. Forest fires and their control. Russian Translation Series 15. A.A. Balkema.

Bater, J.H., and R.A. French (eds.) 1983. Studies in Russian Historical Geography, 2 vols. Academic Press.

Bell, J. 1966. A journey from St. Petersburg to Pekin. 1719-22. J.L. Stevenson, ed. Reprint, Barnes and Noble, Inc.

Berglund, B.E. (ed.) 1991. The cultural landscape during 6000 years in Southern Sweden - the Ystad Project. Ecological Bulletins 41. Munksgaard International Booksellers and Publishers.

Blandon, P. 1983. Soviet forest industries. Westview Press.

Broberg, R. 1989. Finsk invandring till mellersta Sverige. Skrifter Utgivna av Föreningen för Värmlandslitteratur 7. Uppsala.

Campbell, Å. 1982. Från vildmark till bygd. Två Förläggare Bokförlag.

Chekhov, A. 1959. The unknown Chekhov. Stories and other writings. Avrahm Yarmolinsky, trans. Peter Owen Ltd.

Crosby, A. 1987. Ecological imperialism. The biological expansion of Europe 900-1800. Cambridge University Press.

Engelmark, R. et al. 1976. Palaeo-Ecological Investigations in Coastal Västerbotten, N. Sweden. Early Norrland 9.

Erman, A. 1848. Travels in Siberia, 2 vols. London.

Fraser, J.G. 1922. The Golden Bough (abridged). Macmillan Co.

French, R.A. 1960. The Making of the Russian Landscape. Advancement of Science 20, 44-566.

Frenzel, B. 1968. The Pleistocene Vegetation of Northern Eurasia. Science 161, 637-649.

Frierson, C.A. 1993. Flame and conflagration in the daily life of the peasants of European Russia in the Nineteenth Century (unpublished essay).

Grotenfelt, G. 1899. Det primitive jordbrukests metoder i Finland under den historiska tiden. Helsinki.

Heckscher, E.F. 1968. An economic history of Sweden. Göran Ohlin, trans. Harvard University Press.

Heikinheimo, O. 1915. Kaskeamisen vaikutus Suomen metsiin. Helsinki

Huttunen, P. 1980. Early land use, especially slash-and-burn cultivation in the commune of Lammi, Southern Finland, interpreted mainly using pollen and charcoal analysis. Acta Botanica Fennica 113, 1-45.

Högbom, A.G. 1934. Om skogseldar förr och nu och deras roll i skogarnas utvecklingshistoria. Nörrlandskt handbibliotek XIII. Almqvist & Wiksell.

Kardell, L. et al. 1980. Svedjebruk förr och nu. Rapport 20, Avdelning för landskapsvård, SLU.

Levin, M.G., and L.P. Potapov (eds.) 1964. The peoples of Siberia. Stephen P. Dunn, trans. University of Chicago Press.

Lewis, H.T., and T. Ferguson. 1988. Yards, corridors, and mosaics: how to burn a boreal forest. Human Ecology 16 (1), 57-77.

Mead, W.R. 1968. Finland. New York.

Mead, W.R. 1981. An historical geography of Scandinavia. Academic Press.

Montelius, S. 1953. The burning of forest land for the cultivation of crops. Meddelanden från Uppsala Universitets Geografiska Institution, Serie A, No. 87.

Nansen, F. 1914. Through Siberia. The land of the future. Arthur G. Chater, trans. Frederick A. Stokes Co.

Nesterov, V.G. 1945. Protection of forests from fires. English trans. by Canadian Forestry Service. Goslestekhizdat.

Nordström, O. et al. 1989. Skogen och smålänningen, Historiska föreningens i Kronobergs län skiftserie 6. SmpTRYCK AB

Pallot, J., and D.J.B. Shaw. 1990. Landscape and settlement in Romanov Russia 1613-1917. Oxford University Press.

Pyne, S.J. 1885. World fire. The culture of fire on Earth. Henry Holt and Co.

Rasputin, V. 1989. Siberia on fire. Gerald Mikkelson and Margaret Winchell, trans. Northern Illinois University Press.

Raumolin, J. (ed.) 1987. Special Issue on Swidden Cultivation. Suomen Antropologi 4.

Sherbakov, I.P. (ed.) 1979. Forest fires in Yakutia and their influence on the nature of forests. Nauka.

Shostakovitch, V.B. 1925. Forest Conflagrations in Siberia. J. For. 23, 365-371.

Sirén, G. 1973. Some Remarks on Fire Ecology in Finnish Forestry. Proc. Ann. Tall Timbers Fire Ecology Conf. 13, 191-209.

Smith, R.E.F. 1957. Peasant farming in Muscovy. Cambridge University Press.

Soininen, A.M. 1959. Burn-beating as the technical basis of colonisation in Finland in the 16th and 17th centuries. Scandinavian Economic History Review 7 (2), 150-166.

Soininen, A.M. 1974. Vanha Maataloutemme. Historiallisia Tutkimuksia 96. Helsinki [English summary: Old traditional agriculture in Finland in the 18th and 19th Centuries].

Sulimirski, T. 1970. Prehistoric Russia. John Baker

Söderqvist, T. 1986. The ecologists. From merry naturalists to saviours of the nation. Almqvist & Wiksell.

Telitsyn, H.P., and N.P. Kurbatsky. 1994. Essays on history of development of the theory and practice of forest fire control and use in Russia. Unpublished manuscript.

Tolonen, M. 1978. Palaeoecology of annually laminated sediments in Lake Ahvenainen, S. Finland. Parts I-III. Ann. Bot. Fennici 15, 177-240.

Tolonen, M. 1978. The history of agriculture in Sääksmäki traced by pollen analysis. Ann. Bot. Fennici 15, 47-54.

Tänd eld på skogen!. Skog & Forskning No. 4/91 (special issue on fire).

Tweedie, A. 1913. Through Finland in carts. London.

Uggla, E. 1973. Fire ecology in Swedish forests. Proc. Ann. Tall Timbers Fire Ecology Conferences 13, 171-189.

Vakurov, A.D. 1975. Forest fires in the North. Nauka.

Valendik, E.N. 1990. The struggle with big forest fires. Nauka.

Viro, P.J. 1969. Prescribed Burning in Forestry. Communicationes Instituti Forestalis Fenniae 67 (7).

Wein, R.W., and D.A. MacLean (eds.) 1983. The role of fire in northern circumpolar ecosystems. John Wiley and Sons, Inc.

Weiner, D.R. 1988. Models of nature. Ecology, conserevation, and cultural revolution in Soviet Russia. Indiana University Press.

Wik, M. et al. 1986. Sedimentary records of carbonaceous particles from fossil fuel combustion. Hydrobiologia 143, 387-394.

Wood, A. (ed.) 1991. The history of Siberia. From Russian conquest to revolution. Routledge.

Zackrisson, O. 1976. Vegetation dynamics and land use in the lower reaches of the River Umealven. Early Norrland 9, 1-74.

Zackrisson, O. 1981. Forest fire history: Ecological significance and dating problems in the North Swedish boreal forest. In: Proceedings of the Fire History Workshop (M.Stokes and J.Dieterich, techn. coords.), 120-125. USDA For. Serv. Gen. Tech. Report RM-81.

J.G.Goldammer and V.V.Furyaev (eds.), Fire in Ecosystems of Boreal Eurasia, 45-54.

Retrospective Analysis of Natural Fire Regimes in Landscapes of Eastern Fennoscandia and Problems in Their Anthropogenic Transformation

A.N. Gromtsev [1]

1. Introduction

The problem of forest fires has long been extensively dealt with in the scientific literature (Melekhov 1948; Sukachev 1975; Vakurov 1975; Spurr and Barnes 1984; and many others). However, it remains a pressing issue, especially in connection with the hot debate which arose between ecologists and economists, and which has even been discussed in the foreign news media in recent years (Abel 1985; Anonymus 1988). Nowadays, most forest ecologists consider fire to be one of the main ecological factors responsible for the structure and dynamics of the biotic components of taiga landscape, primarily natural forests. An analysis of the literature shows, however, that systematized data on fires were obtained either by dating fire scars on trees or by studying the archives. In the former case, a retrospective analysis of fire regime is usually restricted to 250-300 years (the age of the older tree generation); in the latter, a general analysis is made when only big fires are considered, without specifying areas and habitats. However, to assess the effect of fires on natural forests, detailed data covering a period of thousands of years should be collected with regard for area and the types of growth conditions under which forest-forming species develop. As well, patterns of the pyrogenic factor were discussed, with few exceptions (Furyaev and Kireyev 1979), regardless of forest landscape structure. It is impossible, however, to agree with the above authors, who argue that the problem of forest fires cannot be resolved for an individual biogeocenosis because a landscape basis is needed.

Fire regime was studied as part a comprehensive investigation of Karelian forest landscapes conducted by the Forest Research Institute, Karelian Research Center, Russian Academy of Science (Volkov et al. 1990). We understand the term landscape to represent an ecosystem of a certain order existing within a geographical landscape. It is "an ecological system dominated by structurally and metabolically interrelated and interdependent landforms, Quaternary deposits, soils, microclimates, hydrographic networks as well as phyto- and zoocenoses" (Volkov et al. 1990). Landscapes that are similar in the structure of their constituents, internal relations, metabolism and energy, are collectively defined as "a type of landscape." Our studies were based on a classification of forest landscapes based, in turn, on a zonal-typological principle (Tab.1) and a map of landscapes.

[1] Institute of Forest, Karelian Scientific Center, Russian Academy of Sciences, 185610 Petrozavodsk, Russian Federation

Types of landscapes were differentiated on the basis of genetic types of relief and Quaternary deposits, the extent of waterlogging and predominant primeval forests. For example, type 1m represents landscapes developed on mature plains, with intense soil waterlogging, where primeval forests are predominantly spruce. At the present stage of our investigation, 20 types of landscapes have been classified; 16 of these were differentiated and characterized in the mid-taiga subzone of Karelia and 15 types in the north taiga subzone. The methods and techniques used for the comprehensive study of taiga landscapes have been described in detail (e.g., Volkov et al. 1990).

Karelia is remarkable for having the most varied northern and mid-taiga landscapes existing in the taiga zone of European Russia. About 60% of the classified forest landscape types are encountered here. They occupy over 60% of this vast region. It is, therefore, safe to say that the patterns of fire regime revealed in Karelia can be applied to most of the European taiga zone.

Tab.1. Classification of Karelia's forest landscapes. Numbers in the body of the table refer to the 20 types of landscapes classified.

Predominant primeval forests	Waterlogging (%)		
	Intense (>50)	Medium (15-20)	Weak (<15)
Lacustrine, lacustrine-glacial (lg) and marine (m) plains			
Spruce	1 (m)	2 (lg)	-
Pine	3 (lg, m)	4 (lg)	5 (lg)
Glacial (g) and aqueoglacial (ag) hilly-ridge plains			
Spruce	-	6 (g)	-
Pine	7 (ag)	8 (ag)	9 (ag)
Glacial accumulation plains with rugged topography			
Spruce	-	10	-
Pine	-	11	-
Base-level tectonic hilly and hilly-ridge (bl) plains with glacial deposits (blg) and low mountain topography (l)			
Spruce	-	12 (blg, l)	-
Pine	13 (bl, blg)	14 (bl,blg)	-
Base-level tectonic ridge (selka) plains			
Spruce	-	15	16
Pine	-	17	18
Rocky plains			
Pine	-	19	20

2. Methods

Over 40 landscape profiles totalling about 215 km in length were laid out. Large-scale field studies of taiga areas were conducted. A lot of planning materials as well as archives were studied. Our goal was to reveal the regularities of the landscape structure as well as the natural and anthropogenic dynamics of the biotic components of the taiga landscape, primarily forest cover.

In the course of our mire study, peat samples were taken for botanical analysis. In the samples we noticed distinct fire layers (FL) formed by fires and consisting of ash and coal. Later, such layers were recorded systematically. Holes were drilled along the periphery of all waterlogged sites along the profiles, and in the centre of peat deposits. The number of holes at each site depended on site length (a total of 600 holes were drilled to a depth of 1.0-1.5 m). When studying FLs in core samples, we determined their number, thickness and depth of occurrence. To characterize the periodicity of fires, the following indices were used for a hole in a peat deposit and for its specific depth interval: 1) the relation of the number of holes with at least one FL to the total number of holes (%); 2) the average number of FLs per hole; and 3) the greatest number of FLs based on data for a single hole. Preliminary dating of fires was done using data on average peat accumulation rates in the region (Yelina et al. 1984). Conventional methods were also employed to elucidate some characteristics of fire regimes in dry valleys. The thickness of charcoal layers was determined in ca 50-cm-deep holes made in each type of biogeocenosis in a landscape profile. Dead wood and stumps with traces of fire were evaluated and fire scars (FS) on trees were dated. Similar work was done when studying the most typical central parts of landscape areas.

Our experience has shown that although the above technique is simple and reliable, it could be improved. We tried to overcome some of its disadvantages. Firstly, dating fires based on average peat accumulation rate is very rough. We assumed this rate to be 0.67 mm yr^{-1}, although it varies from 0.46 to 1.2 mm yr^{-1} in different types of peat deposits (Yelina et al. 1984). Furthermore, the vertical growth of peat deposits depends largely on slope angle. Obviously, radiocarbon analysis is required for the precise dating of fires (Swynar 1987) although the conclusions drawn will hardly be substantially affected. Another problem is to clearly detect FLs. They are visible in light-colored peat deposits of transitional and raised types. Doubts, if any, are dispelled by the presence of small, but easily detectable pieces of charcoal. However, recognition of FLs in dark-colored peat deposits of the fen type and in peat-humic soils is often difficult or unreliable. In this case we used so-called "indicator deposits," i.e., waterlogged habitats in this type of landscape (in the same landscape profile) with straw-colored raised peat.

All available data on the periodicity and areal distribution of fires were correlated with the structure and dynamics of forest cover in order to determine the role and characteristics of fires which control the life of forest ecosystems. The results of our studies provide a basis for some general conclusions.

3. Results

Data were obtained on fire regimes which existed in different types of mid- and north taiga landscapes in the past few millenia. Fire records and some landscape characteristics of fires are presented in Table 2. The average age of the peat deposits studied is 1500-2500 years.

Tab.2. Periodicity of forest fires in different types of landscape as shown by stratigraphic analysis of peat deposits.

Index	Fire periodicity index values in different types of landscape (explication)									
	North Taiga Subzones					Mid-Taiga Subzones				
	1m	8ag	12l	14blg	19	6g	6g*	8ag	12blg	14bl
	Occurrence of FLs at various depths (m)									
<0.20	3	14	0	10	3	0	14	10	4	51
0.21-0.50	6	12	0	10	39	14	19	32	5	47
0.51-1.00	0	48	0	2	33	0	25	69	8	74
1.01-1.50	0	35	(0)	3	8	0	12	35	(0)	50
1.51-2.00	(0)	(0)	(0)	5	(+)	(0)	0	17	(0)	18
>2.00	-	(+)	-	0	(0)	(0)	0	(+)	(0)	(0)
	Total occurrence of FLs in drill holes (%)									
	8	67	0	21	49	14	36	86	13	85
	Average number of FLs per hole at various depths (m)									
<0.20	0.03	1.14	0	0.10	0.03	0	0.14	o.14	0.04	0.63
0.21-0.50	0.08	0.16	0	0.09	0.46	0.14	0.33	0.96	0.05	1.39
0.51-1.00	0	0.62	0	0.02	0.44	0	1.15	1.14	0.17	2.96
1.01-1.50	0	0.57	(0)	0.03	0.08	0	0.29	1.35	(0)	1.94
1.51-2.00	(0)	(0)	(0)	0.05	(+)	(0)	0	0.83	(0)	0.18
>2.00	-	(+)	-	0	(0)	(0)	0	(+)	(0)	(0)
	Number of FLs per hole									
	0.08	1.14	0	0.21	0.79	0.12	1.70	3.40	0.17	4.61
	Number of FLs according to record drill hole data (yrs BP)									
<300	0	0	0	0	0	0	1	0	0	1
300-750	1	0	0	1	2	1	2	0	0	7
751-1500		4	0		3	0	5	1	2	11
1501-2200		4	0			0	2	4	0	
2201-2950			0				0	8		
>2950							0			
	Total number of fires according to record drill hole data									
	1	8	0	1	5	1	10	13	2	19

FL = Fire Layer
(0) = no FLs in less than 10 drill holes
(+) = FLs are present in less than 10 drill holes
\- = no data

Well-defined FLs were revealed at different depths. The oldest layer was found in the mid-taiga subzone at a depth of 2.85 m, which corresponds to an age of more than 4000 years. A charred organo-mineral layer was detected at a depth of 3.7 m (over 5500 years).

Consequently, fires were a natural environmental factor at least since the late half of Holocene time. Older FLs were sought for by drilling thick (up to 10 m) peat deposits not indicated on landscape profiles. It should be noted that these are only the traces of the fires which affected waterlogged habitats. The fires which spread over dry valleys alone were not included here and were recorded separately by the presence of FS's on trees.

All fire periodicity indices generally vary over a wide range. For example, the total occurrence of FLs (see the above technique) varies from 0 to 86%. According to record drill hole data, the number of fires varied from 0 to 19 in the past 1500 years; between 1500 and 3000 years ago, although our data is quite sparse, the number of fires recorded in drill holes ranged from 0 to 12. Some zonal characteristics of fire regime are noteworthy. The mid-taiga subzone is more commonly affected by fires than the north taiga subzone. Evidently, this is not a chance phenomenon, but it is difficult to interpret it convincingly at the present stage of our study. It seems that fires broke out more seldom in north taiga forest habitats in the late Holocene time because the degree of their waterlogging was twice as high. On the other hand, the warm period, during which the danger of fire is great, is much shorter in north taiga. Besides, in the past the north taiga subzone was poorly developed and was, therefore, less commonly affected by fires of anthropogenic origin.

It is generally difficult to assess an anthropogenic effect on natural fire regimes. In the post-glacial epoch, the north-western part of the taiga zone was rapidly developed by Mesolithic (10,000-5,000 years BC) and more recent tribes that moved from the south and southwest. This suggests that lightning was not the only cause of fires since then. On the other hand, until the 20th century the extent of fires was limited only by natural factors. Thus, man initially increased the number of fires without affecting their spread. The fires which broke out in the remote past cannot be divided into natural and anthropogenic. In any case, the taiga zone repeatedly suffered from fires for thousands of years regardless of their origin.

According to our data, the landscape patterns of fire regime are very prominent in the taiga. Each type of landscape has its own, more or less distinct, characteristics of fire regime. However, the following tendency is generally observed. FLs are least abundant in spruce-dominated landscapes, (e.g., north taiga: 1 m, 12 l; mid-taiga: 6 g, 12 blg; Tab.2). Fires were most common in pine-dominated landscapes (e.g., north taiga: 8 ag, 14 blg, 19; mid-taiga: 8 ag, 14 blg). In both spruce- and pine-dominated landscapes fire regimes are related to both genesis of vegetation and degree of waterlogging. In the following some fire regime characteristics in these main categories of taiga areas are discussed in detail.

3.1 Spruce-dominated landscapes

Intensely waterlogged marine plain landscape (north taiga 1 m)

Thirty-six holes were drilled, but only three FLs were found within a 50-cm-thick peat unit. Their age is no more than 750 years. A large virgin spruce forest with a maximum spruce age of 250 years and a maximum pine age of more than 300 years is preserved in this area. No traces of fire scars have been found on tree trunks. No traces of fire have been observed in the litter and in upper soil layers. Thus, only one or two fires broke out in this landscape in the past 3000 years. It is a minimum index for the region. Here, fires were very uncommon because open mires and waterlogged forest account for over 80% of the landscape.

Mid-waterlogged hilly-ridge low-mountain landscape (north taiga 12l)

Thirty holes were drilled, but no FLs were found. A large virgin spruce forest is preserved in the landscape. The maximum age of spruce is 270 years and that of pine is 300-400 years. No fire scars are seen on the trunks. The charcoal found in the litter and in upper soil layers suggest that an extensive fire, which affected all non-waterlogged forest habitats, gave rise to a new spruce forest distribution. Here, waterlogged habitats are represented by non-flammable grassy, horsetail-sphagnum, sedge-sphagnum, near-creek and other types. Consequently, in this type of landscape fires broke out no more than two times every thousand years and spread only along dry valleys alone.

Glacial hilly-ridge mid-waterlogged (mid-taiga 6 *g, 6 g*)

A total of 38 holes were drilled, but only a few FLs were revealed in peat deposits within the landscape outlines (6 g). Poorly flammable blueberry, moist blueberry, grassy, horsetail-sphagnum and other types of spruce forests, often cut by rivers and creeks, are predominant here. In the forests not affected by cutting, the trees are 230-250 years old. No traces of fire scars are observed on tree trunks. Charcoal is usually found under the litter. Along the periphery of the landscape areas (6 g*) fires occurred approximately once in 250 years (record drill hole data), sometimes more often (once in 150 years 751-1500 years ago). However, all these fires spread only along esker-like sand ridges which intruded the core of landscape areas. The central part of this landscape type was affected by large-scale fire only one or two times in thousand years.

Base-level-tectonic hilly and hilly-ridge mid-waterlogged landscape with glacial deposits (mid-taiga 12 blg)

Twenty three holes were drilled, but few FLs were revealed. According to record drill hole data, only two fires broke out here in the past 1500-2000 years. Because the forest was cut repeatedly, it appears that man is responsible for all the fires in the last 2-3 centuries. The absence of FLs in "indicator deposits" (see the above technique), including those occurring at a depth of more than 5 m, suggests that here fires occured no more than 1-2 times in 1000 years in the late Holocene.

Thus, spruce-dominated landscapes fall into two major categories. "Spruce" landscapes were partly damaged by fires 1-2 times in 1000 years. Obviously, such disasters were due to weather anomalies, e.g. droughts. They completely destroyed spruce forests over large areas, but the most intensely waterlogged sites with flow type moistening, fens, etc. were unaffected. It is from here that preserved spruce spread after fires toward the adjacent area. These landscapes can be regarded as a standard of pyrogenic invulnerability. The biotic components of the landscape were sporadically, but totally, affected by the pyrogenic factor.

In another category of spruce-dominated landscape the effect of fire was more frequent and selective. Fires spread predominantly toward the periphery of landscape outlines along the most elevated parts of landforms with pine forests growing on dry sandy soils. However, the central parts of these natural complexes were damaged by large-scale fires no more than one or two times in 1,000 years.

3.2 Pine-dominated Landscapes

Aqueoglacial, hilly-ridge mid-waterlogged landscape (north taiga 8 ag)

Twenty nine holes were drilled. FLs were found in every drill hole, except watered peat deposits. The total occurrence of FLs is 67%. According to record drill hole data, 8 fires occurred in the time interval from 751-2200 years ago (one fire in 180 years). Many pine trees have fire scars. Their dating has shown that in dry valleys fires break out no less than once in 100 years. The charcoal layer revealed under the litter and in upper soil layers is ubiquitous. Consequently, a fire-free period in this type of landscape in dry valleys is no longer than 100 years, devastating fire occurring once in 200-300 years.

Base-level-tectonic hilly and hilly-ridge landscape with glacial deposits (north taiga 14 blg)

One hundred and one holes were drilled. The occurrence of FLs is generally low (21%) to very low (0-10%) at certain depths. No more than one FL was observed in each hole. Charcoal was ubiquitous. It was shown by dating numerous FS's on trees that four fires broke out in the past 300 years (ca. 100, 150, 220, and 300 years ago). Fires were most common in rocky and cowberry (i.e., driest) habitats. In all likelihood, these fires were caused by the anthropogenic factor, i.e., more intense economic development of the territory during that period. Available data suggest that the landscape territory suffered from large-scale fires no more than 1-2 times in 1000 years. Fires occurred more often (no less than once in 300 years) in dry valleys. A greater number of fires in the past few centuries is due to more vigorous human activities.

Rocky mid-waterlogged landscape (north-taiga 19)

Twenty-nine holes were drilled. The total occurrence of FLs is very high (79%). FL-free holes are usually watered. Record drill hole data show that the landscape was affected by five fires in the interval from 300-1500 years ago (one fire in 210 years). Charcoal is common under the litter and in upper soil layers. FS's are ubiquitous on tree trunks. Devastating fires broke out here no less than once in 300 years, but in dry valleys they were more common. Only small fragments of spruce forests growing in tectonic faults along creeks with peat-humic soils were unaffected.

Aqueoglacial hilly-ridge mid-waterlogged landscape (mid-taiga 8 ag)

Twenty nine holes were drilled. This type of landscape is a mid-taiga analog of the north taiga type described above. They are almost identical in fire regime characteristics. However, in the mid-taiga this type of landscape is remarkable for a greater number of fires: approximately one devastating fire every 300 years during the last 3000 years. This territory was most often damaged by fires 751-1500 years ago, when one fire broke out in 150 years. Rapidly spreading ground fires are even more common in dry valleys.

Base-level-tectonic hilly and hilly-ridge mid-waterlogged landscape with glacial deposits (mid-taiga 14 blg)

Forty one holes were drilled. Traces of fire are ubiquitous in peat at various depths in the dry valley-mire transition zone and in open forest mires. Here, according to drill hole record data, 18 large-scale fires (1 fire every 70 years) broke out 300-1500 years ago. Here belong only the fires which in dry years spread from dry valleys toward waterlogged areas, including near-creek spruce forests and dwarf shrub-sphagnum pine forests. Fires also influenced open sedge-sphagnum mires where distinct FLs are encountered. No FLs were reported only from sites where groundwater wedges out, as well as from fens and the central parts of extensive open mires. However, the above number of fires does not include small-scale ground fires affecting dry habitats, rather than waterlogged areas. It was shown by dating fire scars on trees, that fires spreading rapidly over dry valleys affect such landscapes once or twice in 100 years. Consequently, there is a time gap of 30-50 years between fires here. This is the maximum fire frequency for the region.

Thus, two major categories are recognized in the pine-dominated group of landscapes. In one category, giant fires occurred 1-3 times every 100 years and rapidly spreading ground fires broke out even more often. These areas provide a standard of pyrogenic vulnerability. The biotic components of these landscapes adapted to frequent fires. Their structure and spontaneous dynamics were controlled by fire for thousands of years. In other types of landscapes large-scale fires occurred once or twice every 1000 years, while in dry valleys ground fires broke out about once every 100 years.

4. Discussion

There are also transitional fire regimes which incorporate thc above characteristics. They are typical, for example, of pine-dominated, base level-tectonic ridge (selka) landscapes (17,18). This is due to the complex structure of the landscape, i.e., rugged topography, complex and fractionated soil cover, and compositional mosaic forests. These, in turn, are responsible for frequent alteration of forest habitats with different types of fire regime. Thus, the upper parts of dry soil pine-dominated crystalline ridges were affected by fires no more than once in 200 years, whereas inter-ridge depressions, in which spruce forests growing on moist and wet soils dominated, were seldom affected by fires.

Generally speaking, each type of forest landscape has its own fire regime, its different variants existing in habitats of the same type, e.g., blueberry-pine forests. It should be noted without considering an anthropogenic effect that the pyrogenic factor is related to both secondary and primary forest plant succession. In the course of primary succession in the Holocene (after glacial retreat 10,000-12,000 years ago) a certain forest cover structure was forming. It depended on the composition of Quaternary deposits as well as tendencies in mire- and soil-forming processes and transformed with changes in climatic conditions. This structure either favored the birth and spread of fires (pine forests on sandy soils) or restricted and even prevented them (spruce forests on loamy wet soils). On the other hand, some types of landscape fires alone created the "pyrogenic" structure of forest cover which ensured their permanent role. They destroyed spruce undergrowth in pine forests at an interval of a few decades, preventing the replacement of pine by spruce, and accumulated litter. Species-specific, readily flammable grass-dwarf shrub cover was formed beneath the fire-thinned canopy on partly mineralized substrate. Depending on fire intensity, part of the standing timber fell down, thereby favouring the accumulation of new "fuel" reserves. Thus, favorable conditions were created for frequent fires and their spread.

Nowadays fire is being gradually excluded from the life of forest ecosystems. At the same time, it would be safe to say that the primeval, virgin forests growing in the region represent different stages and variants of pyrogenic successions. For thousands of years fire regulated forest formation, destroyed dead organic matter, affected the forest growth properties of habitats, contributed to the formation of specific faunistic and ornithological complexes, etc. It seems that in the absence of fires taiga forests "grow decrepit" and degenerate. This is apparent in the accumulation of coarse undecomposed litter, which deteriorates the forest growing properties of soil, the decreased productivity and simpler structure of forest cenoses, the less mosaic forest cover of the landscape, as well as the adequate change of the biota as a whole. We understand fires as a natural powerful factor responsible for the renewal and homeostasis of spontaneous forest ecosystems for thousands of years. In this connection, it is necessary to assess the possible consequences of the exclusion of fires as an ecological factor. One of the most important results of fires is the transformation of plant composition. In southeastern Karelia, the area occupied by virgin pine forests decreased by no less than 40% because of cutting and the gradual elimination of fire (Gromtsev 1988). Spontaneous pine forests are largely "a product" of fires. This problem studied by many researchers (Melekhov 1948; Sukachev 1975) was logically resolved, experimentally supported, and embodied in the "cyclically erosional-pyrogenic theory for natural reforestation of Scotch pine" proposed by Sannikov (1983; see also Sannikov and Goldammer, this volume).

Practical utilization of fires is a problem in applied ecology. In nature reserves, it is possible to preserve pine forests in the original state only if a natural fire regime is maintained, which means that ground fires periodically affect these cenoses. Nowadays, in some protected areas of Karelia, primeval pine forests are being clearly replaced by spruce derivatives. This process will last many decades, and it is psychologically difficult to accept fires in a nature reserve. We presume, however, that there is only one option: either we preserve forest ecosystems in the natural state by letting them be periodically affected by fires, or biocenoses will be gradually transformed or actually degenerated. It would be problematic to admit fires in spruce forests as they destroy stands in large areas once or twice every 1000 years. In any case, spontaneous forests growing in different landscape types can only be preserved if a certain variant of the fire regime which set in during the post-glacial period, is maintained. Using a map of landscapes, we can reliably determine the characteristics of natural fire regime in any part of Eastern Fennoscandia and extensive adjacent territories.

Cosiderable experience in this field has been accumulated in the USA, where many national parks have special programmes on the use of fire for virgin forest preservation (Stottlemyer 1981). In forests used for industrial, protection, recreational and other purposes fires can only be used to a limited extent to facilitate pine regeneration, the utilization of cutting wastes, the cleaning of cutting areas, etc. Experience in taking these measures has been accumulated in domestic and foreign forestry.

5. Conclusions

The problem of forest fires has long been closely studied by researchers and practitioners and is still a pressing issue. The results of our investigations allow us to state unambiguously that fires have been one of the major ecological factors responsible for the structure and dynamics of forest vegetation in the taiga zone for thousands of years. Each type of landscape had its own variant of the fire regime. The number of fires varied from 2-3 in 100 years to 1-2 in 1000 years. The complete elimination of the pyrogenic factor results in a transformed taiga forest structure, the patterns of which are not yet fully determined, and its ecological consequences not completely assessed.

References

Abel, F. 1985. Molniya: i bich, i blago. Za rubezhom, No. 44, pp. 16-17 (originally published in the Frankfurter Allgemeine Zeitung, Frankfurt am Main) <in Russian>.

Anonymus 1988. Ognenny ad v Yellowstone. Za rubezhom [Yellowstone on fire-on the fireline],U.S. News and World Report No.41, p.18. Washington.

Furyaev, V.V., and D.M. Kireyev. 1979. Izuchenie poslepozharnoi dinamiki lesov na landshaftnoi osnove [Studies of postfire forest dynamics on a landscape basis]. Novosibirsk, Nauka, 160 pp. <in Russian>.

Gromtsev, A.N. 1988. Retrospektivnyi analiz antropogennoi dinamiki lesov yuzhnoi Karelii za 1840-1980 [Retrospective analysis of anthropogenic forest dynamics in Southern Karelia from 1840 - 1980]. Lesnoi zhurnal, No.4, 125-127 <in Russian>.

Melekhov, I.S. 1948. Vlyanie pozharov na les [Effects of fire on the forest]. Leningrad, Nauka, 159 pp. <in Russian>.

Sannikov, S.N. 1983. Tsiklicheski erozionno-pirogennaya teoriya yestestvennogo vozobnovleniya sosny obyknovennoi [Cyclical erosion-pyrogeny theory of natural renewal of pine]. Ecologia, No.1, 10-20 <in English>.

Spurr, S., and B. Barnes. 1984. Lesnaya ecologiya [Forest Ecology]: per. s ang. Moscow, Lesnaya promyshlennost, 480 pp. <in Russian>.

Stottlemyer J. 1981. The evolution of management policy in national parks. J. For. 79 (1), 16-20.

Sukachev, V.N. 1975. Izbrannye trudy. 1975. [Selected Works] T.3. Moscow, Nauka, 545 p. <in Russian>.

Swynar L.C. 1987. Fire and the forest history of the north Cascade Range. Ecology 68, 791-802.

Vakurov, A.D. 1975. Lesnye pozhary Severa [Forest fire in the north]. Moscow, Nauka, 100 pp. <in Russian>.

Volkov, A.D., A.N. Gromtsev, G.V. Yerukov, et al. 1990. Ecosystemy landshaftov zapada srednei taigi (structura, dinamica) [Ecosystems of landscapes of the western central structure and dynamics]. Petrozavodsk, Karelskiy Filial AN SSSR, 350 pp. <in Russian>.

Yelina G.E., O.L. Kuznetsov, A.I. Maksimov. 1984. Strukturno-funktsionalnaya organizatsia i dinamika bolotnykh ekosystem Karelii [Structural-functional organization and dynamics of a Karelian bog ecosystem]. Leningrad, Nauka, 198 pp. <in Russian>.

Yelina G.E., O.L. Kuznetsov, A.I. Maksimov. 1984. Strukturno-funktsionalnaya organizatsia i dinamika bolotnykh ekosystem Karelii [Structural-functional organization and dynamics of a Karelian bog ecosystem] Leningrad, Nauka, 198 pp. <in Russian>.

J.G.Goldammer and V.V.Furyaev (eds.), Fire in Ecosystems of Boreal Eurasia, 55-64.

The Impact of Fire on Finnish Forests in the Past and Today

J. Parviainen [1]

1. Introduction

In contrast to many other countries within the boreal region, the occurrence of wildfires in Finland has decreased considerably during the past few decades because of land-use changes and efficient fire prevention and control systems. A hundred years ago, however, forest fires were part of the natural succession in Finland's boreal coniferous forests. In addition to wildfires, man-made fires influenced the forest ecosystems; e.g. in the form of shifting cultivation, and later on, as prescribed burning in connection with the practice of forestry. Today, because of the ecologically favorable impact of fire, steps are being taken to revive the use of controlled fire in Finnish silviculture.

2. Wildfires

In addition to fires lit by lightning, fires were also set by hunters. Hundreds of years ago, the forests of Finland were untouched wilderness areas. However, hunters would roam hundreds of kilometres beyond the outermost settlements. Occasionally they would ignite forest fires deliberately, but mostly by accident. Negligence in putting out campfires was probably the main cause behind fires.

At times, forests would be burnt purposely to attract elk into the smoke, away from the insect pests. Hunters were then in a good position to slay them. Forests were also burnt with the purpose of providing better feeding grounds for elk (Kardell 1984).

The oldest statistics in Finland on the numbers of forest fires and on the areas swept over by them go back some 130 years (Saari 1923). While wildfires were very common on crown land in the 19th century, the areas affected were relatively small. Some 50-70,000 hectares were burned in the worst years 1868, 1888 and 1894. A hundred years ago, there were 150-200 forest fires each summer (Tab.1).

The small area affected in individual fires is explained by the weather conditions and the mosaic-like structure of Finnish woodlands, the alternation of dry heathland forests and wetland sites. One third of Finnish forests grow on peatland sites. Although hundreds of thousands of hectares of forest may have been burned by fire during hot and dry summers by the suitable assistance of the wind, Finland has not been afflicted by extremely large fires to the extent of more continental regions like Siberia or boreal China.

[1] The Finnish Forest Research Institute, Joensuu Research Station, FIN-80101 Joensuu

Tab.1. Occurrence of wildfires on crown land in Finland during the period 1865-1920.

Period	Average Fire Data per Reference Period		
	Area Burned (ha)	Number of Fires	Average Fire Size (ha)
1865-1870	13 764	105	131
1871-1880	8 507	134	63
1881-1890	8 707	127	69
1891-1900	9 335	121	77
1901-1910	3 407	87	39
1911-1920	3 560	127	28

Research results obtained in northern Sweden indicate that wildfires on dry sand and gravel soils ignited in natural conditions by lightning have recurred at an average interval of 50 years (Zackrisson 1977). On moist moraine soils this interval has been 120 years. South-facing slopes have been more prone to wildfires than north-facing slopes; similarly, hill tops were sensitive to fire damage whereas valleys were not. With forest fire statistics and the average interval between fires as the basis, it can be estimated that every hectare of Finnish forest land has been burnt down at least once during the past 400-500 years.

The average annual number of wildfires (400-500) has remained constant during the last decades because of efficient fire prevention. Today, the average area afflicted is a mere 0.5 ha (Fig.1) whereas a hundred years ago it was between 60-80 hectares. The total area burnt during the past 40 years amounts to approx. 80,000 ha (The Yearbook of Forest Statistics 1990-91).

Spruce (*Picea abies*) is usually killed by fires. While old, thick-barked pine (*Pinus sylvestris*) will usually survive, the butt will often become scarred in places where the cambium under the bark is killed by the fire. Silver birch (*Betula pendula*) is another species that has developed a degree of fire resistibility. Pioneer tree species (birch, aspen, alder and pine) are the first to reclaim sites affected by stand-replacement fires. The young forest is dominated by broadleaves of both coppice and seed origin. Gradually, the short-lived light-demanding pioneer species begin to give up and pine begins to take over. Later on, the sites begin to have an increasing proportion of spruce and in the climax stage stands will be composed almost entirely of spruce. The species composition in the climax stand depends on site quality. On dry sites, pine will be the dominant species, but as the site becomes more moist, spruce takes over (Kalela 1945, 1948; Kuusela 1990; Schmidt-Vogt 1991).

The fire has a vital importance in maintaining the nutrient cycle, biological productivity, and biodiversity. In a climax stage forest the nutrients are bound up in the growing stock and in the underlying raw-humus layer. Only a few percent of the total nutrients are actually involved in the cycle between the trees and the soil. The raw-humus layer of an old stand of spruce, for example, will contain 1500 kg ha^{-1} of nitrogen in an unavailable form. At the same time, a mere 20 kg ha^{-1} of nitrogen is actually cycling between the trees and the soil (Kellomäki 1987).

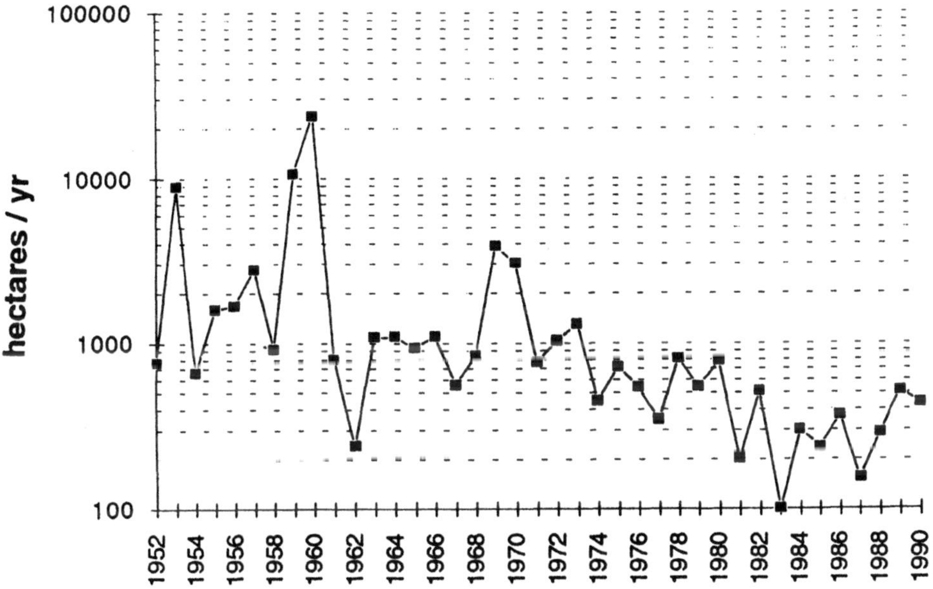

Fig.1. Forest area annually burned by wildfires in Finland during the period 1952-1990.

3. Shifting Cultivation

Although wildfires belong to the succession of boreal forests, the most important impact that fire had on forests in the 18th and 19th centuries in Finland involved the partnership of man. The rural populations of those times relied on the practice of shifting cultivation for their livelihood (Fig.2).

The first phase was to burn down the more fertile forests dominated by broadleaves near villages. In the 16th century, shifting cultivation began to be practised in central Finland - with a technique to cut down and burn mature stands of spruce. The new technique led to the wider adoption of shifting cultivation and also the spreading out of human settlement throughout the country.

Two principal forms of shifting cultivation were practised (Vilkuna and Mäkinen 1988): "Huuhtakaski" technique - strike-down shifting cultivation in the eastern part of Finland was applied in pristine forests while the actual swidden areas close to settlements were subjected to turnip, barley and rye shifting cultivation. "Pykälikkömenetelmä" - ring-barking (girdling) of standing trees (perhaps as many as 300 per hectare) was a preparatory stage in shifting cultivation when practised in pine-dominated woodland areas. Ring-barking killed the standing trees. The drying-up of the trees and their roots made the raw-humus layer more porous. This in turn improved the outcome of the burning. Not all of the ring-barked trees were necessarily cut down.

Fig.2. Swidden agriculture in eastern Finland in the early 1900s - a heavy, dirty and difficult job (original photo from Museovirasto).

The fallowing period in turnip, barley and rye shifting cultivation was 20-40 years. A fresh opening would be made in a Grey Alder woods or a mixed woods that had been established by national regeneration in the area. Fast-growing tree species did not take long to build up a sufficient amount of combustible woody material. Plots would be cropped for 1-2 years and then they would be left fallow and allowed to revert to forest.

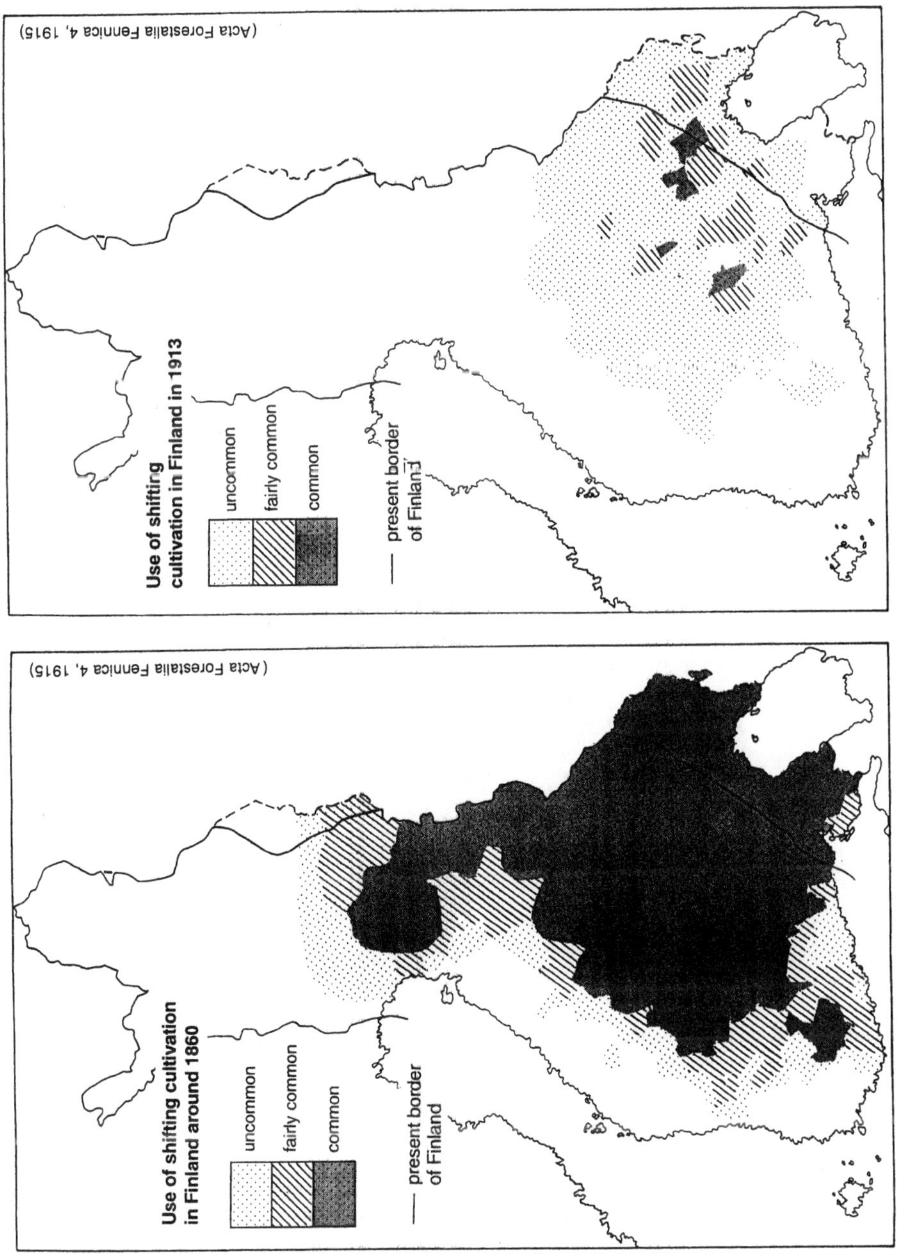

Fig.3. The relative application of shifting cultivation in different localities of Finland during the years 1860 (3a) and 1913 (3b).

During the intervals between consecutive swidden operations, the plots of land would often serve as forest pastures for cattle husbandry, since modern-day meadows had not yet been developed. Grazing of farm animals hindered the forestation, as mere grass was not adequate to satisfy the nutritional requirements of the farm stock. The cows and horses would eat tree seedlings and lower branches and consequently swidden areas would develop into patchy forests (Heikinheimo 1915). These areas would also be accessed for winter fodder for the cattle and horses and conifer branches as bedding. Collecting of conifer branches tended to reduce the survival of spruce. Fields and meadows had to be fenced, which consumed a lot of sapling and thus influenced the species composition of swidden areas (Helander 1949).

Heikinheimo's studies (1915) revealed that shifting cultivation in Finland was practised over more than 4 million hectares per year (Fig.3). By the beginning of this century, some 50 - 75 percent of Finland's forest area had been exploited in this manner. In the eastern part of Finland, shifting cultivation was practised longer and more intensively than anywhere else in the country.

Fig.4. Shifting cultivation during the period 1700-1920 was in particularly widespread use in forests close to settlements. As a consequence, the proportion of broadleaved species rose to a high level (Photo: Olli Heikinheimo).

Shifting cultivation gave rise to conflicting opinions. It was a necessity for the livelihood of the rural population. Nevertheless, the government authorities were worried because of it. On the other hand, already in the 19th century it was feared that loss of forest cover because of shifting cultivation would lead to a cooling of the climate and to Finland becoming unsuitable for human settlement (Leikola 1988). Another, more urgent cause for concern involved the rising economic significance of forests. Sawtimber and timber for mining were becoming increasingly important. Concern over the loss of the country's forest resources culminated in inviting of the German forestry expert E. von Berg in 1858 to come and report on the state of Finland's forests. The report gave a very gloomy view of the forests in areas of intensive shifting cultivation (Fig.4). Moreover, tar distillation, a practice that destroyed the best trees, had created large areas of dismal landscape, especially in western Finland. The forests in the vicinity of villages were all-aged woods of pine, hardwoods and mixed species. Shifting cultivation had already made it difficult to find firewood or timber for building and construction near villages (von Berg 1859a/b).

Legislation was passed in an endeavour to safeguard timber for industrial needs. A bill passed in 1734 made it compulsory for farmers to have a licence before they could burn over forests. But it was not until the end of the century that this regulation began to have a real effect in the heartland of shifting cultivation. The Private Forest Act of 1929 still permitted the practice of shifting cultivation, but only in places where it could be defended as being a sensible practice.

In addition to influencing the availability of timber, shifting cultivation had a major influence on the landscape, the cycling of soil nutrients and regional biodiversity (Linkola 1987; Kuusela 1990). It also influenced the nutritional status of inland waters and the species composition of the forests. The results of the first national forest inventory, conducted in 1921-1923, showed that 50-60 % the forest land in south-eastern Finland was covered by woodlands less than 40 years of age (Ilvessalo 1927). Broad-leaved species accounted for nearly half of the forest area with the share of spruce being only 12-15 %. The most recent inventory conducted in the same part of the country, and completed in 1988, revealed that spruce and pine amount to 40 % each with broadleaves making up the remaining 20 % (Metsäntutkimuslaitos 1992). This shows that, following the demise of shifting cultivation, Finland's forests have experienced a significant increase in the proportion of spruce.

4. Prescribed Burning

With the end of the era of shifting cultivation in the early part of this century, methods derived from this practice began to find use in the regeneration of under-productive forests. Burning of logging waste and the raw humus layer was recommended as a means of promoting the natural restocking of regeneration sites. Broadcast-seeding-on-snow in spring, with prescribed burning preceding it, found widespread use in the 1920s. Prescribed burning in those times amounted to approx. 8,000 ha per year (Fig.5). With time, however, this method's popularity declined; in the 1930's, the annual area burnt in this manner was only a few hundred hectares a year (Kulotustoimikunnan mietintö 1980).

Prescribed burning enjoyed a comeback after World War II and the peak of over 30,000 ha was reached in the mid-1950's. It was particularly useful in assisting with the regeneration of northern Finland's spruce stands to pine, characterised by their thick layer of raw humus. However, this prescribed burning's second coming came to an end in the latter half of the 1960s when it was replaced by mechanized site preparation. The area annually treated fell to 500-1000 ha a year and stayed at that level up to the recent past.

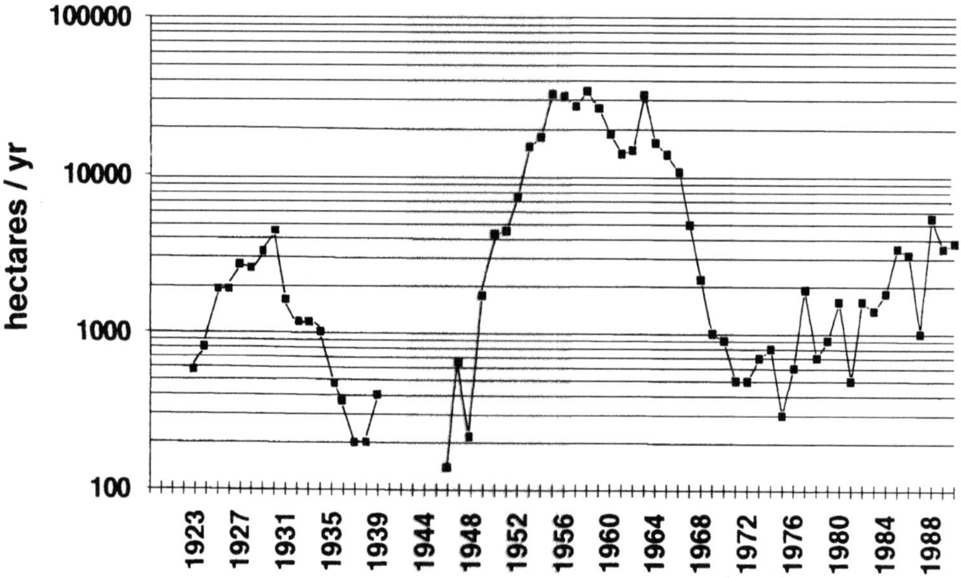

Fig.5. Areas annually treated with prescribed fire in Finland during the period 1920-1990.

The reasons behind the decline in prescribed burning have primarily been technical. The success of prescribed burning depends on weather conditions and this leads to difficulties in organising the operation. The risk of fire getting out of control, the increasing popularity of mechanised site preparation, the risk of nutrients being leached from the soil, and the increased risk for fungal or insect epidemics in the dense young pine stands are the most common forest regeneration problems associated with prescribed burning.

Prescribed burning has many advantageous influences on the physical, chemical and biological properties of the soil. It also affects forest succession and biological diversity (Annila 1993). These advantageous effects depend on site fertility and the intensity of the fire (Mälkönen 1993).

Prescribed burning is recommended for application on moist and dryish mineral soils with sufficient stores of soil water. On more fertile sites, prescribed burning can reduce the availability of phosphorus. The nitrogen released from burning organic matter is lost entirely into the atmosphere and consequently the stores of nitrogen in the forest are reduced. This loss of nitrogen is not, however, fatal for the development of new forest growth, because only a small proportion of the nitrogen stored in forest soils is in a mineralised form. Prescribed burning has an advantageous influence of the pH of the humus layer, which may rise by 1-2 units and stay there for several decades. The rise in pH in the uppermost mineral soil layer has been modest.

Artificially regenerated forests by planting or seeding in Finland currently amount to slightly more than 4 million ha. Since the total productive forest area is approx. 20 million ha, artificially regenerated forests represent 20 % of the whole. Following soil scarification, regeneration sites produce an abundance of naturally arisen conifer and hardwood seedlings, and these develop into mixed woods (Parviainen 1988). Over a period of 80-90 years, prescribed burning has been applied on a total of 0.5 million ha. This corresponds to 12% of Finland's artificially regenerated forests and 2-3% of the total forest area.

5. Conclusions

Every forest stand in Finland has been affected by wildfires at least once during the past 400-500 years. Shifting cultivation (1700-1920) was practised in 50-75 % of Finland's forests while prescribed burning (1920-1990) has been applied to 2-3 percent of the country's forests.

Despite the powerful impact that shifting cultivation has had during the last century on Finland's forests, the volume of the growing stock and species composition, the utilisation of Finland's timber resource has been sustainable. The country's forest area increased somewhat in the 1900s, and the current growing stock and increment are higher than ever before since measurements were began (90 years; Parviainen 1992). Since wildfires, a fundamental element of the natural development of boreal forests, have been subdued, forest structure and biological diversity will change unless fire or its substitutes are implemented as silvicultural measures.

There is a distinct open stage in the fire ecology based developmental cycle of a boreal forests. The long-term preservation of forest productivity and the maintenance of the nutrient cycle inevitably mean that the natural cycle must be simulated by creating an open stage. Nowadays, the likelihood of forest fires is greatly reduced by efficient fire prevention. Prescribed burning is hardly practised because of technical reasons. While its effects are not entirely identical, clear cutting does create an open stage as did wildfires in the past. The cycle in nature can be further simulated through small clear cutting areas, by favoring broadleaves, and by speeding up the soil's nutrient cycle through light site preparation techniques. A fact in favor of prescribed burning is its ecological benefits.

One element of the variation in the biodiversity of Finnish forests lies in the country's remaining natural forests (approx. 10 percent of the area of forest land). Untouched forests are needed to be set aside as national parks and nature reserves representing different site types and regions. Almost all of the current conservation areas consist of old, first-growth climax stands, and they are located mostly in northern Finland. In order that forest environment be enriched with various successional stages, conservation areas should be extended to include pristine forests of all ages. In fact, there is a need to burn areas of standing forest here and there, and thus create natural forests in their early stages of development.

References

Annila, E. 1993. Tuli metsäluonnon monipuolistajana. Julkaisussa: Piri, E. (toim.). Tuli metsän ekologisessa kierrossa. Metsäntutkimuslaitoksen 75-vuotisjuhlaretkeily Kolilla 7.-8.6.1993. Metsäntutkimuslaitoksen tiedonantoja 462, 19-23.

von Berg, E. 1859a. Kertomus Suomenmaan metsistä. Näköispainos. 1988. Helsingin yliopiston metsänhoitotieteen laitoksen tiedonantoja No. 63, 63 pp.

von Berg, E. 1859b. Die Wälder in Finnland. Jahrbuch Königl. Sächs. Akad. f. Forst- u. Landwirthe 13, 1-118.

Heikinheimo, O. 1915. Kaskiviljelyksen vaikutus Suomen metsiin. Acta Forestalia Fennica 4, 1-264, 1-149 (liite) ja 1-59 (Referat).

Helander, A.B. 1949. Suomen metsätalouden historia. WSOY. Porvoo - Helsinki.

Ilvessalo, Y. 1927. Suomen metsät. Tulokset vuosina 1921-1924 suoritetusta valtakunnan metsien arvioimisesta. Summary: The forests of Suomi (Finland). Results of the general survey of the forests of the country carried out during the years 1921-1924. Metsätieteellisen koelaitoksen julkaisuja 11. 421 pp. + 192 pp., taulukot.

Kalela, E.K. 1945. Suomen metsien puulajidynamiikka. Terra 57, 1-19.

Kalela, E.K. 1948. Luonnonmukainen metsien käsittely. Metsänhoitajien jatkokurssi. V. Silva Fennica 64, 16-32.

Kardell, L. 1984. Historieskogen. Julkaisussa: Historiska nyheter No. 30. Statens historiska museum. Stockholm.

Kellomäki, S. 1987. Metsäekologia. Silva Carelica 7.

Kulotustoimikunnan mietintö. 1980. Komiteamietintö 1980:1. Helsinki. 80 p. + 22 p. liitteitä.

Kuusela, K. 1990. The dynamics of boreal coniferous forests. SITRA 112. 172 pp.

Leikola, M. 1988. Esipuhe julkaisussa von Berg, E.: Kertomus Suomenmaan metsistä 1859. Helsingin yliopiston metsänhoitotieteen laitoksen tiedonantoja No. 63.

Linkola, M. 1987. Metsä kulttuurimaisemana. Summary: The forest as a cultural landscape. Julkaisussa Reunala, A., and P. Virtanen. Metsä suomalaisten elämässä. Silva Fennica 21 (4), 362-373.

Metsäntutkimuslaitos 1992. Metsätilastollinen vuosikirja 1990-91 <Yearbook of forest statistics 1990-91>. Folia Forestalia 790, 281 pp.

Mälkönen, E. 1993. Kontrolloitu tulen käyttö maan käsittelyssä. Julkaisussa: Piri, E. (toim.). Tuli metsän ekologisessa kierrossa. Metsäntutkimuslaitoksen 75-vuotisjuhlaretkeily Kolilla 7.-8.6.1993. Metsäntutkimuslaitoksen tiedonantoja 462, 15-18.

Parviainen, J. 1988. Metsänviljely. Perusteet ja sovellutukset. Silva Carelica 9, 177 pp.

Parviainen, J. 1992. The Finnish approach. Long-term and sustainable forestry. Paperi ja Puu - Paper and Timber 2, 78-81.

Saari, E. 1923. Kuloista etupäässä Suomen valtion metsiä silmälläpitäen. Acta Forestalia Fennica 26 (5), 155 pp.

Schmidt-Vogt, H. 1991. Die Fichte. Band II/3. Waldbau - Ökosysteme - Urwald - Wirtschaftswald - Ernährung - Düngung - Ausblick. Verlagsbuchhandlung Paul Parey, Hamburg. 804 pp.

Vilkuna, K., and E. Mäkinen. 1988. Isien työ. 10. painos. Otava. Keuruu. 359 pp.

Zackrisson, O. 1977. Influence of forest fires on the North Swedish boreal forest. Oikos 29, 22-32.

J.G.Goldammer and V.V.Furyaev (eds.), Fire in Ecosystems of Boreal Eurasia, 65-89.

The Role of Paleofire in Boreal and Other Cool-Coniferous Forests

J.S. Clark [1] and P.J.H. Richard [2]

1. Introduction

Two qualities distinguish the paleo record of fire in boreal regions. First is its highly fragmentary nature in light of the importance of boreal forests for carbon storage and emissions to the atmosphere. One might expect boreal forests to be a focus of research on impacts of temperature change and biospheric feedbacks due to C release. Fire can have effects on the C balance of individual stands that persist for decades in boreal regions (Van Cleve and Viereck 1981). Combustion-produced elemental (graphitic) C can have several important environmental impacts, influencing climate (Shaw 1982; Rosen et al. 1984; Ch´ylek et al. 1984; Penner et al. 1991), air quality (Sandberg et al. 1979), human health (Laflamme and Hites 1978), and the concentrations of greenhouse gases in the atmosphere (Seiler and Crutzen 1980; Crutzen and Andreae 1990; Crutzen and Goldammer 1993). Stratigraphic records of several important combustion products (e.g. charcoal, fly ash, reduced organic tracers) exist from some regions, but there is a conspicous lack of synthesis needed to establish baseline (i.e. pre-industrial) biomass burning emissions (Malingreau et al. 1993). Such records are particularly scarce in boreal regions (Clark and Robinson 1993).

Second is the high potential of boreal forest sediments to provide the long-term information required to understand effects of climate change on forests. A lack of sediments suitable for analysis can often frustrate research efforts at lower latitudes, but is rarely a limitation in mid to high latitude glaciated landscapes. Much of the boreal biome is either scattered with lakes of many sizes, some with annual laminations, or it is paludified with extensive mire deposits.

The availability of sites together with the compelling questions those sites might answer makes occasion for review of the subject, with special attention to future directions. We begin with mention of the various methods used to extract and quantify sediment charcoal. This discussion aids interpretation of paleo evidence for fire in the Tertiary, especially since the end of the Pleistocene. For some of the period data are so few as to limit our discussion to potential research directions. Our most extensive data come from presettlement time in North America, which are sufficient to suggest baseline particulate emissions needed for interpretation of modern emissions (e.g. Malingreau et al. 1993).

[1] Department of Botany, Duke University, Durham, North Carolina 27706, U.S.A.
[2] Département de Géographie, Université de Montréal, Montréal, Québec, Canada H3C 3J7

2. The Sedimentary Record of Biomass Burning

Records of biomass burning differ in types of deposits and in the method of analysis. Many factors that contribute to sediment charcoal records are not well understood, so any attempt at synthesis must be proceeded with caution. We outline here some of the important concerns, but more comprehensive reviews of methodology can be sought elsewhere (Patterson et al. 1987; Clark 1988a; Clark and Robinson 1993; Clark and Royall 1994).

2.1 Combustion-Derived Constituents of Sediments

Sediments consist of a heterogeneous mix of organic and inorganic constituents. Some of these constituents derive from the surrounding landscape, having a range of source areas determined in part by vector agent. Others are produced and/or modified in a water column or on a mire surface following deposition. These components can include a host of compounds that result from combustion. A full review of these constituents is beyond the scope of this chapter (Novakov et al. 1996). A few words on the general classes of combustion-derived compounds and ways to identify them should aid in the understanding of the subsequent discussion.

The broad classes of combustion-derived C compounds that can be extracted from sediments include organic (C_o) and inorganic C. C_o includes cellulose, hemicellulose, and lignin. Inorganic C includes carbonate C, which can be precipitated from the water column, and elemental C_e, present in combustion emissions. Char is a heterogeneous and poorly characterized mixture of C_o and C_e, the relative proportion of which depends on combustion temperature, oxygen availability, and initial composition of the fuel. Because chemical, optical, and physical properties of char broadly overlap with those of organic compounds not produced by fire (e.g. lignins), characterization efforts by atmospheric chemists (Cadle and Groblicki 1982) and interpretation by paleoecologists (Patterson et al. 1987; Clark 1988b; MacDonald et al. 1991) has long been difficult. "Charcoal" quantified by optical microscopy (the preferred method of paleoecologists) contains large amounts of cellulose and lignin, constituents that occur in almost all lake sediments. The C_e component of charcoal can vary widely, and graphitic C is 1-5% (Clark et al. in prep). C_e can constitute up to 50% of carbonaceous aerosols (Novakov 1984; Cadle and Groblicki 1982). The "charcoal" reported in marine studies mentioned below is quantified by methods that remove larger fractions of the C_o than is included in the char optical microscopy counts.

Among the organic tracers of combustion in sediments are the highly reduced and rather stable polycyclic aromatic hydrocarbons (PAHs). PAHs are observed to increase dramatically in 20th century lake and marine sediments (Laflamme and Hites 1978) and in soils (Jones et al. 1989) due to fossil fuel combustion, but they have not yet proven valuable for tracing past wildfire. They are adsorbed to aerosol size particulates, thus appearing to represent combustion at regional to subcontinental scales. Particulates produced by fossil fuel combustion (e.g. "soot", "fly ash") can be useful for tracing past distributions of industrial emissions (Clark and Patterson 1984; Wik and Renberg 1991).

2.2 Types of Deposits

The records discussed here include lake, mire, and marine deposits. These records appear to differ in terms of the spatial and temporal scale at which they record past combustion. Mire deposits can include actual lenses of charcoal representing fires that burn across the mire surface. These lenses give evidence for the actual location of a burn. Unfortunately, smoldering "ground fires" in organic deposits can consume, and therefore destroy, evidence of previous fire; they record a minimum number of local fires. Organic deposits also contain charcoal that is blown in from more distant fires, thereby preserving a record of regional burning (e.g. Mehringer et al. 1977; Bradshaw and Zackrisson 1990). Mires are abundant at high latitudes and could be exploited in a more systematic way for analysis of past fire.

Lake sediments record combustion at "local" to "regional" scales, depending on the lake size and on the size range of particles quantified in the analysis. Large particles (10^2 to 10^5 μm diameter) appear to reflect local fires, those that occur within 10^2 m of a lake basin (Clark 1990) (Fig.1). Smaller particles (10^1 to 10^2 μm) appear to reflect combustion at a regional scale (Clark 1988b; Clark and Royall 1994). Theoretical studies (Prentice 1985) suggest that small basins emphasize nearby sources of pollen, whereas large basins average contributions from a wider area. Because pollen grains fall within the diameter ranges of particles quantified by charcoal analysis, it is likely that the same holds true for charcoal. There have yet to be studies done exploring the effect of basin size on the source area of charcoal. High temporal resolution can be obtained in some lake sediments, particularly those having varved (annually laminated) sediments. Such sediments have been used to identify individual fires and so to construct distributions of past fire intervals, e.g. those in Figure 1 (Clark 1990).

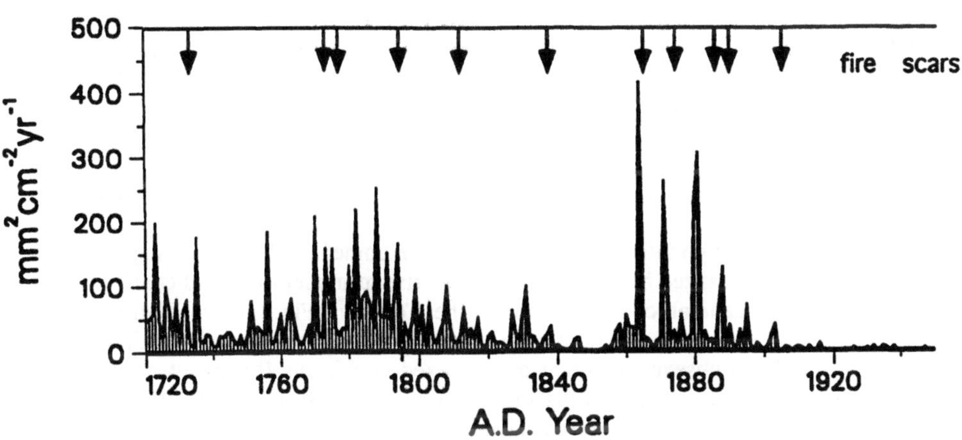

Fig.1. Stratigraphic charcoal data from NW Minnesota and the occurrence of fire within the catchment watershed. The decreased fire importance after A.D. 1910 reflects effective fire suppression in the region (Clark 1990). In general peak accumulation rates exceeding 100 mm^{2}cm^{-2}yr^{-1} correspond to the occurrence of fire within the lake catchment.

Marine sediments are characterized by slow sedimentation and a potentially large source region. Difficulties resolving charcoal source have been partly overcome by the analysis of large networks of cores that show some consistent spatial (Smith et al. 1973) and temporal (Herring 1985) patterns. These patterns are suggestive, as we shall see, but there is great uncertainty concerning how charcoal is eventually incorporated into sediments. These records appear to reflect combustion at broad spatial and temporal scales.

Because the different methods of analysis (e.g. optical microscopy, organic extractions of C_o, wet oxidations, thermal methods, organic tracers such as PAHs), they all measure somewhat different constituents; it is important to acknowledge these differences when attempting to compare across studies and sites. Charcoal analysis of lake and mire deposits have been completed most commonly using optical microscopy and, to a lesser extent, using wet oxidations in nitric acid. Particles quantified by optical methods contains only a small fraction of C_e. Wet oxidations do not cleanly discriminate charred material from lignin (Clark, in prep.). The analysis of carbonaceous compounds in marine sediments has been completed by assays that rely on chemical and optical properties of C_e (reviewed by Goldberg 1985). The different types of deposits appear to record different spatial and temporal scales. In the mapped distributions we present here, we report the methods of analysis, and limit maps to one or several comparable methods of analysis or sediment type.

3. The Long-Term Relationship Between Fire and Boreal Forests

Fire is clearly an important part of modern boreal forest ecology. Vast areas of boreal forest burn annually (Stocks 1993). Fire scars and stand structure attest to the impact of fire in recent decades. Apparent adaptations to fire (e.g. serotinous cones) suggest a long-term relationship between boreal tree species and fire. Surprisingly, there has been little attempt to elucidate the long-term relationship between fire and boreal forests, i.e. from the rise of "boreal-like" assemblages in the Tertiary to the redistributions of boreal species during glacial episodes, to the effects of past human cultures. Here we summarize some of the stratigraphic evidence for these relationships. We restrict ourselves to vegetation and/or climate types that bear some similarity to modern boreal forest in terms of climate and/or presence of species that today dominate these forest types.

3.1 Pre-Quaternary "Boreal" Forest and the Pacific Ocean Record of Combustion

The record of mid to high latitude conifer forest development and charcoal suggests a close relationship has long existed between the two. The rise of modern lowland coniferous forest occurred with the general cooling trend of the Tertiary (Fig.2). Even during the late Cretaceous, coniferous taxa were consistently most abundant at mid to high latitudes (Crane and Lidgard 1990). Subsequent expansion of mixed coniferous forests attended mountain building and cooling, especially during the Oligocene to Miocene (Wolfe 1975). Increased continentality and fluctuating moisture availability saw the demise of subtropical elements in the montane western US (Barnosky 1984). Expansion of coniferous forest and then taiga occurred in central and eastern Siberia by the late Miocene, by which time *Picea mariana* was present in Alaska. The taiga expanded to its modern range by the end of the Pliocene, when tundra was first clearly recognizable. Mixed conifer forests dominated Pliocene floras of the northwestern United States and Europe (Wolfe 1975).

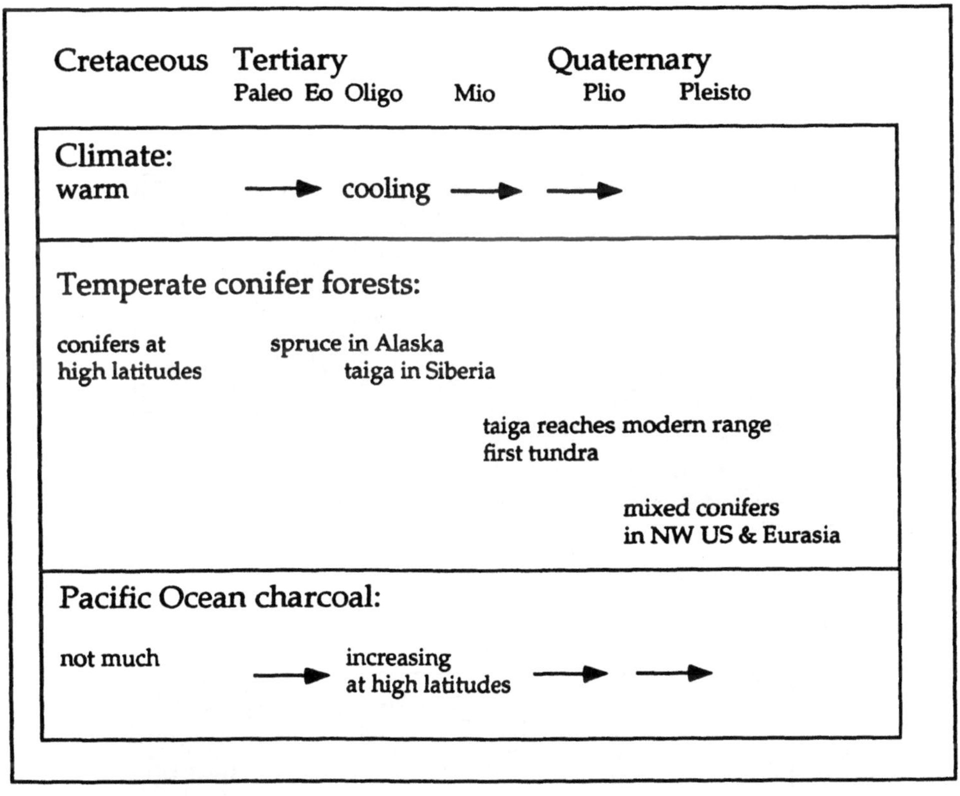

Fig.2. The expansion of cool-conifer forests during the Tertiary reflects global cooling and increasing charcoal accumulation in deep-sea cores from similar latitudes (Fig.3).

Charcoal accumulation in cores of Pacific Ocean sediments suggest that fire may have increased in importance together with the development of high-latitude conifer forests. Smith et al. (1973) found that the sediment "surface" concentration of charcoal showed broad geographic continuity. Their "charcoal" is quantified by infrared spectroscopy of samples subjected to a heating and grinding procedure. One of the absorbance bands that develops during this process is proportional to the mass of C_e. Slow sedimentation rates in these cores mean that samples represent roughly the last 100,000 yr. They found highest concentrations in the Northern Hemisphere at latitudes of temperate and boreal forests (Fig.3a). Concentrations reach a minimum in the equatorial zone and then rise to a second modest peak at 40°S. They interpreted this result as evidence of the greater importance of fire at these mid to high latitudes.

a)

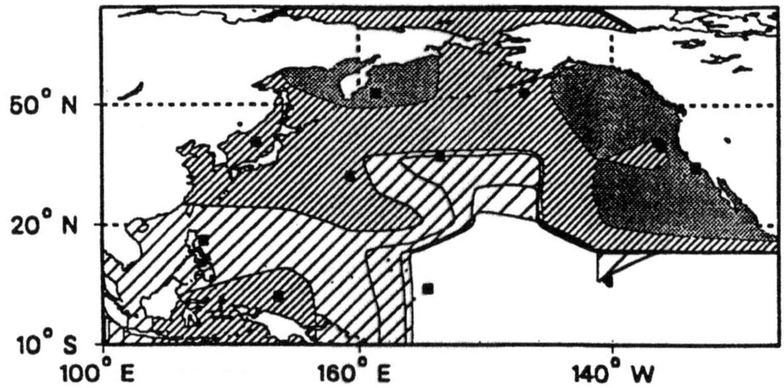

b)

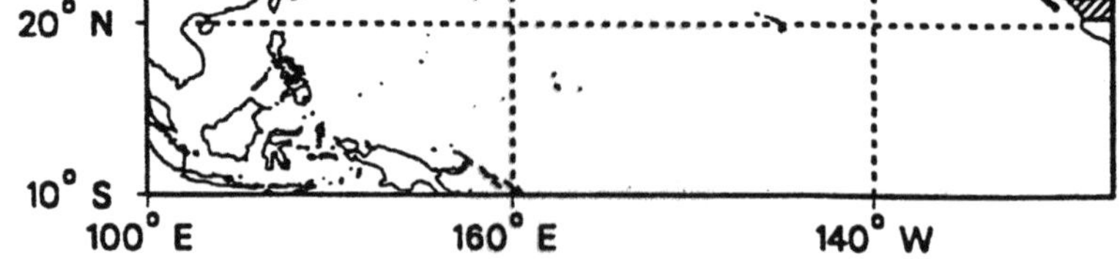

Fig.3. Charcoal accumulation in deep ocean cores (Herring 1985) (a) compared with those of North America (this study) (b). Contours in (a) are 0.1, 0.5, 2, 5, and 50 mg $m^{-2}yr^{-1}$. In (b) they are 1, 5, 20, 50, and 500 g $m^{-2}yr^{-1}$.

Herring (1985) analyzed the charcoal stratigraphy in a smaller number of Pacific sediment cores extending from the present back to the Cretaceous. His "charcoal" is the fraction remaining after the removal of carbonate C, followed by a wet oxidation of C_o. He found that the latitudinal trends observed by Smith et al. (1973) were present since the Cretaceous, but that charcoal accumulation rates increased throughout most of the Pacific over time. Maximum rates are observed in the Quaternary. The highest rates were observed at mid to high latitudes, especially since the Miocene, coinciding in space and time with the expansion of coniferous forests. Charcoal accumulation rates in the deep ocean peak from 50 to 70°N latitude, broadly corresponding with the distribution of temperate coniferous forest. The highest values in coastal sediments may be at somewhat lower latitudes, from 25 to 40°N, especially along coastal North America (Fig.3a).

Although understanding of transport and sedimentation at these core locations is limited, results suggest that conifer forests may long have been among the key contributors to atmospheric loadings of combustion-derived trace gases and particulates. There is a dearth of pre-Quaternary charcoal records from lake sediment that are needed to better resolve charcoal transport and so determine whether cool conifer forests are indeed the source of these particles. The data are yet too few and the understanding of transport and deposition too limited to conclude that biomass burning follows the Tertiary rise of cool-conifer forests. Observation of particle transport over oceans (e.g. Andreae 1983), meso-scale modelling of particle transport, and efforts to resolve the processes that lead to deposition in sediments are all needed to better understand the long-term relationship between fire and the boreal forest.

3.2 The last Interglacial: a Potential for Fire History

The foregoing section outlines the expansion of boreal forest during the late Tertiary, a time of cool climates and glaciations, and perhaps too, fire. The Pacific Ocean record suggests that fire may have been more important in boreal forests, especially during the Quaternary. It leaves open the possibility that charcoal came from other biomes (e.g. grasslands also expanded during the Tertiary), because a source area of deep-sea charcoal is not established. Lake sediment charcoal derives from a more restricted area and so would help to resolve charcoal source. Most lake sediment studies are restricted to recent times, the Holocene or portions thereof. Although the pre-Holocene terrestrial record of fire is meager, we include here a brief consideration of the vegetation changes of the last Interglacial to show potential and to set the stage for a review of the charcoal evidence from Holocene lake sediments.

The last interglacial episode is of interest at this time of concern over climate elevated atmospheric CO_2 and warming, because it may provide an example of such scenarios. The pollen record of vegetation of the last interglacial in Eurasia (Mikulino in Russia, Eemian in Europe) suggests somewhat warmer winters, and precipitation possibly higher than today. Boreal-like forests extended further north than do those of the present, and *Betula* was more abundant everywhere. *Larix* was abundant east of the Lena River and in eastern Europe. *Picea* was a major component of central Siberian forests. Montane areas to the south included *Pinus sibirica* with broad-leaved forests at lower elevations. *Abies* was more restricted than in the Holocene, being confined to montane regions of the Urals, Altay Highlands, and Okhotsk region in Russia (Grichuk 1984) and along forelands of the Massif Central and Vosges (Woillard 1978; Reille and de Beaulieu 1990; de Beaulieu and Reille 1989; 1992; Pons et al. 1992). Forest steppe existed where steppe occurs today (Grichuk

1984; Wright and Barnosky 1984). Pollen evidence from last interglacial in North America (Sangamon) suggest climate much warmer and drier in the Midwest, and similar to or warmer than the Holocene in the East (Mott 1990). Temperate deciduous taxa extended into parts of Canada beyond modern latitudinal limits. Woodlands or forests containing *Picea* occupied the northern Yukon and eastern Beringia, and *Betula*-shrub tundra grew north of its present limit (Heusser and King 1988).

Sediment charcoal from past interglacial deposits represents a potentially important and unexploited source of information on controls of high-latitude vegetation, particularly if the climate warms. The Tertiary records contained in deep ocean cores (see above) suggest a long-term relationship between fire and high-latitude coniferous forests. Those records lack stratigraphic detail needed to establish closer links between fire and climate. Lake sediments provide an opportunity to refine those relationships in space and time for previous interglacials that bear similarities to and differences from the Holocene (Watts 1988). A case in point are the long charcoal and pollen records from Australia that sparked continuing discussions regarding effects of human-induced fire on vegetation (e.g. Singh et al. 1981, Kershaw et al. 1991, Clark 1983). Some of the interesting questions in Eurasia include the order and rates at which different taxa expanded during interglacials (Müller 1974; Woillard 1978; Watts 1988; de Beaulieu and Reille 1989), the changing relative importance of *Picea, Abies, Pinus,* and *Betula* relative to modern times, and perhaps the role of fire in the progressive elimination of many Tertiary elements with repeated glaciation (West 1970; Liu 1988). Sediments suitable for several methods of analysis are available, including some that are varved (Müller 1974), but they have not been analyzed for charcoal. If fire is to be a key contributor to the overall impacts of climate change on boreal forests in the future (e.g. Stocks 1993), then a perspective of past examples might be valuable.

3.3 Fire Seasons Changed With Climates since the last Glacial Maximum

The changes in seasonality since the last glacial maximum at 18 ka, together with an expanding charcoal data set, may soon afford an opportunity to examine in detail cool-conifer fire regimes under decidedly different climate settings. Past fire occurrence may relate to circulation patterns of the time. The general circulation of the atmosphere governs the regional importance of different air streams over continents. The positions of these air streams determine fire hazard and ignition incidence. For example, climates in central Canada are distinguished by Pacific air penetration during the growing season, blocked by Arctic air to the north and tropical air to the south. The tundra/taiga ecotone corresponds rather closely with the median location of the arctic summer front in North America (Bryson 1966) and in Siberia (Krebs and Barry 1970). The southern limit of the boreal biome falls along the average winter position of the Arctic air stream. A west - east gradient of increasing *Abies balsamea* abundance characterizes the boreal forest, attending the increase in precipitation east of Pacific air dominance and the decrease in fire frequency. The fire season in central Canada (e.g. Manitoba and NW Ontario) begins with the retreat of cold, stable Arctic air and the penetration of dry, less stable Pacific air (Johnson 1992). A steep decrease in fire hazard to the east, at longitudes near the Great Lakes, coincides with the approximate summer boundary of the meridinal flow of moist Tropical air. Year-to-year variability in the positions of these air streams can have important effects on annual fire incidence and area burned (Stocks and Street 1983).

Models of the general circulation of the atmosphere predict dramatic effects of continental and sea ice on air mass dominance and on the moisture content of those air masses. The 18 ka seasonality of incoming solar radiation to mid latitudes was not unlike that of today. But the presence of continental ice sheets in North America and Europe, together with lower sea-surface temperatures (including expanded sea ice), lower atmospheric CO_2 and methane, and higher tropospheric particulate loads may all have contributed to climates that differed dramatically from now. General Circulation Models (GCM's) predict that continental ice sheets altered the position and strength of the jet stream, shifting storm tracks in North America well south of their present location. Sea ice over the North Atlantic may have contributed to dry full-glacial conditions in western Europe. Anticyclonic circulation about the high pressure over the Laurentide ice sheet might help explain dry conditions along the ice margin.

The Holocene saw climate changes that might be linked to a maximum in seasonality of incoming solar radiation to mid latitudes about 9 ka and to the retreat of continental ice sheets. Increasing summer insolation up to 9 ka might have seen a strengthened monsoonal flow to interiors of large continents (Kutzback and Guetter 1986), except where blocked by topography (e.g. the Himalayan Plateau). The principal effect might have been increased aridity in some areas (e.g. those distant from moisture such as the Northern Great Plains) or increased rainfall, such as portions of Africa (e.g. Kutzbach and Street Perrott 1985). The early Holocene presence of a large ice sheet in North America may have had the additional effect of steepening temperature gradients, focusing storm tracks, and blocking the incursion of extreme low winter temperatures to areas south of the ice. Wright (1992) invokes this combination of effects to explain the odd assemblages that followed *Picea* in the early Holocene of the upper Midwest. These assemblages include *Ulmus* at abundances that suggest steep moisture gradients, a situation possibly explained by enhanced penetration of Tropical air and its rather direct contact with cold air near the retreating ice.

To what extent might past fire regimes have responded to the combinations of climate variables of the past? And how has changing fire seasons mediated some of the more direct climate effects on vegetation change? The next section will show that the relatively few data do not reveal obvious patterns, but do constitute a beginning.

3.4 Evidence for Vegetation Change and Fire at the last Glacial Maximum

The extent and composition of full Glacial coniferous forests differed in some conspicuous ways from modern boreal forest. The last glacial maximum in Eurasia saw coniferous trees restricted to southeastern Europe and the southern portions of the Ural and Altay Mountains (Wright and Barnosky 1984). Polar desert, steppe, and tundra assemblages spread over much of western Europe and north and central Siberia, likely due to the dominance of dry Westerlies, recently off the ice-covered North Atlantic. *Pinus*, *Betula*, and *Larix* were scattered near the Scandinavian ice margin (Grichuk 1984). Low moisture in southern Russia explains semi-desert and steppe. Treeless assemblages appear to have been continental and rather dry, with frequent occurrence of grasses, chenopods, and *Artemisia* spanning from north of the Alps in western Europe (Huntley 1988) across western Siberia, and along the ice margins in North America (Barnosky 1984; Ritchie 1987; Webb 1988). The dry periglacial climate in North America is attributed to the easterlies that flowed across the southern margin of the ice sheet. Steppe vegetation occupied areas south of the Alps into eastern Siberia. Coniferous forests dominated the middle latitudes of Japan, between herbaceous assemblages to the north and broadleaved deciduous forest to the south (Tsukada 1983). Montane *Picea-Abies* forests descended into lowlands of North China, mixed with *Pinus, Betula*, and herbaceous taxa (Liu 1988).

Open *Picea* woodlands covered much of unglaciated North America extending at least as far south as Kansas (Grüger 1973) and the Ozark Plateau (King 1973), where tree cover may have been more complete, eastward into Tennessee (Delcourt 1979; Delcourt et al. 1980) and North Carolina (Whitehead 1981). Abundant herbaceous taxa indicate an interrupted canopy over much of the range of *Picea*, perhaps with scattered deciduous taxa. *Picea* woodland included increasing *Pinus* eastward to the southeastern coastal plain (Watts 1970; 1980). *Pinus banksiana* mixed with *Picea*, but its range may have extended further south. The southeastern coastal plain was apparently cold and dry (Watts 1980; Plummer 1993), with northern pines (*Pinus banksiana, P. resinosa*), *Picea*, and *Abies* regionally present, but perhaps much less important than herbaceous taxa (Whitehead 1964; Watts 1980; Hussey 1993). Woodlands in unglaciated sections of interior Alaska and northern Yukon contained *Picea, Betula*, and *Larix* (Lamb and Edwards 1988). Eastern Beringia supported herbaceous taxa (Lamb and Edwards 1988).

Few data and different climate conditions make for a poor understanding of the potential importance of fire under glacial conditions. North-south gradients in tree canopy coverage and in relative abundances of *Picea, Pinus*, herbaceous taxa, and hardwoods in eastern North America may have coincided with similar gradients in the importance of fire. If *Picea* woodlands supported low quantities of interrupted fuels, then it is possible that fire regimes were very different from modern ones. Tundra assemblages from Gould Pond, ME appear to have produced little charcoal before 10,000 yr ago (Anderson et al. 1992), suggesting that full-Glacial tundra may not have burned frequently. On the other hand, charcoal accumulation rates for the full-Glacial sediments of the southeastern Coastal Plain show some of highest values observed anywhere (> 2000 g $m^{-2}yr^{-1}$)(calculated from Hussey 1993), when herbaceous vegetation with little *Picea* would appear to have supported low fuel loads. Charcoal data from Nonbara Bog, Japan extend back 17,000 yr and show values too low to indicate an important role for fire in the temperate coniferous forests containing *Abies, Picea*, and *Tsuga* (Sugita and Tsukada 1983).

The few available data suggest that the full glacial importance of fire obviously varied regionally. Differences between Maine and South Carolina suggest that a single fire regime did not characterize herbaceous/woodland assemblages. Whether fire importance was fuel-limited in some of the cool dry environments might he addressed by more attention to pollen productivity and to careful reconstructions of species composition (e.g. Birks 1976; Cwynar 1982). It is likely that different moisture regimes of full glacial times also supported conifer forests having a range of fire regimes. This is certainly true today; more data are needed to better resolve the geographic regions and the forest types where fire was important.

3.5 Late Glacial and Holocene Cool-Conifer Biogeography

The changing vegetation that led to the eventual establishment of modern forest zonations included invasions of treeless and woodland assemblages by boreal and more temperate taxa. Eurasian forests of the late-Glacial contained *Picea* mixed with *Pinus* and *Betula* west of the Urals and *Larix* to the east (Khotinskiy 1984). *Picea* dissappeared from parts of the Japanese Archipelago after 10 ka for reasons probably climatic but poorly understood (Tsukada 1983). Colonization of deglaciated areas in northern Scandinavia by *Betula* followed herbaceous assemblages that included grasses and *Artemisia* (Alhonen 1971; Berglund and Rapp 1988). Treeless landscapes followed ice retreat in the eastern US and Canada. *Picea* invaded with *Juniperus*, *Alnus*, and *Populus* by 13.3 ka at Tannersville Bog, PA (Watts 1979), by 12.3 ka at Rogers Lake, CT (Davis 1969), and from 11 to 10 ka in southern ME (Davis and Jacobson 1985). *Picea* was a dominant canopy taxon for roughly 1000 yr between transitions to hardwoods and pines in the eastern US. However, *Larix*, *Populus*, and *Abies* may also have been abundant, considering that pollen production, transport, and/or preservation of these taxa can be low (Cushing 1965; Richard et al. 1992).

Modern forest zonation patterns began to develop by 9 to 6 ka, although the early composition of those zones differed in many respects from today. *Picea* was initially more important in Siberia than it is at present, attributed to a climate warmer, moister, and less continental during the early Holocene (Khotinskiy 1984). *Pinus* and especially *Betula* dominated from the Urals westward into European Russia. Modern boreal forest expanded to occupy much of its present range in North America by 6 ka. The demise of *Picea* woodland across eastern North America by 10 ka occurred as *Picea* populations followed the retreating ice northward. This was followed by a delayed expansion of boreal forest in eastern Canada, likely due to effects of the remnant Laurentide ice sheet on conditions to the south and east (Kutzbach and Guetter 1986; Richard and Labelle 1989; Webb et al. 1987). *Picea*-dominated forests were short-lived or absent in the Saint-Lawrence River area between 10 ka and 7 ka. Although macrofossils of most boreal tree species were present in the sediments, *Abies balsamea* and *Pinus banksiana* were dominant in the southwest, and *Populus* in the northeast, during a time of maximum charcoal abundance in the sediments. *Picea glauca* expanded first, replaced by *Abies balsamea*, then by *P. mariana* in Labrador 6 ka (Engstrom and Hansen 1985). *Populus* arrived first in western Canada, followed by *P. glauca* and *P. mariana* by 8 ka. *Pinus contorta* and *P. banksiana* arrived later (MacDonald 1987). The early arrival of *Picea* as far north as the Mackenzie River Delta is explained by the early Holocene summer insolation maximum at high latitudes (Ritchie 1987). In eastern Canada where the present zonal treeline occupies the lowest latitude in the Hemisphere, the modern position was attained soon after ice retreat from 7 (coastal) to 5.5 ka (inland), already with the maximum tree abundance (Payette 1993). *Alnus*-rich *Picea* woodlands that contained some *Abies* thrived around the last remnants of the ice sheet in Quebec by 5.5 ka (Richard et al. 1982). In contrast, warm summers of the mid-Holocene apparently caused treelines in central Canada and Siberia to move 200 to 500 km north of modern locations (Sorenson et al. 1971, MacDonald et al. 1993). The southern boreal margin was invaded by broadleaved deciduous trees in Siberia and by *Pinus strobus* in Ontario (Liu 1990). *Thuja* expanded near the northern range limit at several Canadian sites (Bennett 1987; Liu 1990). *Picea* became less important in Siberia at this time. Treelines moved higher in the Alps and in Scandinavia (Huntley 1988). The thermal maximum of the mid-Holocene ended by 3 ka. Forest lines moved southward, and the modern *Picea* taiga of eastern Siberia came to dominate, still with abundant *Larix*. *Picea* increased in abundance in southern Quebec, Ontario, and the eastern United States, particularly at higher elevations (e.g. Davis et al. 1980; Jackson 1989). Vast peatlands formed on the West Siberian Plain after 4500 yr B.P. Paludification in Labrador has been attributed by Engstrom and Hansen (1985) to invasion by *Picea* with attendant effects on soil organic matter, fertility, and hydrology, in part due to the development of permafrost. Alternatively, paludification may have favored the spread of *P. mariana* over wetlands. Continuous opening of the northernmost woodland in Quebec has been attributed to cooling and to increased fires during the past 3000 yr (Payette et al. 1989).

3.6 Fire since the last glacial maximum

To what extent did fire play a role in the broad-scale and rapid vegetation changes since the end of the ice age? That fires occurred has long been apparent from charcoal lenses in peat bogs of western and northern Europe (Tolonen 1983). But, as we shall see, lake sediments show large site-to-site differences in fire importance, as climate and vegetation changed and few sites have been analyzed.

A feature of many pollen diagrams from temperate regions is a transition from conifer to hardwood taxa near the end of the Pleistocene. In some cases the transitions were abrupt and may have involved fire. The 12 ka transition from full-Glacial cool-temperate coniferous forest to broadleaf deciduous forest at Nonbara Bog, Japan does not show any indication for fire, although there exists a possibility of fires between samples (Sugita and Tsukada 1983). The transition appears to occur at a century time scale, with rapid decline of conifers attending an abrupt *Fagus* increase and a more gradual rise of *Quercus*. The *Picea* dominated assemblages from central New York attended high charcoal accumulation rates and peak values indicating local fires at 50 to 100 yr intervals (Fig.4). Highest charcoal accumulation rates occurred after the arrival of northern pines (Clark et al., in prep.). The expansion of hardwoods was attended by an abrupt decline in charcoal accumulation, with no evidence of subsequent local fire. Everitt Lake, Nova Scotia (Green 1981) agrees to the extent that charcoal accumulation increased upon the arrival of pine, but charcoal remains relatively high after increases in mixed forest taxa such as *Tsuga* and *Fagus*. Neither Gould Pond, ME (Anderson et al. 1992) nor Schleinsee, Germany (Clark et al. 1989) suggest decreased charcoal accumulation after hardwood expansions. Although the number of sites is limited, many lakes contain sediments suitable for charcoal analysis that span the late-Glacial period. Analyses of more lakes are needed to better characterize the role of fire across these transitions.

The relatively few charcoal data support the view that fires have been rather common in many boreal forests during the Holocene. Peat deposits in North America, Scandinavia, and Russia may contain several to many charcoal lenses (Tolonen 1983). Charcoal accumulation rates for the available sites appear to reflect gradients in climate and fuels. Accumulation rates in western Canada (MacDonald 1989; MacDonald et al. 1991) are higher than is observed in eastern Canada and in temperate deciduous forest (see below). The establishment of open *Picea/Pinus* woodlands at 9 ka in southwestern Alberta saw an increase in charcoal accumulation to relatively high rates (>300 g $m^{-2}yr^{-1}$, calculated from MacDonald 1989), comparable to some of the largest values yet observed in North America. Unfortunately, too few sites are available to clearly define the geographic patterns of charcoal accumulation in western Canada.

Fire and climate change may have played a role in the spread of *Picea* into northern Europe after 3200 yr BP. Bradshaw and Zackrisson (1990) use stratigraphic pollen and charcoal from an island site in northern Sweden to suggest *Picea* invasion there lagged behind it's establishment on the adjacent mainland, expanding during an extended fire-free interval centered on 600 yr BP. Tolonen (1987) argues for a similar delayed *Picea* expansion due to fire in southwest Finland. But charcoal accumulation rates increased in southern Finland following the replacement of broad-leaved deciduous trees by *Picea* after 6000 yr BP (Tolonen 1987). *Picea* was replaced by *Pinus* and *Betula* 700 to 800 yr BP, perhaps due to cooler temperatures (Khotinskiy 1984; Huntley 1988). Macroscopic charcoal from these *Pinus sylvestris* - *Betula pubescens* stands show several fires during the last two millennia in northern Sweden (Bradshaw and Zackrisson 1990).

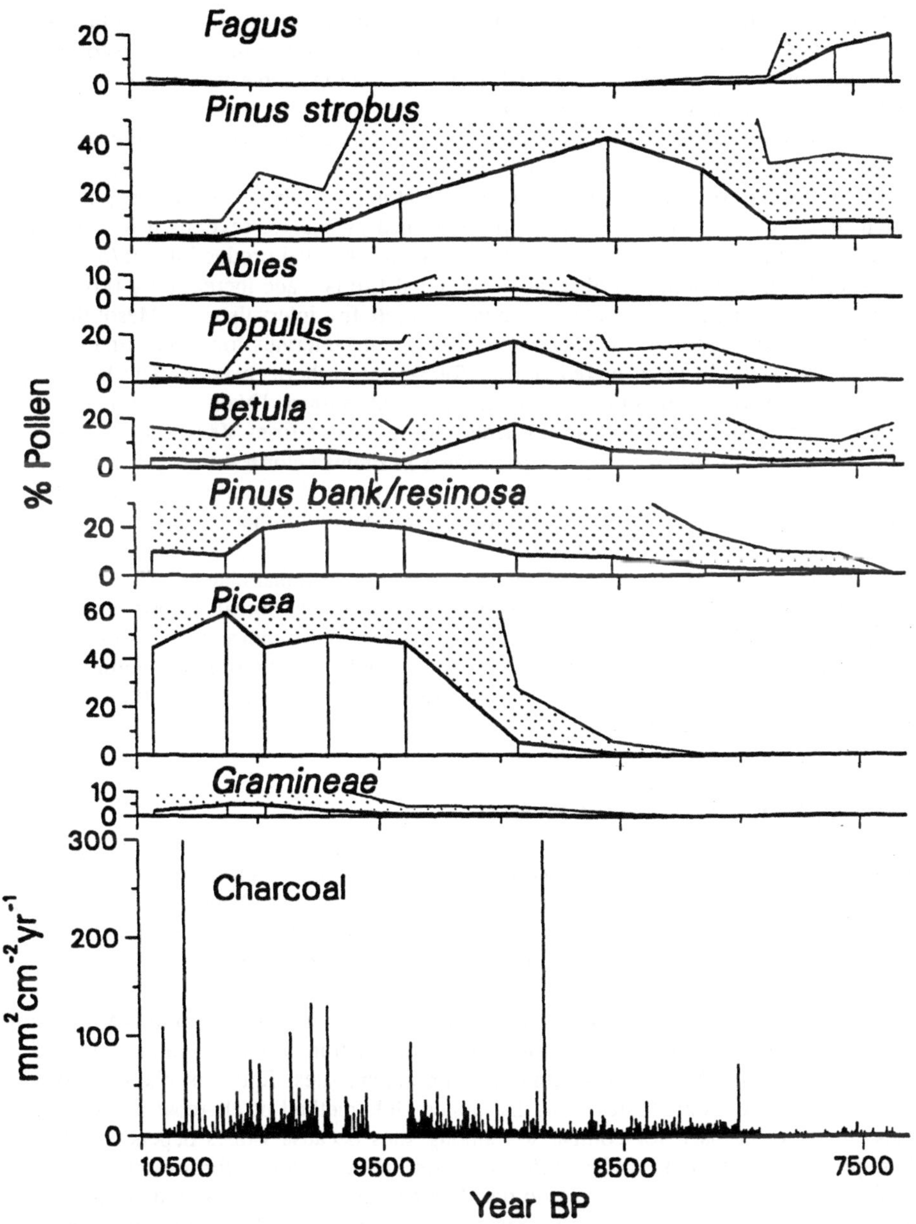

Fig.4. Stratigraphic pollen and charcoal data from western New York (USA) at the transition from *Picea* and *Pinus banksiana* to *Pinus strobus* to hardwoods, including *Fagus*. Charcoal data from this site indicate fires occurred at 50 to 200 yr intervals during the stage of *Picea/Pinus banksiana*, not unlike those of modern boreal forests in central Ontario. Peak accumumulations that would indicate local fire do not appear to occur during the time of *Pinus strobus* dominance. Accumulation rates decline abruptly with expansion of *Fagus*.

Interpretation of European fire is complicated by human activities. Cultural use of fire appears obvious from stratigraphic charcoal data. Dramatic fluctuations in pollen and in charcoal from Lake Ahvenainen (Tolonen 1978) clearly portray influences of Neolithic and later cultures on forests surrounding a small lake. Calculation of intervals between local fires depends on the definition of local fires. Each sample integrates several years of charcoal accumulation; so the series is necessarily smoothed to this degree. Despite use of integrated samples, the series is characterized by high-frequency variability that is difficult to interpret in terms of natural vegetation fires. The investigator must decide on the magnitude and duration of a "peak" in the series that will be defined as one vs. several fires. The erratic behavior of the series may reflect cultural influences on charcoal production and/or transport to the lake. If so, interpretation of local vs. regional fire is made increasingly problematic. Despite these limitations of the data, Tolonen interprets fire intervals since Mesolithic times comparable to those reported by Zackrisson (1977). Fire intervals for recent times average 45 yr in *Pinus* and 120 yr in mixed conifer forests in northern Sweden. The latter figure comes from vegetation most similar to that at Lake Ahvenainen. "Natural" fire intervals are estimated from before heavy human influences to be 240 yr from 3000 to 0 B.C. at Lake Ahvenainen and 80 yr from *Pinus sylvestris* forests in Finnish North Karelia. Tolonen (1983) summarizes many of the important studies of boreal fire with Holocene cultural change in Finland. Fire became more important at Lake Ahvenainen with the onset of Iron Age slash and burn agriculture and cereal cultivation.

Results are far too few to relate the importance of fire to global changes in climate. They are sufficient to establish that fire has been important in boreal forests and that its importance varied from place-to-place. Further progress will require analysis of additional sites and a better understanding of charcoal transport and deposition in lake sediments. This second factor is required to relate charcoal records back to fire importance, a topic addressed in the next section.

5. Implications for Presettlement Emissions

Existing charcoal records from lake sediments were compiled in order to assess i) whether geographic patterns in charcoal accumulation exist and, if so, ii) whether those patterns might suggest how biomass burning in presettlement forests contributed to atmospheric particulate loads. The data were screened according to several criteria:

Dimensions: Charcoal data are reported in the literature usually in three ways. Particle concentrations could not be used for this compilation because values depend on sedimentation rates, which vary within cores and among sites. Ratios of charcoal to pollen are mostly reported from studies using the pollen-slide method of charcoal analysis. Those values could not be used for comparison among sites because values depend on differing pollen production rates of different forest types. *Pinus* forests, for example, generally deposit far more pollen in lakes than do deciduous forests. Charcoal accumulation rates are usually reported as area charcoal per area sediment per year. The studies included in this compilation are those for which accumulation rates are reported, which include times and places dominated by cool-temperate coniferous forests or areas nearby.

Sediment type: Stratigraphic charcoal data come from lake and mire deposits. Depositional processes in lakes might affect accumulation patterns of bouyant charcoal particles. With one exception, we restricted the data set to estimates made on lake sediments, because results may not be comparable to those obtained from mires.

Method of analysis: Different methods of extraction and analysis can impose different biases (Winkler 1985; Patterson et al. 1987; Clark 1988a; MacDonald et al. 1991), particularly with respect to particle size and C_o:C_e. We included for this synthesis only estimates made by optical microscopy, either on pollen slides or thin sections.

For comparison with combustion efficiencies and airborne emissions estimates from modern fires, we transformed charcoal accumulation rates to mass fluxes assuming a density of 0.5 g mL^{-1}. The low density reflects the high porosity of charcoal (Clark and Royall 1994). These conversions were made on charcoal accumulation rates reported from pollen-slide and thin-section methods.

The geographic coverage in the data set is fragmentary. Relatively few records document particle accumulation in boreal forests prior to drastic alterations of native landscapes by humans. Several records from the Pacific Northwest and Rocky Mountains of North America come from montane temperate conifer and *Pinus* forests. South of the boreal forest in eastern North America and eastern Canada are the largest number of sites. They include mixed forests in the Lake States through the St Lawrence River Valley and northern hardwoods mixed with hemlock or other conifers in the northeastern U.S., SW Quebec, and Nova Scotia. Cool temperate mixed forests of Japan are also represented. The boreal records for which accumulation rates are reported are limited to several in Alberta, Quebec, and Finland.

Overall, rates are higher than we anticipated on the basis of net primary production (NPP) estimates reported in the literature (Fig.5a). Our highest values from *Pinus* forests near the prairie/forest border of > 500 g $m^{-2}yr^{-1}$ approach estimates of total aboveground NPP, which fall between 500 to 2000 g $m^{-2}yr^{-1}$ for this region. These high rates would only be reconciled with source strength if sites received substantial charcoal amendments blowing in from the prairies to the west, if charcoal accumulates preferentially in low parts of the landscape, e.g. lake basins, or if biases in analysis lead to overestimates of charcoal accumulation. We view the first possibility as unlikely, because the high values come from thin-section estimates of large particulates with low residence times in the atmosphere. Comparisons between charcoal accumulation rates estimated by the two methods suggest that sources of all particles are closer than the prairies (Clark and Royall 1994).

The relatively high values suggest potential for systematic bias in the method. Only a fraction of aboveground biomass is consumed in a given fire. Assuming for simplicity that production everywhere equals accumulation, we expect accumulation rates comparable to fuel load, scaled by burning efficiencies and by the fraction of a region that burns on an annual basis. ANPP (aboveground net primary production) puts a cap on possible fuel production (only a fraction of ANPP can be consumed by fire), we do not expect charcoal accumulation to exceed ANPP ¥ burning efficiency ¥ fire frequency. Values in our data set exceeded predictions we would make based on these considerations. Sources of bias are several:

Charcoal transport: If charcoal is focused within lakes, due either to post-fire movement by wind (saltation) or in surface flow, then lakes might have higher accumulation rates than the average landscape. Evidence for these processes is almost entirely anecdotal, and no systematic investigations exist. While surface flow is an obvious concern in regions of high rainfall intensity, low soil infiltration rates, shallow soils, and/or slow vegetation recovery, we believe that post-depositional transport is of minor importance beyond a few months in most environments. Worth mention are unexpectedly low values from Duck Pond, MA situated in a kettle on glacial outwash sands. Mixed pine/oak forests at this site should produce large quantities of charcoal, but high infiltration rates through porous sands might limit transport to lakes in surface flow. Fluvial inputs are a potential concern for lakes having inflowing streams. But some of our highest values come from closed basins.

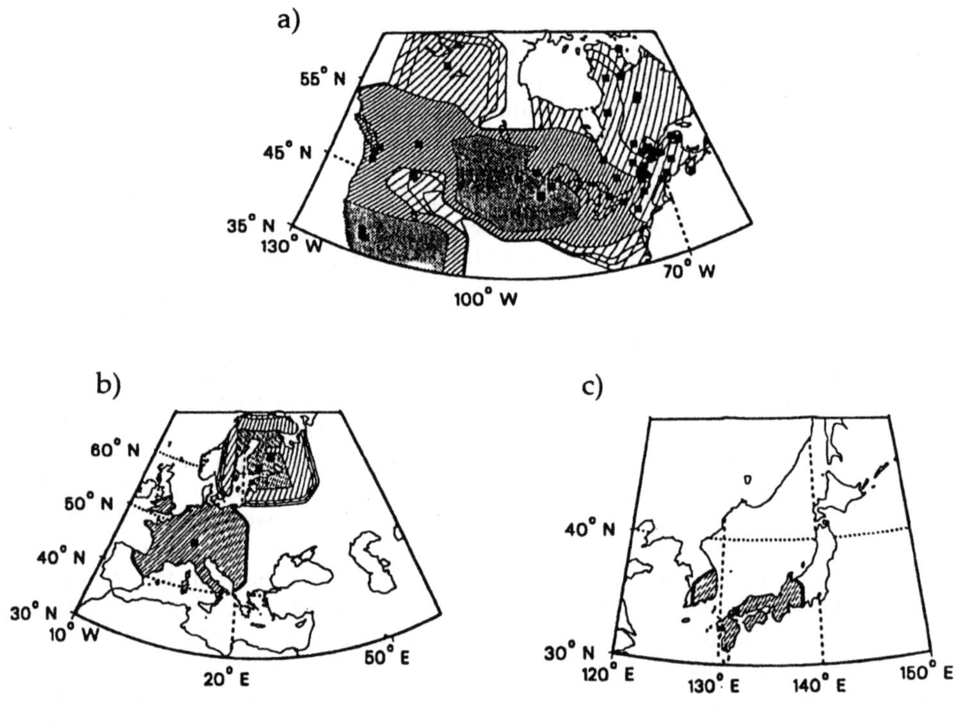

Fig.5. Charcoal accumulation rates contoured from published charcoal data from mid to high latitude forests from North America (a), Europe (b), and Japan (c).

Transport of the relatively large particles (quantified in most investigations of sediments) is poorly understood. Atmospheric chemists tend not to measure them in smoke plumes due to low reactivity and low residence times in the atmosphere. In light of poor knowledge of transport processes, it is possible that movement of charcoal to lakes results in a biased sedimentary record. Sources of systematic (among-site) bias are difficult to identify, as it is for pollen.

Buoyancy and sedimentation: Sedimentation of charcoal is difficult to quantify, because particles are bouyant until pores fill with water they may be grazed by zooplankton and deposited in fecal pellets, and particles can vary greatly in chemical composition, morphology (including porosity), and size. Charcoal might be preferentially focused in the centers of lakes where cores are usually taken (to explain the fact that accumulation rates are higher than we anticipated) (Davis 1972). Alternatively, they might accumulate to a greater degree in littoral zones as a result of wind and wave action on floating particles.

Extraction and analysis: Biases inherent in different extraction and quantification methods are reviewed by Clark (in prep). Nitric acid digestions give erratic results (e.g. Cadle and Groblicki 1982), but more experiments would help determine why this is so (Clark, in prep). Overestimates or underestimates of char appear possible with this method, depending on the duration of the digestion. The two examples where methods were compared include estimates of one showing an order of magnitude higher (Tab.1, Duck Pond) and one 5-fold lower (Tab.1, Rainbow Lake) than obtained by the pollen-slide method. Total C analysis used by paleoecologists to accompany this method (combustion in air at 500°C) misses the graphitic C, which is oxidized at higher temperatures.

Pollen-slide methods probably overestimate char in the fraction counted (see above), but they omit smaller fractions. Overestimates are possible because the wide axis of a particle will be preferentially exposed to view when squashed between the slide and cover slip.

Thin-section methods contain biases in both directions. Downward bias results because small particles are ignored. The bulk of particulate mass is likely in the submicron ranges. Upward bias results because particle volume is estimated from a linear measure of a preferentially long axis using a stereology assumption that all three dimensions are equal (Clark 1988a).

Calculation: Charcoal density can range from 0.4 to 0.7 g $m^{-2}yr^{-1}$. Our use of 0.5 is probably reasonable.

With the exception of ecotonal forests near the prairie/forest border, our charcoal accumulation rates represent < 10% of ANPP. Although perhaps highly relative to expectation of charcoal production, the agreement with published estimates is remarkably good in view of the many sources of bias and error related to charcoal production, transport, sedimentation, extraction, and quantification. Despite different methods used to quantify sediment charcoal and overall high values, we found consistency within the data set (Fig. 5) and clear geographic trends in charcoal deposition. Presettlement charcoal accumulation rates in forests range from 18 g $m^{-2}yr^{-1}$ in high elevation *Pinus* forests in the Rocky Mountains (Lost Trail Bog) to > 1000 g $m^{-2}yr^{-1}$ in mixed hardwood forests near the prairie/forest border in Minnesota (Deming, Arco, and Budd). Fires are known to occur near Lost Trail Bog, albeit at lower frequency than at the Minnesota sites (Clark 1990). Emission rates have not been estimated at either location.

Mixed forests that lie south of the boreal forest show an eastwardly declining trend in charcoal accumulation rates that follows transitions from prairie to mixed oak to eastern deciduous forests. Values decline by an order of magnitude from 2000 to 3000 g $m^{-2}yr^{-1}$ near the prairie/forest border in NW Minnesota to 200 to 300 g $m^{-2}yr^{-1}$ in eastern Minnesota, Wisconsin (Dark Lake and Hell's Kitchen), and central Ontario (Greenleaf Lake). Mixed forests in southern and eastern Quebec and the Northeastern US all show values below 200 g $m^{-2}yr^{-1}$, with most < 100 g $m^{-2}yr^{-1}$.

Low elevation conifer forests in Washington and British Columbia had accumulation rates of 75 g $m^{-2}yr^{-1}$ (Kirk Lake and Hall Lake) to 200 g $m^{-2}yr^{-1}$ (Mineral Lake) at higher elevations. These values agree reasonably well with estimates of presettlement emissions of 4 to 47 g $m^{-2}yr^{-1}$ for the region (Fahnestock and Agee 1983). Given the large uncertainty in estimates of this kind (e.g. Seiler and Crutzen 1980), these values must be considered highly comparable. A study from *Cryptomeria* forests in Japan (Nonbara Bog) showed substantially higher accumulation rates of 1000 g $m^{-2}yr^{-1}$.

The more "boreal" of the sites include values intermediate between these extremes. Alberta sites (Toboggan and Rainbow) have values of 50 to 750 g $m^{-2}yr^{-1}$. The Finnish studies come from areas heavily impacted by agriculture, but all lie near 100 g $m^{-2}yr^{-1}$. These values place boreal particulate emissions near the upper end in the scale of importance in presettlement time. Only mixed hardwood forests near the prairie forest border showed consistently higher values in this data set, and these transitional forests represent a relatively small area. In a separate study we report tropical savanna estimates calculated from Burney's (1987) data (Clark and Stocks, in prep) in the same range as those of these mixed forests near the prairie forest border. Tropical forests show only trace quantities prior to the onset of aboriginal impact, (Clark and Stocks, calculated from Griffen and Golberg 1975). Even *Nothofagus* steppe from Patagonia (Heusser 1990) show substantially lower values (Clark and Stocks, in prep). Taken together with the vast area covered by boreal forests, our results suggest them as one of the principal sources of atmospheric particulates in presettlement times.

6. Conclusions

The paleorecord of charcoal emphasizes both "scarcity" and "potential". Scarcity of data limits our ability to generalize about the role of fire in boreal forests of the past. Analyses are few and inconsistent. The record is now too sparse to evaluate the changing spatial and temporal importance of fire throughout the past.

The record is good enough to provide key insights. First, the big picture "makes sense". The Holocene patterns of charcoal accumulation in eastern North America show geographic continuity and roughly fit our notions of fire importance (Fig.5a). The temporal patterns can also be explained at several scales. The marine records of Tertiary charcoal can be reconciled with the expansion of cool conifer forests at mid to high latitudes (Fig.3).

Changes in many of the charcoal series from lake sediments appear to reflect the fire regimes one might expect (Fig.4). Local occurrence of fires is apparent in the charcoal series from varved lakes (Fig.1).

Potential for progress exists on all fronts. The boreal biome is crowded with sites that could be exploited to understand fire regimes under climate settings that do not exist today, the role of fire during times of rapid climate change, and ways in which fire has contributed to past atmospheric quality, landscape productivity, and carbon balance. Challenges include the resolution of seemingly inconsistent changes that attended the Holocene expansion of different forest types with links to regional climate. Experiments on charcoal production, transport, and sedimentation are underway by Clark (in prep.). Increased ability to interpret stratigraphic data together with better geographic coverage should help resolve the question of past fire importance in the boreal.

ANNEX TABLE

Tab. 1. Sites used to estimate presettlement charcoal accumulation rates in lakes. Explanations and continuation of table on next page).

Site name[1]		Lat (°N)	Long (°W/E)	Method[2]	Reference[3]
Swamp	CA	38	120 W	pollen slides	Smith and Anderson (1992)
Woski	CA	37	120 W	pollen slides	Anderson and Carpenter (1991)
Wawona	CA	37	119 W	pollen slides	Anderson and Carpenter (1991)
Kirk	BC	48	122 W	pollen slides	Cwynar (1987)
Hall	WA	48	122 W	pollen slides	Tsukada et al. (1981)
Mineral	WA	46	122 W	pollen slides	Tsukada and Sugita (1982)
Lost	MT	46	114 W	pollen slides	Mehringer et al. (1977)
Tobaggn	ALB	50	115 W	pollen slides	MacDonald (1989)
Rainb	ALB	60	112 W	pollen slides	MacDonald et al. (1991)
Deming	MN	47	95 W	thin sections	Clark (1990)
Arco	MN	47	95 W	thin sections	Clark (1990)
Budd	MN	47	95 W	thin sections	Clark (1990)
Dark	WI	45	92 W	pollen slides	Gajewski et al. (1985)
Little	WI	45	91 W	pollen slides	Gajewski et al. (1985)
Ruby	WI	45	92 W	pollen slides	Gajewski et al. (1985)
Lake Cld	MN	48	92 W	pollen slides	Swain (1973)
Hells Kn	WI	46	90 W	thin sections	Clark and Royall (1994)
Green	ONT	46	78 W	pollen slides	Cwynar (1978)
Crawfd	ONT	44	80 W	thin sections	Clark and Royall (1994)
Devil Bt	NY	43	78 W	thin sections	Clark and Royall (1994)
Clear	NY	44	74 W	pollen slides	Gajewski et al. (1985)
Ely	PA	42	76 W	thin sections	Clark and Royall (1994)
Conroy	ME	46	68 W	thin sections	Clark and Royall (1994)
Basin	ME	44	70 W	thin sections	Clark and Royall (1994)
Gould	ME	45	69 W	pollen slides	Anderson et al. (1992)
Ange	QUE	48	71 W	pollen slides	Labelle and Richard (1981)*
Atocas	QUE	46	73 W	pollen slides	Richard and Gauthier, unpub.
Boucané	QUE	48	70 W	pollen slides	Richard et al. (1992)
Bride	QUE	7	74 W	pollen slides	Richard, unpub.
Bromont	QUE	45	73 W	pollen slides	Richard, unpub.
Caribou	QUE	48	65 W	pollen slides	Jétte and Richard (1992)*
Castor	QUE	47	73 W	pollen slides	Richard, unpub.
Chavigny	QUE	58	75 W	pollen slides	Richard, unpub.
Daumont	QUE	55	69 W	pollen slides	Richard et al. (1982)
Delorme	QUE	54	70 W	pollen slides	Richard et al. (1982)
Desautl	QUE	50	73 W	pollen slides	Richard (1981)
Diana375	QUE	61	70 W	pollen slides	Richard, unpub.
Diable	QUE	49	66 W	pollen slides	Richard and Labelle (1989)
Dolbeau	QUE	49	66 W	pollen slides	M'Pindy, Gangloff, Richard, unpub.
Flevy	QUE	48	71 W	pollen slides	Richard, unpub.
Fourche	QUE	48	69 W	pollen slides	Richard et al. (1992)
Gabriel	QUE	46	74 W	pollen slides	Richard (1977)*
Geai	QUE	46	74 W	pollen slides	Richard, unpub.

(Tab.1. continued)

Kanaaups	QUE	54	77 W	pollen slides	Richard (1979)*
Léonard	QUE	49	66 W	pollen slides	Labelle and Richard (1981)*
Madelein	QUE	48	71 W	pollen slides	Bussières and Richard, unpub.
Main	QUE	48	71 W	pollen slides	Bussières and Richard, unpub.
Marc	QUE	47	71 W	pollen slides	Labelle and Richard (1981)*
Montagne	QUE	48	69 W	pollen slides	Richard and Larouche (1994)*
Nedlouc	QUE	58	72 W	pollen slides	Richard (1981)
Ormstown	QUE	45	74 W	pollen slides	Richard, unpub.
Ouellet	QUE	48	69 W	pollen slides	Richard et al. (1992)
Perdu	QUE	49	66 W	pollen slides	Richard, unpub.
Pluvieux	QUE	48	69 W	pollen slides	Richard et al. (1992)
St-Franç	QUE	48	72 W	pollen slides	Richard, unpub.
Turcotte	QUE	49	66 W	pollen slides	Labelle and Richard (1984)*
Valin1	QUE	49	71 W	pollen slides	Richard, unpub.
Yamaska	QUE	46	73 W	pollen slides	Richard and Gauthier, unpub.
Yelle	QUE	48	80 W	pollen slides	Richard (1980)*
Duck	MA	42	70 W	pollen slides	Winkler (1985)
Everitt	NS	44	65 W	pollen slides	Green (1981)
Ahvenai	FIN	61	25 E	pollen slides	M.Tolonen (1978)
Laukunl	FIN	63	29 E	pollen slides	K.Tolonen (1983)
Schnsee	GER	48	9 E	thin sections	Clark et al. (1989)
Nonbara	JAP	34	132 E	pollen slides	Sugita and Tsukada (1983)

[1] All abbreviations refer to states (USA) or provinces (Canada) except FIN (Finland), GER (Germany), and JAP (Japan).

[2] Methods are described in the text.

[3] Sites for which the charcoal data are not included in the original publication are indicated by an asterisk.

References

Alhonen, P. 1971. On the early Flandrian stratigraphy and vegetation history of the north Satakunta area, western Finland. Bull. Geological Soc. Finland 43, 39-46.

Anderson, R.S., and S.L. Carpenter. 1991. Vegetation change in Yosemite Valley, Yosemite National Park, California, during the protohistoric period. Madrono 38, 1-13.

Anderson, R.S., G.L. Jacobson, R.B. Davis, and R. Stuckenrath. 1992. Gould Pond, Maine: late-glacial transitions from marine to upland environments. Boreas 21, 359-371.

Andreae, M.O. 1983. Soot carbon and excess fine potas ium: long range transport of combustion derived aerosol. Science 220, 1148-1151.

Barnosky, C.W. 1984. Late Pleistocene and early Holocene environmental history of southwestern Washington State, U.S.A. Can. J. Earth Sci. 21, 619-629.

Bennett, K.D. 1987. Holocene history of forest trees in southern Ontario. Can. J. Bot. 65, 1792-1801.

Berglund, B.E., and A. Rapp. 1988. geomorphology, climate and vegetation in North-West Scania, Sweden, during the late Weichselian. Geographia Polonica 55, 13-35.

Birks, H.J.B. 1976. Late-Wisconsin vegetational history at Wolf Creek, central Minnesota. Ecol.Monogr. 46, 395-429.

Bradshaw, R.H.W., and O. Zackrisson. 1990. A two-thousand-year history of a northern Swedish boreal forest stand. J. Vegetation Sci. 1, 519-528.

Bryson, R. 1966. Air masses, streamlines and the boreal forest. Geographic Bulletin 8, 228-269.

Burney, D.A. 1987. Late Quaternary stratigraphic charcoal records from Madagascar. Quat. Res. 28, 274-280.

Cadle, S.H., and P.J. Groblicki. 1982. An evaluation of methods for the determination of organic and elemental carbon in particulate samples. In: Particulate Carbon: Atmospheric Life Cycle (G.T. Wolff and R.L. Klimisch, eds.), pp.89-109. Plenum, New York.

Ch´ylek, P., V. Ramaswamy, and V. Srivastava. 1984. Graphitic carbon content of aerosols, clouds and snow, and its climatic implications. The Science of the Total Environment 36, 117-120.

Clark, J.S. 1988a. Charcoal-stratigraphic analysis on petrographic thin sections: recent fire history in northwest Minnesota. Quat. Res. 30, 67-80.

Clark, J.S. 1988b. Particle motion and the theory of charcoal analysis: source area, transport, deposition, and sampling. Quat. Res. 30, 81-91.

Clark, J.S. 1990. Fire and climate change during the last 750 years in northwestern Minnesota. Ecol. Monogr. 60, 135-159.

Clark, J.S., and W.A. Patterson. 1984. Pollen, 210Pb, and opaque spherules: an integrated approach to dating and sedimentation in the intertidal environment. J. Sedimentary Petrology 54, 1249-1263.

Clark, J.S., J. Merkt, and H. Muller. 1989. Post Glacial fire, vegetation, and human history of the northern Alpine forelands, south-western Germany. J. Ecology 77, 897-925.

Clark, J.S., and J. Robinson. 1993. Paleoecology of fire. In: Fire in the Environment: The Ecological, Atmospheric, and Climatic Importance of Vegetation Fires (P.J. Crutzen and J.G. Goldammer, eds.), pp.193-214. John Wiley, New York, NY.

Clark, J.S., and P.D. Royall. 1994. Pre-industrial particulate emissions and carbon sequestration from biomass burning in North America. Biogeochemistry 23, 1-17.

Clark, R.L. 1983. Pollen can charcoal evidence for the effects of Aboriginal burning on the vegetation of Australia. Archaeology in Oceania 18, 32-37.

Crane, P.R., and S. Lidgard. 1990. Angiosperm radiation and patterns of Cretaceous palynological diversity. In: Major Evolutionary Radiations (P.D. Taylor and G.P. Larwood, eds.), pp.377-407. Clarendon Press, Oxford, England.

Crutzen, P.J., and M. O. Andreae. 1990. Biomass burning in the tropics: impact on atmospheric chemistry and biogeochemical cycles. Science 250, 1669-1678.

Crutzen, P.J., and J.G. Goldammer. (eds.). 1993. Fire in the Environment: The Ecological, Atmospheric, and Climatic Importance of Vegetation Fires. Wiley, New York.

Cushing, E.J. 1965. Problems in the Quaternary phytogeography of the Great Lakes region. In: The Quaternary of the United States (H.E. Wright, Jr and D.G. Frey, eds.), pp.403-416. Princeton University Press, Princeton, New Jersey.

Cwynar, L.C. 1978. Recent history of fire and vegetation from laminated sediment of Greenleaf Lake, Algonquin Park, Ontario. Can. J. Bot. 56, 10-21.

Cwynar, L.C. 1982. A Late-Quaternary vegetation history from Hanging Lake, northern Yukon. Ecol. Monogr. 52, 1-24.

Cwynar, L.C. 1987. Fire and the forest history of the North Cascade Range. Ecology 68, 791-802.

de Beaulieu, J.-L, and M. Reille. 1989. The transition from temperate phases to stadials in the long upper Pleistocene sequence from Les Échets (France). Rev. Palaeogeography, Palaeoclimatology, and Palaeoecology 72, 147-159.

de Beaulieu, J.-L, and M. Reille. 1992. The last climatic cycle at La Grande Pile (Vosges, France) - a new pollen profile. Quat. Sci. Rev. 11, 431-438.

Davis, M.B. 1969. Climatic changes in southern Connecticut recorded by pollen deposition at Rogers Lake. Ecology 50, 409-422.

Davis, M.B. 1972. Redeposition of pollen grains in lake sediments. Limnology and Oceanography 18, 44-52.

Davis, M.B., R.W. Spear, and L.C.K. Shane. 1980. Holocene climate of New England. Quat. Res. 14, 240-250.

Davis, R.B., and G.L. Jacobson. 1985. Late Glacial and early Holocene landscapes in northern New England and adjacent areas of Canada. Quat. Res. 23, 341-368.

Delcourt, H.R. 1979. Late Quaternary vegetation history of the eastern highland rim and adjacent Cumberland Plateau of Tennessee. Ecol. Monogr. 49, 255-280.

Delcourt, P.A., H.R. Delcourt, R.C. Brister, and L.E. Lackey. 1980. Quaternary vegetation history of the Mississippi embayment. Quat. Res. 13, 111-132.

Engstrom, D.R., and B.C.S. Hansen. 1985. Postglacial vegetational change and soil development in southeastern Labrador as inferred from pollen and chemical stratigraphy. Can. J. Bot. 63, 543-561.

Fahnestock, G.R., and J.K. Agee. 1983. Biomass consumption and smoke production by prehistoric and modern forest fires in western Washington. J. For. 81, 653.

Gajewski, K., M.G. Winkler, and A.M. Swain. 1985. Vegetation and fire history from three lakes with varved sediments in northwestern Wisconsin. Rev. Palaeobotany and Palynology 44, 277-292.

Goldberg, E.D. 1985. Black Carbon in the Environment. John Wiley, New York.

Green, D.G. 1981. Time series and postglacial forest ecology. Quat. Res. 15, 265-277.

Grichuk, V.P. 1984. Late Pleistocene vegetation history. In: Late Quaternary Environments of the Soviet Union (A.A. Velichko, ed.), pp.155-178. University of Minnesota Press, Minneapolis, Minnesota.

Griffen, J.J., and E.D. Goldberg. 1975. The fluxes of elemental carbon in coastal marine sediments. Limnology and Oceanography 20, 456-463.

Grüger, J. 1973. Studies on the Late Quaternary vegetation history of northeastern Kansas. Geol. Soc. America Bull. 84, 239-250.

Herring, J.R. 1985. Charcoal fluxes into sediments of the North Pacific Ocean: The Cenozoic record of burning. In: The Carbon Cycle and Atmospheric CO_2: Natural Variations, Archean to Present (E.T. Sundquist and W.S. Broecker, eds.), pp.419-442. Geophysical Monogr. 32, American Geophysical Union, Washington, D.C.

Heusser, C.J. 1990. Late-Glacial and Holocene vegetation and climate of subantarctic South America. Rev. Palaeobotany and Palynology 65, 9-15.

Heusser, L.E., and J.E. King. 1988. North America. In: Vegetation History (B. Huntley, and T. Webb III, eds.), pp.193-236. Kluwer Academic Publ., Dordrecht, Netherlands.

Huntley, B. 1988. Europe. In: Vegetation History (B.Huntley and T.Webb III, eds.), pp. 341-383. Kluwer, Dordrecht, Netherlands.

Hussey, T.C. 1993. A 20,000-Year History of Vegetation and Climate at Clear Pond, Northeastern South Carolina. MS Thesis, University of Maine, Orono, Maine.

Jackson, S.T. 1989. Postglacial vegetational changes along an elevational gradient in the Adirondack Mountains (New York). Biological Survey Museum Bulletin 465, 29 pp.

Jétte, H., and P.J.H. Richard. 1992. Contribution á l'histoire postglaciaire de la végétation en Gaspésie méridionale, Québec. Géographie Physique et Quaternaire 46, 273-284.

Jones, K.C., J.A. Stratford, K.S. Waterhouse, E.T. Furlong, W. Giger, R.A. Hites, C. Schaffner, and A.E. Johnston. 1989. Increases in the polynuclear aromatic hydrocarbon content of an agricultural soil over the last century. Environmental Science and Technology 23, 95-101.

Johnson, E.A. 1992. Fire and Vegetation Dynamics. Cambridge University Press, Cambridge, England.

Kershaw, A.P., D.M. D'Costa, J.R.C. KcEwen Mason, and B.E. Wagstaff. 1991. Palynological evidence for Quaternary vegetation and environments of mainland southeastern Australia. Quat. Sci. Rev. 10, 391-404.

King, J.E. 1973. Late Pleistocene palynology and biogeography of the western Missouri Ozarks. Ecol. Monogr. 43, 539-565.

Khotinskiy, N.A. 1984. Holocene vegetation history. In: Late Quaternary Environments of the Soviet Union (A.A. Velichko, ed.), pp.179-200. University of Minnesota Press, Minneapolis, Minnesota.

Krebs, J.S., and R.G. Barry. 1970. The arctic front and the tundra-tiaga boundary in Eurasia. Geogr. Rev. 60, 548-554.

Kutzbach, J.E., and P.J. Guetter. 1986. The influence of changing orbital parameters and surface boundary conditions on climate simulations for the past 18000 years. J. Atmospheric Sci. 43, 1726-1759.

Kutzbach, J.E., and F.A. Street-Perrott. 1985. Milankovitch forcing of fluctuations in the level of tropical lakes from 18 to 0 kyr BP. Nature 317, 130-134.

Labelle, C., and P.J.H. Richard. 1981. Végétation tardiglaciaire et postglaciaire au sud-est du parc des Laurentides, Québec. Géographie physique et Quaternaire 35, 345-359.

Labelle, C., and P.J.H. Richard. 1984. Histoire postglaciaire de la végétation dans las région de Mont-Saint-Pierre, Gaspésie. Géographie physique et Quaternaire 38, 257-274.

Laflamme, R.E., and R.A. Hites. 1978. The global distribution of polycyclic aromatic hydrocarbons in recent sediments. Geochimica et Cosmochimica Acta 42, 289-303.

Lamb, H.F., and M.E. Edwards. 1988. The Arctic. In: Vegetation History (B. Huntley and T. Webb III, eds.), pp.519-555. Kluwer Academic, Dordrecht, Netherlands.

Liu, K.-B. 1988. Quaternary history of the temperate forests of China. Quat. Sci. Rev. 7, 1-20.

Liu, K.-B. 1990. Holocene paleoecology of the boreal forest and Great Lakes-St. Lawrence forest in northern Ontario. Ecol. Monogr. 60, 179-212.

MacDonald, G.M. 1987. Postglacial vegetation history of the Mackenzie River basin. Quat. Res. 28, 245-262.

MacDonald, G.M. 1989. Postglacial palaeoecology of the subalpine forest–grassland ecotone of southwestern Alberta: new insights on vegetation and climate change in the Canadian Rocky Mountains and adjacent foothills. Palaeogeography, Palaeoclimatology, and Palaeoecology 73, 155-173.

MacDonald, G.M., T.W.D. Edwards, K.A. Moser, R. Pienitz, and J.P. Smol. 1993. Rapid response of treeline vegetation and lakes to past climate warming. Nature 361, 243-246.

MacDonald, G.M., C.P.S. Larsen, J.M. Szeicz, and K.A. Moser. 1991. The reconstruction of boreal forest fire history from lake sediments: a comparison of charcoal, pollen, sedimentological, and geochemical indices. Quat. Sci. Rev. 10, 53-71.

Malingreau, J.-P., F.A. Albini, M.O. Andreae, S. Brown, J.S. Levine, J. Lobert, T.A. Kulbusch, L. Radke, A. Setzer, P.M. Vitousek, D.E. Ward, and J. Warnatz. 1993. Quantification of fire characteristics from local to global scales. In: Fire in the Environment: The Ecological, Atmospheric, and Climatic Importance of Vegetation Fires (P.J. Crutzen and J.G. Goldammer, eds.), pp. John Wiley, New York.

Mehringer, P.J., S.F. Arno, and K.L. Petersen. 1977. Postglacial history of Lost Trail Pass Bog, Bitterroot Mountains, Montana. Arctic and Alpinus Research 9, 345-368.

Müller, H. 1974. Pollenanalytische Untersuchungen und Jahresschichtenzählungen an der holstein-zeitlichen Kieselgur von Münster-Breloh. Geologisches Jahrbuch A21, 107-140.

Mott, R.J. 1990. Sangamonian forest history and climate in Atlantic Canda. Géographie physique et Quaternaire 44, 257-270.

Novakov, T. 1984. The role of soot and primary oxidants in atmospheric chemistry. The Science of the Total Environment 36, 1-10.

Novakov, T.C., Cachier,H., Macko, S. and Clark, J.S. 1996. Characterization of Particulate Products of Biomass Combustion. In: Sediment records of biomass burning and global change (J.S. Clark, B.J. Stocks, H. Cachier, and J.G. Goldammer, eds.). Springer Verlag, New York (in press).

Patterson, W.A., K.J. Edwards, and D.J. Maguire. 1987. Microscopic charcoal as a fossil indicator of fire. Quat. Sci. Rev. 6, 3-23.

Payette, S., C. Morneau, L. Sirois, and M. Desponts. 1989. Recent fire history of the northern Quebec biomes. Ecology 70, 656-673.

Payette, S. 1993. The range limit of boreal tree species in Quebec-Labrador: an ecological and palaeoecological interpretation. Review of Palaeobotany and Palynology 79, 7-30.

Penner, J.E., S.J. Ghan, and J.J. Walton. 1991. The role of biomass burning in the budget and cycle of carbonaceous soot aerosols and their climate impact. In: Global Biomass Burning: Atmospheric, Climatic, and Biospheric Implications (J.S. Levine, ed.), pp.387-393. MIT Press, Cambridge, Massachusetts.

Plummer, L.N. 1993. Stable isotope enrichment in paleowaters of the southeast Atlantic coastal plain, United States. Science 262, 2016-2020.

Pons, A., J. Guiot, J.L. de Beaulieu, and M. Reille. 1992. Recent contributions to the climatology of the Last Glacial-Interglacial cycle based on French pollen sequences. Quat. Sci. Rev. 11, 439-448.

Prentice, I.C. 1985. Pollen representation, source area, and basin size: toward a unified theory of pollen analysis. Quat. Res. 23, 76-86.

Reille, M., and J.-L de Beaulieu. 1990. Pollen analysis of a long upper Pleistocene continental sequence in a Velay maar (Massif Central, France). Palaeogeography, Palaeoclimatology, and Palaeoecology 80, 35-48.

Richard, P.J.H. 1977. Histoire post-wisconsinienne de la végétation du Québec mérional par l'analyse pollinique. Service de la recherche, Direction générale des Fôrets du Québec. Publications et rapports divers 1 (312 pp.) + 2 (142 pp.).

Richard, P.J.H. 1979. Contribution á l'histoire postglaciaire de la végétation au nord-est de la Jamésie, Nouveau-Québec. Géographie Physique et Quaternaire 33, 93-112.

Richard, P.J.H. 1980. Histoire postglaciaire de la végétation au sud du lac Abitibi, Ontario et Québec. Géographie Physique et Quaternaire 34, 77-94.

Richard, P.J.H. 1981. Paléophytogéographie postglaciaire en Ungava, par l'analyse pollinique. Collection Paléo-Québec 13, 153 pp.

Richard, P.J.H., and C. Labelle. 1989. Histoire postglaciaire de la végétation au lac du Diable, mont Albert, Gaspésie, Québec. Géographie Physique et Quaternaire 43, 337-354.

Richard, P.J.H., A.C. Larouche, and M. Bouchard. 1982. Age de la déglaciation finale et histoire postglaciaire de la végétation dans la partie centrale du Nouveau-Quebec. Géographie Physique et Quaternaire 36, 63-90.

Richard, P.J.H., A.C. Larouche, and G. Lortie. 1992. Paleophytogeographie et paleoclimats postglaciaires dans l'ouest du Bas-Saint-Laurent, Quebec. Geographie Physique et Quaternaire 46, 151-172.

Richard, P.JH., and A.C. Larouche. 1994. Histoire postglaciaire de la végétation au lac du Diable, mont Albert, Gaspésie, Québec. In: Il y a 8000 ans à Rimouski. Paléoécologie et archéologie d'un site de la culture Plano (C. Chapdelaine, ed.), pp.49-89. Collection Paléo-Québec 22, 314pp.

Ritchie, J.C. 1987. Postglacial Vegetation of Canada. Cambridge University Press, Cambridge, England.

Rosen, H., A.D.A. Hansen, and T. Novakov. 1984. Role of graphitic carbon particles in radiative transfer in the Arctic haze. The Science of the Total Environment 36, 103-110.

Sandberg, D.V., J.M. Pierovich, D.G. Fox, and E.W. Ross. 1979. Effects of fire on air. United States Department of Agriculture Forest Service General Technical Report WO-9, Washington, DC.

Seiler, W., and P.J. Crutzen. 1980. Estimates of gross and net fluxes of carbon between the biosphere and the atmosphere from biomass burning. Climatic Change 2, 207-247.

Shaw, G.E. 1982. Perturbation to the atmospheric radiation field from carbonaceous aerosols. In: Particulate Carbon: Atmospheric Life Cycle (G.T. Wolff and R. L. Klimisch, eds.), pp.53-74. Plenum Press, New York.

Singh, G., A.P. Kershaw, and R. Clark. 1981. Quaternary vegetation and fire history in Australia. In: Fire and the Australian Biota (A.M. Gill, R. H. Groves, and I.R. Noble, eds.), pp.23-54. Australian Academy of Science, Canberra, Australia.

Smith, D.M., J.J. Griffen, and E.D. Goldberg. 1973. Elemental carbon in marine sediments: a baseline for burning. Nature 241, 268-270.

Smith, S.J., and R.S. Anderson. 1992. Late Wisconsin Paleoecologic record from Swamp Lake, Yosemite National Park, California. Quat. Res. 38, 91-102.

Sorenson, C.J., J.C. Knox, J.A. Larsen, R.A. and Bryson. 1971. Paleosols and the Forest Border in Keewatin, N.W.T. Quat. Res. 1, 468-473.

Stocks, B.J. 1993. Global warming and forest fires in Canada. The Forestry Chronicle 69, 290-293.

Stocks, B.J., and R.B. Street. 1983. Forest fire weather and wildfire occurrence in the borealforest of northwestern Ontario. In: Resources and Dynamics of the Boreal Zone (R.W. Wein, R. R. Riewe and I.R. Methven, eds.), pp.249-265. Association of Canadian Universities Northern Studies, Ottawa, Ontario.

Sugita, S., and M. Tsukada. 1983. Vegetation history around Nonbara bog in the San'in District for the past 17,000 years. Japanese J. Ecology 33, 225-230.

Swain, A.M. 1973. A history of fire and vegetation in northeastern Minnesota as recorded in lake sediment. Quat. Res. 3, 383-396.

Tolonen, K. 1983. The post-glacial fire record. In: The Role of Fire in Northern Circumpolar Ecosystems (R.W. Wein and D.A. MacLean, eds.), pp.21-44. John Wiley, New York.

Tolonen, M. 1978. Palaeoecology of annually laminated sediments in Lake Ahvenainen, S. Finland. I. Pollen and charcoal analyses and their relation to human impact. Ann. Bot. Fennici 15, 177-208.

Tolonen, M. 1983. Pollen evidence of vegetational change following early European settlement on Monhegan Island, Maine, northeastern U.S.A. Boreas 12, 201-215.

Tolonen, M. 1987. Primary and secondary forest successions reflected in a late Holocene pollen sequence from an upland mire near the SW coast of Finland. Ann. Bot. Fennici 24, 97-109.

Tsukada, M. 1983. Vegetation and climate during the last Glacial maximum in Japan. Quat. Res. 19, 212-235.

Tsukada, M., S. Sugita, and D.M. Hibbert. 1981. Paleoecology in the Pacific Northwest. I. Late Quaternary vegetation and climate. Verhandlungen der Internationalen Vereinigung für Limnologie 21, 730-737.

Tsukada, M., and S. Sugita. 1982. Late Quaternary dynamics of pollen influx at Mineral Lake, Washington. Botanical Magazine (Tokyo) 95, 401-418.

Van Cleve, K., and L.A. Viereck. 1981. Forest succession in relation to nutrient cycling in the boreal forest of Alaska. In: Forest Succession: Concepts and Application (D.C. West, H. H. Shugart, and D.B. Botkin, eds.), pp.185-211. Springer-Verlag, New York.

Watts, W.A. 1970. The full-glacial vegetation of northwestern Georgia. Ecology 51, 17-33.

Watts, W.A. 1979. Late Quaternary vegetation of central Appalachia and the New Jersey coastal plain. Ecol. Monogr. 49, 427-469.

Watts, W.A. 1980. The late Quaternary vegetation history of the southeastern United States. Ann. Rev. of Ecology and Systematics 11, 387-409.

Watts, W.A. 1988. Europe. In: Vegetation History (B. Huntley and T. Webb III (eds), pp.155-192. Kluwer Academic Press, Dordrecht, Netherlands.

Webb III, T. 1988. Eastern North America. In: Vegetation History (B. Huntley and T. Webb III (eds), pp.385-414. Kluwer Academic Press, Dordrecht, Netherlands.

Webb III, T., P.J. Bartlein, and J.E. Kutzbach. 1987. Climatic change in eastern North America during the past 18,000 years; comparisons of pollen data with model results. In: North America and Adjacent Oceans During the Last Deglaciation (W.F. Ruddiman and H.E. Wright, Jr, eds.), pp.447-462. Geological Society of America, Boulder, Colorado.

West, R.G. 1970. Pleistocene history of the British flora. Pages 1-12. in D. Walker and R.G. West, editor. Studies in the Vegetational History of the British Isles. Cambridge University Press, Cambridge, England.

Whitehead, D.R. 1964. Fossil pine pollen and full-Glacial vegetation in southeastern North Carolina. Ecology 45, 767-777.

Whitehead, D.R. 1981. Late-Pleistocene vegetational changes in northeastern North Carolina. Ecol. Monogr. 51, 451-471.

Wik, M., and I. Renberg. 1991. Recent atmospheric deposition in Sweden of carbonaceous particles from fossil-fuel combustion surveyed using lake sediments. Ambio 20, 289-292.

Winkler, M.J. 1985. Charcoal analysis for paleoenvironmental interpretation: a chemical assay. Quat. Res. 23, 313-326.

Woillard, G.M. 1978. Grande Pile Peat Bog: a continuous pollen record for the last 140,000 years. Quat. Res. 9, 1-21.

Wolfe, J.A. 1975. Some aspects of plant geography of the northern hemisphere during the Late Cretaceous and Tertiary. Annals of the Missouri Botanical Garden 62, 264-279.

Wright, H.E. 1992. Patterns of Holocene climatic change in the midwestern United States. Quat. Res. 38, 129-134.

Wright, H.E., and C.W. Barnosky. 1984. Introduction to the English edition. In: Late Quaternary Environments of the Soviet Union (A.A. Velichko, ed.), pp.xiii-xxii. University of Minnesota Press, Minneapolis, Minnesota.

Zackrisson, O. 1977. Influence of forest fires on the North Swedish boreal forests. Oikos 29, 22-32.

J.G.Goldammer and V.V.Furyaev (eds.), Fire in Ecosystems of Boreal Eurasia, 90-104.

Fire and Climate History in the Central Yenisey Region, Siberia

T. W. Swetnam [1]

1. Introduction

Siberian forests are shaped by complex interactions of climate, forest fires, insect outbreaks, and humans. Unravelling the linkages and mechanisms of these interactions will require long temporal and large spatial-scale perspectives. A full account of the past input of Siberian forest fires to global atmospheric composition (e.g., trace gases, aerosols, and particulates), as well as the potential for future input, will require accurate area burned time series and maps (Valendik 1990; Stocks 1991; Goldammer 1992; Dixon and Krankina 1993). Unfortunately, complete and reliable fire activity records are available only for the past one or two decades at best. Furthermore, these data cover only monitored areas, which are less than half of the total area in many remote regions of Siberia (Valendik 1990). Complete meteorological observations extending more than a few decades into the past are generally only available for larger cities. Hence, we must turn to "natural archives", such as lake deposits and tree rings, to reconstruct long-term historical changes and trends in climate and disturbance processes (Valendik and Ivanova 1990).

Dendro-ecological analysis can provide simultaneous reconstructions of past climate and disturbance variations across a broad range of spatial and temporal scales (Fritts and Swetnam 1989). Direct comparisons of these long-term reconstructions can reveal patterns of association, including the strength and lagging relations between climate and disturbances (Swetnam and Betancourt 1990, 1992; Swetnam and Lynch 1993; Valendik and Ivanova 1990). Methods of reconstructing fire histories in forest communities from tree-ring analysis of forest age structures and fire scars are well known (e.g., Heinselman 1973; Zackrisson 1977; Furyaev and Kireev 1979), as are methods for reconstructing past climatic variations from tree-ring measurements (Fritts 1976; 1991; Cook and Kairiukstis 1990). However, combined assessments of well-dated, multi-century fire and climate histories from the same region are rare.

Carefully cross-dated tree-ring reconstructions have two distinct advantages over fire histories developed by simple tree-ring counting of fire scarred specimens or stand age structure analysis. By "cross-dating" I mean the systematic cross-identification of ring-width patterns (or other tree-ring features), within and among all sampled trees, so that all rings are exactly dated (Stokes and Smiley 1968; Swetnam et al. 1985). These advantages include: (1) the ability to accurately determine fire dates to the year, and often the season of occurrence; and (2) the ability to date sub-fossil wood (e.g., stumps, snags, and logs), which

[1] Laboratory of Tree Ring Research, The University of Arizona, Tucson, Arizona 85721, U.S.A.

often increases the temporal length of the reconstructions (Baisan and Swetnam 1990; Brown and Swetnam 1994; Swetnam 1993).

Without seasonal to annual dating resolution it may be difficult or impossible to detect important linkages between environmental processes operating on these time scales (Baumgartner et al. 1989). Distinguishing the effects of climatic factors versus human activities on past fire regimes may also depend on our ability to resolve year-to-year synchrony or asynchrony of fire, climate, and human events. Explicit assessments of disturbance sizes within forest stands, and disturbance extent within regions, are also facilitated by accurately dated events because individual "point" records (e.g., chronologies from trees or stands) can be meaningfully compared and aggregated across space (Swetnam 1993). Disadvantages of crossdated fire history studies include: (1) a typical limitation to time periods of less than 1,000 years (i.e., the maximum life spans of most trees and sub-fossil wood), (2) a "fading" record with increasing time before present because of a progressive loss of older tree-ring material, and (3) additional time and effort may be required to carry out the cross-dating procedures.

In this paper I present a dendro-ecological assessment of fire history and climate changes in pine forests of central Siberia, west of the Yenisey River. I describe and discuss synchronous changes in climate, fire, and human history in this region over the past four centuries. The importance of accurately identifying the temporal and spatial structure of fire regimes and climate (i.e., synchrony) is emphasized. This work is preliminary to a larger assessment of fire and climate changes in central Siberia that will encompass many more tree-ring sites and will link to remote sensing of fire patterns at the landscape scale.

2. Methods

2.1 Study Area

Tree-ring samples were obtained in five locations along the Dubches and Kac Rivers on the west side of the Yenisey River (Fig.1). All fire scarred samples were from relatively dry *Pinus sylvestris* forests with extremely sandy soils (Fig.2). This region is a hilly, glaciopluvial plain formed by glacial moraines and ancient river meanders. Coniferous forests occur in a landscape mosaic on slightly elevated ridges, hills, and terraces above bogs and river meanders. Nearly pure *Pinus sylvestris* forests occur in the driest sites with ground cover dominated by lichens (especially *Cladonia* spp.) and scattered low shrubs (e.g., *Vaccinium* spp. [Fig.2]). Elevations ranged from approximately 100 to 200 meters above mean sea level.

2.2 Sample Collection and Analysis

Increment core collections were obtained in 1989 from one *Pinus sylvestris* stand and one *Pinus sibirica* stand in the region by D.A. Graybill of the Laboratory of Tree-Ring Research, University of Arizona and S. Shiyatov, of the Institute of Plant and Animal Ecology, Yekatarinburg (Sverdlovsk). Two cores were taken from each of ten trees in the *P. sylvestris* stand, and 17 trees in the *P. sibirica* stand (Fig.1). These samples were dendrochronologically cross-dated, measured, and standardized by D.A. Graybill into two ring-width index chronologies (Stokes and Smiley 1968; Graybill 1979). Standardization of the tree-ring width measurements involved fitting negative exponential curves or straight lines to each increment core series, then dividing the ring widths by the values of the fitted lines. Dimensionless tree-ring indices with a mean of approximately 1.0 were produced. All individual core series were then averaged into site chronologies (Graybill 1979; Swetnam et al. 1985).

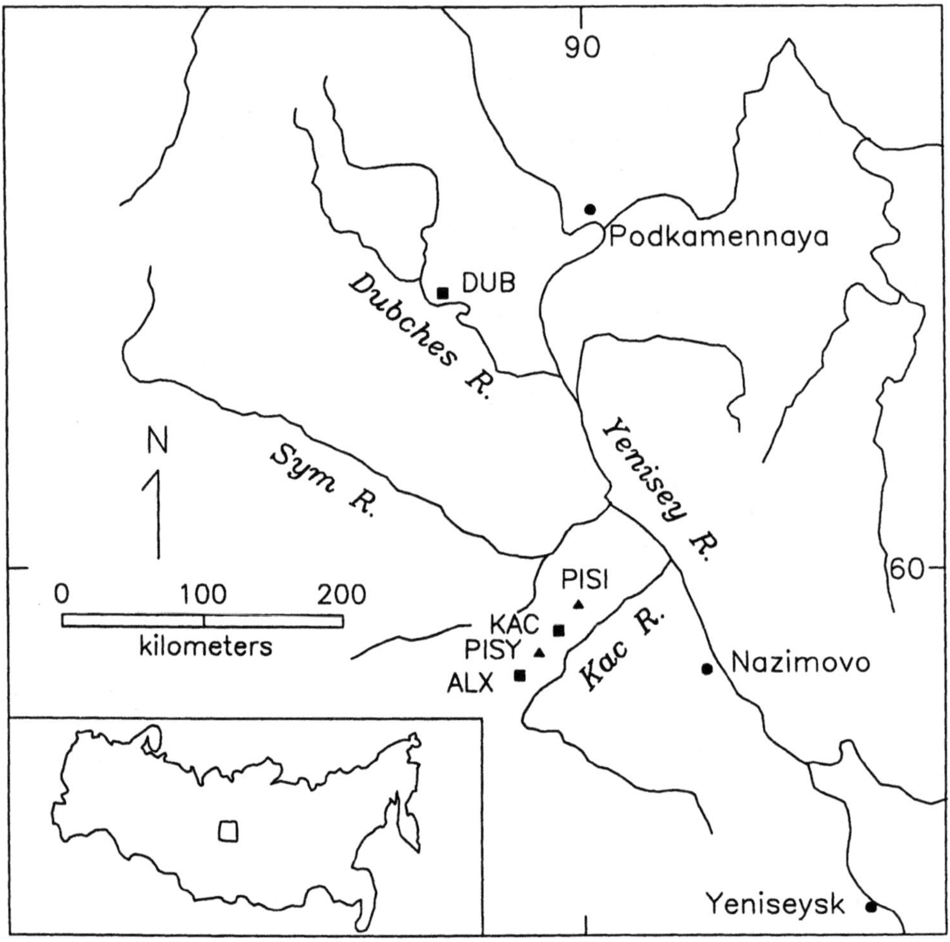

Fig.1. Map of Central Yenisey region with locations of major rivers and sample locations. Fire scar collection sites and tree-ring width chronology sites are shown with letter codes: DUB = Dubches River fire scars; KAC = Kas River fire scars, ALX = Aleksandrovskiy Shlyuz fire scars; PISY = *Pinus sylvestris* tree-ring width chronology; PISI = *Pinus sibirica* tree-ring width chronology.

Development of calibrated climate-tree growth models using these chronologies has not yet been achieved, but cross-dating among the tree-ring samples indicates that a substantial climate signal is contained in the year-to-year ring-width variations. The Pearson correlation (r) between the mean chronologies from the two sites was 0.49 ($p < 0.01$, 1609-1988).

Fig.2. A (left): Old-growth *Pinus sylvestris* stand with *Cladonia* moss ground cover. This stand is located near the Sym River. B (center): *Pinus sylvestris* stand with *Cladonia* moss and *Vaccinium* low shrub ground cover. This stand is located near the Dubches River.

Fig.3. (right) Fire scarred trees up to 400 years old, and containing 10 or more scar dates, are not uncommon in uneven-aged pine stands in the Central Yenisey region. Intense surface fires typically occurred at 20 to 50 year intervals. These old-growth stands are found in small patches on ridges and relatively dry sites and are often surrounded by larger even-aged stands originating from intense, stand-replacement crown fires.

The *Pinus sylvestris* stand was in a well drained site in sandy soils, and so it is reasonable to infer that the ring-width variations were largely controlled by meteorological rainfall and associated soil moisture availability. The *P. sibirica* stand was also growing in sandy soils, but in a more mesic microsite. Thus, the differences between the two species chronologies may be related to different climatic sensitivities. Further studies in this region are needed to determine the dendroclimatic responses of these species.

In summer of 1991 I sampled small clusters of 4 to 5 fire scarred trees with a chainsaw at each of three sites (Fig.3). The sampled trees were dispersed over areas of about five hectares. Trees with well-preserved multiple scars within sites were preferred. A total of 13 living and dead trees were sampled at the sites.

The cross section samples were transported to my laboratory in Tucson, Arizona, U.S.A. All sections were finely sanded with electric beltsanders, and the tree-rings were crossdated (Stokes and Smiley 1968; Swetnam et al. 1985). Years and season of occurrence of past fires were determined by observing the intra-ring position of the fire scars under magnification (30x) (Dieterich and Swetnam 1984).

All fire dates (years) recorded on individual trees were compiled and plotted in composite fire scar chronology charts for inspection of detailed fire history patterns (Dieterich 1980; Swetnam and Dieterich 1985; Baisan and Swetnam 1990). This graphical analysis emphasizes the explicit chronological and spatial aspects of fire history, particularly the synchrony of fire events recorded within and among sites.

All fire dates recorded among the three sites were directly compared with the tree-ring width index chronologies. This was accomplished by (1) plotting the chronologies and overlaying the fire dates for graphical inspection, and (2) by conducting a "superposed epoch analysis" (Lough and Fritts 1987). The superposed epoch analysis was applied to different time periods to assess temporal changes in the climatic sensitivity of the reconstructed regional fire regime. This analysis was carried out by computing the mean tree-ring-width index during all fire years, and lagged years before and after the fire years. The departure of these mean values relative to the mean index value for the entire chronology was plotted to assess the magnitude of tree-growth variations during, prior to, and after the specified fire dates. This provided a tool to assess the average inferred climatic conditions in a time window bracketing all identified fire events. A "bootstrap" method was used to estimate confidence intervals for the observed departures. This was accomplished by randomly selecting the same number of years as the actual fire years, recomputing the mean departures during the randomly selected "pseudo-fire years" and lagged years, and then repeating this process 1,000 times. The standard deviation of the 1,000 random trials were used to compute the statistical significance (p level) of obtaining observed departures simply by chance (Mooney and Duval 1993).

The tree-ring chronologies used in the superposed epoch analysis were "white noise" residuals derived from Box Jenkins modeling of the standard index chronologies (Monserud 1986; Biondi and Swetnam 1987). This procedure removed the persistence (or autocorrelation) in the standard tree-ring index chronologies that may have been due to tree physiological factors (e.g., carryover of stored food materials) (Fritts 1976). This pre-treatment of the tree-ring chronologies was deemed necessary to distinguish lagging patterns due to tree physiology versus climate-fire interactions in the superposed epoch analysis.

In addition to the fire dates determined by fire scar samples collected in this study, fire dates from a landscape-scale fire reconstruction by V.V. Furyaev (Furyaev and Kireev 1979; Antonovski et al. 1992) were also compared with the tree-ring width chronologies. This data set is from the same region - an area of 165,000 ha centered on 59°N - 90°E. Finally, a short time series of annual area burned from the Krasnoyarsk region (1981 to 1991) (Goldammer 1992) was compared with the tree-ring chronologies to assess their potential usefulness as indicators of past fire activity.

3. Results

3.1 Tree-Ring Dating Quality, Characteristics, and Problems

Tree-ring specimens exceeding 400 years old were found in two of the three fire scar collection sites (Dubches and Kac) (Fig. 4). The oldest living tree at the Kac site had an inner ring date of 1574. Also, a sample from a standing dead tree at this site had an inner ring date of 1581 and an outer ring date of 1771. A maximum of 12 fire scar dates were observed on a tree at the Kac site. This demonstrates that ancient fire history material from well preserved living and dead trees in central Siberia can be obtained and cross-dated.

Crossdating quality was fair to good, i.e., year-to-year ring-width variability was sufficiently high that all fire scar specimens, except one, were successfully cross-dated. Few locally absent or completely absent rings were identified (less than 1%). However, because most of the older trees had very slow growth rates, cross-dating or accurate ring counting would not be possible without preparation of very well sanded viewing surfaces. Identification of scar position within rings also required very finely prepared surfaces.

The master fire chronologies (Fig.4) show that recorded fires were usually synchronous among trees within the sampled areas. This is not surprising since the sampled areas were relatively small (less than 10 hectares). However, a few fires were recorded by only one or two trees within sites. With the current small sample sizes from these sites it is not possible to determine with confidence that this differential scarring among trees is related to relative fire extent.

While synchrony of fire dates among trees within sites was high, there was little correspondence of fire dates among the sites. A total of 30 fire years were recorded among all sites from 1585 to 1978. Only the fire dates of 1742 and 1978 were in common among two different sites. However, eight other fire years differed by only one year relative to fires in other sites: 1780-1781, 1869-1870, 1897-1898, 1923-1924. Microscopic examination of the specimens suggests that some of these differences may be due to interpretations of the so-called "dormant season" fire scars. Dormant season fire scars are lesions on the ring boundary between the latewood of one years growth and earlywood of the next years growth. In this study the calendrical dates (years) assigned to dormant season fire scars were the years of formation of the adjacent latewood cells. This assumes that fires occurred after the growing season of the trees in the previous summer or fall. However, it is entirely possible (perhaps probable) that some dormant season fire scars were created by fires that occurred in the spring or summer before cambial growth started. For example, in the case of the 1780 fire scars at Aleksandrovskiy Shlyuz we observed some slight indications of damaged earlywood cells adjacent to dormant season scars on some trees. Hence it is possible that a fire occurred at this site early in the growing season of 1781 just prior to cambial growth. I will return to the implications of this dating problem later in discussing results of the climate-fire comparisons.

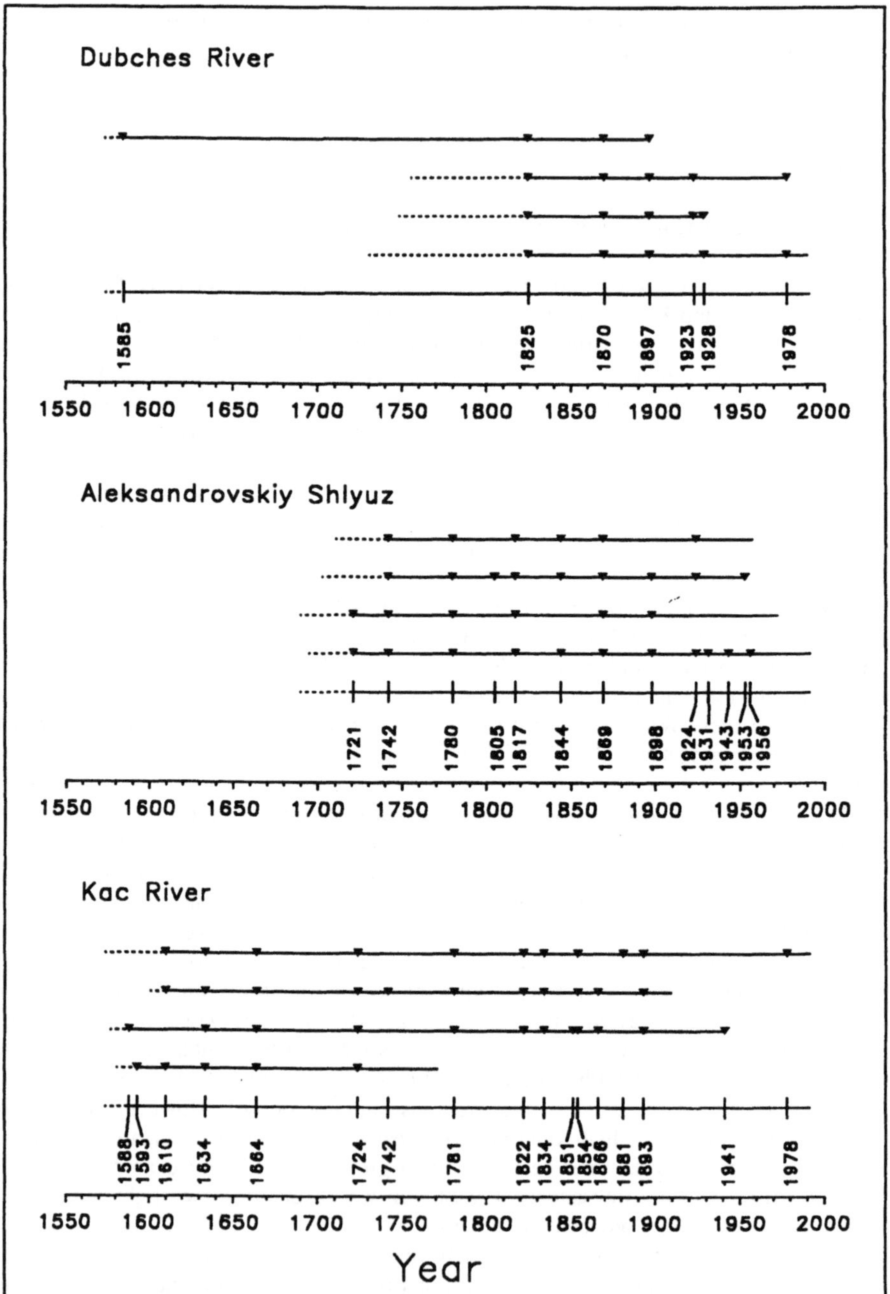

Fig.4. Master fire chronologies for three sites in Central Yenisey region. The horizontal lines are the sampled life spans of trees, and the arrowheads are dated fire scars.

3.2 Synchrony of Fire and Climate Events

Comparison of the tree-ring chronologies with a short record of fire activity in the Krasnoyarsk region (Goldammer 1992) suggests that the tree-ring width chronologies are a fairly good indicator of regional fire patterns (Fig.5), confirming earlier observations of Valendik and Ivanova (1990). Fluctuations in ring widths generally show an inverse relation to fire activity; i.e., decreases in tree growth correspond with high fire activity, and increases in tree growth correspond with relatively low fire activity. However, the overlap of the tree-ring record and observed fire record is quite short (1981 to 1988), and so updated tree-ring chronologies and extended documented fire records are needed.

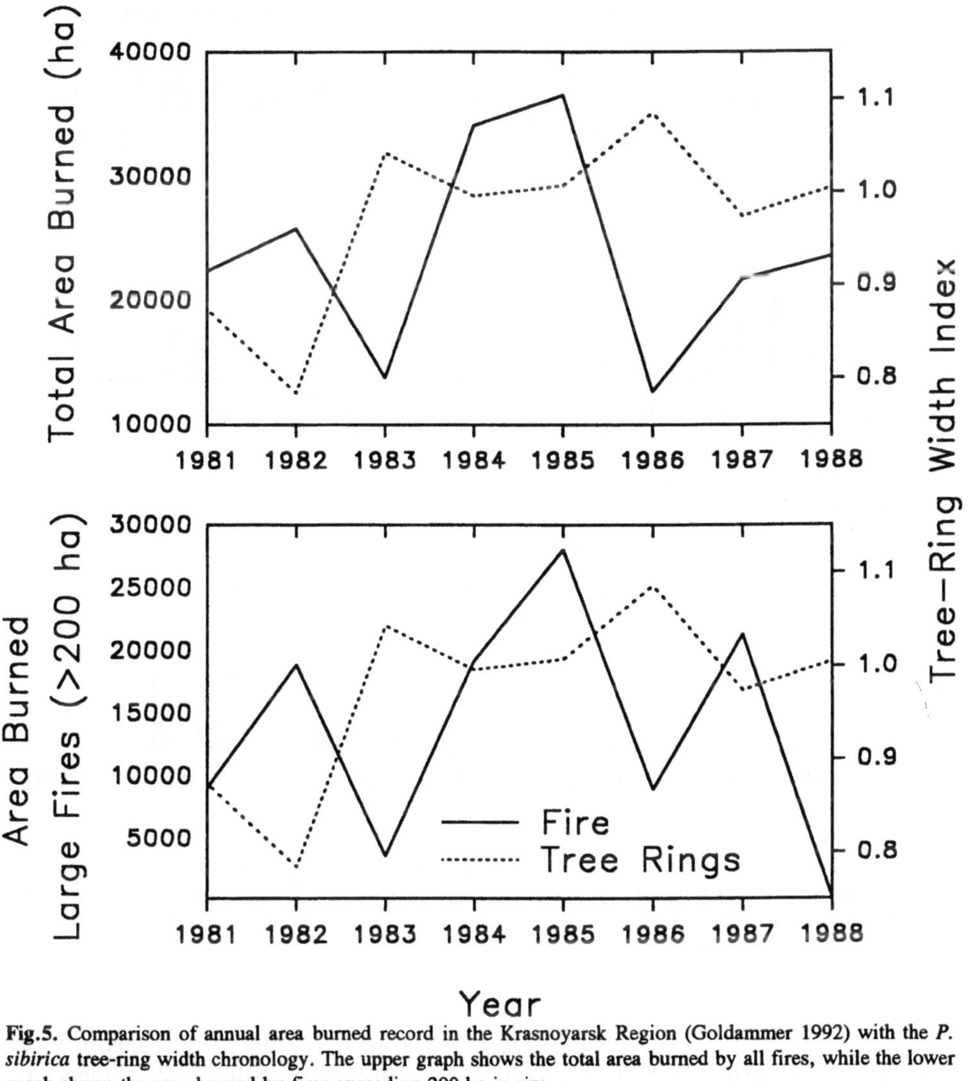

Fig.5. Comparison of annual area burned record in the Krasnoyarsk Region (Goldammer 1992) with the *P. sibirica* tree-ring width chronology. The upper graph shows the total area burned by all fires, while the lower graph shows the area burned by fires exceeding 200 ha in size.

Direct comparison of the combined fire scar dates from all three sites with the tree-ring width index chronologies indicates that fires most consistently occurred during inferred dry years and periods (low tree-ring growth) before ca. 1880 (Fig.6). In the comparison with the *P. sibirica* chronology, 13 out of 18 fire years (72%) from 1610 to 1870 coincided with dry years (tree-ring growth below the long-term mean of the chronology), while 6 out of 12 fire years (50%) coincided with dry years from 1881 to 1978. In the comparison with the *P. sylvestris* chronology, 16 out of 18 (89%) of the fire years from 1610 to 1870 coincided with dry years, while 7 out of 12 (58%) coincided with dry years from 1881 to 1978.

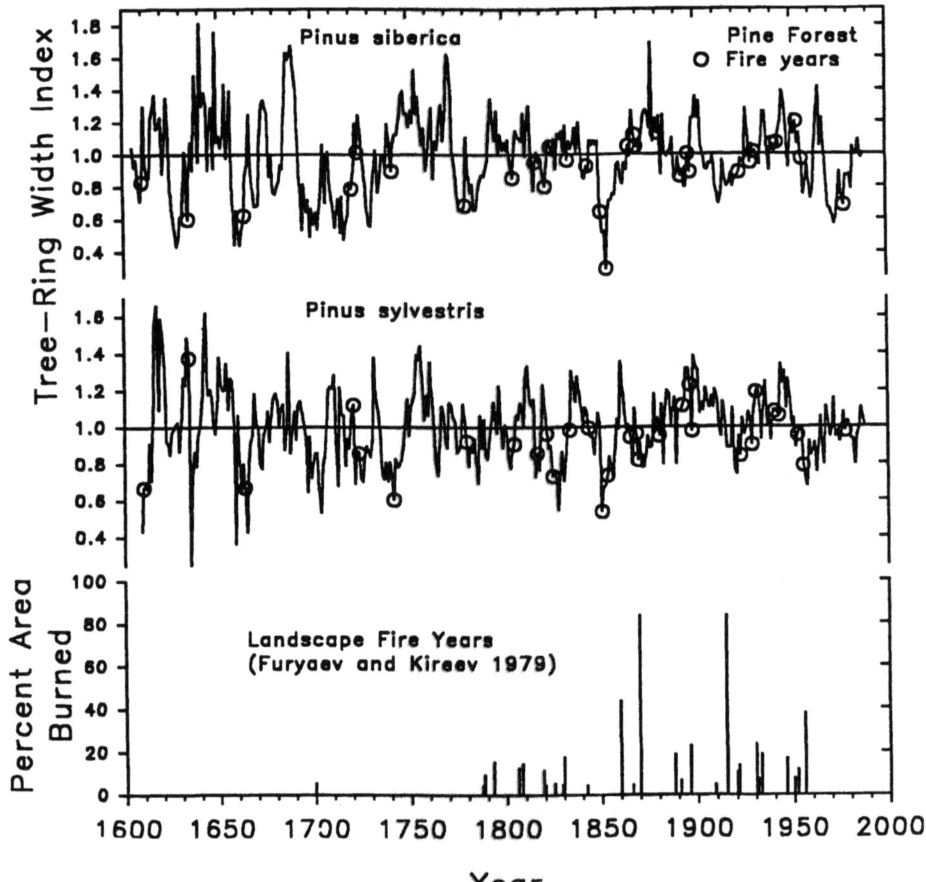

Fig.6. Comparison of fire years (dates) and tree-ring width chronologies. The upper graph shows the comparison of fire years determined in this study (open circles) with the *P. sibirica* chronology, while the middle graph shows the same fire years compared to the *P. sylvestris* chronology. The lower graph shows fire years and percent area burned estimated by Furyaev and Kireev (1979).

Individual fire years compiled by Furyaev and Kireev (1979) do not appear to be consistently associated with dry or wet years inferred from the tree-ring width chronologies (Fig.6) from this same region. In comparison with the *P. sibirica* chronology, 9 out of 15 fire years (60%) coincided with dry years from 1700 to 1870, and 8 out of 14 fire years (57%) coincided with dry years from 1888 to 1956. In comparison with the *P. sylvestris* chronology, 8 out of 15 fire years (53%) coincided with dry years from 1700 to 1870, and 4 out of 14 fire years (29%) coincided with dry years from 1888 to 1956. Although there appears to be no obvious correspondence between the Furyaev and Kireev fire years and dry years, high and low levels of fire activity at decadal time scales seem to correspond better with similar time-scale drought (tree-growth) fluctuations. For example, relatively low fire activity appears in the Furyaev and Kireev chronology during the 1830s and 1840s, and from 1871 to the late 1880s, when wet conditions (larger tree-rings) prevailed (Fig. 6).

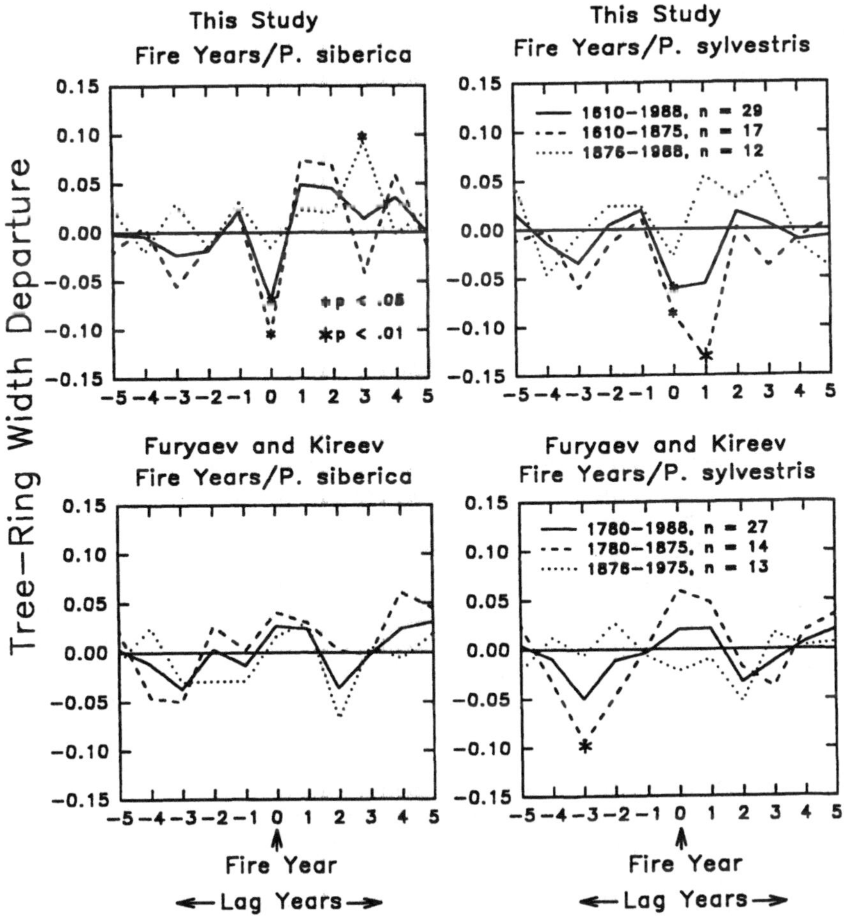

Fig.7. Results of superposed epoch analysis showing mean departures of tree-ring growth during fire years (year 0), and years lagged before and after the fire years. Comparisons of the fire years determined in this study with the tree-ring width chronologies are shown in the upper graph. Comparisons of the fire years determined by Furyaev and Kireev (1979) are shown in the lower graph.

The superposed epoch analysis generally confirmed these synchronous and asynchronous patterns of estimated fire and climate variations (Fig.7). Fire years determined in this study from 1610 to 1870 were significantly drier ($p < 0.05$) than would be expected to occur by chance ($p < 0.05$). However, fire years occurring after 1880 were not significantly drier ($p > 0.05$). The fire years in the Furyaev and Kireev data set were not significantly drier than would be expected by chance either before or after 1880 ($p > 0.05$) (Fig. 7).

Several of the lagged years (i.e., years before and after the fire years) also had significant departures. The fire years determined in this study showed a highly significant ($p < 0.01$) negative tree-growth departure at year +1 (the years following the fire dates), but only in the comparison of the pre-1880 period with *P. sylvestris* (Fig.7). This pattern may be related to the dormant season dating problem mentioned earlier (and discussed in the next section). A significant positive departure ($p < 0.05$) at lag year +3 was also observed in the post-1880 period with the *P. sibirica* and fire dates determined in this study, and a highly significant negative departure ($p < 0.01$) at lag year -3 was observed in the pre-1880 period with *P. sylvestris* and the Furyaev and Kireev fire dates.

4. Discussion

I propose that a change in climatic sensitivity of central Siberian fire regimes occurred after ca. 1880, and this change was related to the increased importance of people as a source of fire in these landscapes. The Trans-Siberian Railroad was begun in the early 1890s. Pyne (this volume) discusses the very important impact on Siberian fire regimes of the railroad, and subsequent increased human populations in formerly remote areas. People had many reasons for burning the taiga, especially for clearing natural vegetation for agriculture and to improve the ease of travel. Siberian settlers primarily burned in the spring, whereas the peak burning period from lightning ignited fires may have been somewhat later in the study area - perhaps June to July (Valendik 1990). Although weather conditions would have had an important influence on the size and intensities of human set fires, people would have burned whenever and wherever it was useful, convenient, and possible to do so. Thus, a decrease in the synchrony of fires and dry years may be expected in the post-settlement era.

If people were important in manipulating fire regimes we might also expect to see changes in fire frequencies coincident with increased populations. A few more fires were documented in the first half of the 20th century than the latter half of the 19th century in both the fire chronology presented here, and Furyaev and Kireev's chronology (Fig. 6). There also appears to be a decrease in fire frequency in the second half of the 20th century.

The superposed epoch analysis indicated that fire years in the three sampled sites were significantly dry, especially before 1880, but some unexpected lagging relations were also observed. The first years following the pre-1880 fire years were also very dry, according to the *P. sylvestris* chronology, but not the *P. sibirica* chronology. Two possible explanations for these observations are: (1) A substantial number of the fire years recorded as dormant season fire scars in the pre-1880 period should actually be dated to the calendar year of the adjacent earlywood cells, rather than the adjacent latewood cells. In other words, many of these fires may actually have occurred in the spring of the next year (lagged year + 1), rather than the year (year 0) they are currently assigned to. (2) Alternatively, the *P. sylvestris* tree-ring chronology is not entirely independent of the fire years - i.e., synchronous fires may also have burned through the stand of trees that were cored, and the negative growth in lagged year +1 is due to detrimental effects of the fires. At this time I have no explanation for the positive response in the *P. sibirica* chronology in lagged year +3.

The lack of significant association between fires and dry years (at year 0, Fig. 7), using the Furyaev and Kireev chronology, could be due to different characteristics of these data. For one, this chronology is based on a "landscape" approach to determining fire history, where both spatial and temporal observations (i.e., using remote sensing and age structure analysis) of extant forests in a large region were used to reconstruct the chronology. Only about 29% of the forest area in this region corresponds to the "pine forests on sands" (Antonovski et al. 1992) that may be represented by the fire scar chronologies that I compiled. Hence, combined fire regimes across several forest types may be less sensitive to the climate signal contained in the tree-ring width chronologies. Second, it is also possible that some or many of the fire dates in the Furyaev and Kireev chronology are in error by one to several years. Such errors are not unexpected in fire chronologies constructed by age structure analysis or by simple ring counting of tree-ring specimens (as opposed to cross-dating) (Madany et al. 1982; Lorimer 1985). The significant correspondence between the pre-1880 Furyaev and Kireev fire dates and the *P. sylvestris* chronology at the -3 year lag (Fig.7) suggests that some fire dates may be about 3 years too late.

These interpretive problems of fire season, calendrical dates, and lagging relations with climate may be resolved with sampling of more fire scarred specimens within sites, and development of additional well-dated tree-ring chronologies. Analysis of intra-annual position of fire scars within many trees from a site (i.e., more than 10) can reveal important temporal patterns of fire seasonality (Baisan and Swetnam 1990). Preferably, new tree-ring chronologies should be developed from sites not significantly affected by surface fires. Resolving the seasonality of past fires, and the temporal relations between fires and climate, could be very useful in distinguishing the influences of humans and climate on fire regimes. For example, if human-set fires tended to occur primarily in the spring and early summer, and larger spreading fires set by lightning occurred later, then temporal shifts to human dominated fire regimes should be reflected in fire scar seasonality.

Collaborative Russian-American studies on modeling cambial phenology and climate-tree growth relations at the cellular level are currently underway (Fritts et al. 1991). This work promises to help us better interpret the seasonal timing of past fires because the climate-growth model predicts the timing and rate of earlywood and latewood formation as a function of daily weather. Hence, we should be able to define the "growing season window" of trees under different climate scenarios, and the probable position of fire scars occurring at different times of the year relative to this window.

Another promising line of investigation would be to link dendrochronologically derived fire histories with remote sensing of spatial fire patterns. The work of Dr. V. V. Furyaev and his colleagues (Furyaev 1980; 1987; 1991; Furyaev and Kireev 1979) along these lines has already accomplished much in documenting spatial and temporal fire patterns. One goal of new studies would be to develop well-dated chronologies of actual amounts of area burned over large regions of Siberia. This could provide quantitative, high-temporal and spatial resolution assessments of atmospheric input of Siberian fires over time periods of centuries. Long-term fire histories from tree-ring networks may also be useful for comparison with, and verification of, proxy indicators of northern latitude fires currently being recovered from the Greenland ice sheet (Legrand et al. 1992).

5. Conclusions

Multi-century tree-ring based fire history and climate reconstructions can be developed for the central Yenisey region of Siberia. Dendro-ecological analysis of ancient living and dead tree-ring samples will extend environmental history back at least four centuries. Experience in developing networks of disturbance and climate histories in the western U. S. indicates that aggregation of well-dated chronologies at landscape and regional scales can result in the discovery of important climate-disturbance linkages (Swetnam 1990; Swetnam and Betancourt 1990; Swetnam and Lynch 1993; Swetnam 1993). These large spatial scale and long temporal-scale patterns may lead to new hypothesis and insights useful in the design and testing of dynamical models for prediction of future climate-driven ecosystem changes.

Changes in historical climate-fire patterns in the central Yenisey suggest that the late 1800s settlement and expansion period in Siberia resulted in a shift in the climatic responsiveness of Siberian fire regimes. This implies that assessment of climate-fire interactions in the past century are probably not a good analog for these interactions in previous centuries, particularly in areas close to human settlements. The role of humans in modifying the fire regimes must be considered in interpreting fire and ecosystem changes during this period. This will require a "historical ecology" approach, where both human cultural and paleoecological histories are compiled and studied in tandem (e.g., Zachrisson 1977). I recommend that networks of well-dated fire chronologies be reconstructed along distance gradients from human settlements in the central Yenisey region. Comparison of fire history patterns at increasing/decreasing distances from settlements should reveal the specific effects of humans versus climate on changing fire regimes. Historical research on human population trends and past land use practices in these same areas will be needed.

Acknowledgements

I thank Drs. Eugene Vaganov, Alexander Shaskin, Gallina Ivanova, and others of the Institute of Forest, Russian Academy of Sciences, Krasnoyarsk for inviting me to Siberia and their generous help in obtaining the tree-ring specimens. I also thank Svenje Mehlert, Chris Baisan, and Dana Perkins in helping to prepare and date the specimens, and Henri Grissino-Mayer for preparing Figure 1. This research was partially funded by the University of Arizona Small Grants Program.

References

Antonovski, M.Y., M.T. Ter-Mikaelian, and V.V. Furyaev. 1992. A spatial model of long-term forest fire dynamics and its application to forests in western Siberia. In: A Systems Analysis of the Global Boreal Forest (H.H. Shugart, R. Leemans, and G.B. Bonan, eds.), 373-403. Cambridge University Press, Cambridge.

Baisan, C.H., and T.W. Swetnam. 1990. Fire history on a desert mountain range: Rincon Mountain Wilderness, USA. Can. J. For. Res. 20, 1559-1569.

Baumgartner, T.R., J. Michaelsen, L.G. Thompson, G.T. Shen, A. Soutar, and R.E. Casey. 1989. The recording of inter-annual climatic changes by high-resolution natural systems: tree rings, coral bands, glacial ice layers, and marine varves. AGU Geophysical Monogr. 55, 1-14.

Biondi, F., and T.W. Swetnam. 1987. Box-Jenkins models of forest interior tree-ring chronologies. Tree-Ring Bulletin 47, 71-96.

Brown, P.M., and T.W. Swetnam. 1994. A crossdated fire history in a coast redwood forest near Redwood National Park, California. Can. J. For. Res. 24, 21-31.

Cook, E.R., and L. Kairiukstis, eds. 1990. Methods of dendrochronology: Applications in environmental sciences. Kluwer Academic Publishers, Dordrecht, The Netherlands.

Dieterich, J.H. 1980. The composite fire interval - a tool for more accurate interpretation of fire history. In: Proceedings of the Fire History Workshop, October 20-24, 1980, Tucson, Arizona. USDA Forest Service General Technical Report RM-81, 8-14.

Dieterich, J.H., and T.W. Swetnam. 1984. Dendrochronology of a fire scarred ponderosa pine. For. Sci. 30, 238-247.

Dixon, R.K., and O.N. Krankina. 1993. Forest fires in Russia: carbon dioxide emissions to the atmosphere. Can. J. For. Res. 23, 700-705.

Fritts, H.C. 1976. Tree rings and climate. Academic Press, London.

Fritts, H.C. 1991. Reconstructing large-scale climatic patterns from tree-ring data. University of Arizona Press, Tucson.

Fritts, H.C., and T.W. Swetnam. 1989. Dendroecology: A tool for evaluating variations in past and present forest environments. Adv. Ecol. Res. 19, 111-189.

Fritts, H.C., E.A. Vaganov, I.V. Sviderskaya, and A.V. Shaskin. 1991. Climate variation and tree-ring structure in conifers: Empirical and mechanistic models of tree-ring width, number of cells, cell size, cell-wall thickness and wood density. Climate Research 1, 97-116.

Furyaev, V.V. 1980. The aerospace photo mapping of after-fire forest dynamics. Reprint from Issledovanie Zemli i Kosmosa (Research of Earth and Space), Moscow, Vol. No. 2, 51-56 <in Russian>.

Furyaev, V.V. 1987. Use of Aerospace photos for study and assessment of forest fire consequences. Nauka, Siberian Branch of Academy of Sciences, USSR, Novosibirsk. pp. 85-98 <in Russian>.

Furyaev, V.V. 1991. Remote sensing detection of patterns of forest disturbance by fire. Mapping Sciences and Remote Sensing 28, 241-245.

Furyaev, V.V., and D.M. Kireev. 1979. A Landscape Approach in the Study of Postfire Forest Dynamics. Novosibirsk: Nauka <in Russian>.

Goldammer, J.G. 1992. Fire statistics from Krasnoyarsk Region - The center of future international fire research and development in boreal Eurasia. International Forest Fire News No. 7 (August 1992), 8-12. ECE/FAO Geneva.

Graybill, D.A. 1979. Revised computer programs for tree-ring research. Tree-Ring Bulletin 39, 77-82.

Heinselman, M.L. 1973. Fire in the virgin forests of the Boundary Waters Canoe Area, Minnesota. Quat. Res. 3, 329-382

Legrand, M., M. De Angelis, T. Staffelbach, A. Neftel, and B. Stauffer. 1992. Large perturbations of ammonium and organic acids content in the Summit-Greenland ice core: Fingerprint from forest fires? Geophys. Res. Lett. 19 (5), 473-477.

Lorimer, C.G. 1985. Methodological considerations in the analysis of forest disturbance history. Can. J. For. Res. 15, 200-213.

Lough, J.M., and H.C. Fritts. 1987. An assessment of the possible effects of volcanic eruptions on North American climate using tree-ring data, 1602 to 1900 A.D. Climatic Change 10, 219-239.

Madany, M.H., N.E. West, and T.W. Swetnam. 1982. Comparison of two approaches for determining fire dates from tree scars. For. Sci. 28, 856-861.

Monserud, R.A. 1986. Time series analysis of tree-ring chronologies. For. Sci. 32, 3449-3472.

Mooney, C.Z., and R.D. Duval. 1993. Bootstrapping: A non-parametric approach to statistical inference. Sage University Paper 95, Series: Quantitative Applications in the Social Sciences, Sage Publications, London. 95pp.

Pyne, S.J. 1996. Wild Hearth. A prolegomenon to the cultural fire history of northern Eurasia (this volume).

Stokes, M.A., and T.L. Smiley. 1968. An Introduction to Tree-Ring Dating. University of Chicago Press, Chicago, 73 pp.

Stocks, B.J. 1991. The extent and impact of fires in northern circumpolar countries. In: Global Biomass Burning, Atmospheric, Climatic, and Biospheric Implications (J.S. Levine, ed.), 197-202. MIT Press, Cambridge, Mass.

Swetnam, T.W. 1990. Fire History and climate in the Southwestern United States in J. S. Krammes, Tech. Coord, Proceedings of Symposium on Effects of Fire in Management of Southwestern U. S. Natural Resources, November 15-17, 1988, Tucson, Arizona. USDA Forest Service, General Technical Report. RM-191, 6-17.

Swetnam, T.W. 1993. Fire history and climate change in giant sequoia groves. Science. 262, 885-889.

Swetnam, T.W., and J.L. Betancourt. 1990. Fire-southern oscillation relations in the southwestern United States. Science 249, 1017-1020.

Swetnam, T.W., and J.L. Betancourt. 1992. Temporal patterns of El Niño/Southern Oscillation - wildfire patterns in the southwestern United States. In: El Niño: Historical and Paleoclimatic Aspects of the Southern Oscillation (H.F. Diaz and V.M. Markgraf, eds.), 259-270. Cambridge University Press, Cambridge.

Swetnam, T.W. and J.H. Dieterich. 1985. Fire history of ponderosa pine forests in the Gila Wilderness, New Mexico. In: J. E. Lotan, B. M. Kilgore, W. C. Fischer, and R. W. Mutch, Tech. Coords., Proceedings-Symposium and Workshop on Wilderness Fire, November 15-18, 1983, Missoula, Montana. USDA Forest Service, General Technical Report INT-182, 390-397.

Swetnam, T.W., and A.M. Lynch. 1993. Multi-century, regional-scale patterns of western spruce budworm history. Ecol. Monogr. 63, 399-424.

Swetnam, T.W., M.A. Thompson, and E.K. Sutherland. 1985. Using dendrochronology to measure radial growth of defoliated trees. USDA Forest Service, Washington, D.C., Agriculture Handbook 639.

Valendik, E.R. 1990. The Struggle with big forest fires. Novosibirsk: Nauka, Siberian Branch <in Russian>.

Valendik, E.N. and G.A. Ivanova. 1990. Extreme fire hazard seasons and their reconstruction. In: S. G. Shiyatov, editor, Abstracts of reports delivered at 5th All Union Conference held in Sverdlovsk.

Zackrisson, O. 1977. Influence of forest fires in the North Swedish boreal forest. Oikos 29, 22-32.

J.G.Goldammer and V.V.Furyaev (eds.), Fire in Ecosystems of Boreal Eurasia, 105-111.

Reconstruction of Summer Temperatures with a Circumpolar Tree Ring Network

F.H. Schweingruber [1] and K.R. Briffa [2]

1. Introduction

Today's research on tree-ring analysis is faced with the question whether growth of forests and climatic conditions have changed under anthropogenic emissions during the industrial period. The base for answering this question is as follows:

◦ All tree-rings are absolutely dated. Therefore they can be compared with dated climatological factors.
◦ Genetical and ecological variability within groups of individual trees can be minimized by building mean tree-growth curves, so-called chronologies.
◦ Ecologically defined tree-ring networks allow direct spatial reconstruction of growth patterns and their climatological interpretation.

Based on a modern radiodensitometric technique (Eschbach et al. 1995) of a tree-ring and temperature network (Schweingruber and Briffa 1996) which covers the northern parts of the northern hemisphere, the time period of the past 300 years remains to be interpreted. It is of crucial importance that the time of most intensive industrialization in boreal forests be compared on a long ranging time scale.

2. Technical and Statistical Methods in Dendrochronology

Measurements of tree-ring widths are fundamental for dendrochronological studies. For 60 years they are carried out by means of special ring width measuring equipments (Anjol 1983). In 1963 Polge developed the X-ray densitometry, a technique used for measuring tree-ring density (Polge 1966). Today the 'densitometer', which is a construction of high technical standard, serves as the measuring instrument. The consequential increase in operating-effort is justified because it allows a climatological interpretation of a tree-ring density profile (Fig.1), as well as several width- and density-parameters.

[1] Swiss Federal Institute for Forest, Snow and Landscape Research, CH-8903 Birmensdorf, Switzerland
[2] University of East Anglia, Climatic Research Unit, School of Environmental Sciences, GB-Norwich NR4 7TJ, United Kingdom

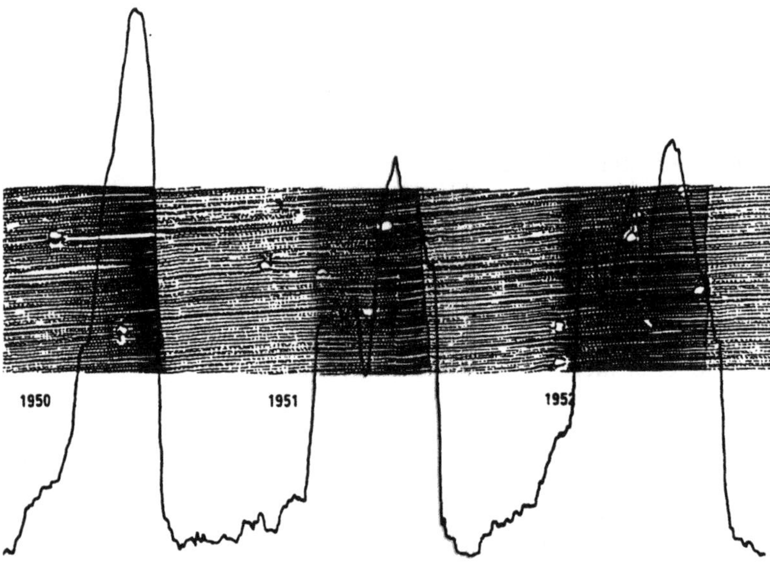

Fig.1. Microphotograph of a thin section of a pine (*Pinus sylvestris*) and the corresponding densitogram. The density curves integrate information on growth conditions during the growing period.

The technical procedure divides into the following steps (Eschbach et al. 1995):

○ Bore- or trunk samples are glued on wooden supports and cut in 1.25 mm thick laths using a double bladed circular saw.

○ In a X-ray chamber the samples are irradiated for 90 min. (20 mA, 11 kV). The medicinal X-ray films produce pictures of strong contrast. They are used especially for distinguishing and measuring early- and late-wood differences.

○ Densities are registered by a 1000 x 30 micron sensor in the densitometer. Of each tree-ring the following parameters are measured or calculated:

- extreme values of density: minimum and maximum densities
- mean values of density: early-wood, late-wood, tree-ring densities
- widths: early-wood, late-wood, tree-ring widths
- relations: early-wood and late-wood densities and widths

In the hemisperic network only tree-ring widths and maximum late-wood densities have until now been considered.

○ Crossdating: Each curve is dated absolutely, e. g. an exact calendar date can be attributed for each tree-ring. This procedure is essential especially for samples of northern origin, where some rings are missing in years of extremely bad climatic conditions. Correct dating is carried out on luminous tables. The calculation of correlations between the individual samples serves as a final control. The dated individual curves represent the basis for calculating mean curves.

○ Each individual curve gets standardized, since a decrease in growth, which is not caused by climate but by competition and aging, becomes apparent as time proceeds. However, these techniques bring along the risk of uncontrolled elimination of long-term oscillations (Briffa et al. 1996).

○ Standardized mean curves are the basis for climatological comparisons during the meteorological measuring period and for interpretation of the time before.

○ Climate-growth comparisons are generally derived by the so-called 'Response functions' (Fritts 1976). This method evaluates the climatological information by computing relationships between meteorological and tree-ring series. The results are expressed in a model. The quality of the model is tested during an independent phase within the meteorological measuring time.

This method contains some problems since for certain areas only very short meteorological measuring series are available and appropriate tree locations are often far from the measuring station (Tessier 1988). These two handicaps diminish the quality of the calibration, especially along the northern timberlines.

3. The Northern Hemispheric Network

The presented network is designed for the interpretation of normal tree-ring sequences which have not been affected by fire (Fig.2). No trees with fire scars have been collected. The fundamentals for the construction of a circumpolar network are the following (Schweingruber et al. 1991, 1993):

○ The network contains only samples from cool and humid areas. In the north, trees are located in the boreal coniferous forest mainly in the extreme northern outposts. The middle geographic latitudes contain samples of the subalpine zone. Those regions, where summer drought also affects the growth of trees on upper timberlines, are not considered, e. g. the Atlas mountains.

○ From these areas, samples of the entire ecological spectrum are included. Comparisons have shown that each year the curves of maximum densities are similar for trees which are located on both dry and humid sites in the mentioned zones.

○ Contemporaneous curves of maximum density of different tree species are similar. This holds especially for double-needled pines *(Pinus sylvestris, P. mugo)*, spruces *(Picea abies, P. obovata, P. mariana, P. glauca, P. engelmannii)*, larches *(Larix decidua, L. sibirica, L. dahurica, L. lyallii)*, and all firs, Hemlocks and *Cedrus* spp. Five-needled pines, e. g. *Pinus sibirica* and *P. longaeva* are less qualified for climatological interpretation.

◦ Dendroclimatological quality varies according to regional conditions. It is excellent in northern Eurasia, good in middle latitudes of Europe, the Rocky Mountains, northeastern America and Asia. The information obtained from material of the central and western parts of North America is, however, less reliable.

◦ The time periods to which the information of the mean curves is related, ranges from single years to decades and even to centuries.

◦ The mean curves contain information on growth conditions and mean summer temperatures. For Central Europe, the Nordic countries and North America it has been demonstrated that maximum late-wood density mainly reflects mean daily temperatures during the summer - on average between April and September (Briffa et al. 1994).

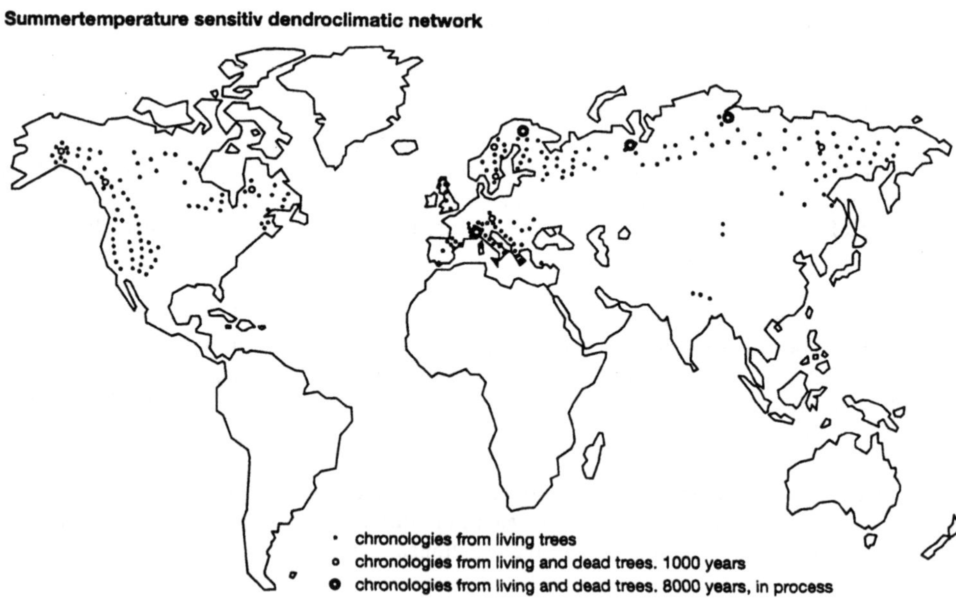

Fig.2. The northern hemispheric densitometric network developed at the Swiss Federal Institute for Forest, Snow and Landscape Research, Birmensdorf. The symbols show the approximative position of chronology sites.

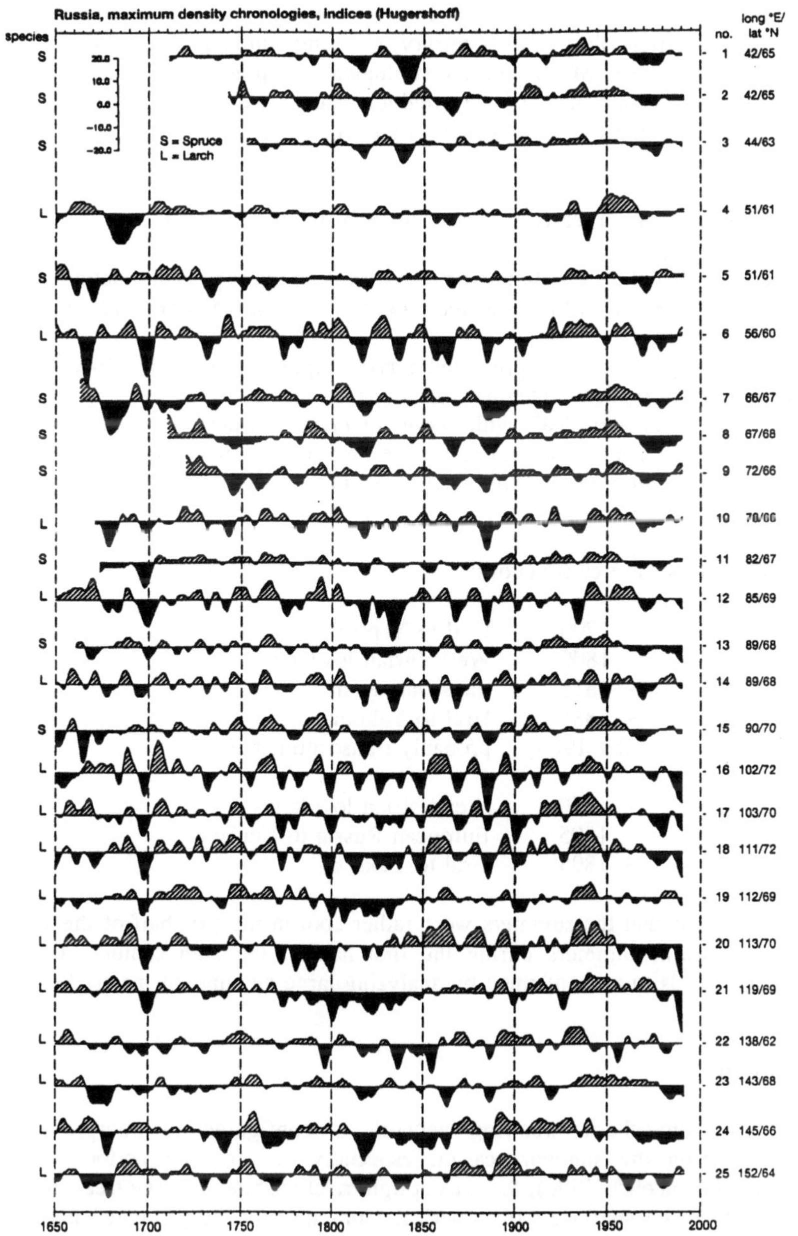

Fig.3. Maximum density chronologies from living spruce and larch growing at the northern timberline in Russia along a west-east transect from longitude 42°E to 152°E. The mean curve has been smoothed with a 13-term low pass filter. The mean of each chronology is computed for the total length of the curve (Schweingruber and Briffa 1996).

Special attention should be payed to the fact that, in most cases, maximum densities react immediately to regional temperatures, show almost no influences from the previous year and are affected only to a minor extent by local environmental factors. Thus an interpretation of the annual values is possible. On the contrary, tree-ring widths primarily reveal the growth potential at a certain site. Mostly the climatological interpretation gets quite complicated, since the ecological occurrences of the previous year as well as the local site- and stand-relationships take influence on the growth of the tree rings.

4. Interpretation of the Tree-Ring Network in Northern Eurasia

With living trees spatial dendroclimatological reconstruction is possible, dating back to 1650 A.D., however those east of the Ural trees count more than 600 years old, when located in fire-protected places.

The tree-ring sequences of the presented curves (Fig.3) include only 340 years, therefore it is impossible to recognize centennial variations.

The following discussion deals with curves of maximum densities derived from Russia, north of 60°N. Here the decadal variations are obvious. We have good reasons to say that phases of high maximum density are equal to warm periods and those of low density equal to cold periods.

The following phases are conspicuous:

cold phases:	~ 1700	Ural to Yenisey
	~ 1800	Westsibirian lowlands
	~ 1815	transcontinental
	~ 1885	Ural to Yakutia
	after 1960	probably transcontinental
warm phases:	~ 1760	Westsibirian low lands
	~ 1805	European Russia to Yenisey
	~ 1895	Ural to Yakutia

In general it seems that the summers were rather cool in the first half of the 19th century, in contrast to warm summers during the first half of the 20th century. However, this statement still needs to be confirmed by analyzing large amounts of material.

5. Conclusions

The northern hemispheric tree-ring network is configured to interpret temperature fluctuations during the summer season, especially the growing period. According to Swetnam and Betancourt (1990), there exist tight relationships between tree-ring widths and fire frequency in the southwest of America. Consequently the presented network can be assumed to assist in estimating the frequency of forest fires. However, in order to confirm proof of this hypothesis, a comparative study has to be made in Siberia, in which the frequency and intra-seasonal location of fire scars within the tree ring (e.g., Ortloff et al. 1995) be compared with densitometric results. The interpretation of the fire situation over the past 50 years is hardly of any use for calibration, since this was the time of intense settlement in northern areas and many forest fires were caused by man. The fire frequency of the recent cold phase in Siberia is probably estimated too high when compared to the pre-industrial period. Given a constant density of human settlement in the boreal coniferous forest, fire frequency will be higher in the approaching warm phase compared to corresponding periods in former times.

References

Anjol R. W. 1983: Tree-ring analysis using Catras. Dendrochronologia 1, 45-53.

Briffa K.R., P.D. Jones, and F.H. Schweingruber. 1994. Summer temperatures across northern North America: regional reconstructions from 1760 using tree-ring densities. J. Geophys. Res. 99, 25835-25844.

Briffa K.R., P.D. Jones, F.H. Schweingruber, W. Karlén, and S.G. Shiyatov. 1996. Tree-ring variables as proxy-climate indicators: Problems with low-frequency signals. In: Climate fluctuations and forcing mechanisms of the last 2000 years (P.D.Jones, R.S.Bradley, and J.Jouzel, eds.). Springer-Verlag, Berlin (in press).

Eschbach, W., P. Nogler, E. Schär, and F.H. Schweingruber. 1995. Technical advances in radiodensitometrical determination of wood density. Dendrochronologia 14 (in press).

Fritts, H.C.,1976. Tree rings and climate. Academic Press. London, New York, San Francisco.

Ortloff, W., J.G.Goldammer, F.H.Schweingruber, and T.W.Swetnam. 1995. Jahrringanalytische Untersuchungen zur Feuergeschichte eines Bestandes von *Pinus ponderosa* DOUGL. ex LAWS. in den Santa Rita Mountains, Arizona, USA. Forstarchiv 66, 206-214.

Polge, H. 1966. Etablissement des courbes de variation de la densité du bois par l'exploration densitométrique de radiographies d'échantillons prélevés à la tarière sur des arbres vivants. Ann. Sci. Forest. 23, 1-206.

Schweingruber, F.H., K.R. Briffa, and P.D. Jones. 1991. Yearly maps of summer temperatures in Western Europe from A.D. 1750 - 1975 and western North America from 1600 to 1982. Vegetatio 92, 5-71.

Schweingruber, F.H., K.R. Briffa, and P. Nogler. 1993. A tree-ring densitometric transect from Alaska to Labrador: comparison of ring-width and maximum density chronologies in the conifer belt of northern North America. Int. J. Biometeorology 37, 151-169.

Schweingruber, F.H., and K.R. Briffa. 1996. Tree-ring density networks for climate reconstruction. In: Climate fluctuations and forcing mechanisms of the last 2000 years (P.D. Jones, R.S. Bradley, and J.Jouzel, eds.). Springer-Verlag Berlin (in press).

Swetnam, T.W., and J.L. Betancourt. 1990. Fire-Southern oscillation relations in the southwestern United States. Science 249, 1017-1020.

Tessier, L. 1988. Analyse spatio-temporelle des relations cerne climat. Publication de l'Association de Climatologie, 213-224.

J.G.Goldammer and V.V.Furyaev (eds.), Fire in Ecosystems of Boreal Eurasia, 112-128.

Analysis of the Distribution of Forest Fires in Russia

G.N. Korovin [1]

1. Introduction

Fire is a significant factor in the formation and maintenance of boreal forests and the structure and dynamics of the Russian Forest Fund. The effective management of natural fires, and the correct evaluation of the scale of its impact on forest ecosystems and environment is impossible without a clear understanding of the distribution and dynamics of forest fires, and the process of forest fire ignition and development. The purpose of this work is to estimate the temporal and spatial distribution of wildfires, and its dependence on natural and economic conditions, based on an analysis of longterm forest fire data from the Russian Territory.

2. Object of the Investigation

The object of this investigation is the aggregation of forest fires that occurred in the actively protected territory of the Russian Forest Fund during the preceding long period. This aggregation occurs under the mutual influences of climate, forest vegetation, and other natural and economic factors. In the basis of its formation lie the process of randomization, that consists in summation the accidental factors that stipulate the existence of the stochastic component in the main parameters of forest burning. The indicated circumstances predetermine the application of mathematical statistics and the use of the theory of random function as the main suitable instruments in analyzing forest fires.

The existence of regular cyclic changes in climatic and weather conditions, the seasonable and daily movement of atmospheric processes, and changes in vegetation phenological phases determine the season and daily periodicity in the process of forest fire occurrence and development (Korchagin 1954; Kurbatsky 1962). The dependence of atmospheric and other natural processes on latitude and subordination to the determined spatial regularities, stipulate the existence of some determinate number of homogeneous microstructures in the forest fire region. The possibility of selecting such microstructures is confirmed by the schemes of forest fire zones, suggested by Melekhov (1946) and Mokeev (1961).

The specifics of natural and economic conditions in various regions of the country brings about the existence of local features in the temporal and spatial structure of forest fires.

[1] International Forestry Institute, Moscow 117418, Russian Federation

Inventory of these features together with the general regularity of the process of forest fire occurrence and growth is necessary in solving the applied tasks of forest pyrology.

3. Information/Data Collection

The problem of collecting, storing, and processing data to analyze the spatial and temporal distribution of Russian forest fires has been solved through the creation of a computer database for all forest fires within the actively protected territory of the Forest Fund (Pokrivailo et al. 1984). This includes a centralized fund of contemporary forest fire records since 1969, a reference book of Forest Fund territories and divisions, and special computer software for database processing. For each fire in the centralized fund database, information is appended on the administrative territory (geographic coordinates, forest economy, forest district, distance to the nearest populated point and transportation routes), the date and time of detection, initial attack, and extinguishment, the means of detection and suppression, forest vegetation conditions (predominant tree species, soil type, natural fire danger class), weather conditions (fire danger class, wind speeds, precipitation), the fire cause and type of fire (ground, surface, crown). For this analysis of the spatial and temporal distribution of fires, more than 2,500 fires from this database, along with fire statistics since 1947, were used.

4. Methods of Forest Fire Analysis

To investigate the spatial and temporal distribution of forest fires it is necessary to consider the whole active protected territory of the Russian Forest Fund as an aggregation of relatively homogeneous parts, and to consider an analyzable period as an aggregation of separate temporal intervals. Fire occurrence and development was analyzed spatially based on latitude zones. The decade, year (fire danger season), month, day and hour were consecutively considered as the main temporal intervals. To solve applied tasks and study the local features of the spatial and temporal structure of Russian forest fires, administrative-territorial and production-territorial units were considered: the areas of forest enterprises, avia districts, administrative regions and areas (territories, republics), and regional aerial protection bases.

The temporal structure of forest fires was investigated by analyzing the temporal series of the number, area, and duration of forest fires, within different sections of the Forest Fund territory. The structure of the temporal series was investigated for the complete fire season, and included estimates of the variability of the main forest burning indicators, analysis and selection of the tendency and periodic (determined) compounds, analysis of the spectral and correlated estimations of filtered-out temporal series, and the building of the models that describe this series (Korovin et al. 1984). The spatial distribution of forest fires was studied by building and analyzing the map of forest fires in different temporal intervals. Investigation of the spatial structure of the forest fires includes estimation of the empirical distribution of forest fires on protected territory, its distance from populated points and from the net of transport roads, and the selection and identification of the corresponding theoretical distribution. Forest fire distribution was also analyzed according to fire causes, types of forest vegetation, fire danger classes, and other indicators that characterized the natural-economic conditions of the process of forest fire occurrence and development.

5. Investigation Results

Annual Dynamics of the Number and Area of Forest Fires

The temporal series (fire seasons between 1947 and 1992) of the area affected by forest fires on the Forest Fund's actively protected territory is characterized by great variability in both the number of fires and the annual area burned (Fig.1). The number of fires ranged from 10,000 to 34,000 annually, with the corresponding area burned varying between 200,000 and 2,700,000 hectares.

The large amplitude of fluctuation, imperfections in the methods of inventory, and the existence of systematic error when determing fire sizes, excludes the possibility of confirming a reduction in annual area burned during this period. The data of the Forest State Inventory may serve as indirect confirmation to the existence of such a tendency, according to which, the areas of cinders and lost plantations were reduced from 44.8 to 29.8 million ha during the 1966-88 period. After excluding trends, the internal structure of the time series (evaluated fire seasons) and the area of forest fires is near random.

About 77% of the total fire area is the result of surface fires, with nearly 22% resulting from crown fires. Less than 1% of the area is burned by underground fires.

Seasonal Forest Fire Dynamics

The visual analysis of the temporal series of forest fire occurrence for ten-day periods in Russia (Fig.2) shows the existence of the, obviously expressed, seasonal motion of fires in each latitude zone, testifying to the nonstationary process of forest fire occurrence. The specific character of the internal dynamics of the number of fires in each latitude zone is an indicator of the homogeneity of the investigated field of burning.

In the lower latitude zone (40°-48°N), the bimodal distribution of the number of fires may be observed. The first peak of the distribution is between the second half of April and the first half of May, while the second peak occurs in October. The amplitude of the first peak is 2-3 times higher than the second and almost an order of magnitude higher than the average number of fires in the other period of the season. The peaks of burning in this latitude zone coincide with the beginning and end of the vegetation period, and with the period of the minimum moisture. In the average latitude zones (48°-64°N) the distribution of the number of fires is near to unimodal with a visible right asymmetric trend that is reduced with movement northward. The peak fire occurrence is displaced to the middle of the fire danger season. In the upper latitude zone (64-72°N) the curve of distribution is unimodal. The peak burning period is in June-July and arranged to the middle of the fire danger season.

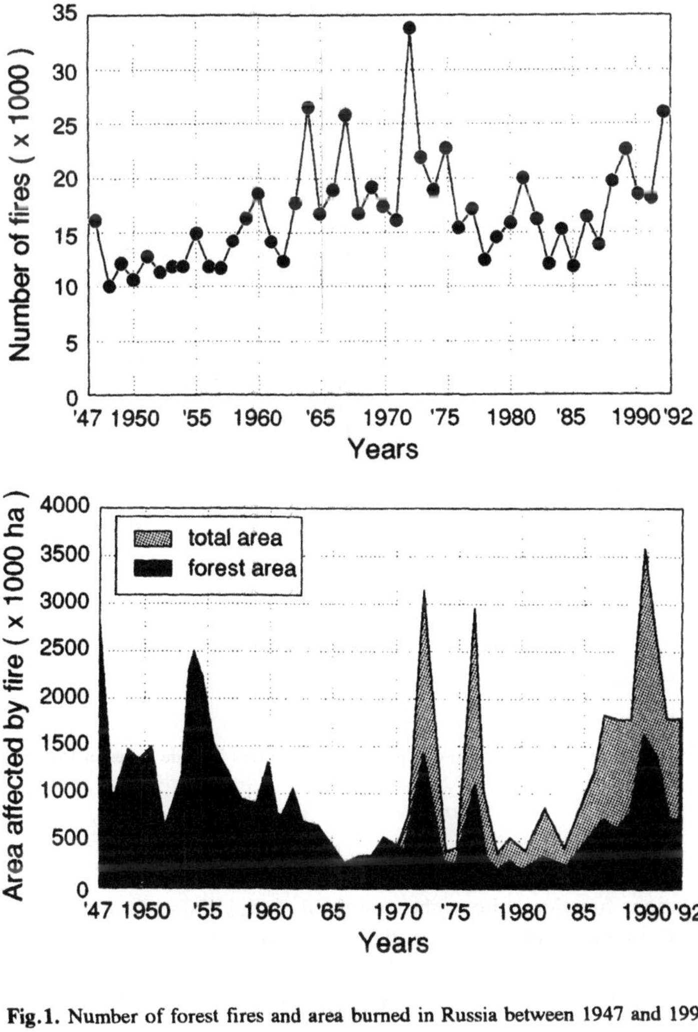

Fig.1. Number of forest fires and area burned in Russia between 1947 and 1992.

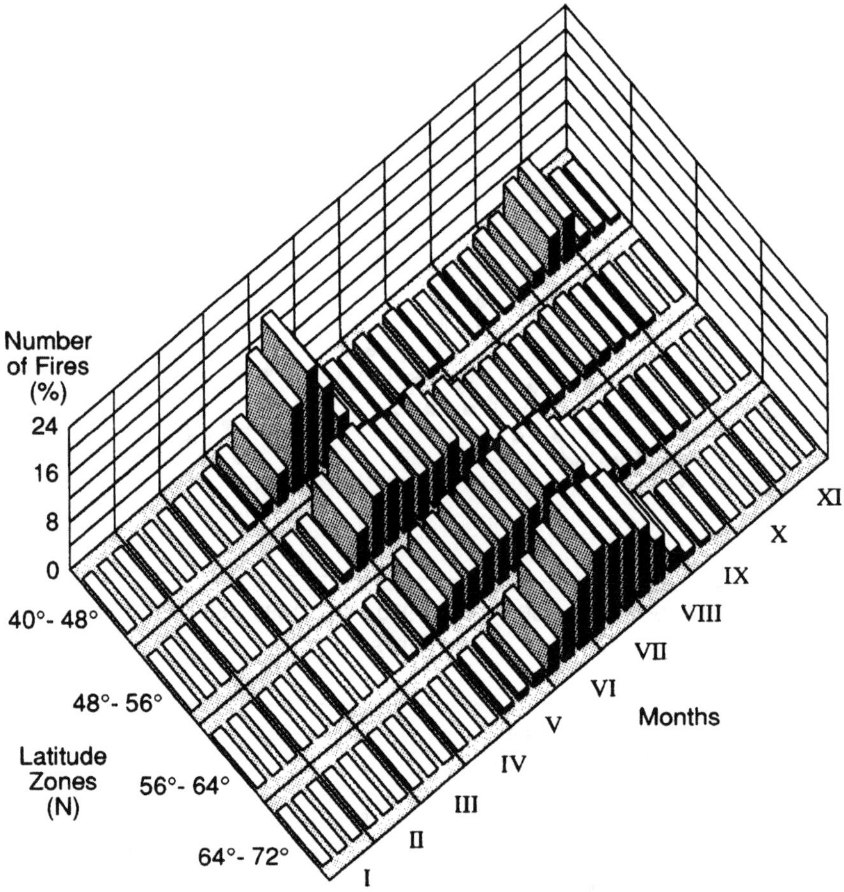

Fig.2. Forest fire occurrence by ten-day periods for different latitude zones in Russia.

The autocorrelation function filters the temporal series (after excluding the periodic component), and in each latitude zone the exponent character tests the possibility of presenting the succession in the form of a Markov process. The temporal lag for both 10-day periods and individual days is shown, for each latitude zone, in Figure 3.

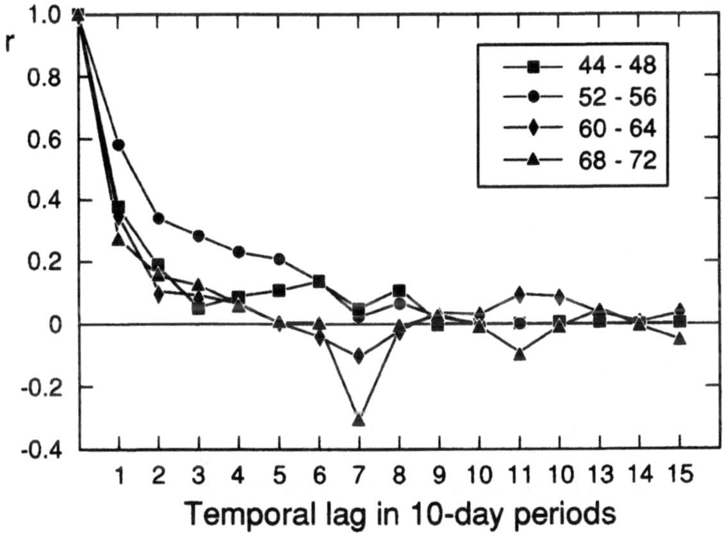

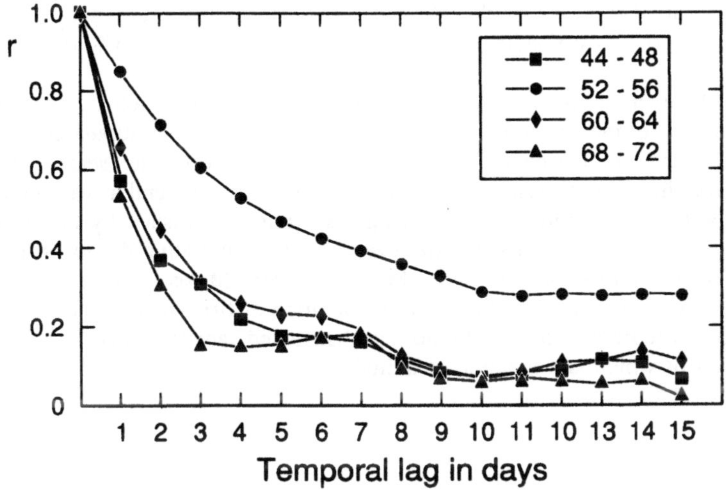

Fig.3. The temporal lag of forest fire occurrence for both individual 10-day periods for various Russian latitude zones.

Autospectrum analysis can be used on the results of such temporal series of the number of fires, showing that the fundamental part of the dispersion of the process is on the low frequency end of the fluctuation. A model of the temporal series of ten-day average number of fires, corresponding to the results of an analysis of selected autocorrelation functions and autospectrums, is given by the equation:

$$\omega(t)=g(t)+\sigma_x(t)y(t) \tag{1}$$

where:

$y(t)$ - the Markov random process;
$g(t)$ - the determined periodic component;
σ^2_x - variance of the random process $x(t)$.

The main difference of the temporal series of the daily number of fires from the ten-day totals is the existence of high fluctuation influences, stipulated by inter day variability of forest fire occurrence. The graphs of selected autocorrelated functions, calculated by initial temporal series of the daily number of fires also has periodic character, but the seasonal motion of burning is expressed in them more weakly then in the graphs of autocorrelated functions of the ten-day number of fires (Fig.3).

The Terms of the Beginning and Duration of the Fire Season

The fire danger season is determined as the part of the calendar year during which 95% of forest fires occur. The terms of the beginning and end of the fire danger season in each latitude zone were correspondingly defined when 2.5% and 97.5% of all fires have occurred.

Essential differences in climatic and forest vegetation conditions, and terms of the beginning and end of the vegetation period in various latitude zones stipulates corresponding differences in terms of the beginning and duration of a fire danger season. The average duration of the fire danger season, in terms of its beginning and end at various latitudes, is shown in Figure 4. The borders of fire danger seasons, established by this method, have better correlation with the results that were received by elaborating the forest fire zones on the territory of the former Soviet Union (Meleckov 1946; Mokeev 1961).

The analysis of a selected distribution of the duration of fire danger seasons illustrated the possibility of its representation as the normal distributed random value, the average mean and dispersion of which is a function of latitude:

$$F_{\varphi}(L)=\frac{1}{\sigma\sqrt{(2\pi)}}\int_{-\infty}^{t}\exp[-\frac{(\tau\sqrt{1-\sin\varphi}-t)^{2}}{2\sigma^{2}}]dt \qquad (2)$$

where:

$F_{\varphi}(L)$ - distribution function of fire danger season duration, corresponding to latitude;
L - duration of fire danger season;
σ - standard deviation of fire danger season duration
τ - the number of calendar days in the year.

Similar results were received for the beginning date of the fire danger season, as a function of territorial latitude. A negative correlation between beginning date and duration of the fire danger season was revealed. The closest relationship between them is at the North latitude, where the coefficient of correlation is 0.75-0.80.

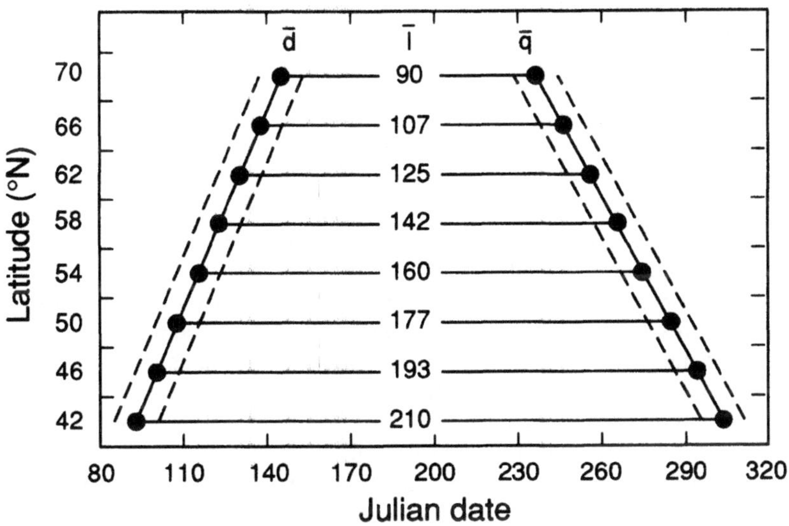

d - Begin of the Fire Danger Season $\bar{d} = \tau/2\,(1-\cos\varphi)^{1/2}$
l - Duration of the Fire Danger Season $\bar{l} = \tau\,(1-\sin\varphi)^{1/2}$
q - End of the Fire Danger Season $\bar{q} = \bar{d} + \bar{l}$

Fig.4. Variation in fire season length with latitude in Russia

Daily Fire Dynamics

The moisture content of forest fuel materials, related to the daily motion of atmospheric processes and analogical influences of the number of forest fire sources determine the time of forest fire occurrence in the daytime hours. An analysis of a selected distribution of the time of the occurrence of an established forest, and the area of each forest fire, if the forest vegetation and weather conditions during fire growth are known, shows that all of them, independent of the territory latitude and the day length, has a unimodal character with a small left asymmetric. The mean of the time of fire occurrence varies in the interval of 13-14 hours, and its standard deviation is monotonic growth from 2-3 hours with the increasing of day length from 10-23 hours (Fig.5).

Selection of the theoretical law shows that the satisfactory approximation gives a normal (Gaussian) distribution and beta-distribution. Parameters of these distributions are the functions of the day length and may be determined by the calendar date and territory latitude. Beta-distribution must be taken concentrated on the interval of 5 -22 hours, and is necessary to give the remainder a zero probability of fire occurrence.

Forest Fire Duration

Selected distributions of the duration of forest fires on the Forest Fund's actively protected territory shows that the majority of the fires were extinguished within 2-3 days after starting (Fig.6). The timeliness of fire extinguishment was best in the zone protected by ground suppression forces in comparion to the aerial protection zone. This is primarily due to better accessibility. In addition, the longer burning fires (>5 days) account for by far the highest percentage of the total area burned (Fig.6).

Spatial Distribution of Forest Fires

Analysis of forest fire maps of the Forest Fund's actively protected territory showed that the number and density of fires was closely correlated with the distance to population centres or transportation routes (Fig.7/8), with the fire incidence decreasing quickly as one moves away from populated or travelled areas.

A comparison of forest fire maps for various years shows that the distribution of the number of fires by latitude zones on country regions is comparatively stable. This is explained by the poor infrastructure in each region, the number of fire sources, and the constant spatial distribution of forest fuel materials.

The distribution of large forest fires within the Forest Fund's actively protected territory is characterized by particularly pronounced instability and irregularity. 50-90 percent of the large fires occur every year in 3-4 regions of the country where extreme fire weather conditions prevail (Fig.9). The number of fires that have occured in these zones are twice as high, and the area burned is up to three times higher than in other zones with comparable weather conditions.

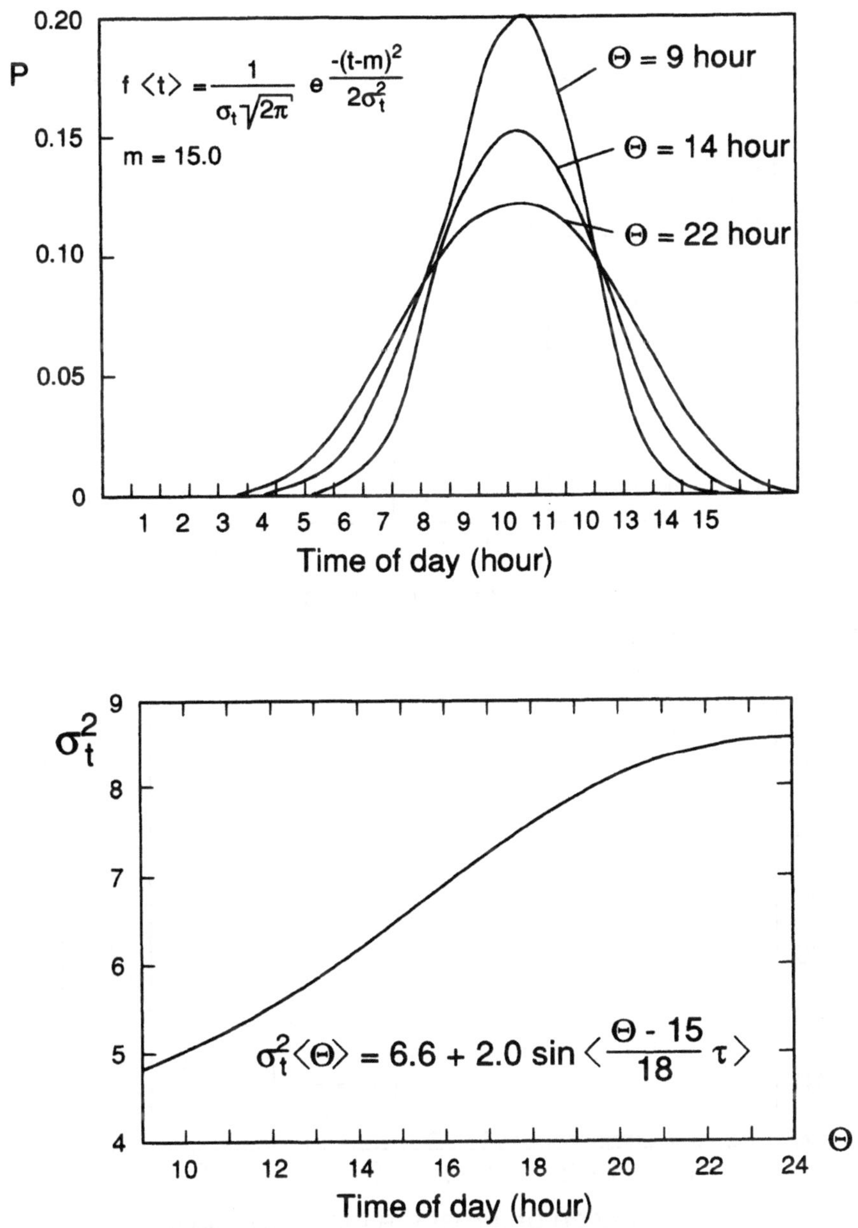

Fig.5. Probability and variance of fire occurrence in Russia throughout the day.

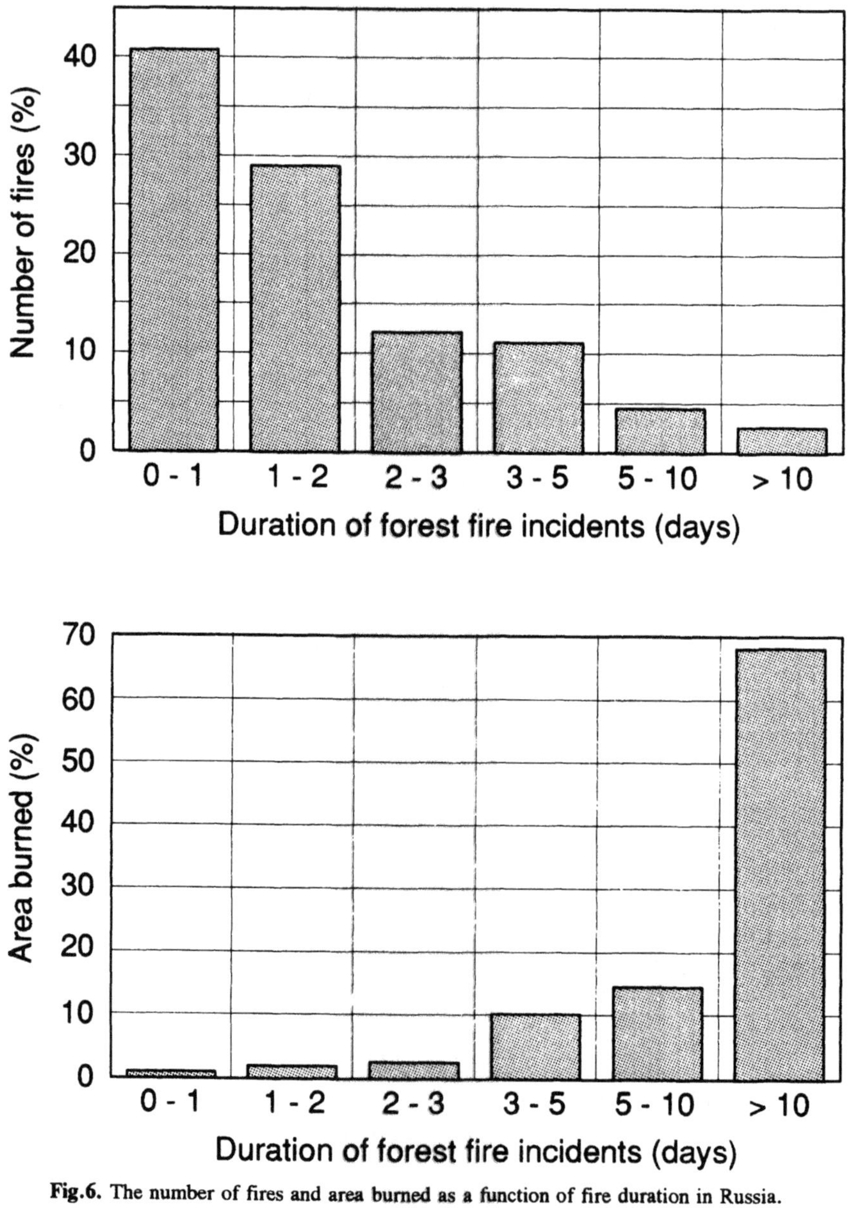

Fig.6. The number of fires and area burned as a function of fire duration in Russia.

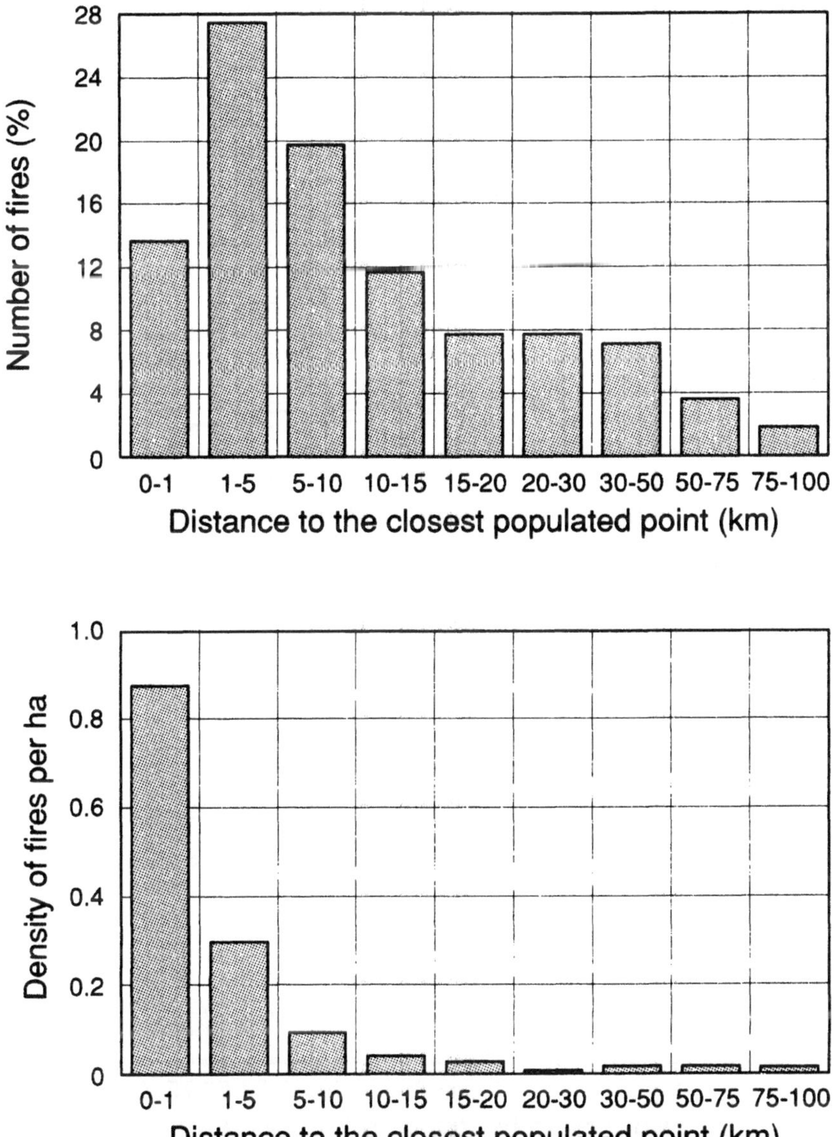

Fig.7. The number and density of forest fires in Russia relative to the distance from population.

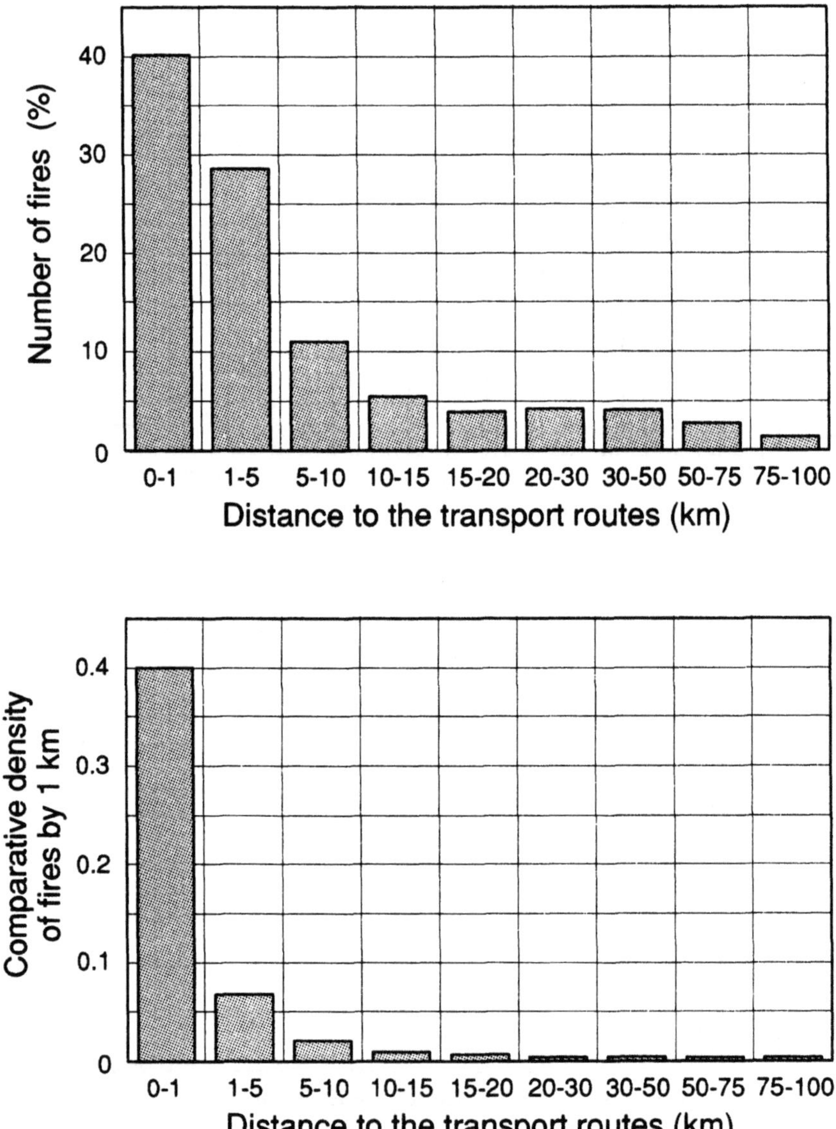

Fig.8. The number and density of forest fires in Russia relative to distance from transportation routes.

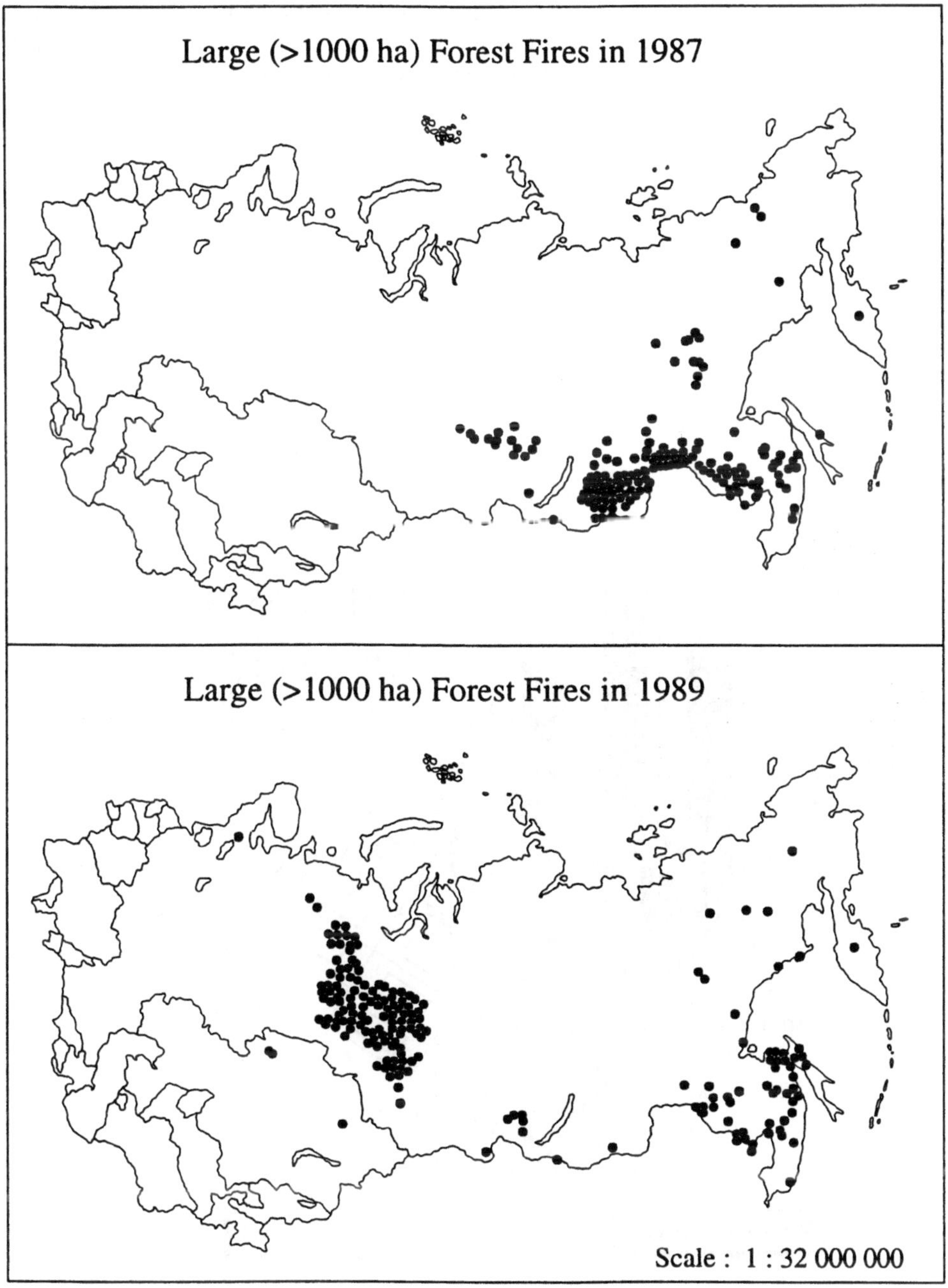

Fig.9. The spatial distribution of large (> 1000 ha) forest fires in Russia in 1987 and 1989.

Distribution of Fires by Forest Types

Most forest fires occur in light coniferous stands (pine, larch) with predominant ground cover of grass, lichen, heather, and green moss (Fig.10). In the dark coniferous stands (spruce, fir) the majority of fires occur in green moss forests, while in hardwood stands most fires occur where grass is the main ground cover.

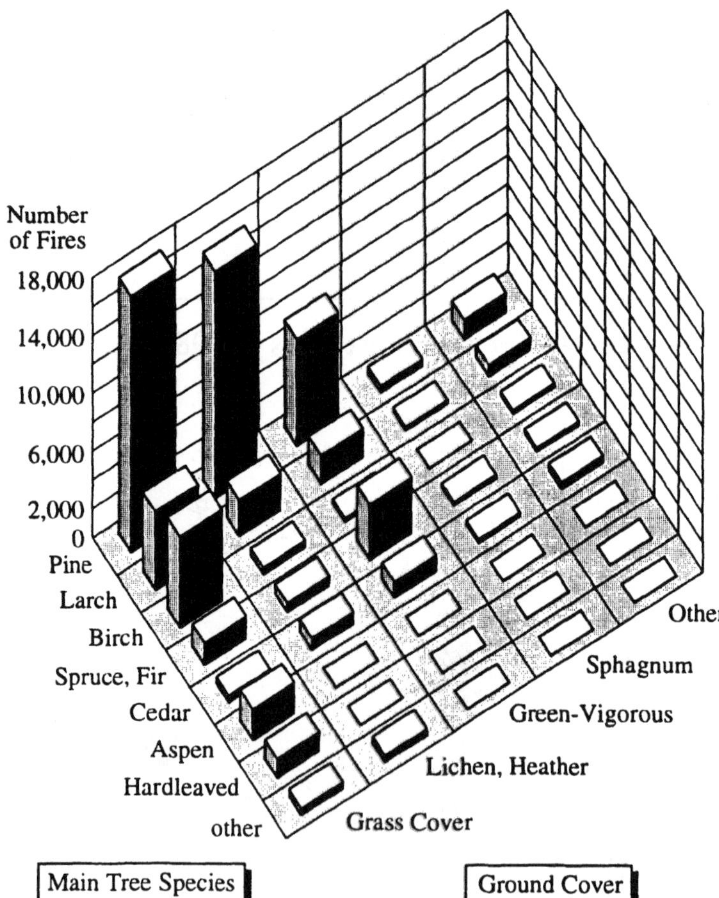

Fig.10. Distribution of forest fires in Russia by forest type.

Forest Fire Causes

Most fires are caused by anthropogenic sources, or negligence (30-50% depending on month). Lightning causes approximately 15% of fires, while the causes of another 20% are unknown. Most lightning fires occur in the summer months while most agricultural fires occur during the spring (Fig. 11).

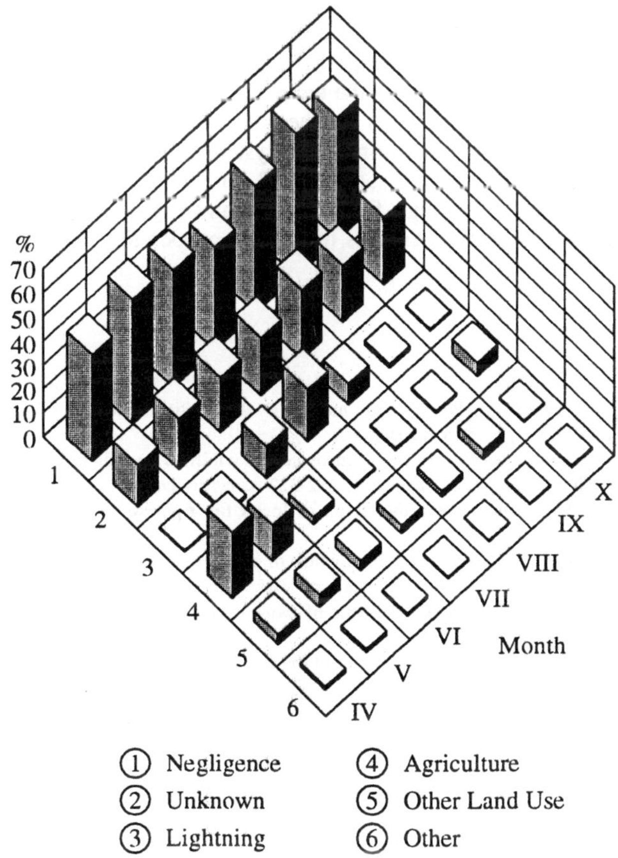

Fig. 11. Distribution of forest fires in Russia by causes.

6. Conclusions

The previous data analysis gives a general understanding of the temporal and spatial structure of forest fires on the Russian Forest Fund's actively protected territory. Gradual accumulation of the accounting and experimental data on forest fires, and the development of the technical base and software for processing and interpretation will permit more detailed analysis of forest fires at the federal, regional and local levels.

To estimate forest burning on the whole territory of the Russian Forest Fund, it is necessary to organize forest fire monitoring in the northern parts of Siberia and the Far East. The absence of a developed infrastructure to organize aerial and ground fire monitoring in these regions make it necessary to use satellites and other remote sensing technology.

To estimate the impact of fires on forest ecosystems and the environment, and to forecast the ecological results of the influence of fire, it is necessary to essentially expand the standard set of registration parameters and to include them in the forest fire computer data bank. To increase the size and reliability of this databank will require better data collection standards, and regular observations of large fire and smoke dynamics.

The most effective means of analyzing the temporal and spatial structure of Russian forest fires, and their ecological consequences on the structure and dynamics of boreal forests requires the use of a Geographical Information System (GIS). The International Forestry Institute in Moscow has developed such a GIS, in cooperation with other scientific organizations. The interest of foreign fire scientists and organizations in obtaining better information on Russian forest fires ensures that future cooperation will expand.

References

Korovin, G.N. et al. 1984. Analysing and modeling the static structure of the field of forest burning-L.:LenNIILKH. 64 pp. <in Russian>.

Korchagin, A.A. 1954. Conditions of forest fire occurrence and wildfires burning of the European North. LGU scientific writings. The Geographical Science 9, 182-322 <in Russian>.

Kurbatsky, N.P. 1962. The techniques and the tactics of forest fire fighting. Goslessbymisdat, Moscow <in Russian>.

Melekhov, I.S. 1946. The fires seasons and construction of a geographical scheme of the forest fires zones. Archangelsk Forestry Institute Scientific Publications VIII, 1-15. Archangelsk <in Russian>.

Mokeev G.A. 1961. Fire danger zones and the time of more powerful development of forest fires. Forestry No. 8, 53-57 <in Russian>.

Pokrivailo, V.D. et al. 1984. Organization and using of the forest fire data banks on the base of IBM-ES-1020. L.:LenNIILKH. 60p.

J.G.Goldammer and V.V.Furyaev (eds.), Fire in Ecosystems of Boreal Eurasia, 129-138.

Temporal and Spatial Distribution of Forest Fires in Siberia

E.N. Valendik [1]

1. Introduction

According to the Operational Information System currently used in the Russian Federal Forest Service, large forest fires are those exceeding 200 ha in area. Although they account for only 1-2% of the total number of fires, and 10-15% in extremely dry years, they are responsible for 50-70% of the annually burned area and 90% of fire damage. These fires are much more intense than smaller fires, and thus their economic and ecological impact is more profound, since an increase in the linear dimensions of a fire leads, logically, to its new qualitative features. For a fire to achieve a large size, several factors, including fire control, are to be involved. These are interrelated factors, and when combined, they favour small fires becoming large.

Environmental factors include vegetation pattern, topography, and weather conditions. The availability of certain fire fighting resources and the possibility of their application in a specific environmental situation is considered as a technical factor. Organization relates to all the problems of forest protection in general and to individual fire suppression organization.

Of recent fires in the unprotected area of northern Siberia, the largest one on record was 850,000 ha in size. This fire was preceded by a long drought period and remained active for 20 days. Large forest fires were reported in the press as far back as 1876. In the early twentieth century, dry years with large forest fires were frequent in Siberia. For example, forests for 200 km around Irkutsk are known to have burned in 1901. In 1908, forests were burned along the Krasnoyarsk-Tomsk part of the Siberian railway. In 1910, mass fires occurred across the Far East. The rate of forest burning in Siberia was especially high in 1915; from May to September, fires burned across the area from Tobolsk to Lena River and their total area was assessed at 12,000,000 ha (Shostakovich 1924). This mass fire outbreak was caused by a long drought period. In a vast area covering eastern Tumen Region, most of northern and eastern Tomsk Region, Yenisei-Angara Region of Krasnoyarsk Territory, the northern and western parts of Irkutsk Region, and the south-western part of Yakutia, a considerable precipitation deficit was observed. Along the perimeter of this area, precipitation was 50-60% and in the central part - only 30% of the norm. During the summer of that year, there were two especially long rainless periods: a 26-day period (mid-June to mid-July), and a 20-day rainless period (late July to mid-August). Two fire activity

[1] V.N. Sukachev Institute for Forestry and Timber, Siberian Branch, Russian Academy of Sciences, 660036 Krasnoyarsk, Russian Federation

peaks fell within these two periods. Over the past two decades, long drought periods and a high rate of forest burning was recorded in 1970, 1973, 1975, 1976, 1981, 1982, 1985, 1986, 1987, 1990, and 1991.

2. Data Analysis

Large fire distribution across Siberia for the 1970-83 period is presented in Figure 1, and the data from these fires is summarized in Table 1. Fires tend to occur in local concentrations, but this is natural since fires recur in certain regions remarkable for high dry year occurrence and anthropogenic stress. Several areas of high fire concentration can be identified: Sverdlovsk Region, the northwestern part of Tumen Region, the eastern part of West-Siberian lowland including Tomsk Region, the northern part of Novosibirsk Region, the Yenisei part of Krasnoyarsk Region, the Angara part of Krasnoyarsk and Irkutsk Regions, the northwestern parts of Yakutia and Chita Region, the southern parts of Amur Region and Khbarovsk Territory, the whole of Primorie, and northeastern and southeastern parts of Magadan Region. The largest number of fires appear to occur in light coniferous stands, with the exception of Far East regions, where fires burned mostly in hardwood stands during spring and autumn periods of the fire season.

The greatest number of fires (83%) was recorded in Eastern Siberia and Primorie, whereas the Trans-Ural Region and Western Siberia (Sverdlovsk, Tumen, and Tomsk Regions) account only for 17% of all fires.

In Tomsk Region, a mass large fire situation occurred in 1982. This was due to extreme weather conditions that prevailed there and in the Yenisei part of Krasnoyarsk Territory. Although almost half of Tomsk Region is comprised of bogs and bogged forests, 10 of the 13 years in question were fire years. 1977 and 1982 are remarkable for the greatest number of fire outbreaks.

For Sverdlovsk Region, the highest rate of forest burning was observed in 1977, with large fires accounting for 50% of the total number. A large fire outbreak also occurred in 1982. It is noteworthy that climatic and site conditions in this area are exceptionally favourable for fire occurrence. The situation is aggrevated by vast areas being covered by young conifers, for which reason fires occur ranging from May to October.

In Tumen Region, large fire years include 1972, 1974, and 1977. In the past few years, the rate of burning is relatively low, and no tendency to large fire number increase has been observed. However, this is true for the protected forest area only. In northern, unprotected, forests the number of large fires keeps increasing. This is obvious from satellite imagery analysis.

In Altai Territory, and Novosibirsk and Omsk Regions, large forest fires, although small in number, occurred annually, begining in 1976. The number of large fires was greatest in 1982, concentrated mostly in the northern part of Novosibirsk Region and in Tomsk Region. On the whole, a high rate of burning in Western Siberia was recorded for 1872, 1974, 1975, 1977, and 1982.

In Angara and southern parts of Krasnoyarsk Territory large forest fires occur annually, but vary in number for each year. For example, in 1970 and 1971, they account for 3.1% and 4.9%, while in the three years to follow, they account for 11.1%, 6.7%, and 6.7% of the total number of fires, respectively. Then a year follows when large fires made up only 1.2% of all fires. Over the next four consecutive years large fire outbreaks were observed. Further, after a year break, multiple large fires occurred again in two years in succession.

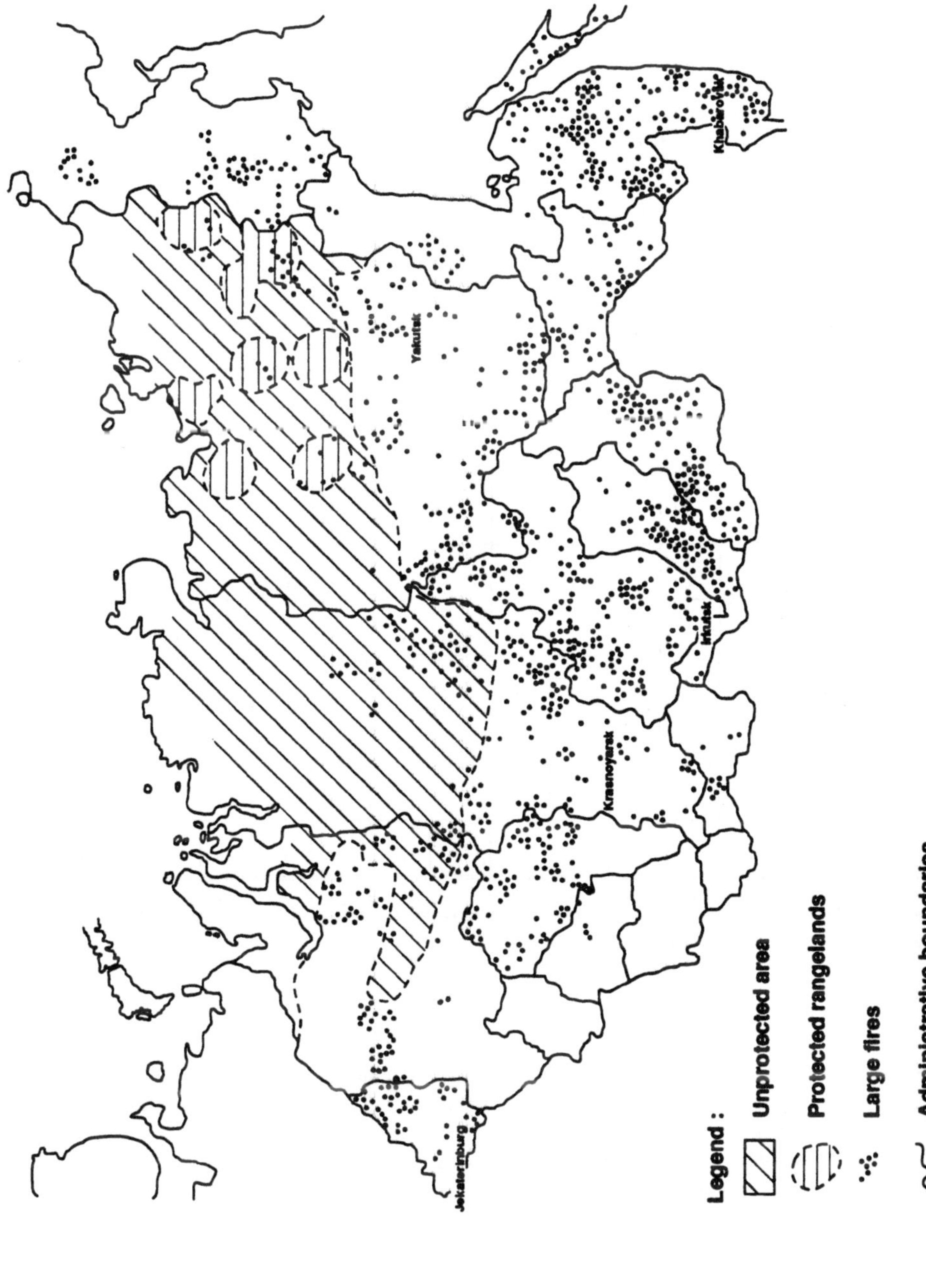

Fig.1. Distribution of large forest fires in Siberia in the period 1970-1983.

Tab.1. Distribution of large forest fires (as a percentage of total number of fires) in Siberia and the Far East for the period 1970-1983.

Region / Territory / Republic	Years (1970-83)													
	70	71	72	73	74	75	76	77	78	79	80	81	82	83
Khabarovsk Territory	19.7	6.2	7.2	3.5	4.5	12.1	10.4	4.8	4.1	4.1	10.8	4.1	4.8	3.5
Yakutian Republic	41.5	12.5	7.8	17.3	0.4	1.6	0.4	4.0	3.1	-	4.7	4.3	1.2	1.2
Irkutsk Region	12.8	18.6	7.0	-	0.8	0.8	1.7	2.9	16.1	23.5	1.7	10.7	1.7	1.7
Krasnoyarsk Territory	3.1	4.9	11.1	6.7	6.7	1.2	15.4	14.1	8.6	8.0	1.8	9.2	7.4	1.8
Buryatian Republic	-	-	2.1	-	2.8	-	-	3.6	32.6	40.5	2.1	14.2	1.4	0.7
Chita Region	-	-	38.7	1.6	-	2.4	12.1	5.6	17.8	17.8	-	2.4	-	1.6
Tomsk Region	-	2.4	-	0.8	5.6	-	2.4	20.8	2.4	0.8	1.6	8.0	45.6	9.6
Primorie Territory	1.0	1.0	-	-	5.0	5.0	-	52.0	2.0	5.0	6.0	1.0	14.0	8.0
Sverdlovsk Region	-	-	-	7.5	1.1	21.5	1.1	49.5	2.2	-	-	1.1	16.0	-
Magadan Region	3.4	6.9	9.2	39.1	3.4	-	1.2	16.1	-	-	-	6.9	1.2	12.6
Tumen Region	-	-	23.9	3.0	41.8	-	-	16.4	-	-	-	6.0	7.4	1.5
Amur Region	19.4	1.6	-	-	-	-	6.4	27.4	11.4	30.6	1.6	1.6	-	-
Novosibirsk Region	-	8.0	-	-	8.0	-	12.0	4.0	8.0	4.0	4.0	4.0	48.0	-
Total	12.1	6.5	8.5	6.3	4.3	4.0	4.9	13.1	8.9	10.5	3.6	6.3	8.0	3.0

Note: horizontal bars indicate the absence of large fires

This situation is particularly interesting in that Krasnoyarsk Territory is latitudinally stretched over some 3,000 km. In that vast an area, weather conditions favourable for large fires occur annually somewhere in the region.

An exceptionally high rate of forest burning was recorded in Irkutsk region in 1979, when 40% of all fires of the period under consideration occurred. Also, multiple fires burned in 1970, 1971, 1978, and 1981, but no periodicity was observed. Fires can occur in two consecutive years to be followed by a 5 or 1 year fire-free period. Fires do not tend to decrease in number.

In the Trans-Baikal Region (Buryatia and Chita Regions), fires are not an annual event. In Buryatia, most serious fire situations were observed in 8 of the past 13 years. Here, 40 % of all fires occur in the spring period of the fire season. Large fire outbreaks occurred in 1978, 1979, 1981, and 1987. The picture is almost similar in Chita Region. The major large fire outbreak occurred in 1972 - that year accounting for 38.7% of all large fires. The next fire outbreak occurred in 1976 and was repeated in the following four years. It should be noted that 1976 to 1979 was a period of extreme forest fire danger throughout Western Siberia, except for its far northeastern part. In Trans-Baikal Region, the majority of fires burn in spring and fall. In summer, fires are not as frequent. Fires are most active in the north of the region.

In Yakutia, the greatest number of fires was recorded in 1970, and by 1976 it decreased 10-fold. Over 1985-86, however, large fires again dominated the region. This fact indicates no improvement of fire control policy in Yakutia; large fire occurrence is fully controlled by weather conditions during fire seasons.

In the southern part of Khabarovsk Territory, fires burn annually. Over the period in question, fire outbreaks occurred twice every 5 years, namely in 1970, 1975, 1976, and 1980. The maximum and minimum number of fires occurred in 1970 and 1973 respectively. The number of fires varies from year to year, which is not to say that it tends to decrease.

Large fires are common in Primorie Region. From 1970 through 1976 they occurred periodically, with a 1-to 2-year interval, and beginning with 1977 burned annually. The greatest number of large fires (49.5% of all fires) was recorded for 1977. From 1978 to 1981 fire was a rare event in the area, and in 1982 a severe spring fire outbreak was recorded.

As is obvious from research study results, large forest fires occur normally in the background of numerous small fires. Their occurrence is determined mainly by environmental and fire control organization factors. Environmental factors responsible for large fires include wind and instability of the atmosphere. Many small fires, when accompanied by strong wind, develop rapidly and escape control (Kurbatsky and Valendik 1974; Kurbatsky and Sheshukov 1978). Organizational factors that make a large fire possible are as follows: poor fire suppression organization, late arrival of fire fighting resources at the fire, inadequate availability of heavy machinery and hand tools for suppression, and many fire bosses inexperienced in operational decision-making.

In the area under consideration, large fires occur very non-uniformly, in both number and time. To estimate the actual rate of burning and to identify regions of the highest fire danger, the area was analyzed as to the number of large fires per 1,000,000 ha. Since these account for not more than 10% of the total number of fires, we estimated the rate of burning according to a four-class system. Table 2 shows large fire frequency per 1,000,000 ha of protected forest area for regions, territories, and republics.

As is seen in the table, the lowest level of forest fire protection is in Primorie Territory and Sverdlovsk Region, areas characterized by a very high rate of forest fires. The same situation is observed in the southern part of Khabarovsk Territory, Novosibirsk and Tomsk Regions, and in the Republic of Buryatia. In other regions, things improve. If it were possible, however, to estimate the rate of forest fires for the whole area under study, one would apparently get results of quite another sort. In this case, the situation would turn out troublesome in the northern part of Tomsk Region, as well as in Krasnoyarsk Territory and Yakutia, since forest protection in these areas is far from being well-organized and reports on fire extent do not give the real picture. The results of METEOR satellite imagery analysis show that, in these areas, individual fires can grow quite large.

For Siberia and the Far East, one can identify fire seasons of two types. One is a long, continuous season, when a high rate of burning is maintained for 5 to 7 months. This is the case with southern taiga forests. Another is a short, continuous fire season, when forest fire danger is at its maximum during 1 or 2 summer months. This fire season type is characteristic of central and northern taiga forests. For certain southern taiga forest areas (Tuva Republic and the southern part of Khabarovsk Territory), a two-period fire season can be observed. This is the case when forest fire danger can achieve its peak in spring and autumn. The short spring period (1 to 1.5 months long) is ended by first summer rains and the beginning of the growing season. The autumn period begins when summer rains stop and grass cover dries out. This period ends as soon as autumn rains begin to fall.

Tab.2. Large fire frequency (recurrence) per 1,000,000 ha for the period 1970-1983.

Region / Territory / Republic	Protected Forest Area (ha x 1,000,000)	Fire Frequency	Rate of Forest Fires *
Primorie Territory	11.9	10.0	very high
Sverdlovsk Region	13.3	6.9	very high
Novosibirsk Region	4.4	5.8	high
Buryatian Republic	25.1	5.5	high
Khabarovsk Territory	56.0	5.2	high
Tomsk Region	26.6	4.7	high
Chita Region	31.5	3.9	moderate-to-high
Irkutsk Region	69.2	3.5	moderate-to-high
Krasnoyarsk Territory	50.0	3.3	moderate-to-high
Magadan Region	27.0	3.2	moderate-to-high
Yakutian Repuplic	110.0	2.3	moderate-to-high
Amur Region	29.7	2.1	moderate-to-high
Tumen Region	52.8	1.3	moderate

* up to 2 - moderate
2.1 to 4.0 - moderate-to-high
4.1 to 6.0 - high
more than 6.0 - very high

From Table 3 it is clear that the number of large fires is the greatest in May, June, and July. In some years, however, the maximum fire activities can shift, in southern taiga forests, either to April or to August. Steady high occurrence of forest fires in July is common only in the north of Krasnoyarsk Territory, as well as in Yakutia, and Tomsk and Magadan Regions.

Tab.3. Distribution (%) of large fires for the period 1970-1983

Region / Territory / Republic	April	May	June	July	Aug.	Sept.	Oct.
Khabarovsk Territory (southern part, up to Amur Latitude)	5	33	22	21	7	7	5
Novosibirsk Region	8	52	20	4	-	8	8
Chita Region	26	46	10	12	-	6	-
Amur Region	-	42	3	43	3	-	8
Sverdlovsk Region	11	27	30	24	3	5	-
Tomsk Region	3	11	46	26	11	3	-
Krasnoyarsk Territory (southern taiga zone)	3	21	43	31	-	2	-
Irkutsk Region	-	22	40	34	3	1	-
Buryatian Republic	8	26	46	18	1	-	1
Tumen Region	3	9	6	70	10	2	-
Magadan Region	-	10	13	64	13	-	-
Yakutian Republic (central and southern parts)	-	6	28	58	16	-	-
Primorie Territory	2	20	11	16	12	25	14

In the Amur Region, the fire season has some specific features. It can be treated as a double-period season since fire activity peaks occur in May and July. The rate of burning is low in June, whereas in October it is somewhat higher. This can be attributed to the fact that in the southern part of the region, high burning rates are observed in spring. Inn June, when fresh green vegetation begins to develop, accompanied by summer rainfall, the rate of burning decreases to nearly zero. In July, fires occur in the northern mountain parts bordering Yakutia. Therefore, the second fire activity peak corresponds to high rates of burning characteristic of forests of northern regions.

In Primorie Territory, there are two distinct fire activity peaks, one in spring (May) and the other in autumn (October). The autumn period of maximum fire activity is even longer than 1 month, covering September and October.

In Chita Region, the fire season is also somewhat peculiar. Fires begin to burn as early as April, and in May, fire activity achieves its maximum. In June, the rate of burning is moderate. In individual exceptionally dry years, the number of fires increases suddenly because of a generally high rate of burning of mountain forests in the north of the region.

Tab.4. Large fire occurrence (%) related to the number of precipitation-free days

Region / Territory / Republic	Number of precipitation-free days up to: *						
	10	20	30	40	50	60	70
				Spring period			
Yakutian Republic	8	62	73	81	88	100	-
Magadan Region	-	62	100	-	-	-	-
Primorie Territory	32	100	-	-	-	-	-
Amur Region	-	15.4	92.3	100	-	-	-
Chita Region	2	14	29	72	94	100	-
Buryatian Republic	11	89	100	-	-	-	-
Khabarovsk Territory	-	31	67	89	100	-	-
Irkutsk Region	-	4	88	100	-	-	-
Sverdlovsk Region	-	100	-	-	-	-	
Tomsk Region	13	100	-	-	-	-	-
Krasnoyarsk Territory	12	50	75	100	-	-	-
				Summer period			
Yakutian Republic	-	-	8	22	66	93	100
Krasnoyarsk Territory	-	10	33	47	71	100	-
Magadan Region	-	3	30	50	69	93	100
Amur Region	-	-	7	93	100	-	-
Tumen Region	-	-	4	30	100	-	-
Chita Region	-	-	11	22	87	100	-
Buryatian Republic	-	-	-	20	98	100	-
Khabarovsk Territory	-	-	43	75	96	100	-
Irkutsk Region	-	-	26	59	70	98	100
Sverdlovsk Region	-	3	58	74	100	-	-
Tomsk Region	-	3	29	76	94	100	-
				Autumn period			
Primorie Territory	-	-	2	9	67	80	100

* Precipitation less than 2.5 mm was ignored

The number of precipitation-free days can be considered as a predictor of large fire occurrence in a region. In Table 4 fire occurrence dynamics are presented relative to the number of rainless days. Three periods of the fire season: spring (April-May), summer (June-August), and autumn (September-October) are covered.

In spring, forest stands of Primorie, Sverdlovsk and Tomsk Regions, and Buryatia attain high flammability very rapidly. 100% of large fires burn here during a 20-day rainless period. This suggests that the spring period of the fire season, with respect to fire weather, is extremely dangerous, which means that for fire suppression to be successful, additional fire fighting resources are needed. As is obvious from the Table 4, 20 precipitation-free days are a critical period, when large fires can burn in forests from southern Siberia throughout the Far East. In summer, large fires begin to occur all over the area after a 30-day drought. A 40 to 50-day dry period is responsible for 66 to100% of all fires. In the autumn period of the fire season large fires are rare. They are mostly concentrated in Primorie Territory, where they follow a 20-30-day rainless period. In autumn, a 60-70-day drought accounts for 100% of large fires.

3. Conclusions

All across Siberia and the Far East, large fires vary widely in both frequency and time of occurrence. In Western Siberia, fires are relatively infrequent, except for Tomsk Region, where they have increased in number during recent years. This area experiences large fires once in the average every two to three years, while severe fire outbreaks occur once or twice a decade. In most parts of Eastern Siberia and the Far East, fires are an annual event, and large-scale fire outbreaks are reported once every 3 to 4 years; often occurring in two consecutive years.

The number of rainless days is the major factor controlling large fire occurrence: a 10-day, 20-day, and 30- to 40-day dry period in spring, summer, and autumn, respectively. Maximum numbers of large fires in spring falls within a 30- to 40-day dry period, and in autumn within a 40- to 50-day drought. In the area of interest, fire seasons of two types can be observed:

- a short, continuous fire season, with thr rate of forest fires being very high during 2 or 3 summer months. Fire seasons of this type are common in the northern parts of Siberia and the Far East.

- a long fire season, when the rate of forest fires remains high for 5 to 7 months. This is the case with southern taiga forests. For some years, a double fire season can be observed in this taiga zone, which is characterized by two short-term maxima of rates of burning (i.e. autumn and spring fire activity peaks). In summer, however, fires do not burn.

This large-scale study of large forest fires in Siberia allowed us to better understand temporal and spatial changes in rate of forest burning and fire occurrence. In the area under study, two completely different types of fire regimes can be found, that are determined by different environmental conditions (i.e, the fire regime in the taiga forests of Western Siberia differs in all respects from that in the forests of Eastern Siberia). Low fire recurrence determines the fire regime in bogged (swamp) forests of Western Siberia, whereas light conifer forests growing on low mountains of Eastern Siberia are characterized by high rates of fires.

The current forest landscape of Siberia is thus the result of the relationship of spatial and temporal dynamics of large fires with physical, geographical, and climatic conditions of the area.

References

Kurbatsky, N.P., and E.N. Valendik. 1974. On past experience. J. Forestry (Lesnoye khozyaistvo) No.3, 65-67 <in Russian>.

Kurbatsky, N.P., and M.A. Sheshukov. 1978. Forest fires in Khabarovsk Territory. J. Forestry (Lesnoye khozyaistvo) No.4, 78-83 <in Russian>.

Shostakovich, V.B. 1924. 1915 forest fires in Siberia. News from West Siberian Department of Russian Geographers Society (Izvestia Vostochno-Sibirskogo otdela russkogo geograficheskogo obshestva) Vol. 47. Irkutsk, p.1-9 <in Russian>.

J.G.Goldammer and V.V.Furyaev (eds.), Fire in Ecosystems of Boreal Eurasia, 139-150.

Major 1992 Forest Fires in Central and Eastern Siberia: Satellite and Fire Danger Measurements

B.J. Stocks [1], D.R. Cahoon [2], J.S. Levine [2], W.R. Cofer III [2], and T.J. Lynham [1]

1. Introduction

The boreal forest biome is a classic fire-dependent ecosystem, capable, during periods of extreme fire weather, of sustaining the large, high-intensity wildfires responsible for its existence. The natural fire cycle in the North American boreal forest averages 50-200 years (Heinselman 1981). Fire is the major disturbance regime in the boreal forest, and tree species have adapted to this form of disturbance over millennia, to the point where fire is required for adequate regeneration. With increased human settlement of the world's boreal zone over the last century, for both industrial and recreational purposes, there has been a concurrent development of fire management programs designed to protect human interests and forest investment. However, total fire exclusion has proven neither economically feasible nor ecologically desirable, with the result that fires are still a major force in the boreal forests of North America and Eurasia, burning over an average of 5-6 million hectares annually during the 1980s (Stocks 1991). In addition, climate change projections indicate significantly higher temperatures across the world's boreal zone within the next 50 years, and forest fire occurrence and impacts would increase significantly under this scenario (Stocks 1993), potentially providing significant positive feedback to global warming, and greatly influencing the carbon budget of the boreal forest zone.

Forest fire activity in the boreal forests of Canada and Alaska has been well documented over the past 70 years, resulting in reliable fire occurrence and area burned statistics that permit a thorough analysis of forest fire trends in the North American boreal zone. Conversely, published forest fire statistics for Russia (formerly the Union of Soviet Socialist Republics) have been sparse and unreliable for political reasons, but qualitative accounts indicate that fire has been, and continues to be, a major disturbance in Russia, particularly in the remote, underpopulated regions of Siberia. Documentary accounts from the early 1900s describe enormous forest fire losses in Siberia, culminating in the particularly dry year of 1915, when an estimated 14 million hectares burned in this region (Shostakovitch 1925). More recently, during the spring of 1987, the Great China Fire burned over in excess of 1.1 million hectares of boreal forest in northeastern China (Cahoon et al. 1991), attracting worldwide attention. Satellite surveillance of the Great China Fire showed that large areas of southeastern Siberia were also burning during this period, and further analysis revealed

[1] Forest Fire Research, Canadian Forest Service, Sault Ste. Marie, Ontario P6A 5M7, Canada
[2] Atmospheric Sciences Division, NASA Langley Research Center, Hampton, Virginia, 23681, U.S.A.

that in excess of 10 million hectares was burned over in this region during the spring of 1987 (Stocks 1991, Cahoon et al. 1994a). A recent study by Dixon and Krankina (1993) quotes Russian sources stating that an average of 6.8 million hectares burned annually in Russia during the 1988-1991 period. Of this total only 1.5 million hectares burned annually on monitored forest lands, indicating that the vast majority of the area burned in Russia is located in unmonitored areas, primarily Siberia, for which fire records are not kept. Since 1987 NOAA/AVHRR (Advanced Very High Resolution Radiometer) satellite imagery has been collected over central and eastern Siberia for the purpose of documenting the occurrence and distribution of large fires in this region where accurate forest fire statistics were not available (Cahoon et al. 1994a). Large fires are easily detected and mapped using AVHRR, and since a small number of large fires account for the vast majority of the area burned annually in the boreal zone, this exercise provides a reasonable estimate of the impact of fires in central and eastern Siberia. As a part of this ongoing investigation, this paper reports on the development of large forest fires in three distinct regions of Siberia during the 1992 fire season, using recent satellite-derived area estimation techniques to determine fire sizes, and using daily weather observations from nearby stations to determine associated fire danger conditions as reflected through the Canadian and Russian fire danger rating systems.

2. Satellite-Derived Area Estimation of Boreal Fires

The Advanced Very High Resolution Radiometer (AVHRR) provides information over large geographical areas (swath width over 3000 km) at medium resolution (1 km at nadir) with a timely sampling frequency (twice daily), and has proven an ideal tool for monitoring the distribution of global biomass burning. Since 1978 the AVHRR instrument has flown on several National Oceanic and Atmospheric Administration (NOAA) polar-orbiting meteorological satellites. NOAA-11, the most recent satellite in this series, provided the imagery used in this study. The AVHRR instrument provides digital imagery in the visible, near-infrared, and infrared wavelengths of the electromagnetic spectrum, and is extremely useful in delineating burned areas in both forest (Cahoon et al. 1991) and savanna (Cahoon et al. 1994b) ecosystems. Methods recently developed for estimating the surface area of forest fires (Cahoon et al. 1992) were applied to estimate the area burned in selected regions of Siberia in 1992. Coverage of eastern Russia by NOAA satellites has been somewhat limited in recent years, due to gaps in ground-receiving stations, but this will be resolved in 1995 when additional downlinks are established in Siberia.

3. Canadian and Russian Fire Danger Rating Systems

The Canadian Forest Fire Danger Rating System (CFFDRS) has been under development by the Canadian Forest Service since 1968. At the present time, the CFFDRS has two modules or subsystems: the Canadian Forest Fire Weather Index (FWI) System (Van Wagner 1987) and the Canadian Forest Fire Behavior Prediction (FBP) System (Forestry Canada Fire Danger Group 1992). The FWI System, which is used in this paper to reflect fire weather conditions in Siberia in 1992, has been in use throughout Canada since 1970. The FWI System consists of six components that individually and collectively account for the effects of fuel moisture and wind on fire behavior in a standardized fuel type. The three moisture codes, the Fine Fuel Moisture Code (FFMC), the Duff Moisture Code (DMC), and the Drought Code (DC), are numerical ratings of the fuel moisture content of fine surface litter, loosely compacted duff, and deep, compact organic matter, respectively. The three fire behavior indexes, the Initial Spread Index (ISI), the Buildup Index (BUI), and the Fire Weather Index (FWI), are intended to represent rate of fire spread, fuel available for combustion, and frontal fire intensity, respectively. The FWI System components depend solely on daily measurements of dry-bulb temperature, relative humidity, wind speed, and precipitation recorded at noon local standard time. The FWI scale is uniform across Canada, although each major jurisdiction in the country has developed its own qualitative fire danger classification scheme. In general, FWI values less than 3 represent low fire danger; values from 4-10 represent moderate conditions; values between 11 and 22 are considered high; and values above 23 represent extreme fire danger conditions.

Although many fire danger rating systems have been developed in Russia, the most widely used is a relatively simple ignition index based on the work of Nesterov (1949). The Nesterov Index (NI) requires daily observations of dry-bulb temperature, dew-point temperature, and precipitation. The difference between daily temperature and dewpoint is multiplied by temperature and cumulatively summed over the number of days since 3mm of precipitation, to provide a general index of ignition potential. NI values between 300 and 1000 are considered moderate, while NI values from 1000-4000 represent high ignition potential. Values above 4000 indicate the potential for ignition is extreme.

4. General Siberian Climatology

Central and eastern Siberia is characterized by the highest degree of continentality on earth. During winter the region is dominated by the Asiatic high, with its core near Lake Baikal, which results in a very stable situation, with very cold temperatures and very little overwinter precipitation (50-100 mm). The breakdown of the Asiatic high in spring intensifies zonal circulation, causing a rapid progression of storms from west to east, resulting in strong winds with very little precipitation while temperatures rise quickly and snow cover evaporates rapidly. Spring and early summer are very dry throughout central and eastern Siberia, with very low relative humidities and very little precipitation. During summer the west-to-east circulation weakens and cyclonic activity strengthens, which, along with increased convective activity, causes increased precipitation (200-400 mm) during late July and August. Although summer is short it can be quite warm. In autumn a zonal circulation pattern returns, causing widespread cooling across Siberia until the Asiatic high becomes established again in early November.

5. 1992 Siberian Fires and Associated Fire Danger Conditions

An analysis of fire weather in Russia during the 1980s (Stocks and Lynham 1996) indicated extreme fire weather conditions occur in southern Siberia in early May, moving northward as spring arrives at higher latitudes. During June and July generally extreme fire danger conditions exist throughout all of Siberia from the Yenesey River east to the Pacific Ocean. Fire weather severity decreases somewhat in August across Siberia, although conditions remain high to extreme on the Central Siberian Plateau between the Yenesey and Lena Rivers. By September fire danger is generally low throughout Siberia. The 1992 fire season in central and eastern Siberia began in typical fashion, with snow cover disappearing early, insignificant precipitation, low relative humidities, and gradually rising temperatures. Three large fire situations developed in central and eastern Siberia during the June-August 1992 period. In the following analysis, NOAA AVHRR satellite imagery is used to monitor the extent of these fires, and weather data from nearby weather stations (acquired from the NOAA National Climate Data Center) is used to determine the fire weather and fire danger conditions associated with each major fire situation.

5.1 Aldan River Fires

The 1992 Aldan River fires (61°N - 135°E) occurred in mid-June near Ust-Maya, an area with rolling topography (200-300 m in elevation) located in the foothills between the Yakutsk Basin and the mountains of Northeastern Siberia. This is a region of extreme continentality, with discontinuous permafrost and a growing season of 100-120 days, where Dahurican larch dominates, often in mixtures with pine, birch and aspen. Korovin (1996) indicates that the fire season in this region is approximately 130 days in length, running from mid-May through the end of August, with most fires occurring between late May and early August.

Weather and fire danger conditions for the 1992 fire season for two weather stations (Ust-Maya and Ust-Judoma) near the Aldan River fires are presented in Figure 1. The Ust-Maya station is located immediately adjacent to the larger fires, while the Ust-Judoma station is located approximately 150 km to the south. The Figure 1 data reflects very dry spring conditions with low relative humidities and rising temperatures, with no significant precipitation at Ust-Judoma and only one major rainfall at Ust-Maya. As a result, extreme fire weather conditions existed throughout the region in mid-June (Julian Days 165-175), when satellite imagery revealed the Aldan River fires to most active. Codes and indexes of the Canadian FWI System reflect extreme fire danger levels during this period, with ISI values indicating good spread potential and extreme BUI values reflecting a high degree of fuel consumption. NI values were also extreme at both weather stations.

The location of the Aldan River fires is shown in Plate 1a, a NOAA AVHRR image for 30 July (Julian Day 212). Although the Aldan River fires were no longer active at this time, this image shows fire activity further east in the area north of Magadan on the Sea of Okhostk. Plate 1b is a closer view of the Aldan River fires, from the same AVHRR scene, also showing the 1991 fires that occurred in this area. The area burned by the 1992 fires shown in Plate 3 amounts to 391,000 ha, while the 1991 fires in this area burned an additional 335,000 ha.

5.2 Kolyma River Fires

The fires shown burning in the extreme northeastern region of Siberia in Figure 2 began in mid-July and continued to burn through most of August. These fires (64°N - 153°E) were located approximately 500 km north of Magadan, along the southern reaches of the Kolyma River, where the topography is mountainous with elevations ranging between 1000 and 2000 m. In this region permafrost and a very short growing season (80-110 days) limit tree growth, and stunted, low-grade Dahurican larch dominates. The fire season is approximately 115 days long, beginning in late May and ending in late August (Korovin 1996). Fire occurrence is highest between late June and early August.

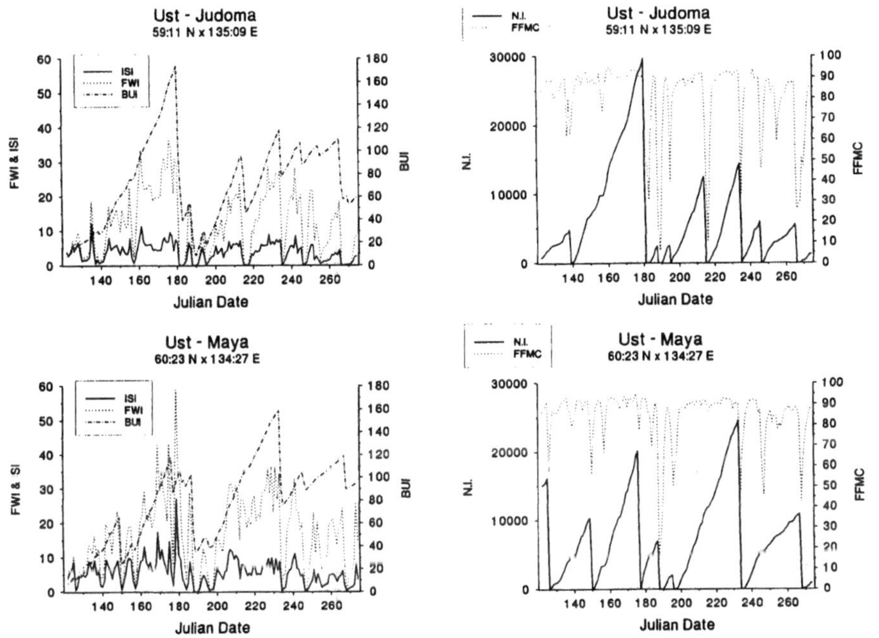

Fig.1. Antecedent and current fire danger conditions near the Aldan River fires, as reflected by the FFMC, ISI, BUI and FWI of the Canadian FWI System, and the NI of the Russian System.

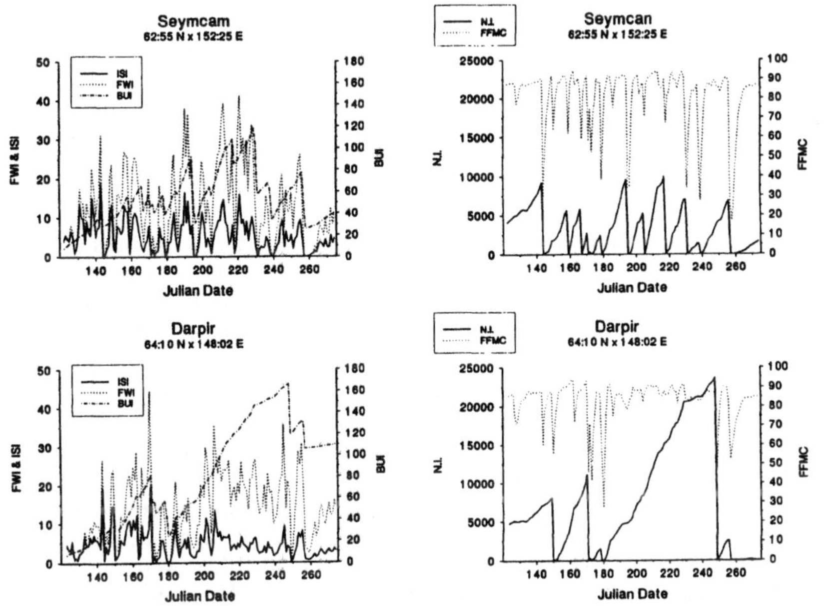

Fig.2. Antecedent and current fire danger conditions in the vicinity of the 1992 Kolyma River fires, as reflected by the FFMC, ISI, BUI and FWI of the Canadian FWI System, and the NI of the Russian System.

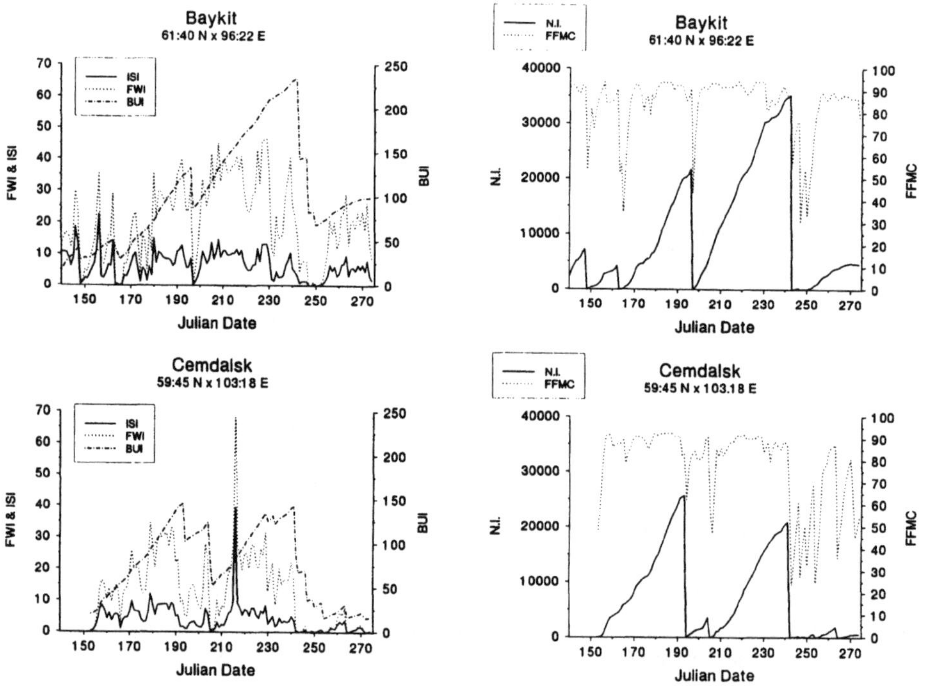

Fig.3. Antecedent and current fire danger conditions near the Chunya River fires, as reflected by the FFMC, ISI, BUI and FWI of the Canadian FWI System, and the NI of the Russian System.

No overhead satellite imagery of the Kolyma River fires was available until September 3, after the fires were over, but oblique views indicated that these fires were burning on 30 July (Julian Day 212) and were still burning on 16 August (Julian Day 229). From Plate 2a, which shows much of northeastern Siberia on 3 September (Julian Day 247), it is evident that the 1992 Kolyma River fires are no longer active, and that large fires also occurred in this region in 1991. Plate 2b provides a closer view of the 1991 and 1992 fire scars along the southern Kolyma River, which amount to a total cumulative burned area of 650,000 ha.

Fire weather in the area of the Kolyma River Fires during the spring and summer of 1992 is summarized in Figure 2 for the two nearest weather stations, Darpir (200 km west) and Semycam (75 km south). As expected at these latitudes, spring arrived late in 1992, with dry but very cool conditions in May, followed by gradual warming and modest precipitation in June. During the months of July and August no appreciable rainfall fell at Darpir and only one significant rainfall event occurred (13 July) at Semycam. As a result, during the active burning period of the Kolyma River fires, fire weather conditions were generally extreme in this region, with extreme FWI and BUI values indicating high intensity fire behavior, excellent spread rates and significant fuel consumption. NI values were much higher during this period at Darpir, as periodic light precipitation kept NI levels at Semycam lower. Rainfall in late August and early September effectively extinguished the Kolyma River fires, with below-freezing temperatures returning to the region by mid-September.

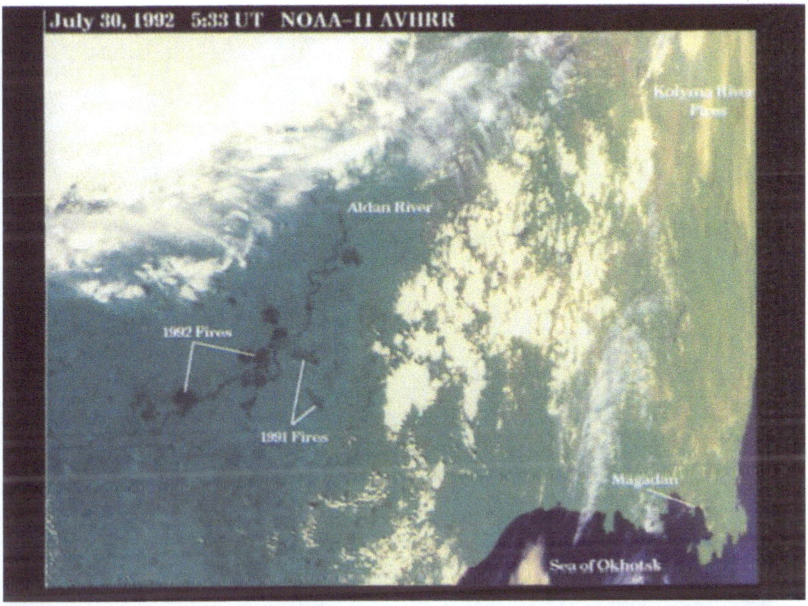

Plate 1a. 30 July 1992 AVHRR image of northeastern Siberia, showing the location of the Aldan River fires and fire activity north of Magadan in the lower Kolyma River watershed.

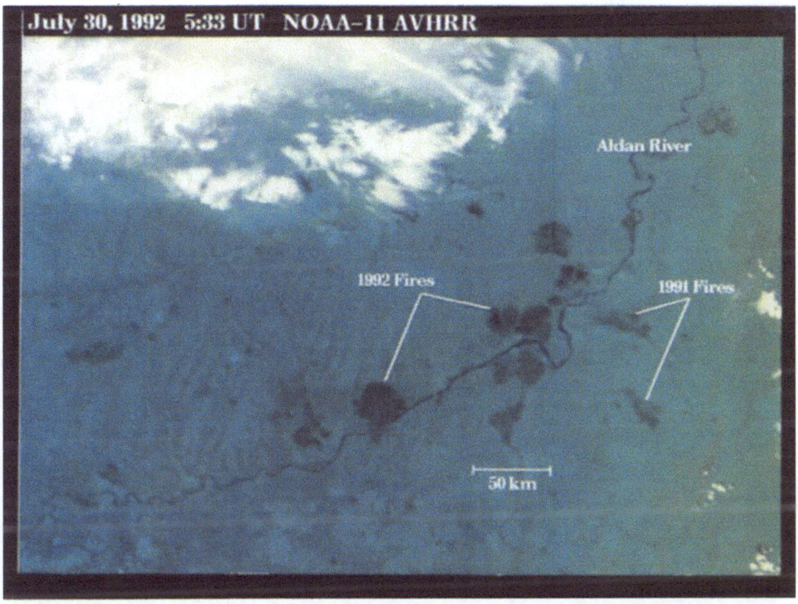

Plate 1b. Larger-scale image (30 July 1992) of Aldan River fire scars showing both 1991 and 1992 fires.

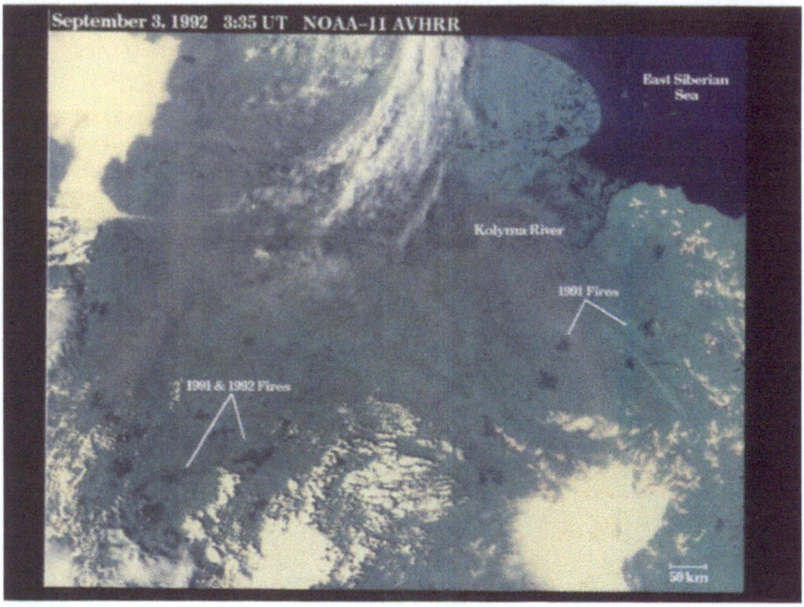

Plate 2a. 3 September 1992 AVHRR image of northeastern Siberia, showing the 1991 and 1992 fire scars in the lower Kolyma River area, and numerous additional 1991 fire scars across the region.

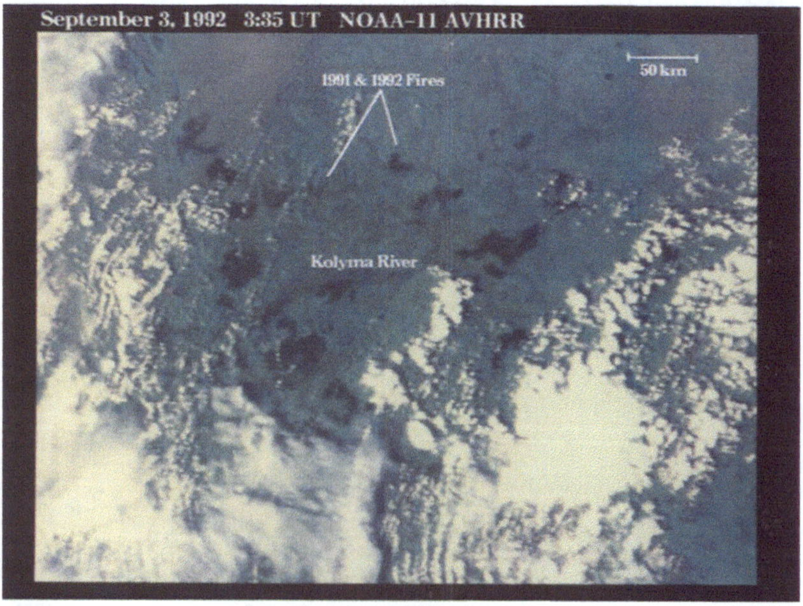

Plate 2b. Larger-scale image (3 September 1992) of the 1991 and 1992 Kolyma River fire scars.

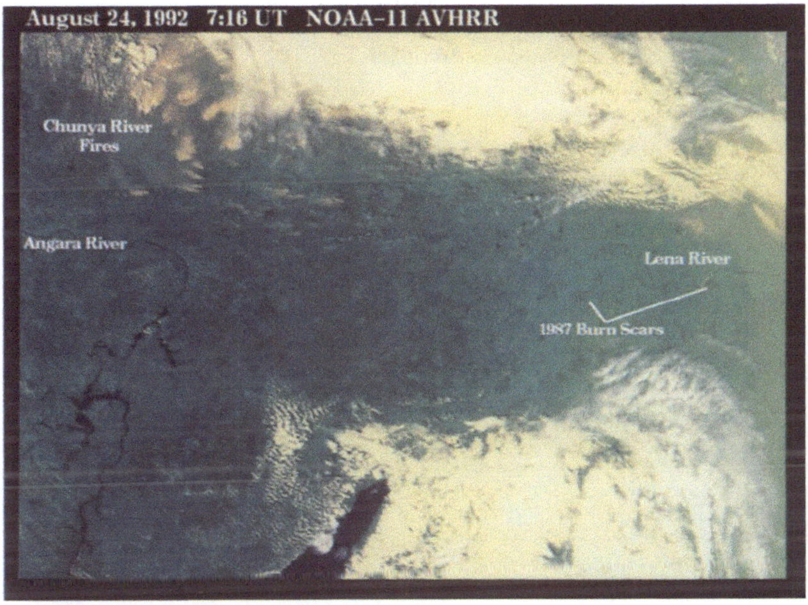

Plate 3a. 24 August 1992 AVHRR image of the active Chunya River fire situation.

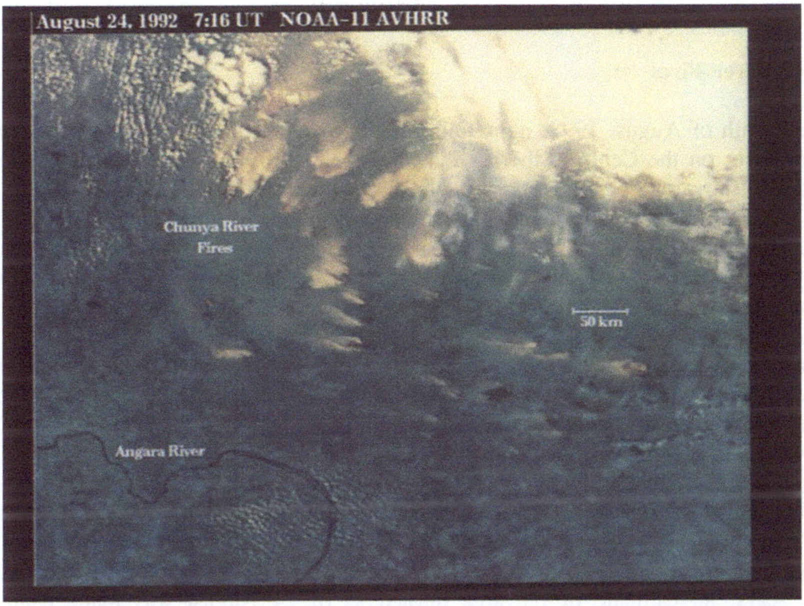

Plate 3b. Larger-scale image (24 August 1992) of the Chunya River fires showing intense fire behavior.

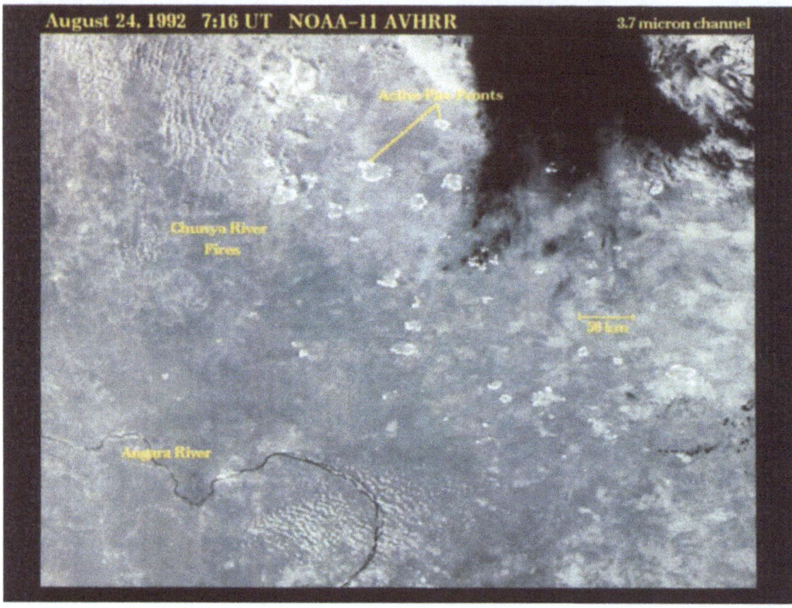

Plate 3c. Infrared image (AVHRR channel 3) of the Chunya River fires on 24 August 1992, showing active fire fronts on all fires.

5.3 Chunya River Fires

During the month of August 1992, numerous fires burned in a large area centered around the Chunya River on the Central Siberian Plateau (62°N - 103°E), approximately 900 km northeast of Krasnoyarsk. This region, with elevations ranging between 600 and 700 m, has a growing season of 120-130 days and supports well-developed forests in which Dahurican larch dominates, usually with mixtures of spruce and pine. Fire season characteristics are similar to those found in the Aldan River area, with the 130-day fire season running between mid-may and late August, and fire occurrence highest between late May and mid-August (Korovin 1996).

The Chunya River fires were first visible on 5 August (Julian Day 218), and were observed obliquely with AVHRR imagery over the next few weeks, remaining quite active during this period. The first directly overhead satellite pass in which the Chunya River fires were visible occurred on 24 August (Julian Day 237). Plate 3a is a long-range view of this fire situation on 24 August, showing numerous active fires in the Chunya River region, but also showing numerous fire scars of various ages along the Lena River to the east. A closer view of the 24 August AVHRR image (Plate 3b) illustrates the intense fire activity occurring over a large area at this time. As no postburn satellite imagery was available of the Chunya River fire scars, an estimate of the area burned as of 24 August was made using the AVHRR infrared channel, which penetrated smoke to permit a clear view of the active fire front (light pixels) and burned area (grey pixels). Plate 3c shows the infrared image for 24 August, in which more than 300,000 ha appear to be burned over.

Figure 3 shows the fire weather prior to and during the Chunya River fires, as measured at Baykit and Chemdalsk, respectively, located approximately 200 km west and 200 km south of the fire area. Warm and dry conditions prevailed in this region throughout May, June and July, with only scattered precipitation. This pattern continued through much of August, creating extreme fire danger conditions during the three weeks that the Chunya River fires were active. Extremely high BUI levels (150-240) indicate very dry fuels and a high degree of fuel consumption, causing the fires to grow steadily with moderate ISI values. The NI also reached extremely high levels during this period. Regular precipitation and cooler temperatures in very late August and early September effectively ended the Chunya River fire situation.

6. Summary

Fire danger conditions over much of Siberia were typically extreme during the summer of 1992, resulting in numerous large forest fire situations. The three described in this study probably burned over close to one million hectares, but likely represent only 30-40% of the total area burned in Siberia that year.

Both the Canadian and Russian fire danger rating systems tracked the increasingly extreme fire danger conditions that developed during these fire events. The calculated extreme fire danger levels are consistent with the intense fire behavior exhibited in each case. Although the NI reflected fire potential well in these situations, an index that returns to zero after very little rain can really only be used to predict ignition potential, and is of limited use in the boreal zone where it is not adequately sensitive to reflect the moisture dynamics of the forest floor, a critical component of forest fire behavior. A multi-index system, such as the CFFDRS, provides much more information (moisture content of various fuel components, potential fire spread rates, fuel consumption, and overall fire intensity) that is useful in both presuppression planning and active fire control activities. Canadian and Russian fire research scientists are now undertaking a cooperative assessment of the suitability of the CFFDRS for use in Russia.

NOAA AVHRR satellite imagery is an excellent and affordable means for the monitoring and documentation of large fire impacts in the boreal zone, particularly in remote regions, where fires often burn undetected. The 1 km resolution is well-suited for measuring large fires, although there is a need for some ground-truthing, perhaps using higher-resolution satellite imagery, to better understand exactly what fire characteristics can and cannot be measured with this approach. With rising concerns over global warming, with its projected significant impacts on the boreal zone in general and boreal fires in particular, remote sensing will be a critical and essential tool in the timely monitoring of boreal ecosystems.

References

Cahoon, D.R., J.S. Levine, W.R. Cofer III, P. Minnis, J.E. Miller, G.M. Tennile, T.W. Yip, P.W. Heck, and B.J. Stocks. 1991. The Great Chinese Fire of 1987: a view from space. In: Global Biomass Burning: Atmospheric, Climatic, and Biospheric Implications (J.S.Levine, ed.), pp. 61-66, MIT Press, Cambridge, MA.

Cahoon, D.R., B.J. Stocks, J.S. Levine, W.R. Cofer III, and C.C. Chung. 1992. Evaluation of a technique for satellite-derived area estimation of forest fires. J. Geophys. Res. 97 (D4), 3805-3814.

Cahoon, D.R., B.J. Stocks, J.S. Levine, W.R. Cofer III, and J.M. Pierson. 1994a. Satellite analysis of the severe 1987 forest fires in northern China and southeastern Siberia. J. Geophys. Res. 99(D9), 18627-18638.

Cahoon, D.R., J.S. Levine, W.R. Cofer, and B.J. Stocks. 1994b. The extent of burning in African savannas. Adv. Space Res. 14(11), 447-454.

Dixon, R.K., and O.N. Krankina. 1993. Forest fires in Russia: carbon dioxide emissions to the atmosphere. Can. J. For. Res. 23, 700-705.

Forestry Canada Fire Danger Group. 1992. Development and structure of the Canadian Forest Fire Behavior Prediction System. For. Can., Ottawa, Ont., Inf. Rep. ST-X-3, 49 pp. + Appendices.

Heinselman, M.L. 1981. Fire intensity and frequency as factors in the distribution and structure of northern ecosystems. In: Fire Regimes and Ecosystem Properties (H. Mooney, J.M. Bonnicksen, N.L. Christensen, J.E. Lotan and W.A. Reiners, eds.), pp. 7-57, USDA For. Serv. Gen. Tech. Rep. WO-26, Washington, DC.

Korovin, G.N. 1996. Analysis of the Distribution of Forest Fires in Russia (this volume).

Nesterov, V.G. 1949. Combustibility of the forest and methods for its determination. USSR State Industry Press.

Shostakovich, V.B. 1925. Forest conflagrations in Siberia. J. For. 23, 365-371.

Stocks, B.J. 1991. The extent and impact of forest fires in northern circumpolar countries. In: Global Biomass Burning: Atmospheric, Climatic, and Biospheric Implications (J.S. Levine, ed.), pp.197-202, MIT Press, Cambridge, MA.

Stocks, B.J. 1993. Global warming and forest fires in Canada. For. Chron. 69, 290-293.

Stocks, B.J., and T.J. Lynham. 1996. Fire weather climatology in Canada and Russia (this volume).

Van Wagner, C.E. 1987. Development and structure of the Canadian Forest Fire Weather Index System. Can. For. Serv., Ottawa, Ont., For. Tech. Rep. 35. 37pp.

J.G.Goldammer and V.V.Furyaev (eds.), Fire in Ecosystems of Boreal Eurasia, 151-167.

Fire Ecology of Pine Forests of Northern Eurasia

S.N. Sannikov [1] and J.G. Goldammer [2]

1. Introduction

Natural fire caused by lightning and volcanic eruptions have occurred in different biomes of the globe long before the appearance of humans and have been recurring to the present day. Like many other types of natural ecological disturbances, they are not a random occasional phenomenon but represent a regular cyclic event. The frequency of lightning-caused fires correlates not only with the climate and type of vegetation but also with magnetic properties of the earth's crust (Novgorodov et al. 1982).

In pristine vegetation types such as forests, savannas, steppes, tundra and bog ecosystems, recurrent fire has been one of the most important ecological factors which had direct and indirect short- and long-term effects on ecosystem properties and dynamics. With the appearance of humans, more frequent anthropogenic forest fires began to act increasingly as an effective agent of transformation of populations and ecosystems, superimposing an imprint on the appearance of biogeocenoses and whole landscapes (Melekhov 1965; Heinselmann 1973; Sannikov 1973, 1981; Kozlowski and Ahlgren 1974; Kimmins 1987; Goldammer 1990, 1993).

A study of diversified ecological consequences of fires in many regions and types of biomes of the Earth show that fire plays a fundamental role in the transformation of habitat conditions and various components of ecosystems. A relatively new branch of modern ecology - fire ecology ("pyroecology") - has been developed in the past three decades. Fire ecology embraces investigations of the manifold effects of fire on the structure, functions, dynamics and evolution of populations, biocenoses and ecosystems. However, in spite of certain attemps to generalize the vast information accumulated on fire ecology of forests in differ ent regions of the globe (Korchagin 1954; Lutz 1956; Mount 1969; Rowe and Scotter 1973; Sannikov 1973, 1981; Sirén 1974; Le Houérou 1974; Kozlowski and Ahlgren 1974; Wein and MacLean 1983; Chandler et al. 1983; Booysen and Tainton 1984; Goldammer 1990, 1993; Goldammer and Jenkins 1990; van Wilgen et al. 1992; Crutzen and Goldammer 1993; Moreno and Oechel 1994; Pyne 1995), we have not yet fully appreciated the role of fire as a comprehensive ecological factor that causes a simultaneous profound transformation of all interrelated components of an ecosystem.

[1] Institute of Forest, Ural Division of Russian Academy of Sciences, 620134 Ekaterinburg, Russian Federation

[2] Max Planck Institute for Chemistry, Biogeochemistry Department, Fire Ecology Research Group, c/o Freiburg University, D-79100 Freiburg, Germany

Assuming a forest fire to be an extrinsic factor influencing ecosystems through a cyclic-pulse action, we consider specific features of fire regimes and empirically generalize the associated pyrogenic changes in the basic component of biogeocenoses with the example of pine forests of Northern Eurasia, which are dominated by common (Scotch) pine (*Pinus sylvestris*).

2. Fire Regimes

During the last 500-600 years fire frequencies in pine forests of the boreal zone have become increasingly influenced by humans. Fire-return intervals vary, depending on region and type of forest, between 15-20 and 60-70 years, thus considerably shorter than in dark-coniferous forests (Swain 1973; Zackrisson 1977; Tolonen 1983).

Regular differences in fire frequency, due to different continentality and climatic aridity, are observed within vast distribution ranges of common pine, the ranges covering almost all of Northern Eurasia. In the transcontinental geographical profile of green-moss pine forests transversing the boreal zone of Eurasia from the West to the East (Fig.1), the mean fire interval decreases from 58-71 years in Scandinavia and the Russian plain (Vakurov 1975; Zackrisson 1977) to 40-42 years in the Trans-Urals, and 14-23 years in Western and Central Siberia (Pobedinski 1965; Furyaev and Kireev 1979). In "dry" lichen pine forests the fire return intervals change in the same manner, but on a somewhat lower level: they decrease from 46 years in Scandinavia to 30 years in the trans-Urals, 13 years in Central Siberia, and 7-8 years in Trans-Baikalia. Like in the boreal zone of North America (Gwynar 1978), in this type of forest located on islands amidst impassable bogs in the Konda basin (Western Siberia), the natural fire interval (60- 70 years) is two or three times longer than in anthropogenically influenced forests.

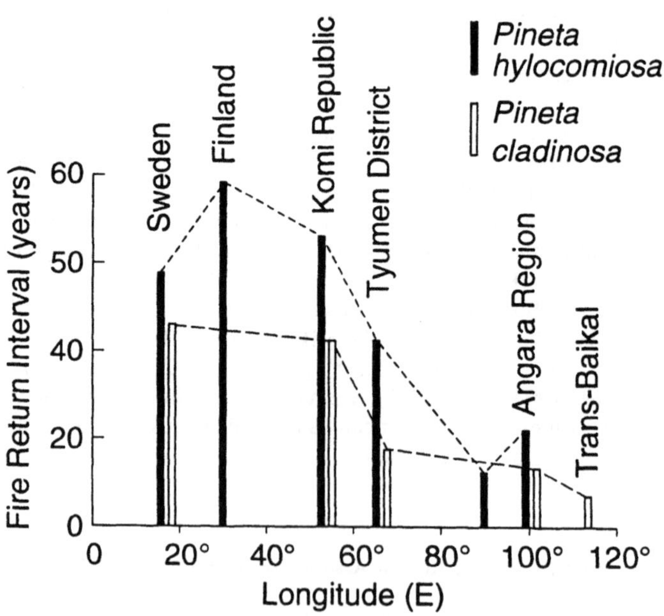

Fig.1. Longitudinal gradient of fire return intervals in pine forest communities of boreal Eurasia.

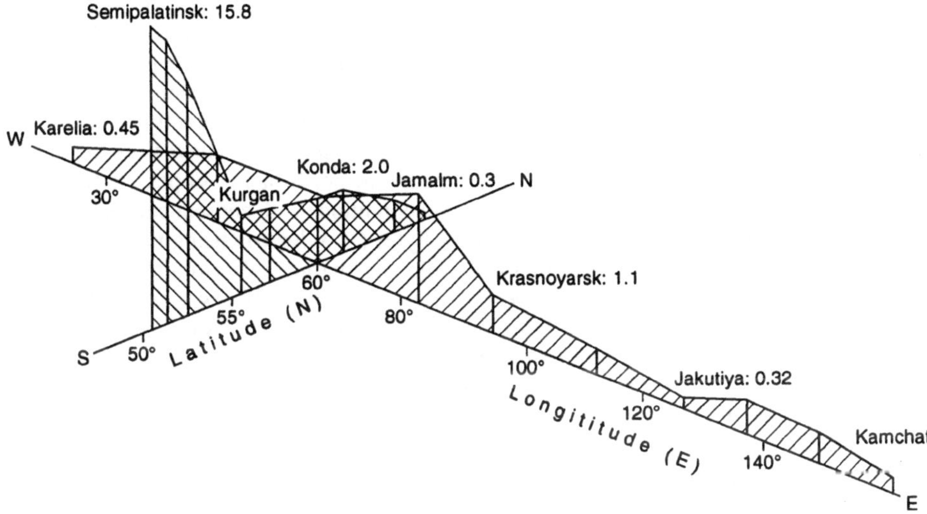

Fig.2. Geographical gradients of lightning fire density (number of fires x 100,000 ha yr[1]) in pine and larch forests of Northern Eurasia. The West-East transcontinental profile is between 55° and 65°N.

From the viewpoint of evolutionary pyroecology, the recurrence of lightning fires is of greatest interest. The mean annual number of fires per 100,000 ha of pine and larch forests can serve as a natural fire occurrence index. From 1981 to 1991 the index of lightning fire density in the forest zone (at 55°-65°N) increased from 0.4-0.5 in the West of the Russian plain (Karelia, Latvia) to 1.6-2.8 in the East of that plain, and 2.1-3.2 in the Trans-Urals and Western Siberia (Fig.2). But farther to the East, the density of lightning fires again lowers rapidly to 1.1 in Central Siberia and 0.3-0.4 in Southern Yakutia and Kamchatka. This is probably connected with a poorer inflammability of litter of the larches which dominate in the forests of Eastern Siberia (Sherbakov et al. 1979).

A still quicker increase of lightning fire density is noted farther East in the arid steppe zone: from Buzuluk pine forests in the Trans-Volga region (4.2) to island pine forests in the Turgai depression (7.3), and further to strip-pine forests of Pavlodar (14.2) and Semipalatinsk (15.8) provinces of Kazakhstan.

In the meridional geographical profile across the distribution range of pine in the Trans-Urals and Northern Kazakhstan, as aridity and lightning intensity of the climate increases, the density of lightning fires increases rapidly along a gradient from forest tundra (0.3) to middle taiga (2.1); then a certain stabilization is observed, and the density increases sharply in the steppe zone (up to 7.3-15.8). Thus, the frequency of anthropogenic and lightning fires in pine forests of Eurasia are becoming higher with increasing continentality of climate and site aridity. It is possible to suggest that fire selection pressure and the effect of fire on the ecological regimes of pine cenoses in steppe pine forests of the Turgai are seven times, and those in strip pine forests of Pre-Irtysh are 15 times higher than in the northern taiga of Western Siberia. On the other hand, the pyrogenic effect is an order of magnitude higher in pine forests of continental regions of Eurasia than it is in coastal zones both in the West and the East of the continent.

Besides the influence of geography and climate at a continental scale, fire regimes are considerably shaped by microclimatic effects and habitats as a consequence of topography (elevation, slope, aspect). Research in the montane-boreal coniferous forests of Northeast China (along the Amur River, Daxinganling mountains, Province of Heilongjiang) shows that south-exposed slopes are generally drier and warmer that north-facing slopes and valleys (Goldammer and Di 1990, and unpubl. data; Goldammer 1993). Consequently, fuels (duff, downed woody material, grass-shrub-tree understory) dry quicker and allow fires to spread across the slopes whereas they usually stop at the edge of north-facing slopes. A feedback loop of topography-determined microclimate and fuel conditions with consequent short-return interval fires, followed by the colonization of pyrophytes (*P. sylvestris* being the dominating species) is a characteristic determinant of small-scale fire regimes. Fire-return intervals on south slopes may range between 5 to 50 years, whereas the generally wet northern slopes may support a fire only in intervals of 30 to 50 years. In the wet depressions, characterized by thick organic layers and biogenic permafrost, fires may burn in short (1-3 yr) intervals in the grass-shrub layer without penetrating into the organic terrain. Fires penetrating and consuming thick organic layers only occur once every few hundred years (ca. 400 to 1200 years according to radiocarbon dates of Goldammer [1993]). As a result of this a variety of micro-habitats are formed by topographic and fire effects.

3. Effect of Fires on Main Ecosystem Components

Tree stand

Thanks to the horizontal canopy closure and height of natural lowland and submontane pine forests, surface fires rarely develop into crown fires. Uneven-aged mountain, bog and anthropogenic pine forests are the only exceptions. Once crown fires occur they do not spread continuously but in jumps, leaving clumps of live mature trees (Sannikov 1985, 1992). This is explained not only by irregular conditions of topography, fuel moisture and fuel loads, but also by the effects of vortex formation in high-intensity fires (see FIRESCAN Science Team 1996).

Intensive surface fires in pine and mixed forests cause destruction to parts of less fire-resistant dark-coniferous (*Picea* spp., *Pinus sibirica*, *P. koraiensis*) and deciduous species, and also of suppressed trees of light-coniferous species (Melekhov 1948; Molchanov and Preobrazhenski 1957; Kolesnikov et al. 1973; Goldammer and Di 1990). Relatively fire-resistant pines and larches (*Larix sibirica*, *L. gmelinii*, *L. czekanowskii*) of older classes and higher growth dominate in the tree stands.

Fire resistance of *Pinus sylvestris* and *Pinus eldarica* stands increase after 20-40 years of life, as soon as the bark thickness in the basal part of the stem reaches 1.5 cm (Sannikov 1973; Sirén 1974; Rego and Rigolot 1990; Goldammer 1979, 1983; Goldammer and Peñafiel 1990; Kolström and Kellomäki 1993). Simultaneously, a more or less regular seed crop and natural regeneration begins in pine forests. Fires stimulate a rapid release of seeds from pine cones without inflicting any major damage of seeds and ensure seeding on the substrate prepared by fire (reduced A_L layer or other surface cover, e.g. lichens). Seeds of *P. sylvestris* partially preserve their germinability even if the cones were charred during a c rown fire (Sannikov 1983). However, this fire adaptation is less pronounced in common pine than in some North American serotinous *Pinus* spp., which retain seeds in closed cones for many years (Clements 1910; Beaufait 1960; Cayford and McRae 1983).

Thinning of tree stands by fire leads to reduction of stem density and canopy closure, productivity and root competition, density falling below 0.3-0.4, and reduction of seed bearing. Moderate fire impact with a less pronounced drop of stand density, however, may lead to a substantial increase of seed productivity during the first 3-5 years after the fire (Sannikov 1973).

Pre-fire mosaic structure of pine forests encourage non-uniform burning and mortality of trees, leading to a higher heterogeneity of tree stands. Recurring fires cause an appearance of new generations of pine regrowth that survives in canopy gaps and in sparse stands, thus leading to the formation of pine stands with group-age (cohort) structure (Shanin 1965; Pobedinski 1965; Zackrisson 1977; FIRESCAN Science Team 1996). As a result, a mosaic of even-aged and uneven-aged (discontinuous "stepped" age-height profiles) stands is formed and maintained, with a time gap between the generations equal to the fire interval (30-60 years). For this reason, even after severe crown fires, fragments of mature tree stands are left on burns and pine populations preserve a continuous ability to self-reproduction (Muller 1929; Sannikov 1973, 1981, 1985). A mosaic-stepped structure of tree stands, which is reproduced by recurrent fires, presents the main mechanism of the "pulsed pyrogenic stability" of a natural population of common pine. The term "pulsed pyrogenic stability" implies the ability of pine populations to successfully survive, reproduce, regenerate and persistently dominate in phytocenoses under conditions of a cyclic fire regime (Sannikov and Sannikova 1985).

Lower layers of phytocenosis

Fires cause an almost complete destruction of regrowth and undergrowth and destroy the above-ground parts of lichen, moss and grass-shrub cover. Hypnum mosses (*Pleurozium schreberi, Dicranum sporarium, Hylocomium splendens*, etc.) are followed by *Politrichum* spp. (*Politrichum piliferum, P. juniperinum*) and other pyrogenic mosses (*Funaria gigrometrica, Marchantia polimorpha, Ceratodon purpureus*), while fruticose lichen cover (*Cladonia rangiferina, C. sylvatica, C. alpestris*) is replaced by their cup-shaped and tubular forms (e.g., *Cladonia botrytes, C. coccifera, C. gracilis*). Shade-tolerant boreal species of herbaceous plants are replaced by "pioneering" meadow-forest species (Sarvas 1937; Korchagin 1954; Pushkina 1960; Viro 1974; Smirnov 1970; Kolesnikov et al. 1973; Sherbakov et al. 1979). The share of "pyrophytes", which are relatively fire-resistant plants, capable of rapid and copious propagation, spreading, reproduction and development on post-fire forest sites, increases. They are represented mainly by the ecobiomorphs of geo- and hemicryptophytes which are regenerated vegetatively from radical or rhyzome buds (Chizhov and Sannikova 1978), e.g. many species of shrubs and grasses which dominate in different parts of the range of *P. sylvestris*. Some species generate their populations from the soil seed bank (Komarova 1986) or by seeding from outside (e.g., *Chamaenerion angustifolium*).

During the first and second year after an intensive fire, the total cover and phytomass of the herb-shrub layer decrease drastically. But after 3 to 10 years these parameters return to the pre-fire level, while the species diversity rises (Malyshev 1957; Kolesnikov et al. 1973; Trabaud and Lepart 1980).

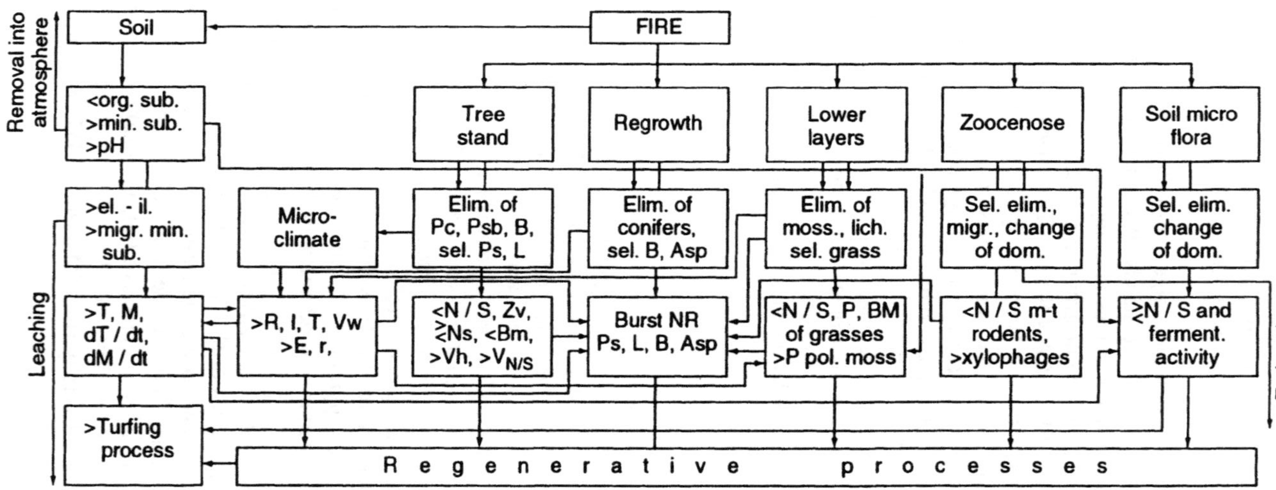

Fig.3. Generalized model of direct and indirect effects of moderate- to high-intensity fire on the basic components of the biogeocenosis of a forest dominated by *Pinus sylvestris*, with intermixed larch and small-leaved species, and with a second tree layer of dark-coniferous species.

Conventional symbols and abbreviations: > - increase; < - decrease; BGC - biogeocenosis; Bm - biomass; d - differential of functions; E - evaporation; I - irradiation; M - soil moisture; N/S - density of populations; Ns - quantity of seeds; NR - natural regeneration; P - percentage cover; R - solar radiation; r - precipitation; T - temperature; t - time; $V_{N/S}$ - variability of the density of populations in space; Vh - variability of trees in height; Vw - wind velocity; Zv - increment of tree stand in volume; Asp - Aspen (*Populus tremula*); B - birch (*Betula pendula, B. pubescens*); L - larch (*Larix sibirica, L. gmelinii*); Pc - spruce (*Picea obovata* a.o.); Ps _ *Pinus sylvestris*; Psb - *Pinus sibirica*; dom. - dominants; el.-il. - eluvial-illuvial process; elim. - elimination; grass. - grasses; lich. - lichens; migr. - migration; min. sub.- mineral substances; moss. - mosses; m-t - mouse type; org. - organic substances; pol.- polytrichous.

Soil

Depending on the region, climate, forest type and burning intensity, fire may change the physical and chemical properties of the soil (see also Mälkonen and Levula, this volume). The thickness of organic layers usually decrease sharply after fire. Very important ecological implications are a rise of the level and stability of the hydrothermal regime of the post-fire substrate, which usually provides optimal conditions for germination of pine seeds and the survival of pine seedlings (Sannikov 1965; Sherbakov et al. 1979; Sannikov and Sannikova 1985), and an increase of moisture and temperature of the mineral soil horizons to a depth of up to 50 cm (Viro 1974; Buzykin and Popova 1978).

A medium- or high-intensity fire (leaving an unburnt layer of duff not more than 1- 2 cm thick) mineralizes part of the organic matter of the duff and plants and, during the initial period of one to three years, causes a rise in pH and the content of mineralized substances available to plant nutrition (Viro 1974). In green-moss and small-herbaceous pine forests of the boreal zone of Scandinavia, the Russian plain and Siberia, the amount of CaO in the upper 10 cm layer of post-fire forest sites increases by 2-3 times; MgO by 1.5 2 times; K_2O and P_2O_5 by 1.3- 1.5 times; and NH_4+NO_3 by 1.5-3 times (Hesselman 1916; Sushkina 1931; Remezov 1941; Arefjeva 1963; Viro 1974; Popova 1979). Losses of total nitrogen are partly compensated by stimulated ammonification and nitrification, and abundant growth of leguminous plants. An exception is the permafrost soil in Siberia, where pyrogenic nitrification is not pronounced (Popova 1979).

Adverse changes in soil properties (compaction, drying, deflation, loss of small soil particles and minerals) take place after frequently recurring fires, not only in pine forests of semiarid zones (Mediterranean, forest-steppes and steppes), but also in boreal forest sites on dry, humus-depleted sandy or gravel soils (Tkachenko 1939; Sherbakov et al. 1979). On the other hand, in waterlogged habitats, especially in the north of the boreal zone, where burning influence is less pronounced, pyrogenic changes in the soil properties are insignificant.

In the far northern boreal and semi-arid regions of Eurasia, where the activity of microflora is suppressed, fires mineralize raw, immobile humus and act as a peculiar reducing agent and a catalyst of biogeochemical and energy cycles in ecosystems (Sirén 1955; Sannikov 1981; Naveh 1990).

Cyclic fires represent a powerful factor of soil formation, the "pyrogenesis" of forest soils (Sapozhnikov 1976). The fires sharply alter chemical properties of the soil and lead to the replacement of mosses by grasses in green-moss pine forests, thereby facilitating the turfing process (Fig.3), which, as the moss cover is regenerated, again leads to podzol formation. Besides exposing the soil and intensifying the heat exchange in the soil, severe fires can greatly alter the hydrological and thermal - specifically permafrost - regimes of the soil, thus stimulating swamp formation, solifluction, thermokarst formation, and water and wind erosion (Pyavchenko 1952; Tyrtikov 1979; Sheshukov et al. 1992). Ancient fires also led to the formation of numerous charred layers in the profile of holocene peatbogs in the forest steppe and boreal zone (Pyavchenko 1958; Tolonen 1983; see Gromtsev, this volume).

Zoocenosis

Effects of fire on the forest fauna may be direct (eliminating individuals or parts of populations) or indirect (by modifying their habitats). The immigration of pyrophilous species which prefer initial stages of pyrogenic successions is favored (see Gossow, this volume). The population density of many species of invertebrates inhabiting the litter and the upper soil layer, reptiles, burrowing mammals, and birds nesting on the surface of soil or in the shrub is drastically lowered immediately after a fire. In the post-fire green-moss pine forests of the Pre-Angara area dominating species in the upper soil horizon are, for instance, annelid worms (Lumbricidae, Enchytraeidae) and insects species of the families *Lycoriidae, Carabidae*, and *Staphilinidae*; while the abundance of molluscs, ticks and collembolans lowers (Gorbachev et al. 1982).

An outbreak of xylophagous insects (bark beetles, longhorned beetles and jewel beetles) is observed in fire-damaged tree stands (Melekhov 1948; Isaev and Girs 1975). In turn, these insects attract insectivorous birds, such as woodpeckers, a nuthatch (*Sitta europaea*), red star (*Phoenicurus phoenicurus*), and others. Well-heated soils of post-fire sites and fire-fallen trees are inhabited by thermophilous insects (*Phaenops cyanea, Melolontha hyppocastani*). One of the most important ecological implications of the reduction of the density of mouse-type rodents, the main consumers of pine seeds shed to the ground, is an abundant appearance of pine regeneration during the first two years after a fire (Sannikova 1977).

Post-fire changes in the species composition and numbers of animals follow the successions of vegetation. As in the forests of the boreal zone of North America, the restoration of the herbaceous grass layer is accompanied by a rapid growth of the population density of mouse-type rodents and common shrews, partridges, blackcocks and sparrow birds; as the tree layer closes up, the number of hare and elk increases (Lutz 1956; Komarek 1969; Scotter 1972; Fox 1983; Kuleshova 1981). The majority of birds and large mammals prefer the post-fire mosaic pattern of the environment.

Microbocenosis

Investigations performed in Scandinavia, on the Russian plain and in Western Siberia show that on relatively humus-rich, wet and sufficiently warm soils of the boreal pine forests, fires cause reproduction and a higher activity of ammonificating, nitrifying, cellulose-fermenting and oligonitrophilic bacteria, and actinomyces (Hesselman 1916; Sushkina 1931; Khrenova 1963; Pietikäinen and Fritze, this volume). This facilitates the decomposition of unburnt organogenic substrate and substantially improves the mineral supply to woody plants.

4. Fires as a Factor of Natural Regeneration of Pine Populations

Fire improves the substrate and ecoclimatic conditions for germination, survival and growth of seedlings of light-coniferous species and reduces the competition of all layers of the phytocenosis and seed-consuming animals. After a sufficiently intensive fire the combination and inter-annual rhythm of main factors of the burn environment correspond, more or less, to the rate and rhythm of the ontogenesis of *P. sylvestris* seedlings (Sannikov, 1973, 1994). Therefore, recurring fires stimulate the outbreaks ("waves") of pine regeneration (Tyurin 1925; Chudnikov 1931; Müller 1929; Sarvas 1937; Korchagin 1954; Sannikov 1983, 1992).

These observations are well in agreement with the fire adaptation or tolerance of most of the ca. 100 species of the genus *Pinus* described in many other parts of the world (e.g. Le Houérou 1974; Klötzli 1975; Naveh 1990; Goldammer and Peñafiel 1990; Goldammer 1993; Agee 1996).

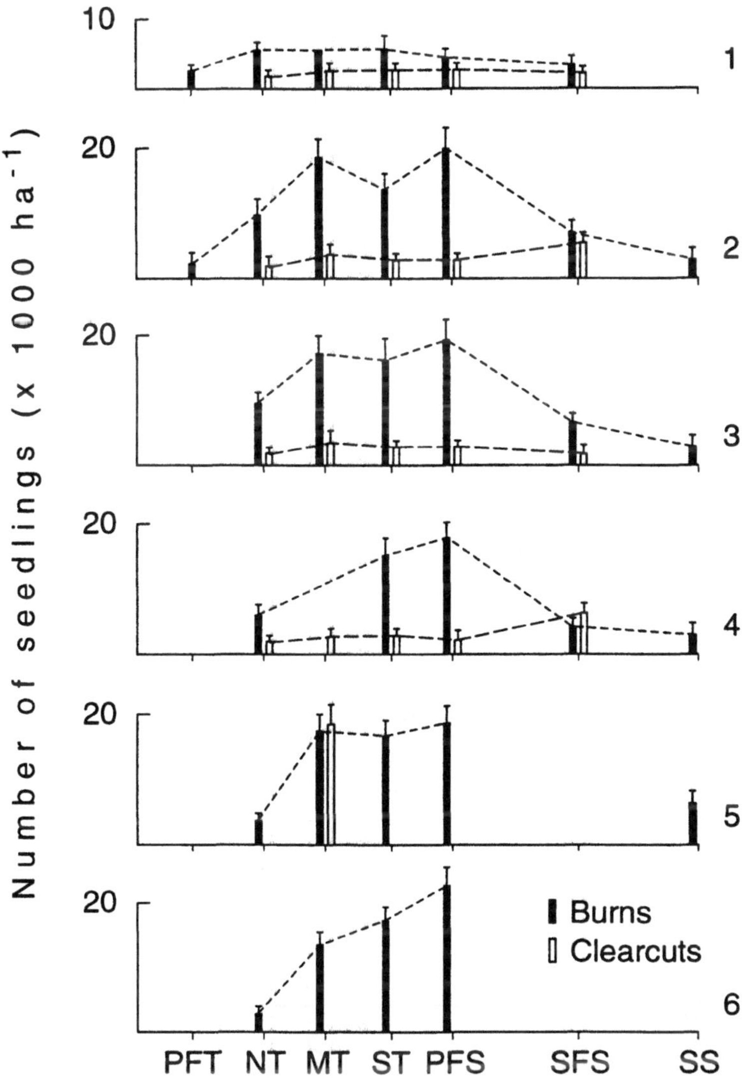

Fig.4. Transzonal meridional profiles of natural regeneration of *Pinus sylvestris* cenopopulations on stand-replacement burns and and clearcuts (with seed trees left nearby) in pine forests of the Transurals and Northern Turgai. PFT - pre-forest tundra; NT - Northern taiga; MT - middle taiga; ST - south taiga; PFS - pre-forest steppe; SFS - south forest-steppe; SS - south steppe. Numbers indicate a series of analogous forest types on: (1) tops of high sand dunes; (2) low tops of hills; (3) middle parts of slopes; (4) foot of slopes; (5) edges of bogs; (6) sphagnous peat-bogs.

With the seeding conditions being close in all the subzones and forest types of the boreal and forest-steppe zones of Western Siberia, the settling density of the self-sown common pine is an order of magnitude higher on post-fire sites than on the raw humus of clearcut sites (Sannikov 1992; Fig.4). The density of self-sown pine varies inversely with the thickness of the unburnt layer of the duff (Fig.5). During the first 10 to 15 years after a fire the survival, parameters, and the cenotic role of the juvenile pine are higher than compared to the same characteristics of unburnt sites (Sannikov 1973, 1992; Sirén 1974; Uggla 1974).

Recurrent fires produce cyclic "regeneration waves" of populations of common pine and other light-coniferous species of the genera *Pinus* and *Larix*, leading to an interruption of the age successions by dark-coniferous species (Fig.6), and to the formation of a stepped age-height structure of tree stands. Regeneration waves can also be observed at regional scales. Large-scale sampling of fire history (fire scar) data in the montane-boreal coniferous forests of Northeast China were combined with the age dating of all sampled trees, revealing a large (long-oscillation) regeneration wave as a consequence of increasing fire activties in the 19th Century (Fig.7).

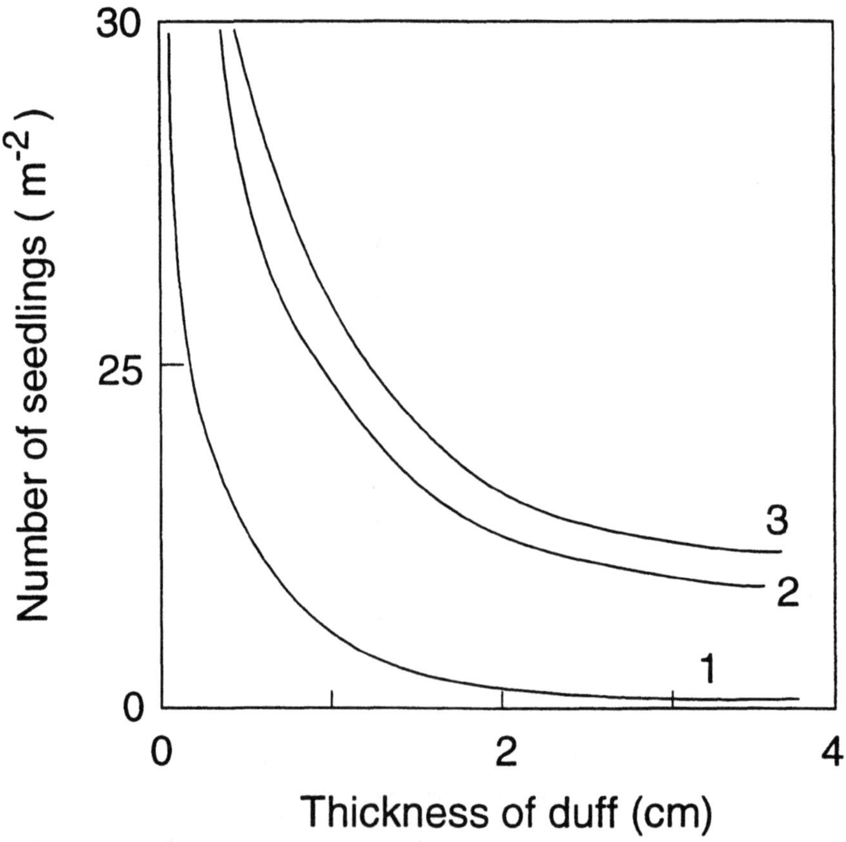

Fig.5. Density of naturally regenerated *Pinus sylvestris* on sites with different thickness of non-burnt duff layer. (1) *Pinetum cladinosum*; (2) *Pinetum vaccinioso-hylocomiosum*; (3) *Pinetum vaccinioso-myrtilloso-hylocomiosum* (after Sannikov and Sannikova 1985).

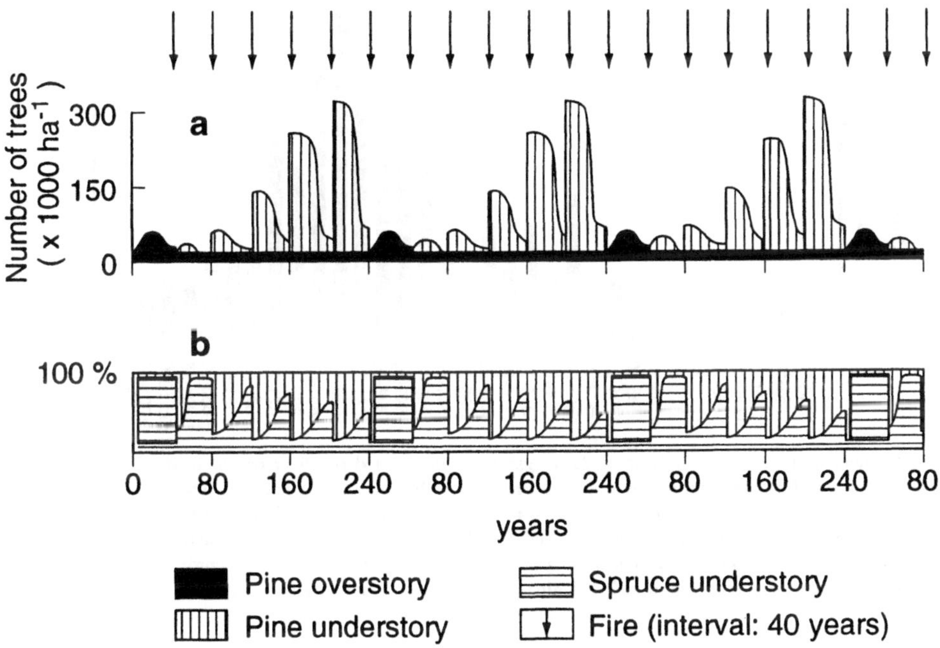

Fig.6. Dynamic model of (a) cyclic pyrogenic regeneration waves of *Pinus sylvestris*, and (b) of changes of pine and spruce (*Picea obovata, P. abies*) with undergrowth in pine forests of boreal zone of Eurasia

5. On Evolutionary Consequences of Fires

Recurrent fires were and remain to be one of the extrinsic factors determining micro-evolution and the genesis of phytocenoses by inducing regeneration, formation of population generations, population waves, isolation, a stringent selection of individuals, and probably mutations, thereby at once changing all the "elementary evolutionary factors" (according to Timofeev-Resovskii et al. 1977). As other natural local ecological disturbances (droughts, floods, hurricanes, avalanches, etc.), fire can probably, lead to irreversible qualitative changes in the gene pool of pine populations and other plant species, animals and microorganisms living in pine forest ecosystems (Sannikov and Sannikova 1985; Sannikov 1991, 1992).

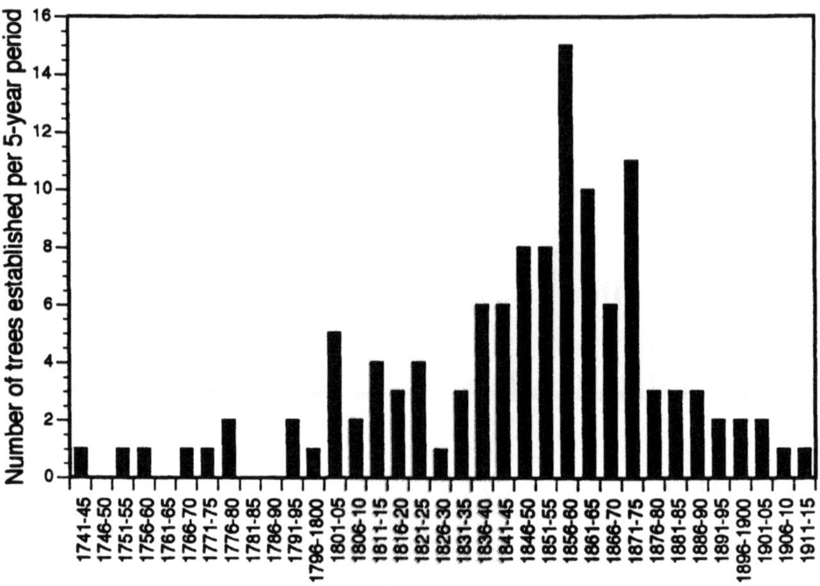

Fig.7. Number of trees (*Pinus sylvestris, Larix gmelinii*) established per 5-yr period in the montane-boreal coniferous forest of Daxinganling (Heilongjiang Province, NE China). The dates of tree origins were taken from a set of samples of fire-scarred trees delivered to a sawmill from the Forest Districts Amuer, Talin, Tuqiang and Walagan where they were analyzed for the establishment of the regional fire history (mean composite fire interval of all four districts: 35 years). Source: unpubl. data of Goldammer and Di, on file at Northeast Forestry University (Harbin, P.R.China) and Freiburg University (Germany).

By eliminating the least fire-resistant individuals from populations, an event occurring repeatedly during the lifespan of each single generation, the "fire selection" constitutes a particular form of directional natural selection in many types of vegetation. On an evolutionary time-scale fire affected and continues to considerably affect the composition of species, biotypes and ecobiomorphs (Korchagin 1954; Sannikov 1973, 1981, 1983; Rabotnov 1978; Trabaud and Lepart 1980). Forests which are regularly affected by fires are inhabited by a wide variety of pyrophytes (Kuhnholtz-Lordat 1939, as quoted by Le Houérou 1974; Sannikov 1973; Mutch 1970; Gill 1981; Chandler et al. 1983). Evidence to support the pyrophytic characteristics of *Pinus sylvestris* is the above-mentioned correspondence between the inter- and intra-annual (seasonal) rhythms of its ontogenesis and the dynamics of post-fire site conditions.

An analysis of the spectra of the ecobiomorphs of plants growing in the grass layer of pine forests in the geographical profile from the north taiga to the steppe zone shows that the share of fire-resistant species along this North-South gradient rises from 55-60 to 80-90 %. Species that are not fire resistant disappear completely. It seems very likely that this fact is due to the density of lightning fires which increases by 16 times along this gradient.

The study of diversified environment-transforming micro- and macro-evolutionary implications of fires in different biomes of the globe at different biochorological levels represent a promising lead in the modern evolutionary ecology, which can be termed as "evolutionary pyroecology" (Sannikov 1984).

Conclusions

1. Pine forests of Northern Eurasia are subjected to periodic fire events which have a strong impact on their evolution and ecology.

2. The frequency of fires on the whole and the density of lightning fires in pine forests are higher with increasing climatic continentality and aridity.

3. Fires represent a powerful exogenic local disturbance factor responsible for the transformation of structures, functions and dynamics of pine-forest ecosystems. The factor is of a cyclic pulsed action, which has a more or less continuous, strong direct or indirect effect on the most critical site conditions and components of biogeocenoses and landscape.

4. The degree and duration of the pyrogenic transformation of an ecosystem depend on abiotic (soil, climate) and biotic (biocenosis) factors of the region and fire regime.

5. In natural green-moss and herbaceous pine forests of the forest and forest-steppe zones of Eurasia dominated by *Pinus sylvestris*, fires recurring in 40-80 yr intervals substantially improve all principal environmental factors for the regeneration and survival of pines and ensure the long-term stability of these ecosystems.

6. Pyrogenic regeneration waves and the patchiness of fire effects lead to a mosaic structure of biogeocenoses and landscapes, and to the characteristic stepped age-/height structure of natural pine forest communities ("pulsed pyrogenic stability ").

7. By sharply changing the environment, and probably enhancing the action of the micro-evolution factors, cyclic fires can lead to irreversible qualitative changes in the gene pool of pine populations and their associated species.

References

Agee, J.K. 1996. Pines and fire. Community-level perspectives. In: Ecology and biogeography of *Pinus* (D.M.Richardson, ed.). Cambridge University Press (in press).

Arefjeva, Z.N. 1963. Influence of fire on some biochemical processes in forest soil. In: Proc. of the Institute of Biology Vol. 36, 39-55. Ural Branch of the Academy of Sciences of the USSR, Sverdlovsk <in Russian>.

Beaufait, W.R. 1960. Some effects of high temperatures on the cones and seeds of jack pine. For. Sci. 6, 194-199.

Booysen, P.de V., and N.M. Tainton (eds.) 1984. Ecological effects of fire in South African ecosystems. Ecological Studies 48, Springer-Verlag, Berlin 426 pp.

Buzykin, A.I., and E.P. Popova. 1978. Influence of fires on forest phytocenoses and properties of soil. In: Productivity of pine forests, pp. 5-44. Nauka, Moscow <in Russian>.

Cayford, J.H., and D.J. McRae. 1983. The ecological role of fire in jack pine forests. In: The role of fire in northern circumpolar ecosystems (R.W.Wein and D.A.MacLean, eds.), 183-199. SCOPE-18, John Wiley, Chichester, 322 pp.

Chandler, C., P. Cheney, P. Thomas, L. Trabaud, and D. Williams. 1983. Fire in forestry. Vols.III. John Wiley, New York, 450 & 298 pp.

Chizhov, B.E., and N.S. Sannikova. 1978. Fire resistance of plants in herbaceous-undershrub layer in pine forests of the Transurals. Lesovedenie No. 5, 68-76 <in Russian>.

Chudnikov, P.I. 1931. Influence of fires on regeneration of Urals forests. Selkolchozgiz, Moscow-Leningrad, 160 pp. <in Russian>.

Clements, F.E. 1910. The life history of lodgepole forests. U.S. Forest Serv. Paper No. 79, Washington, 96 pp.

Crutzen, P.J., and J.G.Goldammer (eds.) 1993. Fire in the environment: The ecological, atmospheric, and climatic importance of vegetation fires. Dahlem Workshop Reports. Environmental Sciences Research Report 13. John Wiley & Sons, Chichester, 400 p.

FIRESCAN Science Team. 1996. Fire in ecosystems of boreal Eurasia: The Bor Forest Island Fire Experiment, Fire Research Campaign Asia-North (FIRESCAN). In: Biomass burning and global change (J.S.Levine, ed.). The MIT Press, Cambridge, MA (in press).

Fox, J.F. 1983. Post-fire succession of small-mammal and bird communities. In: The role of fire in northern circumpolar ecosystems (R.W.Wein and D.A.MacLean, eds.), 155-180. SCOPE-18, Wiley, Chichester, 322 pp.

Furyaev, V.V., and D.M Kireev. 1979. Study of post-fire dynamics of forests on a landscape basis. Nauka, Novosibirsk, 160 pp. <in Russian>.

Gill, A.M. 1981. Adaptive responses of Australian vascular plant species to fires. In: Fire and the Australian biota (A.M.Gill, R.H.Groves, and I.R.Noble, eds.), 243-272. Austral. Acad. Sci., Canberra, 582 pp.

Goldammer, J.G. 1979. Der Einsatz von kontrolliertem Feuer im Forstschutz. Allg. Forst- u. J. Ztg. 150, 41-44.

Goldammer, J.G. 1983. Sicherung des südbrasilianischen Kiefernanbaues durch kontrolliertes Brennen. Hochschul-Verlag Freiburg, Forstwiss. Bd. 4, 183 pp.

Goldammer, J.G. (ed.). 1990. Fire in the tropical biota. Ecosystem processes and global challenges. Ecological Studies 84. Berlin-New York, Springer-Verlag, 497 pp.

Goldammer, J.G. 1993. Feuer in Waldökosystemen der Tropen und Subtropen. Birkhäuser-Verlag, Basel-Boston, 251 pp.

Goldammer, J.G., and S.R.Peñafiel. 1990. Fire in the pine-grassland biomes of tropical and subtropcal Asia. In: Fire in the tropical biota. Ecosystem processes and global challenges (J.G.Goldammer, ed.), 45-62. Ecological Studies 84, Springer-Verlag, Berlin-Heidelberg-New York, 497 pp.

Goldammer, J.G., and M.J. Jenkins (eds.) 1990. Fire in ecosystem dynamics. Mediterranean and northern perspectives. SPB Academic Publishing, The Hague, 199 pp.

Goldammer, J.G., and X. Di. 1990. Fire and Forest development in the Daxingangling montane-boreal coniferous forest, Heilongjiang, Northeast China - a preliminary model. In: Fire in ecosystem dynamics: Mediterranean and northern perspectives (J.G.Goldammer and M.J.Jenkins, eds.), 175-184. SPB Academic Publishing, The Hague, 199 pp.

Gorbachev, V.N., V.K. Dmitrienko, E.P. Popova, E.P., et al. 1982. Soil-ecological investigations in forest biogeocenoses. Nauka, Novosibirsk, 185 pp. <in Russian>.

Gwynar, L.C. 1978. Recent history of fire and vegetation from laminated sediment of Greenleaf Lake, Algonquin Park, Ontario. Can. J. Bot. 56 (1), 10-21.

Heinselman, M.L. 1973. Fire in virgin forests of the Boundary Water Canoe Area, Minnesota. Quat. Res. 3, 329-382.

Hesselman, H. 1916. Om vara skogforyngringsatgardens inverkan pa salpeterbildningen i marken och dess betydelse for barrskogens foryngring. Medd. Stat. Skogsforsoksanst. Vol.2 (13), 923-1076.

Isaev, A.S., and G.I. Girs. 1975. Interaction of tree and insect-xylophages. Nauka, Novosibirsk, 344 pp. <in Russian>.

Khrenova, G.S. 1963. Influence of fire on the microflora of forest soils in Pripyshminsk bors of Transurals. In: Soils and hydrological regime in forest of Urals, Proc. Inst. Biol., Ural Branch, Acad. Sci. USSR, Sverdlovsk, Vol. 36, 151-163 <in Russian>.

Kimmins, J.P. 1987. Forest ecology. Macmillan, New York, 531 pp.

Klötzli, F. 1975. Ökologische Besonderheiten *Pinus*-reicher Waldgesellschaften. Schweiz. Z. Forstw. 126, 672-711.

Kolesnikov, B.P., N.S. Sannikova, and S.N. Sannikov. 1973. Effect of surface fire on the structure and regeneration of tree species in bilberry and cowberry-bilberry pine forests. In: Burning and fires in forest. Sib. Div. Acad. Sci. USSR, 301-321 <in Russian>.

Kolström; T., and S. Kellomäki. 1993. Tree survival in wildfires. Silva Fennica 27, 277-281.

Komarek, E.V. 1969. Fire and animal behaviour. In: Proc. Ann. Tall Timbers Fire Ecol. Conf. 9, 161-241. Tall Timbers Research Station, Tallahassee, Florida.

Komarova, T.A. 1986. Seed regeneration of plants on the latest burns (Forests of South Sikhote-Alin). Far-Eastern Sci. Centre of Acad.Sci USSR, Vladivostok, 224 pp. <in Russian>.

Korchagin, A.A. 1954. Effect of fires on the vegetation and its regeneration on after fire in European North. Proc. Biol. Inst. Acad. Sci. USSR, Geobot. Ser. 3, Vol. 9, 75-148. <in Russian>.

Kozlowski, T.T., and C.E. Ahlgren (eds.) 1974. Fire and ecosystems. Acad. Press, New York, 542 pp.

Kuhnholtz-Lordat, G. 1939. La terre incendiée. Edit. Mais. Carrée, Nimes, 361 pp.

Kuleshova, L.V. 1981. Ecological and zoogeographical aspects of influence of fire on the forest birds and mammals. Zool. J. 40 (10), 1542-1553 <in Russian>.

Le Houérou, H.N. 1974. Fire and vegetation in the Mediterranean Basin. In: Proc. Ann. Tall Timbers Fire Ecol. Conf. 13, 237-277. Tall Timbers Research Station, Tallahassee, Florida.

Lutz, H.I. 1956. Ecological effects of forest fires in the interior of Alaska. U.S.D.A. Tech. Bull. 1133, 121 pp., Washington.

Malychev, L.I. 1957. Influence of fires on the forests of Northern Baikal. Proc. East. Sib. Div. Acad. Sci. USSR, Biol. Ser., 43-53 <in Russian>.

Melekhov, I.S. 1948. Effect of fires on forest. Goslestekhisdat, Moscow-Leningrad, 127 pp. <in Russian>.

Molokhov, I.S. 1965. Forest pyrology and its problems. In: Modern problems of fire prevention and suppression. Lesn. Prom., Moscow, 5-25 <in Russian>.

Molchanov, A.A., and I.F. Preobrazhenski. 1957. Forests and forestry in Arkhangelsk district. IZD, AN SSSR, Moskow, 238 pp. <in Russian>.

Moreno, J.M., and W.C. Oechel. 1994. The role of fire in Mediterranean-type ecosystems. Ecological Studies 107, Springer-Verlag, Berlin, 201 pp.

Mount, A.B. 1969. Eucalypt ecology as related to fire. Proc. Ann. Tall. Timbers Fire Ecol. Conf. 9, 75-109. Tall Timbers Research Station, Tallahassee, Florida.

Müller, K.M. 1929. Aufbau, Wuchs und Verjüngung der südosteuropäischen Urwälder. Schaper, Hannover, 302 pp.

Mutch, R.W. 1970. Wildfires and ecosystems - a hypothesis. Ecology 51, 1047-1051.

Naveh, Z. 1990. Fire in the Mediterranean - a landscape ecological perspective. In: Fire and ecosystem dynamics. Mediterranean and northern perspectives (J.G. Goldammer and M.J. Jenkins, eds.), 1-20. SPB Academic Publishing, The Hague, 199 pp.

Novgorodov, V.D., L.G. Smolnikova, and A.I. Zakharov. 1982. Method of exposure of fire danger on the locality. VNIIGPE, author's certif. USSR, cl. A 62 C 3 02, N 902763, Bull. No. 5., Moscow <in Russian>.

Pobedinski, A.V. 1965. Pine forests of Middle Siberia and Trans-Baikal. Nauka, Moscow 268 pp. <in Russian>.

Popova, E.P. 1979. On the duration of pyrogenical effect on the forest soil properties. In: Burning and fires in forest, 110-116. Inst. Forest and Wood., Sib. Div. Acad. Sci. USSR, Krasnoyarsk <in Russian>.

Pushkina, N.M. 1960. Natural regeneration of vegetation on burned forests. Proc. State National Park, Lappland, No. 4, 5-125 <in Russian>.

Pyavchenko, N.I. 1952. Causes of soil swamping on forest burns. LesnoeKhozyaistvo, No.12, 39-40 <in Russian>.

Pyavchenko, N.I. 1958. Peat bogs of the Russian forest-steppe. Inst. Forest, Acad. Sci. USSR, Moscow, 191 pp. <in Russian>.

Pyne, S.J. 1995. World fire. Henry Holt, New York, 379 pp.

Rabotnov, T.A. 1978. Significance of pyrogenic factors in formation of vegetation cover. Bot. J. 63 (11), 1605-1611 <in Russian>.

Rego, F., and E. Rigolot. 1990. Heat transfer through bark - a simple predictive model. In: Fire in ecosystem dynamics. Mediterranean and northern perspectives (J.G. Goldammer and M.J. Jenkins, eds.), 157-162. SPB Academic Publishing, The Hague, 199 pp.

Remezov, N.P. 1941. Ammonification and nitrification in forest soils. In: Investigations about forest soil science, Vol.1, 43-57. All-Union Research Inst. Forestry, Pushkino <in Russian>.

Rowe, J.S., and G.W. Scotter. 1973. Fire in the boreal forest. Quat. Res. 3, 444-464.

Sannikov, S.N. 1965. Ecological characteristic features of main types of the microenvironment of natural regeneration of pine on clearcuts. Proc. Inst. of Biol., Ural Branch, Acad. Sci. USSR No. 43, 231-242. Sverdlovsk <in Russian>.

Sannikov, S.N. 1973. Forest fires as evolutionary and ecological factor of regeneration of pine populations in Transurals. In: Burning and fires in forest, 236-277. Inst. Forest and Wood, Sib. Div. Acad. Sci. USSR, Krasnoyarsk <in Russian>.

Sannikov, S.N. 1976. Age biology of common pine in Transurals. In: Restoration and age dynamics of forests on the Ural and in Transurals. Proc. Inst. Biol., Ural Branch, Acad. Sci. USSR, No. 101, 124-165. Sverdlovsk <in Russian>.

Sannikov, S.N. 1981. Forest fires as a factor of the transformation of the structure, regeneration, and evolution of biogeocenoses. Ekologiya No. 6, 23-33 <in Russian>.

Sannikov, S.N. 1983. Cyclically erosional-pyrogenic theory of the natural regeneration of common pine. Ekologiya No.1, 10-20 <in Russian>.

Sannikov, S.N. 1984. Evolutionary pyroecology: problems, principles, hypotheses. Burning and fires in forest. Thes. of Rep. Interrepubl. Conf., 35-37. Inst. Forest and Wood, Sib. Div. Acad. Sci. USSR, Krasnoyarsk <in Russian>.

Sannikov, S.N. 1985. Hypothesis of pulsed pyrogenic stability of pine forests. Ekologiya, No.2, 13-20. (in Russian>.

Sannikov, S.N. 1991. Impulsed stability and microevolution of populations. In: Ecology of populations. Nauka, Moscow <in Russian>.

Sannikov, S.N. 1992. Ecology and geography of natural regeneration of common pine. Nauka, Moscow, 264 pp. <in Russian>.

Sannikov S.N. 1994. Evolutionary pyroecology and pyrogeography of the natural regeneration of Scotch pine. In: Proc. 2nd Int. Conf. Forest Fire Research, Vol.II, pp.961-968. Coimbra.

Sannikov, S.N., and N.S. Sannikova. 1985. Ecology of pine regeneration under forest canopy. Nauka, Moscow, 149 pp. <in Russian>.

Sannikova, N.S. 1977. The surface fire as a factor of appearance, survival and growth of pine seedlings. In: Detection and analysis of forest fires, 110-128. Sib. Div. of Acad. Sci. USSR, Krasnoyarsk <in Russian>.

Sapozhnikov, A.P. 1976. The role of fire in formation of forest soils. Ekologiya, No.1, 42-46 <in Russian>.

Sarvas, R. 1937. Über natürliche Bewaldung der Waldbrandflächen: Eine waldbiologische Untersuchung auf trockenen Heideböden Nord-Finnlands. Acta Forest. Fenn. 46 (1), 73-76.

Sherbakov, I.P., O.F. Zabelin, and B.A. Karpel. 1979. Forest fires in Yakutya and their influence on the forest nature. Nauka, Novosibirsk, 224 pp. <in Russian>.

Scotter, G.W. 1972. Fire as an ecological factor in boreal forest ecosystems of Canada. In: Fire in the environment, Symp. Proc., 15-24. USDA For. Serv. Publ. FS-276, Denver, Colorado.

Shanin, S.S. 1965. The structure of pine and larch tree stands in Siberia. Lesnaja promyshlennost, Moscow, 106 pp. <in Russian>.

Sheshukov, M.A., Savchenko, A.P., and Peshkov, V.V. 1992. Forest fires and its control on the North of Far-East. Dal'NIILKH, Khabarovsk, 96 pp. <in Russian>.

Sirén, G. 1955. The development of spruce forest on raw humus sites in Northern Finnland and its ecology. Acta Forest. Fenn. 62, 1-408.

Sirén, G. 1974. Some remarks on fire ecology in Finnish forestry. Proc. Ann. Tall. Timbers Fire Ecol. Conf. 13, 191-209. Tall Timbers Research Station, Tallahassee, Florida.

Smirnov, A.V. 1970. Change of components of forest vegetation of South Central Siberia under impacts of anthropogenic factors. Author's abstract of doctoral dissertation. Krasnoyarsk, 48 pp. <in Russian>.

Sushkina, N.N. 1931. A contribution to microbiology of forest soils in connection with the effect of fire on them. In: Studies in forestry, 137-169. Selkhozgiz, Moscow-Leningrad <in Russian>.

Swain, A.M. 1973. A history of fire and vegetation in Northeastern Minnesota as recorded in lake sediments. Quat. Res. 3, 383-396.

Timofeev-Resovskii, N.V., N.N. Vorontsov, and A.V. Yablokov. 1977. Short outline of the theory of evolution. Nauka, Moscow, 297 pp. <in Russian>.

Tkachenko, M.E. 1939. General silviculture. Goslestekhizdat, Leningrad, 745 pp. <in Russian>.

Tolonen, K. 1983. The post-glacial fire records. In: The role of fire in northern circumpolar ecosystems. (R.W.Wein and D.A.MacLean, eds.), 21-44. SCOPE-18, Wiley, Chichester, 322 pp.

Trabaud, L., and J. Lepart. 1980. Diversity and stability in garrigue ecosystems after fire. Vegetatio 43, 49-57.

Tyurin, A.V. 1925. Fundamentals of silviculture in pine forests. Novaya Derevnya, Moscow, 112 pp. <in Russian>.

Tyrtikov, A.P. 1979. The dynamics of vegetation cover and development of frozen relief forms. Moscow, 115 pp. <in Russian>.

Uggla, E. 1974. Fire ecology in Swedish forests. Proc. Ann. Tall Timbers Fire Ecol. Conf. 13, 171-190. Tall Timbers Research Station, Tallahassee, Florida.

Vakurov, A.D. 1975. Forests fires in the North. Nauka, Moscow, 98 pp. <in Russian>.

van Wilgen, B.W., D.M.Richardson, F.J. Kruger, and H.J. van Hensbergen (eds.) 1992. Fire in South African mountain fynbos. Ecological Studies 93, Springer-Verlag, Berlin 325 pp.

Viro, P.J. 1974. Effect of forest fire on soil. In: Fire and ecosystems (T.T.Kozlowski and C.E.Ahlgren, eds.), 7-45. Academic Press, New York, 542 pp.

Wein, R.W., and MacLean, D.A. 1983.An overview of fire in northern ecosystems. In: The role of fire in northern circumpolar ecosystems (R.W.Wein and D.A.MacLean, eds.), 1-21. SCOPE-18, Wiley, Chichester, 322 pp.

Zackrisson, O. 1977. Influence of forest fires on the North Swedish boreal forest. Oikos 29, 22-32.

J.G.Goldammer and V.V.Furyaev (eds.), Fire in Ecosystems of Boreal Eurasia, 168-185.

Pyrological Regimes and Dynamics of the Southern Taiga Forests in Siberia

V.V. Furyaev [1]

1. Introduction

Investigations of the role of fire in the taiga forest formation are required for estimating and predicting forest resources. Our system of complex study of the role of fire in the forest formation process involves the determination of fire recurrence and the rate of forest fires in natural landscape complexes (NLC) of different complexity ranks (Furyaev 1988). In this system, combinations of factors are analyzed that are responsible for the pyrological regime of a facies, geographical forest blocks (GFB), geographical localities (GL), and landscape. The mean fire recurrence (= mean fire return interval [MFRI]) within a NLC over the past three centuries is considered as the major index of the pyroregime. Fire recurrence is an objective indicator of multi-year drought occurrence, particular NLC ecological regimes, intensity range of stand-replacement fires, as well as repeated fire occurrence depending on vegetation succession and the availability of fire sources. Fire-return intervals correspond to the mean frequency of extreme fire seasons in different NLCs, which is very important for the understanding, interpretation, and prediction of the forest formation process. Average multi-year rates of forest fires (i.e. area covered by fires) is used as a subsidiary index for characterizing a pyroregime.

In the taiga forest zone, fires are an important ecological factor affecting large forest areas every year (see statistical database provided by Korovin, this volume). In every site that was subject to a high-intensity surface or crown fire, climax coniferous forest communities are gradually formed, often through species replacement or age succession. Repeated fires have strong bearings on forest formation by often changing the trend of them. Regular fire influence over two to three centuries may result in a complete change of the general forest type. However, only few researchers have attempted to study, estimate, and predict long-term post-fire forest formation (Kolesnikov and Smolonogov 1960; Reimers and Malyshev 1963; Popov 1967; Komin 1982). The publication of basic works of Clements (1905, 1916) stimulated Canadian and USA scientists to investigate forest formation in relation to fire.

---

[1] V.N. Sukachev Institute for Forestry and Timber, Siberian Branch, Russian Academy of Sciences, 660036 Krasnoyarsk, Russian Federation

2. Study Area and Methods

Investigations were conducted in landscape sample sites established in the southern taiga forest subzone of the West-Siberian Plain, the Yenisei Mountain Ridge, and the Central-Siberian Plateau. The study area ranges from 80-100°E to 54-62°N. This paper presents the analysis of landscape, vegetation, and post-fire forest dynamic maps, which were created for the study area using aero-space imagery. The objects of investigation are the dark coniferous and mixed larch-pine stands that have been regularly affected by fires over the past three centuries. The specificity and complexity of the addressed problematic nature of the research required some new approaches and methods of investigation, apart from currently used ones.

For example, the system of complex investigation of the role of fire in forest formation assumes that consecutive observations be conducted to reveal the complex temporal and spatial interactions between forest formation and fire. The system provides estimates of fire activity in the past, present, and future and applies to facies, GFB, GL, and landscapes, i.e. to long existing NLCs of different taxonomic ranks. Estimation of fires in the past is actually a retrospective identification of past fires. Fire recurrence calculations and mapping are considered when estimating and predicting the role of fires in forest formation.

Fire dating was performed using Melekhov's (1948) methodology. Model trees with fire scars were selected in sample sites established within landscape and topo-ecological profiles. Fire frequency was determined for a 270-year period, from 1700 to 1970. Average multi-year forest rate of burning was determined by dividing the total area of post-fire secondary communities by 200 years. A 200-year time period was assumed to be the average period needed for post-fire regeneration and formation of dark conifer forests.

The complex investigation system is also aimed at constructing dynamic community series formed within a NLC after fires. We interpret this aspect of investigation as an integral estimation of fire consequences reflected in specific features of the post-fire forest formation process.

Principles and methods of the landscape-based study of fire influence on forest formation using aerospace images were developed to fit forest analysis methodology which considers forest areas as NLCs of different taxonomic ranks composing a system of morphological units (Solntzev 1949). Applying this methodology and the methods of ecological analysis ensuing from it, fires of any intensity occurring in any NLC were found to result in a set of inter-related events (Furyaev and Kireyev 1979; Furyaev 1988). Partial or complete consumption of overstory, understory, regrowth, living surface cover, and forest floor organic layers by fire leads to an increase of insolation and precipitation reaching the soil. Both soil temperature regime and surface air humidity change. Salt and small particle penetration into the soil is promoted, and soil formation processes change its trend. Surface and in-soil drainage redistribution is observed, as well as changes in ground water level. All these factors eventually transform NLC ecological regime parameters. Post-fire forest regeneration is treated as a result of changes in all inter-related NLC components, leading to the formation of successive biogeocoenoses, which in turn form dynamic forest regeneration series (Popov 1967; Furyaev and Kireyev 1979).

The methodology allows a better understanding of post-fire vegetation succession trends and changes occurring in NLCs after fires of various intensity. Our experience so far shows that systematized investigation of the role of fire, with respect to long-existing NLCs, is quite feasible since a possibility is provided for predicting forest formation processes and the retrospesctive analysis of its relationship with past fire recurrence.

3. Results and Discussion

3.1 Fire Intensity

Our fire intensity studies show that surface fires in pine facies, which are most widespread in the study area, and well known for the highest fire danger, result for example in a maximum energy release of 27.300, 39.900, and 50.400 $kJ \cdot m^{-2}$ respectively as well as in the complete consumption of surface fuels. The amount of heat released per unit area is controlled by the available fuel, whose load varies considerably with forest type; forest types are considered to be dynamic regeneration stages within the forest facies. A relationship was established between forest-type related fuel load and stand age, canopy closure, and fire interval. Fuel load measurements and surface fire intensity calculations revealed a feature specific for pine stands of the study area. In this area, recurrent past fires periodically reduced the fuel load thereby preventing, in many cases, large fires to result in complete stand destruction. In dark conifer forests, high fuel load and low fire resistance of tree species are the two factors accounting for complete stand consumption by fire followed by community replacement (Furyaev 1970a). Of all ecological regime factors within facies, ground water level has the strongest influence on fire occurrence, intensity, and recurrence in plain forests of Western Siberia (Furyaev 1970b).

3.2 Fire Recurrence

Potential fire recurrence in an elementary NLC facies is determined by its ecological regime, intensity of initial fires, and load dynamics of both living and dead surface fuels. Potential fire recurrence is normally higher than actual fire recurrence, the latter being dependent on the availability of fire sources and relative facies location within the structure of more complex NLCs.

Over the past 270 years, fires occurred in most facies, but their recurrence varied (Tab.1). The highest fire recurrence was found for well-drained facies, whereas the lowest fire recurrence is characteristic of facies comprising of river flood plains. All other facies are intermediate.

In more complex NLCs, GFBs, GLs, and landscapes fire recurrence depends on the ecological regimes of less complex NLCs; their location relative to one another within the morphological structure of NLCs of higher rank, and the percentage of area they occupy. Relative location and area percentage of NLCs differing in ecological regimes determine probability of fire occurrence and behavior.

The highest fire recurrence is observed in GFBs represented by pine stands on terraces and hill slopes. The lowest fire recurrence is common in bogged flood plains of small rivers covered by dark conifer taiga forests, and in overwet flat interfluves occupied by spruce and *Pinus sibirica*. Data on actual multi-year fire recurrence confirm its relationship with NLC ecological regimes. The highest fire recurrence is observed for NLCs whose ecological regime is characterized by low ground water level. Ecological regime factors combined with current weather conditions and availability of fire sources determine fire cycles within an NLC. Fire regimes can be reconstructed by analyzing forest fire history for decades or centuries. Fire recurrence increases consistently along a gradient between river flood plains and river terraces and decreases towards flat interfluve sites. This suggests different fire danger rates within NLCs and non-uniform fire-starting agent distributions across the area.

Tab.1. Duration of fire-return intervals in various types of facies. The ecological regime of facies is classified in accordance with Kireyev (1979a)

Facies Index	Ecological Regime of Facies *	Number of Fires over 270 Years	Fire Interval (years)		
			Mean	Min.	Max.
1.	Pine stands growing on sand under dry-to-moist site conditions: Dr0, Al0, F0, G0, A	23	11	1	88
2.	Pine stands on sand in moist sites: Dr0, Al0, F0, G1, A	20	11	1	93
3.	Pine stands on peat soils in wet sites: Dr0, Al0, F0, G2, A	4	13	1	87
4.	Bogged pine stands. Dr0, Al0, F0, G3, A	3	22	4	51
6.	Larch subors** on sand underlaid by loams in dry-to moist sites: Dr0, Al0, F0, G0, 1B	3	10	2	20
7.	*P.sibirica* subors on sand underlaid by loams in dry-to-moist sites: DR0, A10,F0, G0, B	3	26	6	45
13.	Fir stands on surface loams in dry-to-moist sites: Dr0, Al0,F0, G0, 1D	8	11	5	16
14.	Fir stands with spruce on surface loams in moist sites: Dr0, Al0, F0, G1,1D	4	33	5	103
15.	Spruce stands with *P.sibirica* on peat-loamy soil in wet sites: Dr0-1, Al0, F0, G2, 1D	5	11	3	20
16.	Bogged *P.sibirica* stands on peat-loamy soil: Dr0-1, Al0, F0, G3, 1D	3	19	6	27
17.	Pine stands occupying old river beds on sand (considerable peat layer) in wet sites: Dr1, Al0, F0, G2, 1B	3	31	22	41
18.	Bogged pine stands occupying old river beds (considerable peat layer) in overwetted sites: Dr1, Al0, F0, G2, 1B	3	15	4	26
19.	Bogged *P.sibirica* stands in old river beds (overwetted sites): Dr1, Al0, F0, G2, 1B	3	33	24	40
22.	Spruce stands comprising flood plains growing in dry-to-moist sites: Dr1-2, Al1, F1, G0, 1D	3	26	20	17

* Dr is rate of drainage; Al is alluviality; F is flooding by melted snow; G is ground water level; A,B,C,or D indicate soil richness; 0,1,2,or 3 indicate low, moderate, and high rate of strength of these ecological factors.

** Subor is a stand on transitional, relatively poor soils (USSR classification).

Our observations support Melekhov's (1948) statements that fire behavior is closely related to relief elements. Therefore, the revealed fire recurrence pattern is apparently of a general character.

Fire recurrence is directly related to the fire-return interval (FRI). FRI varies widely with ecological regime and complexity rank of NLC. It increases gradually proceeding from NLCs with dry and dry-to-moist site conditions to those with overwet site conditions (Tab.1).

Forest floor organic fuel load is directly dependent on FRI. Fire recurrence, FRI, and ecological regime, including forest floor organic fuel load and state, are the major factors determining the post-fire forest formation process and should be considered for prediction and modeling.

Tab.2. Fire recurrence and rate of forest fires in plain landscapes of western and central Ob-Yenisei interfluve. Landscape names are classified in accordance with Kireyev (1979a).

Landscape Index	Landscape Name and Description	Fire Recurrence over 270 Years	Rate of Forest Fires (%)
I	Highly elevated, slightly-bogged lacustrine-alluvial glacial Kas-Yenisei Plain represented by eutrophic bogs, southern taiga spruce and pine stands	20	0.32
II	Elevated lacustrine-alluvial loess glacial Ket Plain represented by oligotrophic moors and mesotrophic bogs and by southern taiga *P.sibirica* and *P.sylvestris* stands	110	0.25
III	Highly elevated well-drained lacustrine-alluvial loess Ulu-Yul-Chulym Plain comprised by southern taiga spruce, fir and *P.sibirica* stands	70	0.41
IV	Lacustrine-alluvial loess and alluvial, heavily bogged Lower-Chulym Plain covered by eutrophic swamps, mesotrophic bogs, and spruce and *P.sibirica* stands	60	0.24
V	Highly elevated lacustrine-alluvial loess and alluvial-deluvial well-drained Tom-Yaya Plain dominated by southern taiga spruce and fir stands	50	0.35
VI	Elevated well-drained lacustrine-alluvial loess Chet Plain covered by southern taiga spruce and fir stands	52	0.34
VII	Slightly elevated Chulym-Ket-Ob Flood Plains represented by eutrophic swamps, meadows, and *Salix* and *Populus* stands	185	0.31
Landscape Class Average		79	0.31

3.3 Multi-year Rate of Forest Fires

Aero-space imagery analysis supported by ground data from sample sites provided much information on past fire areas in plain, mountain, and plateau-like landscapes of the study area. The rate of forest fires was found to depend on NLC morphological structures (Fig.1). It was established that NLC vegetation patterns, which were shaped by the ecological regime and previous fires, determine the area of the fires to follow.

In plain landscapes of the western and central Ob-Yenisei interfluve, the rate of forest fires varies with landscape, ranging from 0.15% to 0.41% (multi-year average: 0.31%; Tab.2).

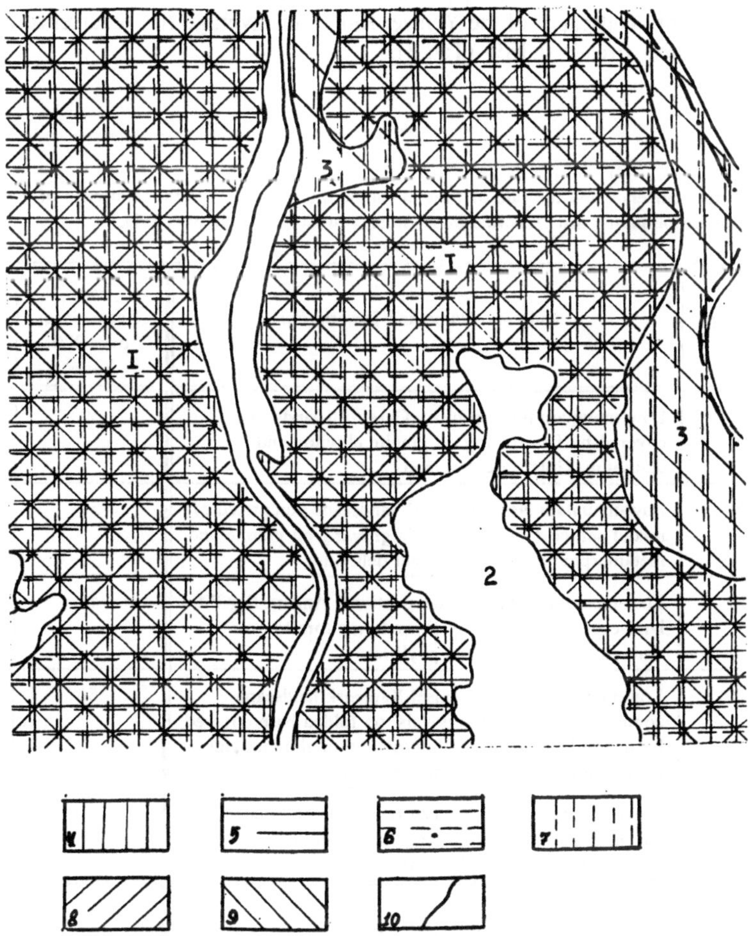

Fig.1. Fragment of a map showing forest types and fire history within GLs. Overwet terraces characterized by soft-slope-hummock relief occupied by spruce and bogged *P.sibirica* stands. Shallow bogs are common around terraces. GFB types: (1) soft hummock terraces covered by spruce and bogged *P.sibirica* stands; (2) mesotrophic bogs in isolated flat-bottom depressions; (3) terrace slopes comprised of fir and spruce stands. Fire dates and area boundaries: (4) 1870, 1915; (5) 1886; (6) 1920; (7) 1933; (8) 1946; (9) 1956; (10) GFB boundary.

The highest fire recurrence is the common elevated Kas-Yenisei Plain, whereas the lowest fire recurrence is characteristic of the flood plain-dominated slightly elevated Chulym-Ket Plain. Weight-average fire recurrence suggests that, in GLs prevailing in the Kas-Yenisei Plain landscape, fires recurred, on the average, every 20 years over several past centuries, while in the Chulym-Ket-Ob Plain they repeated every 185 years. As for the rate of fires, the highly elevated Ulu-Yul-Chulym Plain presents an interesting case. This plain is dominated by well-drained sites occupied by fir stands killed by Siberian gipsy moth (*Dendrolimus superans sibiricus* Tschetv.) outbreaks followed by fires. The landscapes of elevated Kas-Yenisei, Tom-Yaia, and Chet plains have an almost similar, relatively high rate of forest fires. Each of these is remarkable for the domination of well-drained NLCs experiencing periodical Siberian gipsy moth outbreaks.

For mountain landscapes of the Yenisei Ridge, the average fire recurrence is 50 years and the rate of forest fires is 0.18% (Tab.3).

Tab.3. Fire recurrence and the rate of fires in southern taiga forests in the landscape of Yenisei Mountain Ridge. Landscape names are classified in accordance with Kalashnikov (1987).

Landscape Index	Landscape Name and Description	Fire Recurrence over 270 Years	Rate of Forest Fires (%)
5	Low-mountain upwarp-block Garev-Pit landscape covered by fir growing on granite eluvium and deluvium. Golets are common.	50	0.13
6	Low-mountain Novoerudninsk landscape comprised of mixed spruce-*P.sibirica* stands growing on granite and metamorphic slate eluvium and deluvium	50	0.39
10	Low-mountain structural denudational Angara-Big-Pit landscape represented by fir and pine forests on eluvium and deluvium of phyllite	50	0.22
11	Low-mountain South-Yenisei-Kamensk landscape comprised of soft-slope uovals* on folded base covered by mixed larch-pine stands growing on eluvium and deluvium of sandy clay and limestone	50	0.13
30	Low-mountain South Yenisei landscape dominated by fir and spruce forests growing on eluvium and deluvium of metamorphic slates.	50	0.18
Average for the Landscape Class		50	0.18

* Uoval is a stretched upland site up to 200 m high with soft slopes, a flat top, and without a clear foot.

It is noteworthy that fire recurrence appeared to be the same for all the landscapes of this class. In the Novoe Rudnino landscape, however, the rate of forest fires is considerably higher. High rates of forest fires in this landscape can be attributed to the past prevalence of mixed pine-larch low brush-moss forest types which have already been managed for a very long time. As a result of regular cutting and fires, these forests were replaced by secondary birch and aspen-herb-moss communities.

The highest fire recurrence is characteristic of the plateau-like landscapes of the Central-Siberian Plateau, being on the average 47 years (Tab.4).

Tab.4. Fire recurrence and the rate of fires in southern taiga forests of plateau-like landscapes of the Central-Siberian Plateau. Landscape names are classified in accordance with Kalashnikov (1987).

Landscape Index	Landscape Name and Description	Fire Recurrence over 270 Years	Rate of Forest Burning (%)
17	Plateau-like soft-hilled Angara-Yensei landscape dominated by pine and fir stands supported by loamy sand and loam deposits	25	0.25
12	Plateau-like Kozhim-Sygar landscape dominated by larch and pine forests growing on eluvium and deluvium of traps	57	0.02
25	Plateau-like Kova-Chuna landscape represented by mixed spruce-fir, pine, and larch stands on eluvium and deluvium of traps	50	0.16
24	Plateau-like structural Central Angara landscape comprised of pine and fir stands growing on eluvium and deluvium of loamy sand and limestone	50	0.19
32	Plateau-like Chuna-Birusa landscape dominated by pine and mixed fir-spruce forests supported by eluvium and deluvium of traps	50	0.25
Average for the Landscape Class		47	0.16

3.4 NLC Classification by Pyrological Regime

This classification is based on data of actual multi-year fire recurrence. The latter is determined by NLC ecological regime, weather, availability of fire sources, and post-fire vegetation succession patterns. The dependence of the pyrological regime of a facies on its ecological regime is presented in Figure 2.

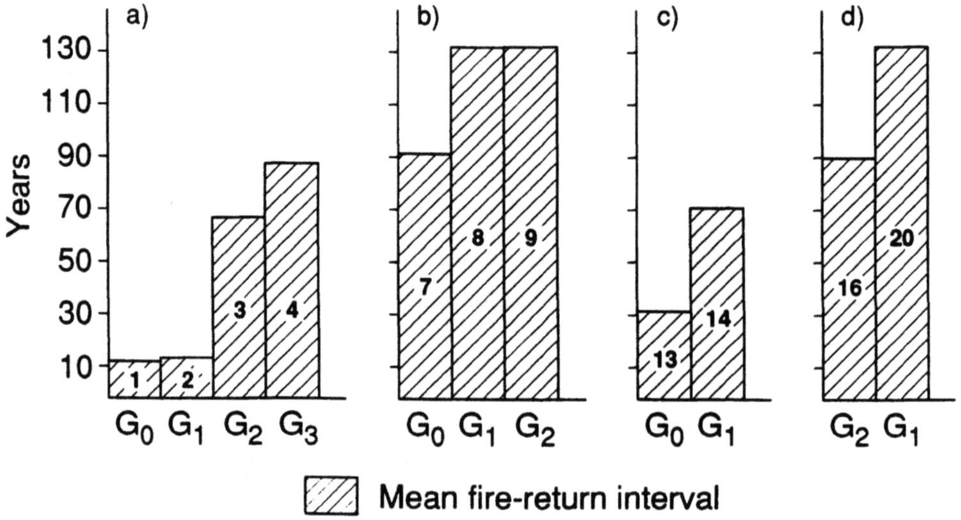

Fig.2. Fire recurrence for various facies:

a - 1-4: Pine stands growing on sands under dry-to-moist, moist, wet, and overwet site conditions
b - 7-9: *P.sibirica* subors on sand underlaid by loams growing under dry-to-moist, wet, and overwet site conditions
c - 13+14: Fir stands on surface loams in dry-to-moist and moist sites
d - 16-20: Spruce stands (with *P.sibirica*) growing on loams in wet and bogged sites

G0 and G1-3 are ground water level indices

As is obvious from the figure, in facies comprised of pine (*P.sylvestris*) a decrease in ground water depth of location from G0 to G3 is directly related to decreasing fire recurrence and increasing fire interval, provided all the other ecological regime parameters are the same. A similar relationship is observed for facies occupied by *Pinus sibirica* stands growing on sand underlaid by loams and for those represented by fir stands on surface loams. In facies comprised by mixed fir-Siberian pine forests and pure *P.sibirica* stands growing on loamy soil, fire recurrence decreases consistently depending on several eco-regime parameters (such as alluviality, flooding, and ground water level).

All facies, GFB, and GL types fall within four classes of the pyrological regime (I - high fire recurrence; II - moderate fire recurrence; III - low fire recurrence; and IV - extremely low fire recurrence). NLCs where fires recurred every 5 to 30, 31 to 70, and 70 to 100 years fall within the pyroregime classes I, II, and III, respectively, while those where fires recurred in more than 100 years are included into pyroregime class IV.

Spatial ecological regime patterns of NLCs can be well demonstrated by GFBs comprising the Kas-Yenisei erosion plain (Fig.3). The data presented show that the lowest fire recurrence is characteristic of GFBs of young and old flood plains. Fire recurrence increases noticeably in GFBs located on pine-dominated slopes and terraces. As one proceeds from terraces to the watershed, fire recurrence declines again. We have estimated pyrological regimes of all the GLs and landscapes identified by Kireyev (1979a, 1979b, 1986) and Kalashnikov (1987) in the area of interest. These estimates were based upon constructed pyroregime maps. Map analysis shows that more than 50% of GLs and landscapes are characterized by moderate fire recurrence, 13% of GLs and 25% of landscapes have low fire recurrence, whereas extremely low fire recurrence is observed in 21% of GLs and 9% of landscapes. Each landscape has a specific pyroregime that is closely related to weather and the occurrence of extreme dry seasons characterized by large fires.

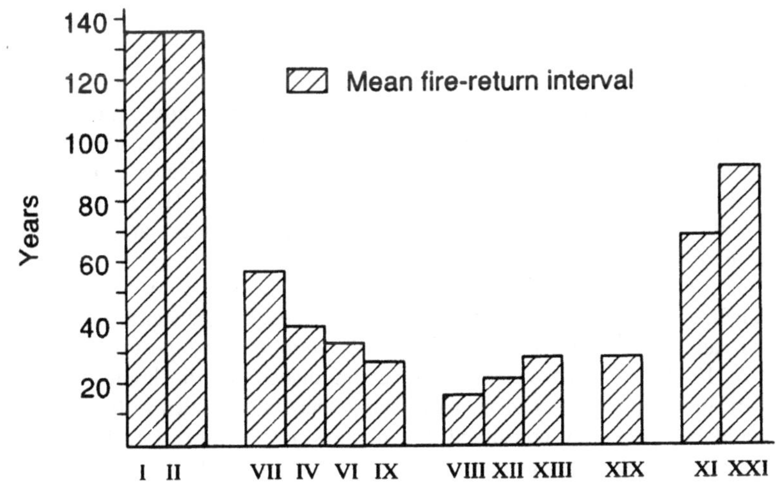

Fig.3. Fire recurrence in various Geographical Forest Blocks (GFBs):

I-II:	GFBs of young and old flood plains
VII, IV, VI, IX:	GFBs represented by dark conifer forest occupying soft-hummock terrace slopes
VIII, XII, XIII:	GFBs represented by pine stands on slopes and flat surface of terraces
XIX:	GFBs represented by fir and spruce stands occupying crest-like flat interfluves
XI, XXI:	GFBs represented by pine and spruce stands growing in flat interfluves and soft slopes, respectively.

Pyroregime class I (high fire recurrence) covers NLCs where forests rapidly attain high flammability. These NLCs are characterized by the absence of alluviality and flooding over the whole fire season, low ground water level, occurrence of surface fires of relatively low intensity, low tree mortality, and the absence of stand replacement.

Pyroregime class II (moderate fire recurrence) is comprised of NLCs whose ecological regimes are less favorable for fire occurrence and spread. These are usually limited to moist sites where ground water level is higher and forests attain high flammability at a lower rate. In these NLCs, fires are noticeably less frequent, but being of high intensity they lead to complete overstory destruction followed by stand replacement. The NLCs included into this pyroregime class, fire recurrence increases in burned areas at initial stages of forest regeneration and decreases during the following stages.

Pyroregime class III (low fire recurrence) is assigned to NLCs whose ecological regime is characterized by a periodical increase in ground water level, with forests rarely attaining high flammability. In these, steady humus or peat fires are common resulting in complete stand destruction followed by stand replacement.

Pyroregime class IV (extremely low fire recurrence) includes NLCs characterized by low or moderate flooding and alluviality and permanently high or very high ground water level. Here, forests attain high flammability very slowly and steady humus and peat fires completely kill the overstory thereby leading to stand replacement.

3.5 Dynamic Series of Dark Conifer Forest Regeneration

In Russia, studies on post-fire forest dynamics were conducted for many geographical regions. Our landscape approach is not only of interest for the study area, but is also an attempt to identify forest formation stages within facies which are considered NLCs with similar ecological conditions. The division of the land surface into facies allows one to more accurately relate vegetation dynamics series to given NLCs within large burned areas; this in turn permits the estimation of potential post-fire ecoregime changes and the prediction of vegetation formation trends.

Dark conifer forests of the study area are represented primarily by facies of fir stands growing on surface loams under dry-to-moist site conditions, spruce stands on surface loams in moist sites, and mixed spruce-*P.sibirica* stands supported by peat soils in wet sites (Kireyev 1976, 1979b). The above facies are distinguished for the highest forest rate of burning and are the major control of post-fire stand succession scales in the study area.

When a stand is completely consumed and no new fires occur, forest regeneration in facies comprised of fir stands growing on loams under dry-to-moist site conditions involves, according to short term major tree species replacement, the following stages (Tab.5):

1 Recently burned area;
2 *Chamaenerium angustifolium* communities;
3 Aspen-mixed herb forest;
4 Aspen-small herb-sedge forest with mixed dark conifer-fir regrowth;
5 Aspen-small herb-sedge forest with the subordinate tree layer composed of dark conifers;
6 Aspen-small herb-sedge forest with the overstory partly contributed by conifers;
7 Fir-small herb-green moss (*Pleurozium schreberi*, *Helacomium splendens*, *Dicranum scoparium*, etc.) forest with birch and aspen partly contributing to the overstory;
8 Climax fir-small herb-green moss forest mixed with spruce and *P.sibirica*.

In fire-killed facies of spruce stands growing on loams in moist sites, climax spruce-small herb-green moss forest regeneration includes the following stages:

1 Recently burned area;
2 *Calamagrostis*, *Rubus idaeus*, *Spiraea*, and *Chamaenerion angistifolium* communities or their combinations.

Mixed herb-low shrub communities vary widely with site conditions and fire intensity (Furyaev 1966).

In areas burned by fires resulting in low forest floor consumption and mineral soil exposure, sedge-*Calamagrostis* communities develop. High-intensity fires promote *Chamaenerion angustifolium* and mixed *Chamaenerion angustifolium-Rubus idaeus* communities. These are succeeded by:

4 Birch-mixed herb forest with dark conifer regrowth;
5 Birch-sedge-mixed herb forest with the subordinate tree layer represented by dark conifer species;
6 Birch-sedge-small herb forest with the overstory partially contributed by dark conifers;
7 Spruce-small herb-green moss forest with a portion of birch and aspen in the overstory;
8 Climax and secondary spruce-small herb-green moss forest with a portion of *P. sibirica* in the overstory.

Regeneration series for facies comprised of spruce stands with *P.sibirica* supported by peat soils in wet sites consists of the following stages:

1 Burned area with completely burned tree layer;
2 Herb-low shrub associations dominated by *Chamaenerion angustifolium*, *Rubus idaeus*, *Calamagrostis*, and *Spiraea* communities;
3 Birch-*Equisetum*-mixed herb forest;
4 Birch-*Equisetum*-green moss forest with *P.sibirica* and spruce regrowth;
5 Birch-*Equisetum*-green moss forest with the subordinate tree layer composed of spruce, *P.sibirica*, and fir;
6 Birch-green moss forest with *P.sibirica*, fir, and spruce contributing to the overstory;
7 *P.sibirica-Equisetum*-green moss forest with a portion of birch in the overstory, and
8 Climax *P.sibirica-Equisetum*-green moss stand.

The established series can be considered a community regeneration pattern within ecologically similar NLCs. Further community development, in a non-repeated fire situation, involves age succession of climax generations presented by Falaleyev (1964).

When burned areas are large conifer seeds are not transported by the wind into the burned sites and repeated fires occur; climax community regeneration is characterized by a long-term major species succession (replacement). Climax tree species do not prevail in the overstory until the first generation life span is over.

Tab.5. Parameters of the overstory, regrowth, and understory for the dynamics of stage series of regeneration of fir stands growing on surface loams under dry-to-moist site conditions.

Species composition and major parameters	Forest Type: Fir-Small-Herb-Green Moss Forest with Spruce and *P.sibirica*							
	Stage / Forest Type							
	1	2	3	4	5	6	7	8
Fir (*Abies obovata*)	-	-	-	-	0.3	0.5	0.5	0.6
Siberian Pine (*P.sibirica*)	-	-	-	-	0.1	+	0.1	0.1
Spruce (*Picea sibirica*)	-	-	-	-	0.2	0.2	0.2	0.3
Aspen (*Populus tremula*)	-	-	0.8	0.7	0.3	0.2	+	-
Birch (*Betula verrucosa*)	-	-	0.2	0.3	0.1	0.1	0.2	-
Climax species age (yr)	-	-	30,0	60,0	80,0	120,0	160,0	180,
Mean stand height (m)	-	-	11,0	18,0	22,0	25,0	24,0	0
Site index	-	-	11,0	11,0	11,0	111,0	111,0	25,0
Canopy closure	-	-	0.6	0.7	0.8	0.8	0.8	111,
Dominant regrowth species	-	-	11,0	11,0	11,0	11,0	11,0	0
Total regrowth (x10^3 ha)	-	-	0.5	4.6	0.8	0.9	5.4	0.7
Understory density	-	-	0.2	0.2	0.1	0.2	0.1	11,0
								5.5
								0.1

4. Fire-caused Forest Disturbance in Relation to Landscape

The use of landscape maps in combination with aero-space imagery analysis allows us to estimate forest disturbances induced by fire. Disturbance is defined as a part of an NLC area occupied by post-fire (secondary) communities. Fire-caused forest disturbance actually describes the forest area which has experienced fires, including recurrent fire over the past 200 years. In most cases, fire influence was manifested through the replacement of conifers by small-leaved species or through climax species generation succession resulting in a young or middle-age conifer stand formation. The above fire-caused forest disturbance definition implies that it is of great importance for fire management and forestry since it provides general information on long term (multi-century) fire influence on forest formation.

Fire-caused forest disturbance varies widely with GL, ranging from 100% sound climax trees species (no disturbance) to 100% replacement by secondary species (complete disturbance). In 88 GLs of the study area, post-fire communities were found to dominate, whereas 222 GLs (72%) are occupied mostly by climax species. Disturbance was established to vary depending on initial climax species composition, ecological regime and structure of GFBs comprising GLs, fire recurrence, and post-fire forest dynamics direction.

In the landscapes of the Ob-Yenisei interfluve, average fire-caused forest disturbance is 66%, ranging from 30 to 84%. In the landscapes of the better drained Yenisei part of the West-Siberian Plain, disturbance ranges from 5 to 68%, with the average being 48%. For the landscapes of the Yenisei Mountain Ridge, a 35% average disturbance is observed, varying from 16 to 78%. For the plateau-type landscapes of the Central-Siberian Plateau, average fire-caused forest disturbance was established to be 33%.

Landscapes of the study area can be divided into three groups by fire-caused rate of forest disturbance. Group I includes landscapes where forests are only lightly disturbed by fire. In these, secondary post-fire communities account for less than 30% of the total forest area. Light disturbance is characteristic of depressions and plains with elevations of 50 and 50-100 m a.s.l., respectively. In these landscapes, bogs and bogged forests cover 30-60% of the area.

Landscapes where forests are moderately disturbed by fire (31-60% of the area is occupied by secondary communities) fall within Group II. These are dominated by elevated plains (100-200 m a.s.l.), plateaus, and low mountains. Bogs account for 10-30% of the area. Heavily disturbed (more than 61% of post-fire forests) landscapes included in Group III are represented by highly elevated plains (elevation more than 200 m a.s.l.) and low mountains. Bog areas do not normally exceed ten percent.

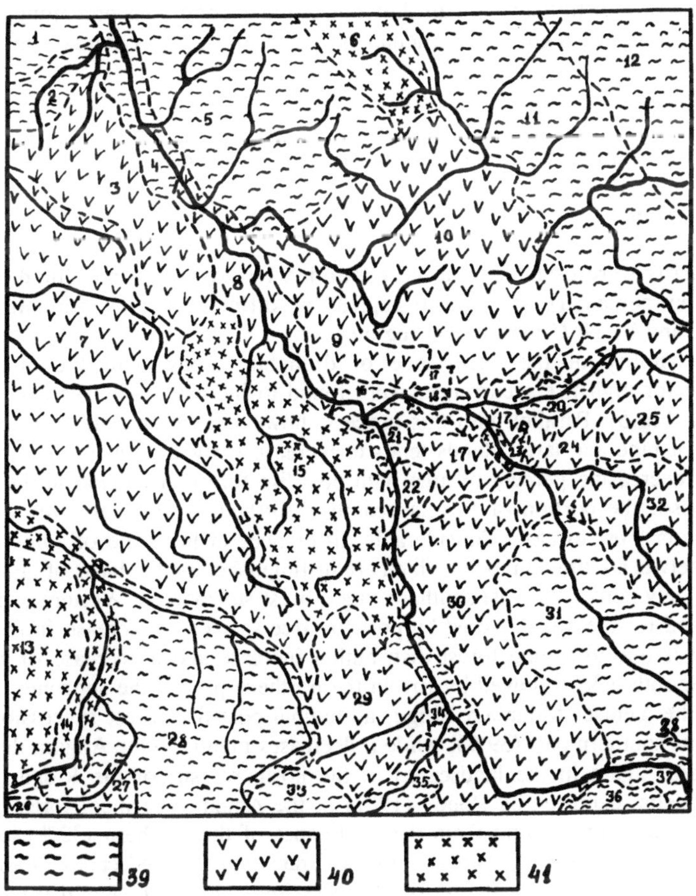

Fig.4. Fragment of a fire-caused forest disturbance map.

1-38:	Individual landscapes
39:	Landscapes with lightly disturbed forests
40:	Landscapes comprised of moderately disturbed forest
41:	Landscapes represented by heavily disturbed forest

Using this classification, a fire-caused forest disturbance map was created (Fig.4). The analysis of the map has revealed that light, moderate, and heavy forest disturbance is characteristic of 50, 37, and 13% of landscapes, respectively. It was established that Siberian gipsy moth outbreaks and the following fires are the two major factors accounting for forest disturbance in the southern taiga zone of Western Siberia. These factors, however, manifest themselves differently in different landscapes. In landscapes represented by highly elevated lightly bogged plains with a very rough relief, disturbance is caused by fires occurring in pine stands and dark coniferous taiga forests in years with extremely long drought periods. In landscapes of moderately bogged elevated plains having very rough topography, forests are disturbed mainly by Siberian gipsy moth outbreaks followed by fire. In the southern taiga zone of Western Siberia, fire is the major factor responsible for disturbance, except for the landscapes of the Yenisei Mountain Ridge. Differences in fire-caused forest disturbance are attributed to climatic conditions prevailing over landscapes and to the ecological regime of particular NLCs of lower ranks composing landscapes.

5. Provincial and Altitudinal Patterns of Post-Fire Forests

5.1 Succession

These patterns were established to be controlled by province- and altitudinal belt-specific climate, the percentage of area occupied by different climax formations, stand fire resistance, and by NLC pyrological regimes.

The probability of the occurrence of droughts, along with the number of NLCs remarkable for high fire recurrence, were found to increase from West to East across the study area (Fig.5). Parallel with this process, however, the percent area covered by more fire resistant pine and larch stands tends to increase eastward. Integrally, the above trends result in a decreasing percent of area with post-fire forest succession in the direction West to East. Also, the portion of forests heavily disturbed by frequent fires was observed to increase from west to east. Considerable differences in the scale of forest succession are observed between altitudinal belts which are distinguished on the basis of climatic conditions varying widely across the vast southern taiga forest subzone. In Western Siberia, there is a distinct trend for the area covered by post fire tree species to decrease from south to north. This trend is due to a pine forest distribution increase and a decrease of areas subject to gipsy moth outbreaks followed by fire in the northern part of the subzone.

As for the Yenisei Mountain Ridge, altitudinal post-fire forest succession is seen most clearly in its southern part (39%), while it decreases greatly in the Angara depression (16%) and increases again in the landscapes of the northern part of the Ridge (78%). For the altitudinal belts identified by Popov (1982) in the Angara province, forest succession was found to increase proceeding from the landscapes of the southern (33%) to central (50%) Angara region, whereas it decreases in the Trans-Angara landscapes (4%). Considerable differences in the scale of post-fire forest succession is due to the fact that the southern parts of the Angara and Trans-Angara regions are dominated by fire resistant climax pine and larch stands, while the central region is prevailed by dark coniferous taiga forests which were earlier widely distributed in watersheds (Vasiliev 1933; Popov 1967) and was replaced, over time, by secondary small-leaved, small-leaved-dark conifer, and small-leaved-pine forests as a result of fire.

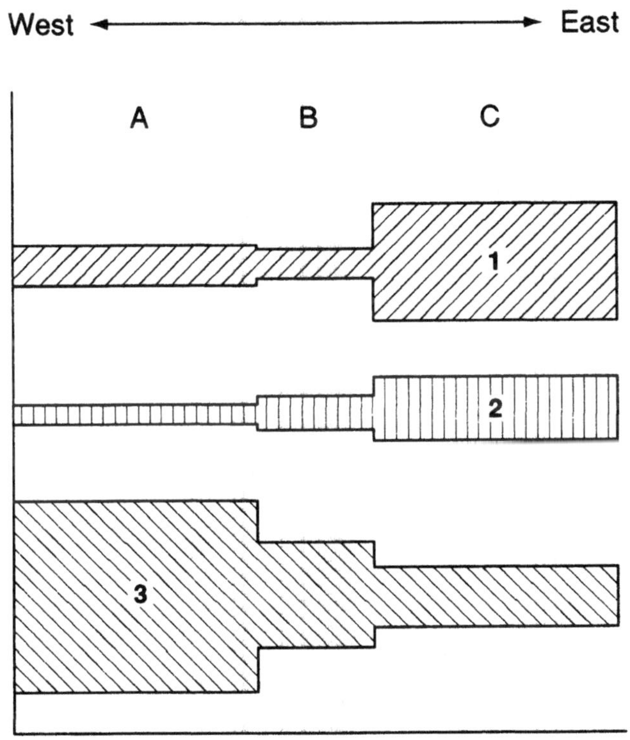

Fig.5. Stand succession scale in the provinces of Western and Central Siberia.

A - Chulym-Yenisei province
B - Yenisei Mountain Ridge
C - Angara province

1 - linear scale corresponding to pine and larch covering 20% of total forest area
2 - linear scale corresponding to the weight-average pyrological regime class
3 - linear scale corresponding to post-fire communities (complete stand replacement) accounting for 10% of the total forest area.

6. Conclusions

In the study area, fire occurrence, intensity and distribution are determined by the NLC ecological regime and morphological structure, post-fire vegetation patterns, and by fire-return intervals. Each landscape is characterized by a particular pyrological regime that is closely related to its climatic conditions.

Most GLs and landscapes of the study area experienced fires of different periods of time. Post-fire communities form ecology-dependent series of stages of forest regeneration dynamics differing in phytocoenotic and morphological parameters.

Forest disturbance, as a result of multi-year fire influence on forest formation, varies with GL and landscape. Landscapes are divided into three groups by the fire-caused rate of disturbance. Differences in disturbance are attributed to the ecological regime and morphological structure of NLCs of lower ranks composing landscapes. Disturbance is greatest in highly elevated well-drained plains and moderate in mountains, while plateau-like landscapes are even less disturbed.

In the study area, the scale of post-fire forest succession is observed to gradually decrease eastward, while the portion of pine and larch communities as well as fire recurrence increase. Provincial and altitudinal patterns of post-fire succession are determined by climatic differences responsible for fire recurrence increase and variations in the contribution of pine and larch, which are more fire resistant than dark conifers. In the study area, post-fire forest succession occurs in four main directions: dark conifers are replaced by hardwood species; dark conifers are replaced by pine and larch stands; pine and larch stands are replaced by dark conifers. Of these, the replacement of dark conifers by hardwood species is of the largest scale, whereas the replacement of pine stands by hardwood species is less common. The scale of the replacement of dark conifers by pine stands is relatively small, while the reverse replacement of pine stands by dark conifers is even less manifested. The NLC pyrological regime is considered the major factor accounting for the non-uniformity of post-fire forest succession.

Data on pyro- and eco-regimes of NLCs of different ranks, post fire forest regeneration stages related to fire recurrence, and calculated values of multi-year forest rate of burning and fire-caused disturbance are believed to be most useful for forest management in the southern taiga zone of Western Siberia. It is recommended that this data be used for forest monitoring by applying remote sensing methods for detecting and monitoring early post-fire forest transformation.

References

Clements, F.E. 1905. Research methods in ecology. University Publ. Co., Lincoln, Nebraska, 334 pp.

Clements, F.E. 1916. Plant succession: an analysis of the development of vegetation. Carnegie Inst. Publ. 242, Washington, 512 pp.

Falaleyev E.N. 1964. Fir forests of Siberia and their complex use. M. Lesprom., 166 pp. <in Russian>.

Furyaev, V.V. 1966. Control Burning of gipsy moth-affected forest areas. Nauka. Moscow. 80pp. <in Russian>.

Furyaev, V.V. 1970a. Fires in taiga forest of Ket-Chulym interfluve. In: Problems of Forest Pyrology (Problemy lesnoy pirologii), pp.273-330. Krasnoyarsk <in Russian>.

Furyaev, V.V. 1970b. The influence of ground water level on the flammability of bogged forests of Ket-Chulym interfluve. In: Problems of Forest Pyrology (Voprsy lesnoy pirologii), pp.155-186. Krasnoyarsk <in Russian>.

Furyaev, V.V. and D.M. Kireyev. 1979. Landscape approach in the study of post fire forest dynamics. Nauka. Novosibirsk. 160 pp. <in Russian>.

Furyaev, V.V. 1988. Forest fire effect analysis in estimating forest formation process. Sylviculture (Lesovedenie) No.1, 59-66 <in Russian>.

Kalashnikov, E.N. 1987. Forest landscape investigation using methods of remote sensing (Angara-Yenisei region case study). In: Forest Investigation With Aero-Space Methods (Issledovanie lesov aerokosmicheskimi metodami), pp.10-34. Nauka Publ. Novosibirsk <in Russian>.

Kireyev, D.M. 1976. Principles of natural landscape complex imagery analysis. In; Landscape Method for Forest Estimation Based on Areal Image Analysis (Landshaftny metod deshfrirovania lesov po aerfotosnimkam), pp.45-72. Nauka <in Russian>.

Kireyev, D.M. 1979a. Taiga forest landscape structure and methods of remote sensing for its study. In: Investigation of Taiga Landscapes Using Methods of Remote Sensing (Issledovanie tayeznih landshaftov distancionnymi metodami), pp.3-38. Nauka. Novosibirsk <in Russian>.

Kireyev, D.M. 1979b. Morphological landscape structure. In: Landscape Approach to Post Fire Forest Dynamics Investigation (Izuchenie poslepozharnoy dinamiki lesov na landshaftnoy osnove), pp.54-67. Nauka. Novosibirsk <in Russian>.

Kireyev, D.M. 1986. Landscapes of the southern part of Central Siberia and their identification in space images. In: Methods of Remote Sensing for Natural Resource Estimation in Siberia (Distancionnie issledovania prirodnyh resursov Sibiri), pp.147-170. Nauka. Novosibirsk <in Russian>.

Kolesnikov, B.P. and E.P. Smolonogov. 1960. Some trends of the regeneration dynamics of Siberian pine forests of Trans-Ural-Ob region. In: Problems with *Pinus sibirica*. Forestry in Siberia (Trudy po lesnomu hozaistvu Sibiri), Vol.6, pp.21-31. USSR Acad. Sci. Publ. Novosibirsk <in Russian>.

Komin, G.E. 1982. Regeneration succession in uneven-aged forests and possibilities for its prediction. Silviculture (Lesovedenie) No.4, 49-54 <in Russian>.

Melekhov, I.S. 1948. The influence of fire on forest. Goslestehizdat. 126 pp <in Russian>.

Popov L.V. 1967. Southern taiga forests of Central Siberia. In: Siberian geographical collection, pp.151-196. Nauka Publ., Moscow-Leningrad <in Russian>.

Popov, L.V. 1982. Southern taiga forests of Central Siberia. Irkutsk University Publ. 360 pp. <in Russian>.

Reimers, N.F. and L.I. Malyshev. 1963. Forest disturbance in Central Siberia. In: Seasonal and Secular (100-year period) Dynamics of the Environment in Siberia (Sezonnaia i vekovaya dinamika prirody Sibiri), pp.74-105. Irkutsk <in Russian>.

Solntzev, N.A. 1949. On the morphology of natural geographical landscape. Problems of Geography (Voprosy geografii), Vol 16, pp.61-86 <in Russian>.

Vasiliev, Ya.Ya. 1993. Forest and its regeneration in Bratsk, Ilim, and Ust-Kut regions. Trudy SOPS, USSR Acad. Sci. Siberian Series Vol.2. 111 pp. <in Russian>.

J.G.Goldammer and V.V.Furyaev (eds.), Fire in Ecosystems of Boreal Eurasia, 186-190.

The Role of Fire in Forest Cover, Structure, and Dynamics in the Russian Far East

A.S. Sheingauz [1]

1. Introduction

The influence of fire on forest status and development is covered extensively in the literature. Much valuable research has covered this at the level of forest type groups and smaller units (e.g., Telitsyn 1988; Valendik 1990; Sheshukov et al. 1992). Fire influences on forest cover units larger than forest stands are described in less detail in terms of burning mechanisms, general trends, and evaluation of losses. At the level of large forest cover units, the effects of fire are more diverse than on the levels of biogeocenosis, parcel, or individual plant (Furyaev 1978). At the level of the biogeocenosis and below, forest fires are generally episodic and catastrophic events causing total removal of vegetation. At the level of the forest tract and higher, at least in the Far East region of Siberia, fire should be considered an inevitable factor in forest development and structure.

The Far East is unique in the respect that researchers and other eyewitnesses have been able to record practically the whole history of natural forest development, so that patterns of vegetation change and disturbance can be reconstructed more completely than in other regions.

2. Regional Fire Patterns

Before the onset of mass colonization in the Far East (seventeenth century in the north and mid nineteenth century in the south) there was clear evidence in the forest cover of large-scale forest fires over several centuries (Krasheninnikov 1949; Middendorf 1860; Korshinsky 1982; Kolesnikov 1968). An increase in fire occurrence as the territory was settled and developed is confirmed by both direct and indirect data. While there are multiple reasons for this increase, one of the major factors was an increase in anthropogenic fires. Although the hypothesis is popular that much of this fire increase is due to lightning fires, there is little evidence at this point for an increase in lightning from historic times.

Currently, the average area burned annually equals or exceeds the area of commercial harvesting. However, there is little local correlation between the area burned and the area harvested. In some regions of the Far East, the leading disturbance factor affecting forest

[1] Institute for Economic Research, Far East Branch, Russian Academy of Sciences, 680042 Khabarovsk, Russian Federation

dynamics appears to be fire, in others, logging operations. In addition, while harvesting is reasonably uniform from year to year, the extent and severity of fires fluctuate widely. Catastrophic fires that occur every 12-15 years affect areas many times larger than those which are burned in a typical year. In general the influence of forest fires on forest structure and dynamics in the Far East area appears greater than that of logging. There are many difficulties, however, in obtaining quantitative data to support this contention. Reliable forest resource coverage is incomplete; fire records are not only poor, but may deliberately misrepresent areas burned; and records of fire effects are lacking, especially for long-burning fires. However, long-term large-scale data bases are available. Although they may not contain sufficient information for numerical models, they do allow development of adequate estimates. Such estimates of the proportion of forest land affected by fire and logging (Tab.1) have been calculated using methods of the author (Sheingauz 1979; Karyakin and Sheingauz 1988). This approach more likely underestimates the areas burned rather than overestimates them. The correlation between fire area and logging activity in these regions is high (r=0.82).

Tab.1. Percent of land in the Russian Far East affected by fire or logging.

Region (Krai/Oblast)	Vegetation of pyrogenic origin (% of area)	Forest cover destroyed by logging (% of area)
Prymorsk	39	31
Khabarovsk	29	36
Amur	37	42
Kamchatka	8	9
Magadan	19	17
Sakhalin	25	39
Russian Far East (Total)	26	30

3. Forest Resource Dynamics

Methods

Understanding the implications of data on the status of forest resources requires the analysis of resource dynamics. Considerable information is available to provide indices of rates and scales of pyrogenic transformations. For the Russian Far East, sufficiently reliable forest cover inventory data start in 1966 (five sets of observations; inventory data of 1993 are not yet available). In addition there are many individual forest stand descriptions available with detailed data on forest resource dynamics. However, this information needs to be organized into some form of classification.

Although there are a number of classification systems that have been developed in forest pyrology, most of these relate to the classification of fires rather than to the classification of post-fire dynamics of forest cover. Some forest cover dynamics classifications have been developed for the Russian Far East (Solodukin 1965; Kolesnikov 1968; Manko 1987), but they are not specifically pyrogenic. On the level of the forest tract and higher it is very likely that purely pyrogenic classifications are not possible, for on this level there are multiple factors in addition to fire, which influence patterns and processes. Of the 16 types of general forest resource dynamics classification schemes in the Russian Far East only two were defined as purely pyrogenic: fire and post-fire. However, pyrogenic elements exist in all of the other anthropogenic types (Sheingauz 1976).

Classification effectiveness is enhanced if each description type includes quantitative features for determining rates of change. Such information is important to include in a forest resource data bank. Unfortunately, despite numerous efforts, lack of funding has prevented the establishment of this type of data bank. Therefore, dynamic types are determined based on the analysis of data from forest tracts. The approach used to isolate fire influences in these data sets involves comparing pairs of tracts, in which all conditions except for fire occurrence are equal. Selecting such pairs is complicated but possible. These analyses are highly labor intensive and require specific statistical approaches (Sheingauz 1986).

In the Evoron Forest Management zone, in the Khabarovsk Region (middle taiga subzone), we identified a 49,000 ha area in the Siroky River basin where repeat inventories of similar accuracy (1:25,000 scale) had been carried out before the catastrophic fires of 1954 and two and 10 years after the fires. This unintentional experiment provided data for the following analysis.

Results

In the spring of 1954 (before the fires), forests covered 86.7% of the area of the basin, and the area was essentially untouched by harvesting. The 1954 fire burned through 54% of the forest, including high severity stand-replacement fires on 8% of the area. As a result, the forest area decreased by 14%, reducing forest cover to 74.9% of the area, while the standing timber volume decreased by 16%. Conifer stands were impacted three times as much as hardwood stands. The greatest damage was to mature and overmature forests. During the next eight years, logging operations were carried out and 3 fires of average intensity occurred. The last of these burned 34% of the study area. A 51% decrease in forest cover (59% decrease in standing timber volume) over this eight-year period reduced the percentage of forest area cover to 36.7%. Only 20% of the volume decrease was due to logging; the rest was destroyed by fires. This situation is typical due to forest industrial development in the Far East Region.

At the larger scale of forestry regions (FR) other patterns are evident. From 1966-1978 in the Russian Far East as a whole, the forested area decreased annually by 0.1% percent (105,000 ha yr^{-1}). In the subsequent 10 years, it increased 1.1% annually (1,275,000 ha yr^{-1}). This reflects changing approaches to forestry, which began early in the 1970s, with increasing emphasis on forest management and reductions in the levels of commercial harvest. The result was increases in forested areas, first in FR where earliest reductions in harvesting occurred, and eventually throughout the Far East Region. The spatial patterns of spruce-fir and larch stands, the main objects of commercial harvest in the region, were quite diverse. This diversity reflected the various patterns of logging of mature and overmature stands, the development of young stands, and stand destruction in all age groups caused by fire.

In all coniferous forests the area of mature and overmature stands decreased by 30 million ha between 1966 and 1988, and 6 million ha were harvested commercially. There was little change in overall stand ages during that time. About 24 million ha of the decrease can be attributed to fire as well as to the effects of natural mortality in spruce-fir stands. Over the same 22-year period, timber volume has decreased by 2.7 billion m^3; only 0.7 billion m^3 were harvested; and ingrowth is estimated at 2.6 billion m^3 over the same period. Another 0.2 million m^3 of lost volume were placed in miscellaneous categories. Based only on timber harvest, miscellaneous losses, and ingrowth figures, we would predict an increase in timber volume of 1.7 billion m^3, rather than the observed decrease. The difference of 4.4 billion m^3 (about 0.2 billion m^3/yr) can be attributed to losses due to fire and spruce mortality, or 5 times more than stock reduction resulting from harvesting and other miscellaneous activities. This relationship is similar to the one observed above for the Siroky basin.

4. Conclusions

Although data show similar sized areas for forest fire and commercial harvesting activities in the Russian Far East, fires have a 4-5 times greater impact than harvesting. The loss is most evident in terms of timber volume. The situation is less clear in terms of forested area. On the one hand forest fires often leave vast wastelands that may last for many years. On the other, total forested areas are increasing due to the emergence of young stands and decreases in logging activity. Two factors help to explain this apparent discrepancy. One, is that long-term efforts at fire control have led to decreases in the average areas burned over the last 20 years. The other, is that the two major years of catastrophic fire in the Russian Far East were in 1954 and 1976. Extensive regrowth of young stands in these fire areas will continue to add to the forested land base, as long as these areas are not reburned.

According to current estimates, only about half of the biological potential of forest lands in the Far East is being utilized, and only one third is utilized in areas with strong anthropogenic pressure (mainly from forest fires).

Forest fires are clearly the dominant factor influencing the status and dynamics of forest resources in the Russian Far East, and have a large impact on determining forest management strategies.

References

Karyakin, V.P., and A.S. Sheingauz. 1988. The urgency of the problems related to effective natural resource use. Geography and Natural Resources 3, 14-21 <in Russian>.

Kolesnikov, B.P. 1968. On the classification of forest cover dynamics forms. In: Materials on Vegetation Cover Dynamics, pp. 33-36. Vladimir Teaching College <in Russian>.

Korshinsky, S.I. 1982. Report on studies of Amur region as as a land use area. Reports of the East-Siberian Branch of Russian Geographers Society 23 (4-5), 73-138 <in Russian>.

Krasheninnikov, S.I. 1949. Description of Kamchatka area. Glavcevmorput, Moscow-Leningrad, 7 pp. <in Russian>.

Manko, Yu.I. 1987. *Abies ajanskii*. Nauka, Leningrad. 280 pp. <in Russian>.

Middendorf, A. 1860. Travelling in the North and East of Siberia. Part 1. North and East of Siberia; Natural and Historical Aspects. St. Petersburg, 812 pp.

Solodukhin, E.D. 1965. Forestry concepts behind management activities in mixed *Pinus sibirica*-broad-leaved forests of the Far East. Vladivostok. 367 pp. <in Russian>.

Telitsyn, G.P. 1988. Forest fires, prevention and control in Khabarovsk Territory. Khabarovsk. 96 pp. <in Russian>.

Furyaev, V.V. 1978. Forest fires as an ecological factor of taiga formation. Forestry Problems in Siberia 4, 210-218. Nauka, Novosibirsk <in Russian>.

Sheingauz, A.S. 1986. Methodical recommendations for analyzing forest area dynamics. DalNIILH, Khabarovsk, 41 pp. <in Russian>.

Sheingauz, A.S. 1976. Forest recourse dynamics classification (an example of the Far East). Forestry 6, 11-20 <in Russian>.

Sheingauz, A.S. 1979. Methods for evaluating actual area rate of burning using forest inventory data. Combustion and Fires in the Forest. In: Proc. All-Union Scientific Meeting. Part 3: Forest Fires and Their Effects, pp. 66-69. Inst. Forest. Sib. Br. Russ. Acad. Sci., Krasnoyarsk <in Russian>.

Sheshukov, M.A, P. Savchenko, and V.V. Peshkov. 1992. Forest fires and their control in the north of the Far East. Khabarovsk. 96 pp. <in Russian>.

Valendik, E.N. 1990. Large forest fire control. Nauka, Novosibirsk, 193 pp. <in Russian>.

J.G.Goldammer and V.V.Furyaev (eds.), Fire in Ecosystems of Boreal Eurasia, 191-196.

Importance of Fire in Forest Formation under Various Zonal-Geographic Conditions of the Far East

M.A. Sheshukov [1]

1. Introduction

When estimating the influence of fire on boreal forests, it must be taken into account that here fire, after climate, soil, and relief, is the most important natural exogenous factor, determining in many respects biological nutrient cycling, soil-forming processes, and forest development (reproduction, growth, composition, dynamics, expansion and destruction). Partial or full destruction of forests by fire may deeply change the entire forest environment and initiate a series of inter-connected effects of far reaching significance. Because of its scale and depth of influence on forest ecosystem (biogeocoenoses) dynamics and their ecological balance, especially in the taiga zone, the effects of fire can considerably surpass technogenic effects. Depending on the type and severity of the fire, vegetation and relief peculiarities, and presence or absence of permafrost and of permafrost type, the influence of fires on forest ecosystem development is highly variable. In some cases, irreversible processes may take place: soil erosion and degradation; formation of stony deposits or thermo-karst events. In others, long-term, negative changes during which degressive successions occur, are accompanied by the strong development or disappearance of organic layers and bogs. "Positive" ecological transformations of the forest environment may occur, determining cyclic initiation of forest forming processes and stimulating the formation of highly productive crops. Some ecosystems, however, do not experience any prominent changes in their development.

However, in considering and evaluating the influence of fires on the dynamics of forest phytocenoses, it is necessary to take into account that in different zonal-geographic conditions the qualitative consequences of fire (especially long-term), even under equal burning conditions, may be manifested in opposite effects on forest ecosystem development. Under some conditions, effects may be fully negative, under others, obviously positive.

2. Effects of Fire in Humid Zones

In zones with a warm and humid climate, for example in the conifer broadleaved forests of the Far East, forest fires negatively influence various aspects of natural forest development. One of the main reasons for this negative effect is the high speed and intensity of

[1] Far East Institute of Forest Economy, 71 Volochayevskaya St., 680020 Khabarovsk, Russian Federation

biogeochemical cycling in these forest ecosystems. This is because favorable climate and soil conditions (sufficient availability of heat, light and moisture) cause such important elements of biological rotation as the assimilation by plants of nutrient elements (carbon, nitrogen and ash elements) from the soil and atmosphere, and the decomposition of dead organic matter to take place with greater speed. In such regions, leaf fall from trees and bushes is almost fully decomposed in a year, forming the richest part of the soil in the form of humus, the organic nutrient substances and mineral elements of which are actively assimilated by the plants, thus accelerating their rotation in the ecosystem. Consequently, under the natural (biological) oxidation of dead vegetation remnants, almost all nutrient substances that form during decomposition, gradually (as litter mineralizes) are absorbed by roots of the plants and included in cycling. Increasing intensity and speed of nutrient cycling (turnover), in its turn, sharply accelerates plant growth and raises the general productivity of forest ecosystems. As a result of mineral element extractions by roots of timber species from lower soil levels and their consequent transformation into litter on the soil surface, a gradual increase of soil fertility occurs as soil enriches with organic and mineral nutrient substances. On the contrary, during fires, as a result of fast flame oxidation, a definite share of valuable nutrient elements such as nitrogen is volatilized into the atmosphere, while a large part of the mineral elements remaining in the ash is leached and washed out by precipitation. In addition to the destruction of forest mull litter and humus, the negative effect of fire in such regions is manifested in the destruction of or damage to forest overstory, juveniles and seedlings, which leads to the replacement of coniferous species by broadleaved trees and, with high fire frequency, by brushwoods. Fires here also promote soil turfing and the disruption of soil or forest forming processes. Thus in humid regions, fires in general have a great and diversified influence on forest formation and development.

3. Effects of Fire in the Subarctic Zone

In the subarctic zone, including tundra and forest-tundra zones and the north part of the northern subzone of the taiga, the effects of fire on forest ecosystem development is manifested particularly negatively. The nature of this effect is determined by the characteristics of frozen soils, which differ greatly among various permafrost zones, and determine to a large extent the impacts of fire. The main indicators of differences among zones could be expressed in terms of the peculiarities of the areal extent and thickness of permafrost, ice concentration in the soil, location of permafrost relative to the soil surface, and ground temperature. It is customary to divide soil and frozen sub-soil thickness into two large groups: 1) cryogenic (dry cryosolic) low ice concentration with many years of long-seasonal pseudo-permafrost--characteristic of the taiga zone, 2) cryolitogenic (true cryosolic) high ice concentration, located close to the surface, saturated long-term permafrost--peculiar to the subarctic. In cryogenic soils because of the lack of moisture chiefly physical-mechanical changes are possible without deep chemical transformation. In cryolitogenic soils physical-chemical reactions occur actively even in cold periods (Vtorushin 1992).

In the subarctic zone the permafrost layer is continuous, solid, and strong, containing a high concentration of subsoil ice close to the soil surface. Under such cryolitic conditions after fires there is a sharp increase in heat exchange between the soil and near-surface atmosphere, and intensive melting of underground ice begins, which raises soil moisture content so that soil sink, landslide and thermocarst mixtures occur (Kruychkov 1973).

Thermocarsts develop most actively in flat areas having loamy, hydromorphous silt soils with a high concentration of ground ice in so called ice wedge areas. After soil sinks develop, the depressions, everywhere formed in burns, are filled with water and may have depths up to 2-3 meters. Flowing together they can form, depending on the local relief, a series of small thermocarst lakes or a network of carst-erosion ravines with craters and gaps. The process of thermocarst events in cryolithogenic regions may continue for 30 to 40 years. Under these deep pyrogenic disruptions of soils and vegetation cover in the subarctic it is possible not only for the northern border of forest patches to shift to the south, but also for land ecosystems to be replaced by water ecosystems, e.g., with formation of thermocarstic lakes.

It is necessary to take into account that, independent from zonal geographic conditions, forest fires fatally influence the development of such phytocoenoses as *Pinus pumila* brushwood, fir-spruce and cedar stands and various associations with lichen cover.

4. Effects of Fire in the Middle and South Taiga Subzones

In contrast, in the middle and especially south taiga subzones, in the regions with discontinuous long-term pseudo-permafrost or long seasonal freezing of soils, the effect of infrequent fires (fire frequency about equal to harvesting turnover) on development and productivity of forest phytocoenoses is evidently positive. This is due to the fact that in regions like these, climate conditions are optimally suited to bio-ecological requirements for growth of the main forest forming species (*Larix* spp., *Pinus* spp.). On the other hand, especially under conditions of moisture surplus, the most intensive development of sphagnum mosses occurs and hence active bog development. Though annual phytomass production by boreal forest ecosystems decreases with higher geographic latitude (advancing to the north through taiga, forest tundra and tundra zones), total stock of dead organic matter (mainly in the form of peat layers) in many biotypes when there are no fires greatly exceeds the stock of annual leaf fall. Peat, peaty and coarse humus layers that form here differ by higher levels of carbon and nitrogen storage. In peat bedding the carbon content could be up to 24-36 t ha^{-1} and nitrogen up to 1.4 t ha^{-1} (Melekhov 1980). Accumulation of great quantities of nitrogen and other important nutrient elements in dead organic matter and their exclusion from biological rotations is one of the most prominent tendencies in the development of northern ecosystems and leads to gradual decreases in their general productivity.

Progressive spread of sphagnum moss and gradual accumulation of peat layers with great insulation ability lead to bog formation, deterioration of the hydrothermal soil status, increase of soil acidity, inhibited activity of microorganisms, slow-down of heat exchange between soil and atmosphere, and thus to the raising of the permafrost level and increased seasonal freezing. The increased seasonal freezing in its turn amplifies bog formation, so that the inter-amplification of all processes occurs. The general result of this is the transformation of red soils into bog soils with related reduction of forest productivity, decreased growth, and degradation (Sheshukov 1978, 1984). Without the repeated influence of fire, the progressive development and strengthening of sphagnum dominance, activation of bogging processes, and the associated long-term shifts and degradation of post-fire phytocoenoses follow this general pattern:

- First 40 years after fire: eutrophic woodgrass phytocoenoses (larch, birch-aspen, larch-aspen-birch; forest site classes I, II);

- 40 to 100 years after fire: eutrophic woodgrass-green moss (a lower canopy layer of fir and spruce is formed; forest site classes II, III);

- 100 to 350+ years after fire: mesotrophic woodbush-green moss-sphagnum (larch-fir-spruce forest, with birch disappearing; forest site classes III, IV);

- After natural decay of vegetation from the first post-fire generation: mesotrophic woodbush-sedge-sphagnum forest phytocoenoses; low productivity larch trees; forest site classes IV, V, Va).

In valley conditions, and especially on clayey hydromorphous soils, mesotrophic forest phytocoenoses in absence of fire may degrade to heterotrophic bush-sedge-sphagnum bog phytocoenoses (fens) or even to oligotrophic sphagnum swamps. The scale of these long-term processes is measured by centuries, they occur over a large geographic area, and their regulation has prominent ecologic-economic importance.

In such naturally developing ecosystems forest fires are the only decisive natural factor able in a very short period to radically and positively alter the ecological environment, the direction and intensity of cycling of soil-biological matter, the age and spatial structure of vegetation, and the conditions that affect tendencies of further development of a biogeocoenosis. Fire effect is expressed also in preventing bogging, releasing, converting, and re-distribution of nutrient substances, in improving the hydrothermal mode of soils and lowering their level of acidity (Sheshukov 1992).

Evdokimenko (1984, 1989) found and estimated on the basis of biomass information the positive influence of understory fires on wood increase in larch trees on permafrost sites. According to his opinion, increased thawing of permafrost as a result of soil thermal mode changes activates root system function, makes extra soil resources that impact productivity of soil environment available, and hence results in an increased accumulated growth of trees.

In general, successful post-fire forest reproduction on burns is observed when: (1) there is good fire mineralization of bedding (when upper and particularly middle little-decayed organic horizons are burned), (2) there are sufficient male cones and high seed production in the first or second season after fire with adequate precipitation during seed germination and early growth. When these factors coincide in time and place, the forest forming processes on burned areas are most successful. In cases when the mineralization of organic layers is weak during fires, non-burned layers prevent seedling regeneration, since in the summer, organic layers often undergo strong drying, which negatively influences root extension into the mineral soil for larch and other species. Fast reproduction of live ground cover (e.g.*L. middendorffii*) on burned areas in subsequent years sharply worsens ecological conditions for the appearance and growth of seedings.

On the basis of his long-term research Sannikov (1973, 1981, 1983) in his original prominent novelty works found that in light conifer forests behind the Urals, in various subzones of moderate zone and forest types with existence of seeding, the density of pine seeding in natural and near-natural habitat types (burned and burned-cut areas) was far higher than in anthropogenically modified habitats (clear-cut areas). He found that population density of pine seedlings on burned areas with a greater extent of organic layer combustion and in areas of bare mineral soil usually increased 5-20 times, seedling survival increased ten times, and growth rates and development increased several times compared with unburned substrate in the same type biogeocoenosis. Repetition of such intense fires at a 40-60 year interval actively induces cyclic waves of population dynamics of undergrowth and cenopopulations in general, forming a vertical mosaic of age-height class composition of stands with intervals of 30-60 years between generations, leading to a situation of "pulse-pyrogenic stability." The important evolutionary-ecological and zone geographic characteristics of natural reproduction of pine cenopopulations described here can provide a reliable scientific basis for purposeful and widescale control of natural forest forming processes not only in flat pine forests behind the Urals and in Western Siberia, but in taiga forests throughout the moderate zone of Eurasia and North America as well (see also Goldammer and Sannikov, this volume).

Research by Utkin (1965), Pozdnyakov (1975), Tsvetkov (1990) and others, testifies that under cryogenic conditions with low ice permafrost the forest forming processes on burned areas with mineralization of surface organic matter by flaming combustion are quite satisfactory in forests of Central Yakutia and Evenkia as well.

5. Conclusions

Thus under various zonal-geographic conditions the effect of fires on forest formation and development is quite different. It could be purely negative, as in humid regions and the subarctic zone (in crylitogenic high icy regions), or evidently positive, as in the taiga zone in regions with long seasonal soil freezing or patchy, poorly developed weak ice permafrost. In the moderate zone, periodic forest fires (with 80-100 year fire rotation) promote radical renovation and rejuvenation of forest phytocoenoses. Hence the pyrogenic factor here could be considered as a natural exogenic rejuvenating force, which rapidly returns climax forest phytocoenoses to a younger (initial) stage of their development cycle.

References

Evdokimenko, M. 1984. Vliyanie lesnykh pozharov na productivnost drevostoev. In: Productivnost lesnykh phitotsenozov, pp. 56-65. [Effect of forest fires on productivity of trees.] Krasnoyarsk <in Russian>.

Evdokimenko, M. 1989. Rol pirogennogo factora v productivnosti drevostoev. In: Factory productivnosti lesa, pp. 53-90. [Role of pyrogenic factors in tree productivity.] Novosibirsk <in Russian>.

Kruychkov, V. 1973. Kraini Sever problemy ratsionalnogo ispolzovaniya prirodnykh resursov. [The far north: On the problem of rational use of natural resources.] Moscow, 182 pp. <in Russian>.

Melekhov, I. 1980. Lesovedenie. Moscow, 405 pp <in Russian>.

Pozdnyakov, L. 1975. Daurskaya listvennitsa [Dahurian larch.] Moscow, 312 pp. <in Russian>.

Sannikov, S. 1973. Lesnye pozhary kak evolutsionno-ecologicheski factor vozobnovleniya populyatsii sosny v Zauralie. In: Gorenie i pozhary v lesu, pp. 236-277. [Forest and Fire as an evolutionary ecological factor in population renewal in Zauralie.] Krasnoyarsk <in Russian>.

Sannikov, S. 1981. Lesnye pozhary kak factor preobrazovaniya structury vozobnovleniya i evolutsii biogeotsenozov. [Forest fire as a transforming factor in renewal structure and evolution of biogeocoenoses.] Ecologiya No. 6, 24-33 <in Rssian>.

Sannikov, S. 1983. Tsiklicheski erozionno-pirogennaya teoriya estestvennogo vozobnovleniya sosny obyknovennoy. [Cyclical erosion-pyrogenic theory of natural renewal of pine.] Ekologiya No.1, 10-20 <in Russian>.

Sheshukov, M. 1978. Vliyanie pozharov na razvitie taezhnykh biogeotsenozov. In: Gorenie i pozhary v lesu, pp. 99-100. [Effect of fires in development of taiga biogeocoenoses.] Krasnoyarsk <in Russian>.

Sheshukov, M. 1984. Pirogenez-vazhneishii factor formirovaniya lesov. In: Gorenie i pozhary v lesu. [Pyrogenic-Important factor in formation of forests.] Krasnoyarsk <in Russian>.

Sheshukov, M. et al. 1992. Lesnye pozhary i borba s nimi na Severe Dalnego Vostoka. [Forest fires and their control in the northern Far East.] Khabarovsk. 94 pp. <in Russian>.
Tsvetkov, P. 1990. Vozobnovlenie na garyakh v listvennichnikakh tsentralnoi Evenkai. [Renewal in burned areas in Central Evenkii.] Lesovedenie No.1, 62-67 <in Russian>.
Utkin, A. 1965. A. Lesa tsentralnoi Yakutii. [Forests of central Yakutia.] Moscow, 208 pp. <in Russian>.
Vtorushin, V. 1992. Kriogenez i merzlotogenez v pochvakh. [Cryogenesis and permafrost development on soils.] Geographiya i Prirodnye Resursy No.3, 38-42 <in Russian>.

J.G.Goldammer and V.V.Furyaev (eds.), Fire in Ecosystems of Boreal Eurasia, 197-210.

Fires in Ecosystems of the Far Northeast of Siberia

G.V. Snytkin [1]

1. Introduction

The forests of the far North East of Siberia are considered a part of the West Siberian light coniferous forest subzone, a part of the Eurasian taiga zone (Kolesnikov 1955). Extreme environmental conditions (average annual temperatures below zero, the presence of permafrost and shallow soils with high peat content) account for a small proportion of forest area (17.3%), belt or island patterns of forest distribution - primarily in river valleys - and the absence of continuous forest cover, particularly in the Magadan region.

From site conditions, two geomorphological complexes can be identified in the area: the complex adjacent to the valley, and the mountain complexes. Each complex is characterized by its intrinsic forest fuel load and fire behavior.

In the whole of the study area, coniferous species - larch and Siberian dwarf pine (*Pinus pumila*) cover 92% of the total forested area. Mature and overmature (declining) stands dominate (82%), and young larch stands make up only 4%. More than half of the area is occupied by larch stands of low productivity (site indices IV to V) with a low degree of stocking (0.3-0.4). Light larch-lichen stands are very common (70%).

Under extreme environmental conditions of the far northeastern part of Siberia, fire has a major influence on ecological and resource potential. This is due to the fact that forest and tundra vegetation grow at the limits of its areal extent and form fragile ecosystems which are very susceptible to fire. Fire danger is very high in light larch stands and *P.pumila* thickets where surface cover is dominated by lichens (Snytkin 1971, 1972). Fires tend to crown almost immediately and burn across large areas. They often turn into catastrophic fires which only abundant rainfall can extinguish (Snytkin 1974). For instance, 32 large fires in the Magadan region affected 400,000 ha in 1971.

Fires result in considerable damage to reindeer farms since lichens, mosses, and low shrubs - the major food for reindeer - are consumed by fire. After fire, regeneration of this vegetation occurs very slowly. Mosses and lichens are replaced by less valuable vegetation, such as *Chamaenerium angustifolium*, *Calamagrostis* spp., *Euphorbia* spp., and other plants that reindeer cannot consume in the wintertime. In the second year following fire, *Vaccinium vitis-idaea* and then *Ledum* invade burned areas, and 5-10 years after fire *Cladonia rangiferina* and *C.silvatica* can be observed. It is not until 10-15 years following fire that *Cladonia alpestris* and *Cetraria* spp. begin to grow. Olds burns are not suitable as reindeer

[1] All-Russian Research Institute for Chemization in Forest Management (VNIIKhleskhoz), Pushkino, 141250 Ivanteevka, Moscow Region, Russian Federation

grazing areas. Almost an entire absence of lichens and overstocked shrubs do not allow these areas to be used as pastures. Lichen pastures recover only 40-50 years after fire. By its natural conditions, vegetation pattern, fire danger, and forest rate of burning, the area in question (i.e. Magadan region) can be divided into five subregions, namely: Arctic, Anui, Anadyr, Kolyma, and Okhotsk (Tab.1).

Tab.1. Meteorological conditions during the fire season in the subregions Anui, Anadyr, Kolyma and Okhotsky

Climatic index	Subregions			
	Anui	Anadyr	Kolyma	Okhotsky
Average air temperature (°C)	2.9	7.2	7.5	7.6
Average relative humidity (%) at 13:00 hrs	78	74	68	82
Average quantity of precipitation (mm)	50	36	39	56
Number of days with wind speed >15 m s^{-1}	4	7	2	10
Average date of appearance and disappaerance of snow cover	27 August 20 June	26 September 18 May	13 September 16 May	10 October 22 May

The arctic subregion has a cold climate and is characterized by tundra vegetation in which mosses, lichens, and grasses dominate (Alisov 1956). Like in the arctic tundra of Eurasia, vegetation cover is not abundant (up to 5 t ha^{-1}) (Bazilevich and Rodin 1971). This can be attributed to long-term influence of low air temperatures and weathering of the nonuniform soil layer. Significant amounts of vegetation are burned by tundra fires. Since too little information is available for this subregion, we will focus on the rest of the four subregions.

In the Anadyr subregion, treeless sites in mountains alternate with *P.pumila* thickets (25%) and open larch stands (4%). Climatic conditions vary: the costal zone has a maritime climate which gradually becomes more continental further inland. In this region, tundra sites and *P.pumila* thickets experience fires. In the sea coastal zone, where strong winds are common, fires can spread at very high rates. Kolyma and Anui fire subregions are characterized by extremely continental climate. Vegetation covers a relatively small area here: forest and *P.pumila* thickets account for 14 and 8% of the total area, respectively.

The Kolyma subregion includes the Kolyma river basin which is under management, while the Anui subregion covers the unmanaged part of the Korkodon-Anui interfluve. In these areas, high-intensity fires occur in *P.pumila* thickets and in larch stands whose second (subordinate) wood layer consists of *P.pumila*. The Okhotsk fire subregion suffers from a negative influence of cold sea and strong winds. Larch stands and *P.pumila* thickets here account for 22 and 14% of the total area respectively. Severe fires occur at the sea coast and are usually accompanied by strong winds. The above division of the area into forest fire subregions corresponds, to a large extent, with both climate (Alisov 1956) and vegetation (Starikov 1958).

Like other territories, the number of fires in the study area increases with increasing forest management activities. High fire occurrence is characteristic of the Anui subregion where wildfires tend to cover large areas. Average fire occurrence and fire size are less in the Okhotsk subregion which is most intensively managed.

The duration of the fire season is defined by the dates of the onset of permanent snow cover and snow melting, the growing season, as well as the first and last fire dates. In the north of the area under consideration (67°N), the fire season lasts 92 days, whereas in the south (60°N) the season is as long as 125 days. Fires occur mostly in July and August, but their quantity varies to some extent by fire subregion. In the Anadyr and Kolyma subregions, spring and summer periods of the fire season tend to merge (1 June to 15 September). Sometimes fires occur in early May and in late September, but these are very rare events. In the Okhot subregion, the largest number of fires occurs in the spring-summer period (late May - early June), while in the summer-autumn period (late July - early September) much less fires are observed; this indicates a distinct difference in the severity between these two fire danger periods. In general, the fire season starts on 25 May and ends on 30 September. Fires occurring in early May and in October are considered an exception.

Especially intensive fires, which are difficult to control, occur in *P.pumila* thickets. The history of these fires goes far back, and they have influenced *P.pumila* and surface (ground) cover for centuries (Tikhomirov 1933). This is supported by the presence of charcoal in soil and in deluvial deposits within the *P.pumila* area. Fire often consumes both the organic soil and *P.pumila*, including the root system which grows in the top soil. After fires in the upper parts of slopes, small gravel and residual vegetation downdrift is observed and soil horizons start to destruct. On steep (30-40°) slopes, pre-fire *P.pumila* thickets are replaced by detritus, which is clear from the presence of charred *P.pumila* stems. Regeneration of such burned areas occurs at a low rate and depends on the available sod layer depth, fire intensity, and the distance to the unburned productive *P.pumila* area.

Sites occupied by *P.pumila* communities are characterized by the highest fire danger. From the viewpoint of pyrological (flammability) properties, *P.pumila* communities are beyond any existing classification, and fire danger in them is at least twice as high as in other conifer communities. A surface fire jumps immediately to the *P.pumila* crown layer and spreads onward as a crown fire.

Under severe climatic conditions of the far northeast of Siberia, both surface and aerial forest fuel loads not only prove that the most important biological processes regulated by environmental and anthropogenic factors are intensive, but also determine to a large extent forest and tundra fire development (growth), intensity, and periodicity.

2. Fuel Inventory Methods

Surface fuel loads were determined by sampling plots of 20 x 25 cm. The fuel load in larch stands were defined by 10 model trees. In the course of the investigations new methods were tested to define the quantity of the organic mass (fuel load) of *P.pumila* stands. For this purpose the diameter of all shrubs at the bottom as well as the total length were measured on investigation plots 1 ha in size. After that ten "middle shrubs" were defined by mean diameter and length. On these middle shrubs the mean number and length of branches of different orders were determined. After that, the middle shrub was cut at its base, its length was measured and the age defined by counting year rings.

The organic mass of branches was calculated in two ways. In the first method the branches of the middle shrub were cut off, needles were gathered, weighed and the fuel moisture determined. In the second method, 20 branches of every order (live and dead) were taken from the middle shrubs. This allows the calculation of the biomass of the middle shrubs and the total shrub fuel load per hectare.

3. The Fuel Types

By fuel load and structure, *P.pumila* communities can be classified into three belts of distribution; upper, middle, and lower. The upper *P.pumila* belt occupies steep mountain slopes, the upper parts of round hilltops and borders the golets zone (mountain rocky area without wood and shrub vegetation). The vegetation pattern is fragmentary. Single creeping *P.pumila* individuals account for up to 15 % of the total belt area. Also, *Betula fruticosa* not more than 1 m high occurs in small numbers. The *P.pumila* layer is in some places invaded by *V.vitis-idaea* and *Ledum*. In the surface cover, suppressed lichens are present. In the upper *P.pumila* belt the average load of absolutely dry fuels is 20 t ha^{-1} (Snytkin 1973a,b; Fig.1).

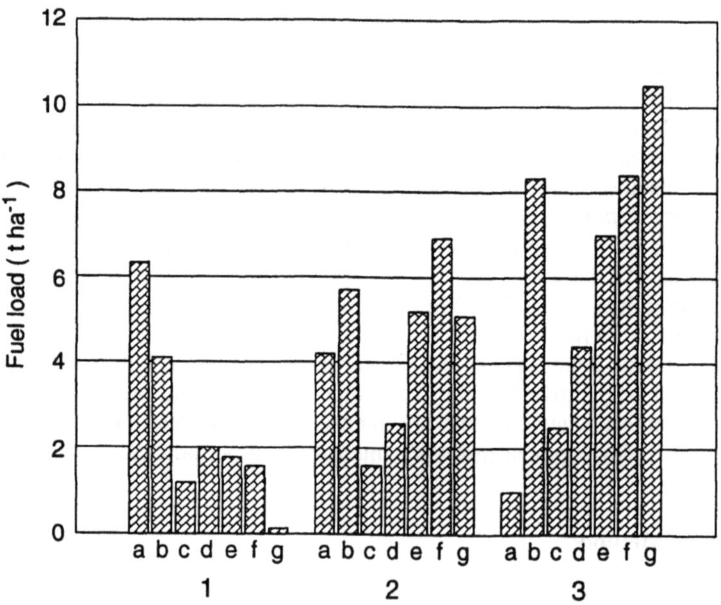

Fig.1. Fuel load in *Pinus pumila* thickets (Okhotsky subregion).

1 - upper slope
2 - middle slope
3 - lower slope

a - lichen, moss
b - leaf litter
c - grass, shrub
d - downed wood
e - needles
f - living branches (diameter <1 cm)
g - dry (dead) branches

The middle *P.pumila* belt is represented by *P.pumila* thickets with lichen surface cover. It occupies the middle parts of mountain slopes and rounded tops of low hills. Lichens dominate the surface cover. *P.pumila* is mixed with *B.fruticosa*. The average height of this wood layer is 1.5-2.0 m. Canopy closure (density of canopy) varies from 20 to 70%. The low shrub layer consists of *Ledum*, *V.vitis-idaea*, *V.uliginosum*, and *Empetrum sibiricum*. The cover of these species is 10-15%. A well developed lichen layer is the peculiarity of this belt. Lichen cover varies from 25 to 90%. The belt is dominated by taiga and permafrost-taiga soils with different podzol contents underlaid by rock debris occurring from the depth of 40-60 cm. In the middle *P.pumila* belt the total forest fuel load is 30 t ha^{-1}.

The lower belt comprised of *P.pumila* thickets with low shrub-moss surface cover includes all aspects of the lower parts of slopes, and the middle parts of north-facing slopes. *P.pumila* canopy cover is 70-100%. *P.pumila* is mixed with *Alnus kamchatica* and *B.fruticosa* not exceeding 2.0-3.5 m in height. Under this tree canopy, a well-developed low shrub layer is composed of *Rododendron*, *Ledum*, *V.vitis-idaea*, *V.uliginosum*, and *Empetrum sibiricum*. Low shrub cover varies from 20 to 60%. The grass cover is thin and discontinuous (patchy). In glades, mosses form compacted cushions. The belt is represented by permafrost-taiga, illuvial-humus, and peat soils. Permafrost is located at the depth of 80-90 cm. In the lower *P.pumila* belt the total forest fuel load is 40 t ha^{-1}.

The differences in fuel accumulation along the altitudinal gradient are explained by soil and climatic conditions, low soil temperature being the major factor in limiting phytomass production (Ignatenko et al. 1976).

Intensities of fires occurring in *P.pumila* thickets are determined by forest fuel load and moisture content. The moisture content of the growing parts of plants is relatively constant during the fire season: 80-124% for small twigs and 100-120% for grass and low shrubs. In drought periods, dead organic matter, lichens and mosses have a low moisture content: for dead branches, lichens, and litter, the moisture content was measured as 14-20, 20-28, and 23-30%, respectively. Moisture content of these forest fuels increases when proceeding from the upper to the lower *P.pumila* belt. This is due to varying *P.pumila* canopy density and, consequently, the various levels of surface cover shading. The live *P.pumila* needle moisture content does not vary much during the fire season and is never below 98%. Since *P.pumila* needles are rich in volatiles, they are easily ignited (Filippov 1977). *Betula fruticosa*, which often accompanies *P.pumila*, is also highly flammable.

Fire intensity is controlled not by the total fuel, but largely by the proportion of available fuels. The largest amount (90-100%) of the total fuel is consumed by fire in the upper *P.pumila* belt. In the middle belt this value is 60-80%, while it is the lowest (20-50%) in the lower belt.

Fire spread rate and shape are determined by wind speed and slope. In *P.pumila* thickets, crown fires can occur in a no-wind situation because of the low crown layer location. Linear fire spread rate varies with slope, topography, and wind speed (Tab.2).

In open larch stands, like in *P.pumila* thickets, forest fuels are confined to certain landscape elements, so they should be considered in relation with geomorphological complexes (i.e. a mountain and valley complex) (Snytkin 1968, 1973b). The valleys are comprised mostly of forest types where surface cover is dominated by either green mosses, sphagnum or sedge. Mountain forests are represented by larch stands with lichen surface cover (Fig.2). An understory with a 0.1-0.5 canopy closure index grows in openings and consists of *B.fruticosa*, *Salix fruticosa*, and *P.pumila*.

Surface cover composition and distribution vary with forest type and dates of occurrence of past fires (if any). Open mountain larch-lichen stands account for 70% of the total area and have surface fuel loads of 10-17 t ha^{-1} (Fig.3). This fuel load is similar to that found for open pre-tundra forests (Basilevich and Rodin 1971). Like in *P.pumila* thickets, fuel load in larch stands increases downslope and is 10-15 and 15-25 t ha^{-1} for the upper and lower parts of slopes, respectively. In larch-green moss and larch-*V.vitis-idaea* stands, occupying areas above flood plain sites account for 28% of the total forest area, fuel load is 25-50 t ha^{-1} (Fig.4).

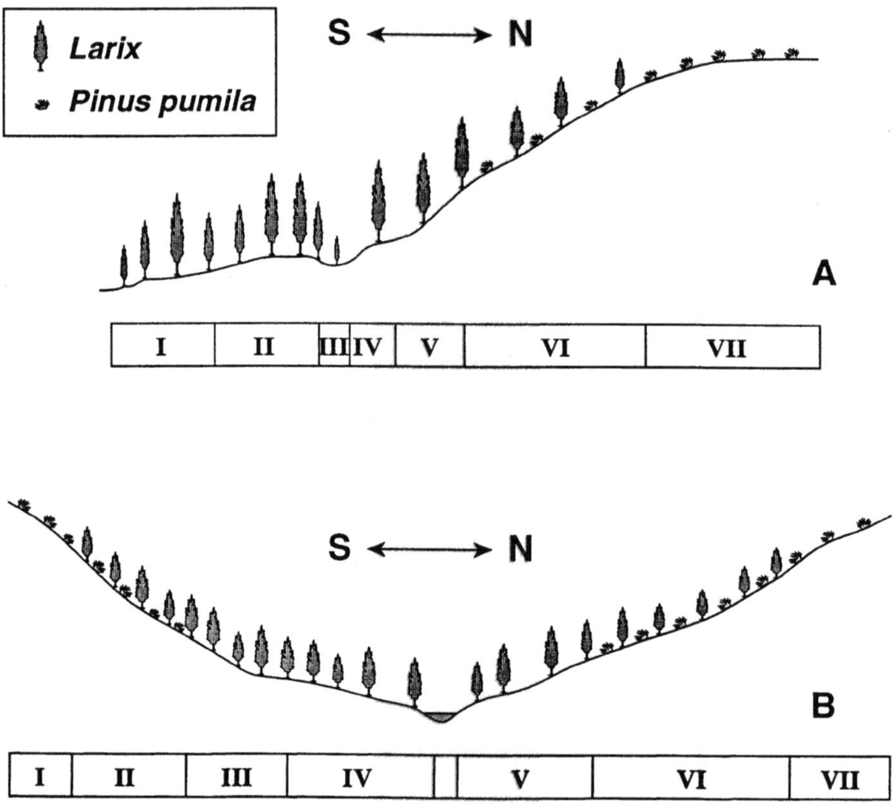

Fig.2. Distribution of larch stands and *Pinus pumila* thickets in basic groups of forest types.

A. South slopes (Tanon river valley, Okhotsky subregion)

- I - grass
- II - moss-lichen
- III - sphagnum
- IV - shrub-sphagnum
- V - moss-shrub
- VI - moss-lichen
- VII - *P. pumila* thickets

B. North and South slopes (Mayorych river valley, Kolyma subregion)

- I - *P.pumila* thickets
- II - moss-lichen
- III - *P.pumila* - *A.fruticosa* stands
- IV - moss-sphagnum
- V - green moss
- VI - lichen
- VII - *P.pumila* thickets

Tab.2. Rate of fire spread ($m\ min^{-1}$) in *Pinus pumila* forests

Slope	Wind Speed ($m\ s^{-1}$)		
	1 - 5	6 - 10	11 - 15
	Rate of Fire Spread ($m\ min^{-1}$)		
0°	3	6	12
10°	5	9	18
20°	6	12	24
30°	9	18	36

In mountain larch stands, fuel load is directly related with slope and aspect. On north-facing slopes, fuel load is twice as high as that on south-facing ones. This large difference in fuel load is mainly due to abundant green mosses on north-facing slopes. For example, in the north-facing slope, near the Maiorich spring, the fuel load is 19 $t\ ha^{-1}$, whereas in the south-facing slope it is only 12 $t\ ha^{-1}$.

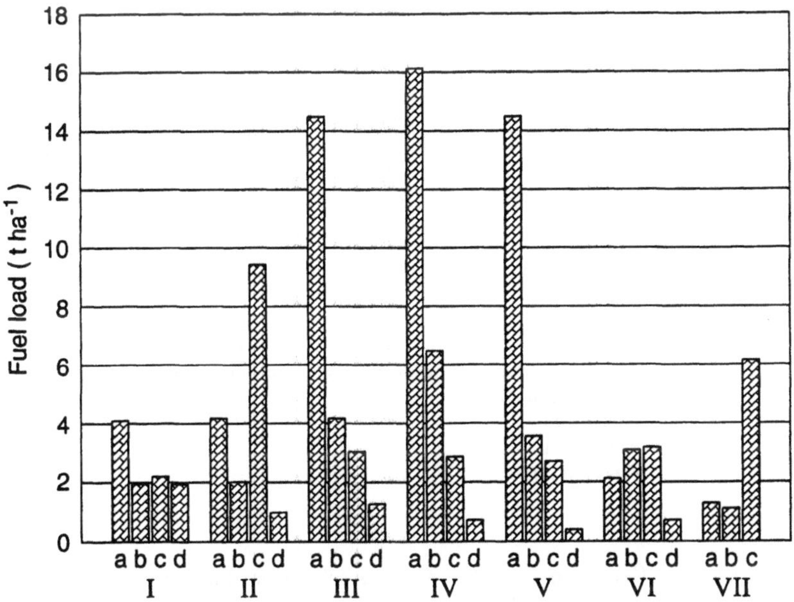

Fig.3. Surface fuel loads in the Kolyma subregion. Groups of forest types:

I - *P.pumila* thickets
II - moss-lichen
III - *P.pumila* - *A.fruticosa* stands
IV - moss-sphagnum
V - green moss
VI - lichen
VII - *P.pumila* thickets

a - lichen, moss
b - leaf litter
c - grass, shrub
d - downed wood

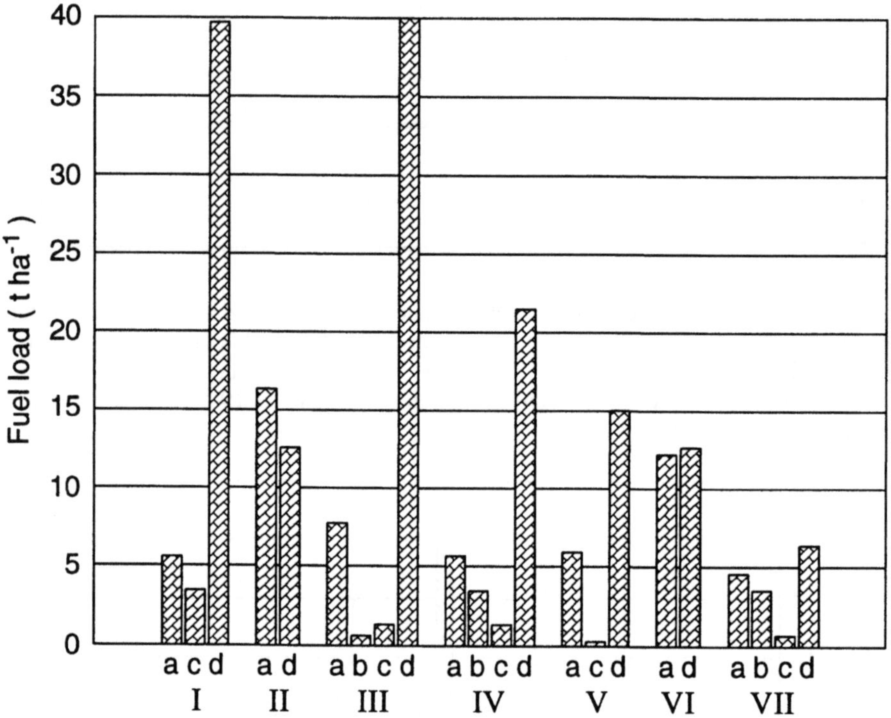

Fig.4. Surface fuel loads in the Okhotsky subregion. Groups of forest types:

I	- grass
II	- moss-lichen
III	- sphagnum (upper part of layer only)
IV	- shrub-sphagnum
V	- moss-shrub
VI	- moss-lichen
VII	- *P.pumila* thickets

a - lichen, moss
b - leaf litter
c - grass, shrub
d - lower decaying part of surface cover (A_F)

In the upper parts of 10-15° slopes in the Kolyma subregion, the surface fuel load is 1.5-2 times less than in the lower parts. This difference increases with increasing slope. When the slope does not exceed 10°, no large differences in fuel load are observed.

Fuel load increases from 10-20 t ha^{-1} under dry site conditions (larch-lichen and larch-*V.vitis-idaea* forest types) to 30-50 t ha^{-1} under hydromorphic conditions (larch-grass, larch-sphagnum stands); these measurements agree well with data obtained for other regions.

Apart from altitudinal vegetation belt and geomorphological characteristics fuel composition varies within fire subregions. For example, in mountain larch stands of the Kolyma subregion, surface fuel load is less than in those of the Okhotsky subregion - 10-25 and 25-50 t ha^{-1}, respectively. This can be attributed to more severe ecological conditions in the Kolyma subregion. Litter and forest floor organic layers account for the largest amount (50-80%) of the total surface fuel load; this indicates an extremely low rate of organic residue decomposition, suggesting insufficiency and low activity of the microflora.

Aerial fuel load (branches, needles) vary significantly in larch stands (from 0.8 to 113 t ha^{-1}). In stands located on slopes near the Maiorich spring (Kolyma subregion), the average fuel load is 4.6 t ha^{-1} (Fig.5). Coastal larch stands at the sea of Okhotsk reach 1.3 t ha^{-1} (Fig.6). The largest proportion of fuel load falls on tree needles (0.6-1.4 t ha^{-1}), then, in order of decrease, thin branches (0.4-1.2 t ha^{-1}), and thick branches (0.3-1.3 t ha^{-1}).

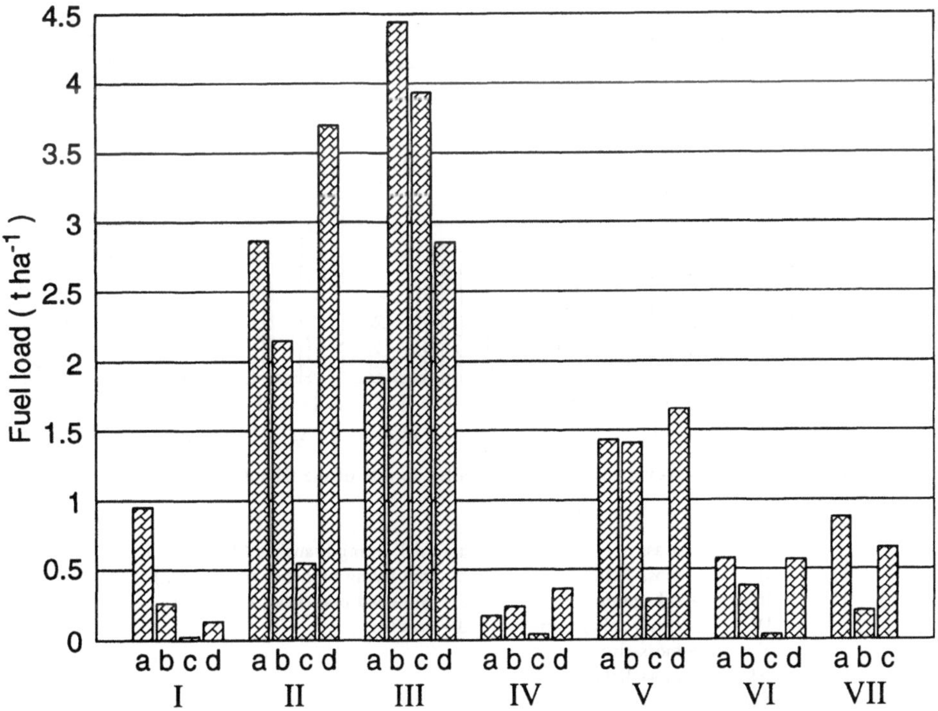

Fig.5. Aerial fuel loads (tree crowns) in mountain forests of the Kolyma subregion. Groups of forest types:

I - *P.pumila* thickets
II - moss-lichen
III - *P.pumila - A.fruticosa* stands
IV - moss-sphagnum
V - green-moss
VI - lichen
VII - *P.pumila* thickets

a - thick living branches (d > 3 cm)
b - thin living branches (d < 3 cm)
c - thin dead branches (d < 3cm)
d - needles

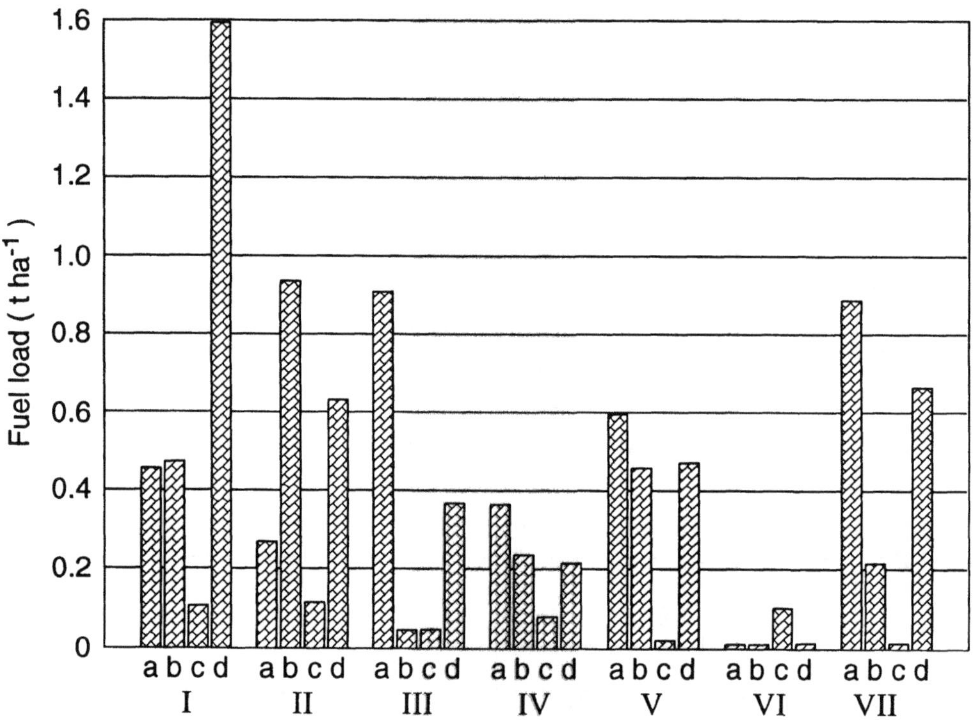

Fig.6. Aerial fuel loads (tree crowns) in valley forests of the Okhotsky subregion. Groups of forest types:

I	- grass	a - thick living branches (d > 3 cm)
II	- moss-lichen	b - thin living branches (d < 3 cm)
III	- sphagnum	c - thin dead branches (d < 3 cm)
IV	- shrub-sphagnum	d - needles
V	- moss-shrub	
VI	- moss-lichen	
VII	- *P.pumila* thickets	

The load of needle biomass, which largely determines crown fire intensity in open larch-lichen stands, varies from 0.1 to 2.8 t ha^{-1}. It is noteworthy that, in larch stands of the Kolyma subregion, needle load is more than in those of the Okhotsky subregion (0.1-1.4 and 0.2-0.6 t ha^{-1}, respectively). Needle load is directly related to stand degree of stocking. Minimum needle load is characteristic of low-stocked stands growing under poor site conditions. For example, in a larch-green moss-sphagnum stand with 0.3 degree of stocking, needle load was found to be 0.4 t ha^{-1}.

Young larch stands on the Okhotsk sea shore are remarkable for the greatest accumulation of dead thin branches which promote crown fires. Near-mature and young stands adjacent to the Maiorich spring have considerable amounts of needle biomass and thin branches (1.4 and 2.6% of the total aerial fuel, respectively). Despite these stands being open, crown fires occur frequently because of the mountain topography (i.e. the effect of slope). The amount of dead and living thin branches is much less in overmatured stands of the Okhotsk subregion (0.1 and 0.3 t ha^{-1}, respectively). Open larch stands occupying valleys rarely experience crown fires.

Load percentage of different aerial fuel groups is dependent on stand inventory parameters. In stands consisting of small trees, the proportions of the total aerial fuel load accounted by different fuel groups are as follows: 40% tree trunks, 37% living branches, 1% dead branches, and 36% needles. In well-stocked, high-tree stands, these values are equal to 28, 37, 16, and 19%, respectively. With increasing stand height and age, relative proportions of needles and small branches increase, while canopy density decreases. Therefore, in old stands with high trees, crown fires are less probable than in young stands. However, the probability of crown fire occurrence increases if larch regrowth is present in these old stands. This is due to fuel load in the regrowth. In most cases, however, fires consume primarily needles, dead and living branches, which account for 18.5, 6.4, and 22.0% of the total biomass, respectively. The total biomass, as well as the amount of biomass accounted for by different tree parts, vary considerably.

In young larch stands, surface fuel load is 10-15% less than in mature and overmature decaying stands. In young stands, mosses and lichens have not recovered yet and, litter is the major fuel.

Fuel load (excluding downed wood) is in a certain way dependent on the forest floor organic layer depth. The studied forest types were classified into 5 groups by organic layer depth. Five forest types (mostly *Larix-V.vitis-idaea* stands) where the forest floor organic layer is 1-2 cm deep, are assigned to Group 1. For this group, surface fuel load varies from 10.7 to 16.3 t ha^{-1}, with the average value of 14.1 t ha^{-1}. Group 2 includes larch-lichen-green moss stands and is characterized by a 3 cm organic layer and a relatively high surface fuel load (19.5-20.5 t ha^{-1}; 20.1 t ha^{-1} on average). Group 3 covers *Larix-Polytrichum* stands. The forest floor organic layer is 5-10 cm deep, and the average surface fuel load is 25 t ha^{-1}. Group 4 (*Larix-sphagnum* stands): organic layer depth is 20 cm and surface fuel load is 35 t ha^{-1}. Larch-grass stands are included into Group 5. This forest type is remarkable for a relatively small (3-4 cm) organic layer depth and high surface fuel load (48 t ha^{-1}). This can be attributed to high organic layer compactness.

Vegetation residue and surface fuel loads were found to correlate with aerial fuel load; crown weight increases with increasing tree diameter. This correlation is observed for larch stands growing in the basin areas of the At-Uryakh and Dukcha rivers.

From the forest fire viewpoint, the distance between stand canopy and surface cover is of special interest since this distance is important for crowning. The smallest distance from surface cover to canopy was found for the youngest (15-year old) of the stands studied. The distance between surface cover and crown layer increases with increasing stand height. For example, at similar stand age and stocking, the branchless part of tree trunks is 0.37 and 1.21 m, at a tree height of 3.31 and 4.91 m, respectively. Lower tree branch removal depends on the stand degree of stocking. At the same stand age, the higher the stocking, the more intensive the branch removal. For example, in a stand with 0.5 stocking, the branchless tree part is 0.16-1.21 m long, whereas it equals to 2.45-4.08 m at 0.7-1.0 degree of stocking. Crown fires, however, tend to occur more frequently in dense rather than in open (light) stands. This can be due to the fact that, despite more intensive branch removal processes, forest fuels are more continuous in dense stands. For this reason, fire spreads faster through a dense larch stand canopy. In young larch stands the average crown length is ca. 66% of the tree trunk height.

Apart from fuel load, forest fire danger is directly related to the fuel moisture content since the latter influences the probability of fire occurrence, burning efficiency, and fire spread rate and intensity. Depending on microclimatic conditions, seasonal and daily fuel moisture content varies with stand characteristics. Lichen moisture content varies widely during a 24-hour period; under drought conditions, it was measured to be 10-11% from noon to 6:00 p.m., 14-16% from 8:00 to 10:00 p.m., and 26-30% from midnight to 2.00 a.m. In dry sites, litter moisture content was found to be almost similar to that of lichens. In moist and wet sites (sphagnum bogs and sedge marshes), where litter is represented by dry tall grass, it forms a compacted layer, and fire danger remains high throughout fire season. Growing grass cover moisture content varies inconsiderably (85-130%) and depends largely on the cover composition. Moisture content of both low shrubs and grasses changes slowly and in a small range. Low shrub moisture content, however, is lower than that of grasses. For instance, *Empetrum sibiricum* and *Ledum* moisture content varies from 60 to 80 and from 70 to 100%, respectively. Downed wood moisture content varies considerably - from 11 to 63%. Moisture content tends to decrease when proceeding upslope, with decreasing canopy closure. Some discrepancy in data on downed wood moisture content can be attributed to great non-uniformity of this fuel group and differences in microclimatic and site conditions.

In mountains, where forest sites are located at various altitudes, fuel moisture content is highly affected by air temperature gradient and inversion. Also, fuel moisture content is closely related with slope aspect: on south-facing slopes, it is lower than on north-facing ones. High moisture content is characteristic of mosses dominating surface cover in *Alnus fruticosa* thickets that often occupy northern slopes. Moreover, grasses and low shrubs cast shadow on mosses promoting thereby a high moisture content in the latter. As for moisture content, mosses covering southeastern slopes are intermediate between those growing on south- and north-facing slopes. Mosses distributed in the middle and lower parts of slopes are remarkable for an especially high moisture content. Top soil here is maintained wet due to melting frost substrates, which is never lower than 20-25 cm throughout the fire season. Moss absorbs water from soil and so its moisture content remains high.

Because of fuel diversity and differences in fuel moisture content, fires of low, moderate, and high intensities can occur in larch-lichen stands widely distributed in the far North-East of Siberia (Tab.3).

Tab.3. Fire characteristics in larch-lichen stands

Fire Intensity	Wind Speed ($m\ s^{-1}$)	Flame Length (cm)	Depth of Fire in Soils (cm)	Width of Fire Seam (cm)	Fire Spread ($m\ min^{-1}$)
Low	1 - 3	3 - 5	2 - 3	4 - 5	1 - 1.5
Medium	5 - 10	6 - 10	5 - 8	8 - 12	2 - 3
High	11 - 15	30 - 50	10 - 12	30 - 50	4 - 10

Low-intensity fires consume the surface portion of the lichen including leaves and litter. Larch trunks are charred to the height of 5-10 cm. Low-intensity fires do not result in tree mortality. Fires of moderate intensity consume the whole lichen layer. Parts of trees are often injured by fire, and tree mortality occurs. 2-3 years after fire, trees start falling down. High-intensity fires consume the whole of the surface cover and forest floor organic layer, as well as such aerial fuels as needles and small live and dry branches. In 3-4 years following a high-intensity fire, all trees fall down. The sites burned by high-intensity fires experience repetitive fires in 10-15 years. Fire danger increases with developing vegetation cover.

In *Larix-V.vitis-idaea* stands, the main surface fuels are dry needles. *Larix-V.vitis-idaea* and larch-lichen stands suffer from periodical fires every 30-50 years which either kill them immediately or lead to their full mortality within 3-10 years. Other forest types, such as larch-green moss stands, are subjected to less frequent high-intensity fires in intervals of 50-100 years. All young larch stands are of fire origin.

Fire danger of forests and other land categories is directly related to the differences in surface fuel composition, as well as to stand structure.

The highest fire danger is characteristic of *P.pumila* thickets and larch stands with *P.pumila* understory and lichen-dominated surface cover (Fire Danger Class I). The litter layer consisting of needles and dry small *P.pumila* branches favors fire spread.

In sites designated as Fire Danger Class II, a certain decrease in fire danger rate is due to the high moisture content of the moss layer and the presence of *Empetrum sibiricum* which hampers fire spread. In young larch stands and *P.pumila* thickets whose surface cover is composed of mosses, crown fires occur. In the remaining sites of fire danger Class II, surface fires are common. From the fire danger viewpoint, open larch stands do not differ from corresponding cut areas, so both of these land categories are included in the same fire danger class.

Fire danger in the sites of Class III decreases because of the occurrence of less flammable plants in the surface cover. Crown fires occur in stands where surface cover consisting of mixed broad-leaved grasses is of moderate density and abundance. The remaining sites are characterized by surface fires of various intensities.

In sites of Fire Danger Class IV, fires are limited to small locations and often burn deep into compacted organic layers. In sphagnum bogs, long-term drought results in peat fires. These fires of low intensity are self-extinguishing as a result of the frozen layer melting and moistening of sphagnum.

For the successful development of methods for protecting forests and tundra from fire under severe conditions of the far North East of Siberia, it is of crucial importance to intensify the study for forest and tundra fire behavior, improve methods of fire danger prediction, and develop a system of fire prevention measures and effective fire control activities.

References

Alisov, B.P. 1956. The Climate of the USSR. Moscow University Publ. <in Russian>.

Bazilevich, N.I. and L.V. Rodin. 1971. Productivity and element cycling in natural and planted phytocoenoses. Biological Productivity and Element Cycling in Vegetation Communities. Leningrad <in Russian>.

Filippov, A.V. 1977. Thermal regime of fires burning through understory and brush layer. In: Combustion in Heterogenic and Gas Systems, p.33-36. Chernogolovka <in Russian>.

Ignatenko, I.V., I.I. Kotlyarov, A.I. Nesterenko and A.A. Pugachev. 1976. The total amount and structure of organic matter in *Pinus pumila* thickets of the Okhotsk sea shore. In: Biology and Productivity of the Vegetation Cover of North East, p.138-158. USSR Acad. Sci. DVNC Publ. Vladivostok <in Russian>.

Kolesnikov, B.P. 1955. Vegetation cover zoning in the Far East and problems of forest replanting and protective forest belt formation. In: Problems of forestry and wood industry Development in the Far East, p.81. USSR Acad. Sci. Publ. Moscow-Leningrad <in Russian>.

Snytkin, G.V. 1968. Vegetative residue and surface cover fuel loads in forests of the Kolyma river basin. Botanical Journal 53, 689-692 <in Russian>.

Snytkin, G.V. 1971. Fire danger rating in in forest of the far North East. J. Forest Management No.4, 67-69 <in Russian>.

Snytkin, G.V. 1972. Fire danger in high mountain forests of the Kolima river basin. Forest Journal No.2, 159-160 <in Russian>.

Snytkin, G.V. 1973a. Primary productivity of *Pinus pumila* thickets in the Okhotsk sea costal zone. In: Topological aspects of studying behavior of substances in geosystems, p.151-153. Irkutsk <in Russian>.

Snytkin, G.V. 1973b. On primary productivity of forest communities in the far North East. In: Investigation of the taiga biota (problems and prospects), p.93-99. Proc. V Meeting of the Board of Scientists on Complex Management of Taiga Areas. Irkutsk <in Russian>.

Snytkin, G.V. 1974. The reasons of large fires in Magadan region and method of their control. Biological Problems of the North, Issue 5, p.168-172. Magadan <in Russian>.

Starikov, G.F. 1958. Forests of Magadan Region. Magadan, 221 pp. <in Russian>.

Tikhomirov, B.A. 1933. Fires in *Pinus pumila* thickets of Penzhin region. Botanical Journal 18 (6), 28 <in Russian>.

J.G.Goldammer and V.V.Furyaev (eds.), Fire in Ecosystems of Boreal Eurasia, 211-218.

Fire-Induced Transformations in the Productivity of Light Coniferous Stands of the Trans-Baikal Region and Mongolia

M.D. Yevdokimenko [1]

1. Introduction

Light coniferous species (*Pinus* spp., *Larix* spp.) dominate the forest area of the Trans-Baikal region and Mongolia. These species are of crucial importance for ecological stability in this region of south-eastern Siberia and a predominant feature of the Baikal Lake biosphere complex. The extreme continental climate (low annual precipitation) is associated with high fire risk. Environmental importance and high fire danger (as compared to adjacent forested areas) are the two factors accounting for the long-standing scientific interest to the productivity of light conifers in this region. From all anthropogenic and technogenic factors, forest fires cause the largest forest disturbance in the region of interest.

This paper presents the results of a study on post-fire forest productivity in pine (*Pinus sylvestris*) stands of the southern and central parts of the Trans-Baikal region and larch (*Larix sibirica*) forests of Mongolia and the northern part of the Trans-Baikal region.

2. Methods

In 37 sample plots, the radial increment (tree ring widths) of all living trees was measured before and after fire. Trunk biomass increment (in cubic meters) was determined from selected model trees representing different trunk diameter classes.

Most observations were started in fresh burns, i.e. in the year of fire or the year following fire. To observe long term fire effects, several plots were established where fires burned some 14-18 years ago. The research program, apart from observations on burns resulted from wildfire, included experimental surface fires of various intensity. These experiments provided for a more accurate comparison of plots subject to "experimental fires" with control plots, since the pre-fire situation in both control and fire plots was the same.

Post-fire stand thinning was observed on permanent sample plots including fresh burns and experimental fire sites. Post-fire changes of pine stand productivity were studied in the regions where annual precipitation is 250-300 mm.

---

[1] V.N. Sukachev Institute for Forestry and Timber, Siberian Branch, Russian Academy of Sciences, 660036 Krasnoyarsk, Russian Federation

3. Results

For highly insolated mountain slopes, surface fires of any intensity were found to have negative bearing on tree increment. Decrease in tree increment depends on fire intensity, tree viability, and also on tree age and thickness since these two latter parameters determine both the role of bark for protecting the tissue against heat and tree crown height above ground.

In middle age stands, tree increment tends to decrease after low-intensity surface fires; although these stands comprise most of the study area, accurate values for tree increment decrease cannot be obtained because of high variability. Increment decrease over 2-3 years was observed most often for thin trees which are most probable to die from fire. In stands growing on relatively moist soils at foothills, a 10-15% increase in increment was observed mainly for thick trees. Similar observation results were obtained for north-facing slopes.

In Table 1, data are summarized on post-fire tree increment by trunk thickness class for a 73-year old mixed larch-pine stand located on the Khamar-Daban macroslope. The burned area was large enough for sample plots to be established close to stand outskirts so that they could represent various fire damage levels. Thus burned plots and unburned plots selected in a stand situated on a 15-20° west-facing slope are characterized by similar pre-fire conditions. The results of model tree growth analysis suggest that the stand showed no signs of growth suppression prior to fire.

Judging by the adjacent forest site, brush layer with a density index value of 0.5 was prominent in the stand before fire. It consisted of *Rododendron dahurica* and single *Alnus fruticosa* Rupr. individuals. Average understory height was 1.5 m. A well-developed ground cover was dominated by *Vaccinium vitis-idaea*. Total dead ground fuel load was 32.7 t ha^{-1}. A general description of the phytocoenoses affected by fire is given below.

In the plot subject to a surface fire of moderate intensity, average trunk char height was 2.8 m. The brush layer was killed by fire except for trunks and I class branches (i.e. initial branches coming from the trunk itself). Over two years following the fire, mortality was observed mainly for trees less than 12 cm in diameter. In the plot that experienced a high-intensity fire the mean tree char height was 4.1 m. The fire completely consumed branches of the understory. The understory trunks were completely charred. Post-fire tree mortality resulted in considerable openings in the overstory. Only trees whose diameter exceeded the average value survived the fire. The third plot was established where a fire of very high intensity burned. In this case local crowning occurred. The understory was completely killed by fire. The overstory became extremely thin. Only large larches and single large pines survived the fire. In many of the surviving larches, the needles of the fire damaged branches recovered. Secondary needles are very dense. The length of the regenerated needles is 7-8 cm, which is twice as long as the norm.

Pre-fire similarity of forest inventory parameters in burned and unburned (duplicate) plots permitted one to attribute the changes in stand productivity to fires of various intensities. The ratio of living-to-dead trees decreases with increasing fire damage. In the first two plots, tree mortality is contributed mainly to the prevailing of thin trees, while large tree crowns are only slightly damaged. In the third plot where surface fire locally went into overstory crowns, mortality of both thin and thick trees were found to be relatively equal. In larch portions of the stand, this trend in tree mortality is not that clear since larch makes up a relatively small percentage of the stand. Furthermore, the capability of larch needles to regenerate hampers the mortality of larches even when their crowns were scorched.

Tab.1. Radial tree increment (mm) for a 2-year period / radial increment changes (%) following fires of various intensity

Fire Intensity	Trunk Thickness Class (cm)									
	4	8	12	16	20	24	28	32	36	40
	Pine									
Unburned Plot (Control)	0.35	0.69	0.97	1.23	1.42	1.50	1.55	1.58	1.60	1.61
Moderate	0.09/-74.2	0.37/-45.0	0.65/-33.0	0.93/ -24.2	1.17/ -17.6	1.36/ -9.3	1.43/ -7.8	1.44/ -8.9	1.46/ [illegible] 8	1.48/ -8.0
High	–	–	0.73/-24.9	0.84/ -13.7	0.92/ -35.1	0.99/ -34.0	1.03/ -33.6	1.07/-32.3	1.10/-31.0	1.11/-31.0
Very high	–	–	-	1.60/+30.1	1.73/+21.8	1.80/+20.0	1.84/+18.7	–	–	
	Larch									
Unburned Plot (Control)	0.28	0.48	0.64	0.79	0.92	1.04	1.13	1.19	1.24	1,28
Moderate	0.11/-60.7	0.28/-41.6	0.43/-32.8	0.58/ -25.6	0.71/ -23.8	0.84/-19.1	0.94/-16.8	1.02/-14.3	1.07/-13.7	1.12/-12.0
High	0.12/-57.0	0.31/-35.4	0.46/-28.1	0.58/ -25.6	0.70/ -23.9	0.78/-30.8	0.80/-29.2	0.81/-32.0	0.82/-33.9	0.83/-35.0
Very high	-	-	-	0.99/+25.3	1.09/+18.5	1.12/+7.7	1.13/0	-	-	-

Surface fires of moderate intensity primarily affect thin trees, while thick tree response to fire is less manifested. The level of crown yellowness and radial increment change suggest that, in this case, stand recover processes can be promoted by cutting all trees less than 8 cm in diameter and through the partial cutting of trees in the 12-16 cm thickness class, depending on the rate they were damaged by fire.

High-intensity fires considerably damage crowns and trunks of the overstory and thus has a strong effect on tree increment. Observed deviations from the main trend for some thin individuals are occasional and can be attributed to unsteady combustion; for this reason, some thin trees were only slightly damaged by "merciful" fire.

It is noteworthy that, in the plot that experienced a very high fire intensity, the trees showed an increase in increment. This is especially interesting because these trees were very small in number, while the major part of the stand was killed by fire completely. Single large trees whose crowns and bark were not mortally damaged by fire benefited much from increased insolation and made effective use of additional amounts of nutrients from soil which became available due to fire-induced forest floor mineralization.

In plots where high-intensity fire caused intensive stand thinning, the survival trees cannot form a closed canopy. However, high viability of some large trees suggest that seed sources are sufficient in the area.

Absolute decrease in current tree increment, relative to the pre-fire increment, characterizes immediate forest fire effects to which trees respond differently depending on their pre-fire state, which in turn describes the relationship between tree trunk current and mean increment. For middle aged pine trees damaged by fire, this coefficient was shown to be related to two factors, namely: tree growth class (according to Kraft's classification, c.f. Tkachenko [1952]) and the length of living crown. This relationship is described by the following equation:

$$Y + 41,75 + 9,75x_1 - 21,25x_2 - 7,2x_1 x_2 ,$$

where Y - is the ratio of current to mean increment; x_1 - is tree growth class (Kraft's classification); and x_2 - crown length. This equation is used to evaluate the viability of pine stands of age classes IV-V in old burns.

In order to get the full picture of fire-caused stand productivity disturbances, increment data were compared with wood biomass losses due to partial tree mortality (Tab.2). In a plot that experienced an experimental fire, the forest thinning process was observed carefully over 10 years. In a 65-yr old pine stand, the number of trees that died over a period of two years following a low-intensity fire was found equal to those dying during a 5-year period under normal conditions. For the same period of time after a moderate intensity fire, a decade of normal mortality was observed. A high-intensity fire led to almost a 3-fold increase in normal decade mortality during the next two years.

From a tree diameter viewpoint, stand structure is strongly affected by high-intensity fire. Apart from thin trees, some moderate diameter and thick individuals also die. The series of tree distribution by diameter show noticable right side asymmetry. Stand thinning was found to be 12, 17, and 21 percent for 2-, 5-, and 10-year periods, respectively. For other experimental fires, the values are considerably lower: 3.4% and 7% in the case of fire of moderate intensity and 1.3 and 5% in the case of low-intensity fire.

Over a 5-year period, tree mortality in the plot subject to high intensity surface fires was 83% of the normal wood increment in unburned plots over the same period. Mortality decreases with time for which reason only half of tree increment is lost over a decade.

Tab.2. Changes in the woodstock (m^3) of a pine stand subjected to experimental fires.

Five years after fire			One decade after fire		
Increment (m^3)	Mortality (m^3)	Mortality-to-increment ratio	Increment (m^3)	Mortality (m^3)	Mortality-to-increment ratio
Unburned Plot (Control)					
8.70	0.06	0.070	17.40	0.67	0.038
Low Intensity Fire					
9.00	0.38	0.042	18.00	1.12	0.062
Moderate Intensity Fire					
9.00	0.56	0.062	18.00	1.65	0.092
High Intensity Fire					
8.70	7.20	0.830	17.40	8.44	0.485

High-intensity surface fires, especially those covering large areas, usually develop into crown fires, at least in a part of the area. In this case, stands either die completely or become open woodlands. This is characteristic of the region under study and is clearly manifested in present pine forest patterns which, in fact, is a mosaic of relatively small (up to 1-5 ha) forest stands. The difference in age of stands of adjacent sites corresponds to the time since fires originated them. A distinct and widespread phenomenon in the typical picture is the replacement of climax species by hardwoods; this is common around Baikal lake, on the Khentey-Chikoy upland, and in other regions characterized by moderate precipitation amounts.

The above post-fire tree increment variations are observed also in larch stands growing under non-permafrost conditions in the Trans-Baikal region.

In the north of the Trans-Baikal region, post-fire tree increment changes in larch stands reflect the typical situation that occurs under permanent permafrost conditions. These are pure larch-dwarf birch stands (age: 135 years) with 19 m mean tree height and 24 cm mean tree diameter; in the unburned plot, stocking density is 0.6.

Radial tree increment observations were performed in three sample plots situated in the same forest inventory unit crossed by an automobile road and a railroad. The control is an unburned stand separated from other plots by roads. The other plots, in fact, are a part of a large burned area. In one of these two plots, a low-intensity fire burned, while the other was subject to a fire of extremely high intensity showing signs of local crowning. In the latter case, the char height of 7.2 m was observed for the trees that remained viable three years following the fire. Many living trees still have signs of crown scorch, obvious from an unusual crown needle distribution and density. In the lower parts of the crowns of the majority of trees needles are mainly concentrated on short small shoots and in very high density. Visual comparison with the unburned control suggests that the needle biomass of lower branches exceeds the pre-fire amount.

As is obvious from the data in Table 3, surface fire promoted radial tree increment. Moreover, a noticeable increase in tree radial increment was observed in the plot subject to a fire of very high intensity. It is noteworthy, however, that in the latter case, 40% of trees died from heavy thermal injuries during the three years following fire. For this reason, wood biomass losses greatly exceed its increment observed for surviving trees. The fire of such an intensity resulted in a considerable stand disturbance. Data that were obtained in a 42-year old larch stand situated in Eastern Khentey (Mongolia) illustrate well the relationship of post mortality with fire intensity. In this stand, characterized by uniform pre-fire conditions, sample plots were established to represent the entire variety of fire effects; for each burnt sample plot, an unburned control plot was selected. The errors of the post-fire tree mortality estimates were small because the fire had occurred only three years earlier and the factors leading to the death of the trees were easy to determine.

Observational results (Tab.4) show that tree mortality is directly related to fire intensity. Tree mortality increases with increasing fire intensity. Also, consistent qualitative changes are apparent. Low-intensity fire only affects suppressed trees, and the mortality pattern is largely the same as was in the control. Fire of moderate intensity kills the majority of thin trees and a considerable number of middle-thick trees. High-intensity fire leads to catastrophic tree mortality involving many trees of Class I and Class II (Kraft's classification).

Tab.3. Radial tree increment (mm) in a 135-year old larch stand.

Unburned Plot (Control)	Low Intensity Fire	High Intensity Fire
Pre-Fire Increment		
0.50 + 0.03	0.35 + 0.04	0.35 + 0.02
Post-Fire Increment		
0.50 + 0.03	0.47 + 0.05	0.43 + 0.03

Tab.4. Post-fire tree mortality (%) by stem diameter class in a 42-year old larch-mixed-herb stand.

Diameter class (cm)	Fire Intensity			Unburned Plot (Control)
	High	Moderate	Low	
2	100	100	57	32
4	100	100	32	15
6	90	54	17	2
8	84	40	7	-
10	53	28	-	-
12	37	10	-	-
14	25	5	-	-
16	18	-	-	-
18	12	-	-	-
20	5	-	-	-
22	-	-	-	-
24	-	-	-	-
Entire Stand	55	29	16	4

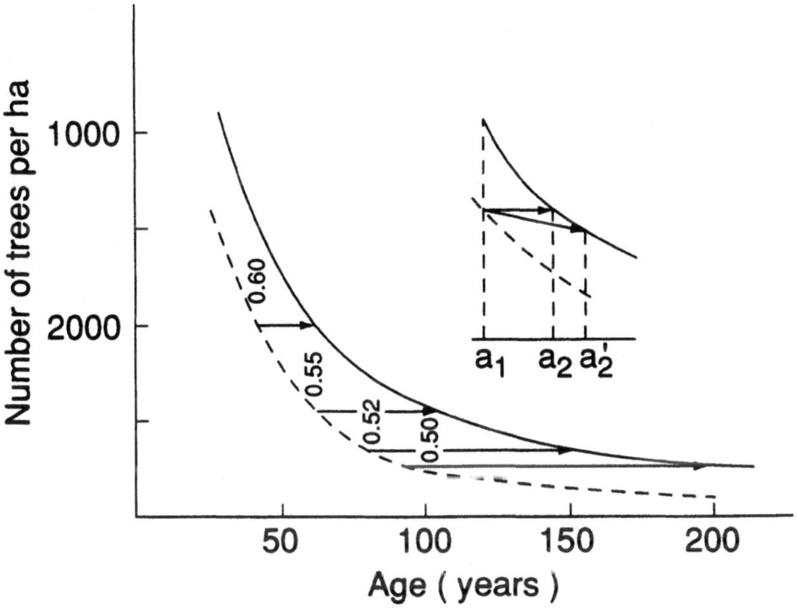

Fig.1. Dynamics of tree numbers in relation to tree age in burned (dashed line) and unburned/ normal (solid line) larch stands. 0.60-0.50 is relative stocking density. Arrows indicate the potential time period required for woodstock and stand density to recover to the norm.

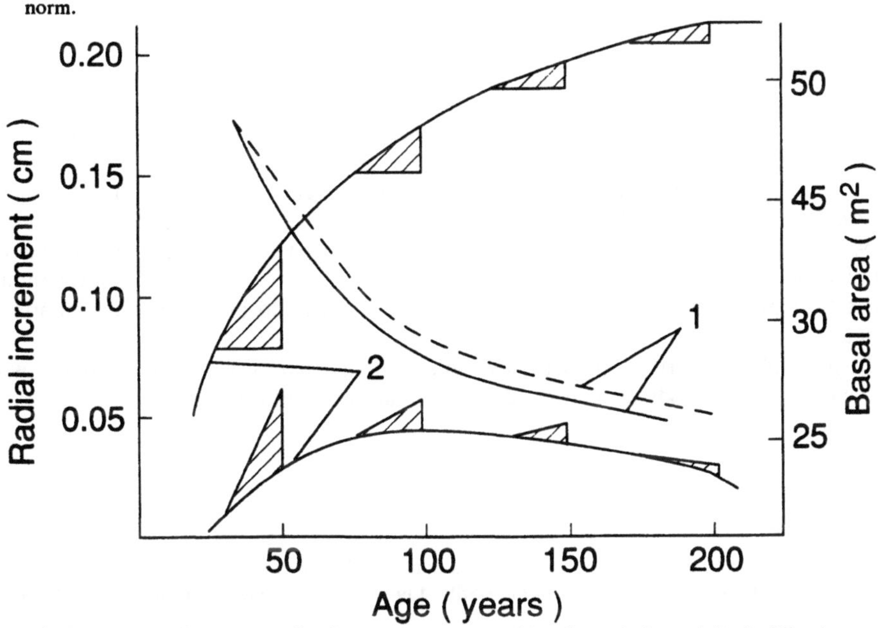

Fig.2. Age-related dynamics of radial tree increment and basal area in burned (dashed lines) and unburned (solid lines) larch stands. 1 - radial increment; 2 - basal area. Shaded segments indicate current increase in basal area by tree group.

Histograms of the number of trees in siberian larch stands taken from tables for tree growth are presented in Figure 1. The curve corresponding to the dynamics of a stand with a normally closed canopy was obtained using the table of Tikhomirov (1967), while its pyrogenic variant is based on the table of Gusev (1967). The second variant can be considered, in a way, as an age development of the static variant in Table 4 which corresponds to fire of moderate intensity.

When analyzing the time periods needed for stand density of stocking to regenerate to the norm, one can conclude that they increase with age (Fig.2). Twenty years are required by young stands to regenerate stocking to the norm, whereas pre-mature stands (aging 60 to 80) need a 100 year period. Thus, a larch stand's capability "to cure" fire damage is exhausted when it reaches middle age. High-intensity fires burning in older, formerly thinned stands can cause tree mortality anomalies that endanger the entire phytocenosis stability. A general description of the time period needed for stand stocking and density to regenerate back to the norm is given by:

$$i = a_1 - a_2 + \Delta a$$

where a_1 is the age at which the stand was damaged by fire; a_2 is the age of a given stand density under natural (no-fire) thinning; and Δa is the correction for post-fire natural thinning.

High-intensity fires, which are still common in the region in question is due to poor forest fire protection, often kill large larch stands and inhibit the following colonization of burned areas by hardwoods (mainly by birch and brush species).

Periodic surface fires also result in low wood quality and trunk rot. The commercial value of such forest stands decreases to 3rd class or lower.

4. Conclusions

To sum up, the following conclusions can be made in regard to forest fire protection for the region in question.

Frequent forest fires occurring in the Trans-Baikal region and Mongolia have negative bearings on the productivity of light coniferous forests growing under water-stress conditions. High-intensity fire-caused forest disturbances endanger the entire natural complex around Lake Baikal. To stabilize the ecological situation in the region, efforts should be made to improve forest fire protection making it adequate to the biospheric importance of Baikal.

In principle, prescribed fires of low and moderate intensity can be used for wildfire prevention, but methods of their conduction should be carefully chosen.

References

Gusev, N.N. 1967. Growth of Larix sibirica stands in the southern part of Central Siberia. In: Growth of the major forest forming woody species of the USSR, 139-140. Lesnaya Promyshlennost Publ., Moscow <in Russian>.

Tikhomirov, B.N. 1976. Growth of Larix sibirica stands in Khakassia. In: Growth of the major forest forming woody species of the USSR, 138-139. Lesnaya Promyshlennost Publ., Moscow <in Russian>.

Tkachenko, M.E. 1952. General forestry. Goslesbumizdat Publ., Moscow-Leningrad, 600 pp. <in Russian>.

J.G.Goldammer and V.V.Furyaev (eds.), Fire in Ecosystems of Boreal Eurasia, 219-226.

Forest Fires in the Eastern Trans-Baikal Region and Elimination of Their Consequences

V.F. Rylkov [1]

1. Introduction: Forest Fire Dynamics

High fire danger in the forests of the Eastern Trans-Baikal Region can be attributed to the following factors:

a) Specific climatic conditions

The non-uniform distribution of precipitation by season (the largest amount is observed in the second half of summer) results in the occurrence of long drought periods in autumn and spring. The spring period is characterized by high solar activity, windy weather (for example, in May 1987, wind speeds in excess of 24 ms^{-1} were measured for 18 days), and extremely dry air with relative humidity values as low as 7-10%. Observational data acquired by Yevdokimenko (1977) for the Chita Region, showed relative humidity was less than 30% every other day in spring.

b) Peculiarities of the mountain taiga vegetation pattern

Conifer species (*Pinus pumila, Larix czekanowskii, L.sibirica, Pinus sylvestris, Pinus sibirica* and others) distributed in the zone of permanent or discontinuous permafrost account for 76.4% of the total forest area. Conifer needles dominate the litter layer and downed dead branches and trees accumulate on the surface, requiring several decades to decompose. In the Chita Region, hill slopes are much steeper than those in Krasnoyarsk and Khabarovsk Territories or the Irkutsk and Amur Regions. Only in Buryatia do hill slopes compare with those in the Chita Region. In the Chita Region, 16-25 degree slopes account for 26% of the total forest area, while slopes exceeding 25 degrees make up another 7%. A combination of steep open slopes covered by dried grass, with north-facing slopes dominated by *Rododendron dahuricum* favors high fire danger. Forests represented mainly by high mountain forest types (Tab.1), with surface cover dominated by lichens, *Rododendron spp.*, *Vaccinium vitis-idaea*, mixed herbs, *Betula nana*, or green moss, are characterized by high fire danger and account for 61.4% of the total forest area.

---

[1] Institute of Natural Resources, Russian Academy of Sciences, Siberian Branch, 672000 Chita, Russian Federation

Tab.1. Forest areas of the Eastern Trans-Baikal Region with respect to forest type

Forest Type	Area (hax10^3)	Land Cover (%)
Mountain forest type	2303.6	8.5
Lichen-dominated surface cover	304.3	1.1
Rododendron-dominated suface cover	4253.7	15.8
Vaccinium vitis-idaea-dominated surface cover	5886.4	22.2
Mixed herb-dominated surface cover	4590.6	17.0
Larix-Alnus forest type	594.1	2.2
Larix-B.nana forest type	1259.2	4.7
Larix-Ledum forest type	5647.2	21.0
Broad-leaved herb-dominated surface cover	61.4	0.2
Forests of flood plains	152.0	0.6
Forests adjacent to brooks	460.4	1.7
Green moss and *V.vitis-idaea*-dominated surface cover	158.6	0.6
Green moss-dominated surface cover	473.2	1.8
Carex and *Sphagnum*-dominated surface cover	177.4	0.6
Sphagnum-dominated surface cover	366.1	1.4
Total	26945.8	100

c) Increasing human activity in sparsely populated areas

According to fire statistics for the forests adjacent to the Chita section of the Baikal-Amur Railway (BAR), forest areas partially or completely burned by fires increase with increasing population density. For example, if 1971-1975 fire damaged forest area levels (before BAR construction) are taken as 100%, these levels increased to 125% for the following 5-year period (BAR construction) and 591% for the 1981-85 period (Kotelnikov and Rylkov 1986). In 1986 alone, the forest area completely destroyed by fires was twice as large as the total 1971-1975 fire-damaged forest area. This is partially due to insufficient forest protection personnel in the BAR zone. For example, two forest exploitation areas in the Chita Region that total 5,600,000 ha are the responsibility of only 26 foresters who are very poorly provided with firefighting equipment (Kotelnikov and Rylkov 1987a). 127 to 1,018 fires occur annually in the forests of the Eastern Trans-Baikal Region. Years of high rate of burning, when both fire frequency and area burned exceed the average annual norm, are followed by 1 to 4-year periods with weather conditions hampering fire occurrence (rainy summer and autumn and low-snow winter).

A considerable decrease in the rate of burning during such periods make forest management authorities a bit too self-confident and they start to focus on forest planting, forest resource use, and other activities. In general, forest fire protection measures are planned with the assumption of a moderate rate of forest burning. As a result, when a year of high rate of burning suddenly occurs, forest fires achieve catastrophic scales. Such years can be predicted. A dry August and September the previous fall, and low snow cover in winter should be considered by the State Forest Protection Service as a warning of high rate of burning in the period to follow. In Table 2, data are presented on rates of forest fires for the past 39 years, which show that a high rate of burning was recorded in 1954, 1955, 1956, 1959, 1961, 1965, 1966, 1967, 1969, 1971, 1972, 1975, 1976, 1978, 1979, 1986, 1987, and 1992. The annual area burned in the 1954-1992 period varied from 0.01 to 153,700 ha. It should be noted that these data do not represent the real picture. Forest protection authorities, in trying to persuade higher officials that everything is fine in fire protection, intentionally reduce the values of annual forest area burned as a sort of bureaucratic approach.

Tab.2. Rate of forest fires in the Eastern Trans-Baikal Region

Year	Number of Fires	Burned Area ($ha \times 10^3$)	Year	Number of Fires	Burned area ($ha \times 10^3$)
1954	538	132.1	1975	696	7.6
1955	410	153.7	1976	820	21.7
1956	251	62.1	1977	461	3.5
1957	302	20.7	1978	549	19.1
1958	392	37.2	1979	660	9.2
1959	581	61.0	1980	513	0.8
1960	313	7.3	1981	514	4.7
1961	511	125.7	1982	390	1.4
1962	308	4.4	1983	295	0.9
1963	280	2.7	1984	127	0.1
1964	572	6.0	1985	395	1.5
1965	888	125.9	1986	791	4.8
1966	890	48.9	1987	938	93.8
1967	1018	28.6	1988	177	1.9
1968	518	1.5	1989	202	0.9
1969	641	2.8	1990	426	11.8
1970	381	1.0	1991	281	1.1
1971	621	0.9	1992	622	10.5
1972	912	81.8			
1973	461	5.4	Total =	20,226	1137.4
1974	583	2.4	Average =	519	29.2

Despite efforts to put a good face on the matter, the fact that the real situation is quite different is confirmed by a number of indirect factors. For example, *Larix dahurica* and *Pinus sylvestris*, being the main woody species in Eastern Trans-Baikal Region, appear, when compared to other regions, to be remarkable for producing cones every or almost every year (Pobedinsky 1991). Larch seed crop varies with year from 2 to 16 kg ha^{-1} (Danilov 1952; Pobedinsky 1962; Pravdin 1963). According to the data obtained by Panarin (1966) and Kruklis and Milyutin (1977), the amount of seed produced annually by *L.dahurica* and *L.czekanowskii* is even greater, 36 and 120 kg ha^{-1}, respectively. The amount of pine seeds is considerably less, varying with forest type from 3.5 to 0.5-1.5 kg ha^{-1}. This crop is sufficient for numerous self-sown seedlings to occur either under the forest canopy or in burns and cut areas. In the Eastern Trans-Baikal Region, about 50,000 ha of forest area are annually subject to harvesting. The total 5-year logged area is 250,000 ha. According to the recent forest inventory data (for 1983-1987), forest regeneration is in progress in a 172,000 ha part of the total area logged over in this period, while in the remaining 78,000 ha part this process has already been completed. The case is different with burned areas. One could expect that the total 5-year burned area would be 150,000 ha. Inventory data summarized by the end of the 1983-1987 period show, however, that burns cover 716,000 ha, open woodland makes up 707,000 ha, and fire-caused forest glades account for 66,000 ha of the total area. This suggests two conclusions: either 1) forest regeneration is slow in burned areas, or 2) the values for forest area burned are reduced by forest protection personnel on purpose. Comparison of official information with the few data available in the literature confirms the validity of the latter. For example, in 1987, 94,000 ha of forest area were destroyed by fires in Chita Region. At the same time, forest fires burned in the Big Hingan - a zone in China bordering Russia (Chita and Amur Regions). The fires were active for 26 days, spreading across over 1,000,000 ha. Forests accounted for 70% of this area. The fires killed about 650,000 ha of forest. Also, fires burned for 22 days in the Ingodinsk forest exploitation area in the Chita Region. Even very rough calculations yielded a 40,000 ha area burned. The official value, however, is only 1,500 ha. In 1987, about ten 20-day forest fires occurred in the Chita Region. Productivity of the forests in this region is equal to that of the forests in NE China. Using this analogy, the total forest area burned in the region should not be less than 5,000,000 to 6,000,000 ha, whereas officially this value was only 94,000 ha.

2. Forest Fire Consequences

a) Accumulation of standing snags and downed wood

Forest fires, whatever their intensities, result in the occurrence of snags and debris. In normal forests of the Eastern Trans-Baikal Region where dry climate prevails, natural mortality represented by snags and downed wood does not exceed 10% of the total standing crop per hectare. Low-productivity forests are widespread in the region, and, on average, have standing crops of 100 m^3 ha^{-1}. Snags and downed wood can thus not make up more than 10 m^3 ha^{-1}. According to 1971-1982 forest inventory data 4.1×10^6 ha of forests were burned, partially or completely, by fires in the region, with snags and downed wood totaling 132×10^6 m^3. This leads to a decrease in commercial forest area (Kotelnikov and Rylkov 1987b).

b) Promotion of young stands

The forests of the Eastern Trans-Baikal Region, especially pine stands, are remarkable for their high regeneration capability. In most burns, forests regenerate completely, without vegetation succession, in 3 to 10 years after a fire. This results in the total forest area remaining the same, with young stands being promoted over stands of other ages. However, forests previously planned for commercial harvesting turns into useless snags and downed wood.

c) Short-term forest succession

A part of fresh burns are invaded by aspen. Aspen stands account for 355,000 ha (i.e. 1.3% of the total forest area). There was a long-standing opinion among forest protection personnel that aspen is an undesirable woody species, and it was a common practice to remove aspen when conducting salvage cutting. Investigations have shown that aspen seedlings hamper grass layer development and that conifer regrowth (pine, larch, and *P.sibirica*) occurs under the aspen canopy. Normally, 10-20 yr old aspen stands are killed by fungi and replaced by young conifer stands (Rylkov 1975).

d) Replacement of larch by *Betula nana*

About 1,000,000 ha of forests (4% of the total forest area) are represented by larch stands with *B.nana*. Forest fires result in larch being completely replaced by *B.nana*-dominated dwarf shrubs. In the Eastern Trans-Baikal Region, *B.nana* is distributed across an area of more than 1,000,000 ha. Most *B.nana* stands replaced larch stands after fires. This is confirmed to be in the presence of clustered and single larch trees; in the clusters, snags and downed wood volume is 5 to 20 m^3 ha^{-1}.

e) Replacement of conifers by *Betula pendula*

Apart from aspen and *B.nana*, burns are often colonized by *B.pendula* stands. These stands make up 3,982,600 ha. Most of them, especially those adjacent to *P.sibirica* cone harvesting areas, are of fire origin and replaced *P.sibirica* stands. Replacement of *P.sibirica* by *B.pendula* takes a long time.

3. A Technology for *Larix / B.nana* Forest Restoration

In *Larix / B.nana* communities, fires kill both regrowth and parent stands. Fire-injured larch die and become snags or downed wood. Fires and active sun rays, previously scattered by canopy, are responsible for an increase in permafrost depth of 0.3-0.4m. Conditions are thus created that do not favor forest regeneration. The *B.nana*-dominated low-shrub layer does not burn out. Due to its high capability to spread, *B.nana*, a low-value species, forms stands in burns, with the standing crop being 10-20 m^3 ha^{-1}. Surviving clusters of larch trees do not contribute considerably to forest regeneration, since larch seeds are prevented from penetrating the soil by continuous moss-*Ledum* surface cover. Even in cases when a few larch seeds succeed in germinating, larch seedlings fail to compete with a well-stocked, 0.8-2.0 m high *B.nana* layer.

B.nana occurrence in place of fire-killed larch forests is favoured by soil swamping. Thus, to promote climax larch forest regeneration artificial soil drainage (reclamation) is needed. This is a very expensive procedure that would cost much more than the forest itself and an additional 100-years of forest protection expenditures. In 100 years, i.e. by the time trees in drained areas reach maturity, the expenses will of course, be partially compensated. However, the sum of money expended for the drainage will be 17.4 times larger than the expected income. This does not make sense. Considering this, a money-saving method was proposed, *Larix / B.nana* forest type restoration put into practice in place of fire-induced *B.nana* thickets (Rylkov and Gerasimov 1987, 1988).

The method is as follows: in swampy areas (25-40 ha in size) covered by well-stocked *B.nana* stands, 3.5-4 m spaced lines are ploughed downslope. Many scientists are of the opinion that since permafrost is located very near the soil surface, these plough lines would stimulate soil erosion and ravine occurrence. However, this does not happen. Water accumulated in the lines permanently streams downwards to result in gradual soil drainage. Favourable conditions are thus created for larch regeneration, with larch clusters being the starting point of this process. Larch regrowth and seedlings appear to concentrate in spaces between the lines. In 7 to 9 years, closed larch canopy (17,500-28,000 trees ha^{-1}) occur in the ploughed area. So the job - the replacement of low-value *B.nana* by larch, is completed with much less money required. Ploughing should be conducted in early spring (late March/early April) when the soil surface has melted to permit the plough to plunge. In cases when the area to be ploughed is crossed by streams or small rivers, the lines should be constructed at right angles to them.

4. Reconstruction of Birch Stands with *P.sibirica* Undergrowth

The post-fire replacement of *P.sibirica* by birch is a long-term process. Under the canopy of birch stands, about 12 km away from *P.sibirica* forest, *P.sibirica* regrowth can be found. The regrowth is distributed in groups due to *Nucifraga carycatactes* activity. *P.sibirica* regeneration is a complex process requiring a lot of time. The portion of *P.sibirica* nuts laid in and not consumed by *N.carycatactes* germinates. *P.sibirica* regrowth, increasing gradually under the birch canopy, forms highly viable biogroups. Over time, *P.sibirica* penetrates into the canopy and begins to dominate the overstory. *P.sibirica*, being a shade-tolerant and slow-growing species, makes up the most part of the overstory by the age of 100-120 years (Parfenov 1979).

P.sibirica trees growing at forest outskirts and in open woodlands start to produce nuts at the age of 30-40 years. The better the site conditions, the earlier the start of nut production. Uneven-aged *P.sibirica* stands with a closed canopy begin to produce nuts at the age of 90-120 years. The latest start of nut production is observed for *P.sibirica* developing under a hardwood canopy. In this case *P.sibirica* trees begin to produce nuts only after the hardwood canopy is completely eliminated. The stronger the hardwood canopy influence, the slower the development of *P.sibirica* reproductive organs. Under strong long-term suppression by hardwoods, *P.sibirica* nut production is delayed up to 140-160 years of age, with some individuals beginning to produce nuts as late as at the ages of 200-220 years (Iroshnikov et al. 1963).

Since more than 90.000 ha of the *P.sibirica* nut harvesting zone are made up of fire-induced birch stands with *P.sibirica* regrowth, our job was to develop a method of *P.sibirica* restoration. The method was expected to result in the earliest possible formation and rapid distribution of *P.sibirica* stands through cutting birch overstory. In this case, optimal conditions were believed to be created for steady *P.sibirica* height increment, that in turn, would lead to an earlier start of nut production. This method is not new: Polikarpov

(1966), who applied the method in the Sayan and Altai mountains, recommends that cutting be divided into several stages, with about 60-80% of the total birch growing stock being removed at the first stage. Other researchers (Bekh 1974; Bekh and Taran 1979) who used this method in the forests of the Southern Trans-Baikal Region, suggest that the complete hardwood overstory should be clearcut at one time. We tested the method in the Eastern Trans-Baikal Region, where *P.sibirica* grows on heavily insolated mountain ridges. We found that after 60-80% of the total birch overstory is removed, the adaptation period is 10 years for a single *P.sibirica* tree and 7 years for a cluster of individuals. The adaptation period is the period from the time when post-cutting environmental conditions begin to influence *P.sibirica* regrowth to the time when a steady height increment (not less than 10 cm yr^{-1}) is recorded (Volosevitch 1968). However, during 7 to 10 years of *P.sibirica* adaptation to changed site conditions, birch appears to develop much more rapidly and begins, over time, to resuppress *P.sibirica*.

The results of our 1982-1985 tests suggest that in order to reduce *P.sibirica* regrowth adaptation time to a minimum, restoration of this species should be a two-stage procedure. The first stage, called "illumination", is the preparation of *P.sibirica* previously suppressed by birch overstory for new environmental conditions. At this stage, birch growing within 2.5 to 3 m around clusters or a single individual of *P.sibirica* are removed. The second restoration stage is cutting the whole of the overstory suppressing *P.sibirica* regrowth. Thus the time needed for restoration totals 4 years. In the first year, initial cutting is conducted, while during the next two years *P.sibirica* regrowth adapts to changed site conditions ("illumination" period). In the fourth year, all the overstory is cut. This results in young *P.sibirica* stands being promoted over birch. From the fifth year on, a steady 10-25 cm annual height increment is recorded for most (85%) *P.sibirica* individuals. At present this method, developed by Rylkov (1991b) as "A Treatment of *P.sibirica* Regrowth Under Birch Canopy" enjoys wide practice.

References

Bekh, I.A. 1974. *Pinus sibirica* stands of southern Trans-Baikal region. Nauka Publ. Novosibirsk. 212 pp. <in Russian>.

Bekh, I.A. and I.V. Taran. 1979. Siberian miraculous tree. Nauka Publ. Novosibirsk. 126 pp. <in Russian>.

Danilov, D.N. 1952. Periodicity of conifer seed production and geographical distribution of the seeds. Goslesbumizdat. 58 pp. <in Russian>.

Iroshnikov, A.I., V.F. Lebkov, and Yu.S.Tcherednikov. 1963. Nut production of *P.sibirica* stands of Lena-Ilim interfluve. In: Nut Production of *Pinus sibirica* in Eastern Siberia. Transactions of Institute of Forest Sib. Br. Russ. Acad. Sci. Vol.62, 34-75. Moscow <in Russian>.

Kotelnikov, A.M., and V.F. Rylkov. 1987a. Problems of forest management in the Udokan forest exploitation area. In: Economical and Geographical Problems of Exploitation, 87-95. Nauka Publ. Novosibirsk <in Russian>.

Kotelnikov, A.M., and V.F. Rylkov. 1987b. Problems of forest resource use and improvement of industrial complexes. In: Geographical Aspects of Economy Policy in Chita Region. Western Trans-Baikal Branch of the USSR Geographers Society, Vol.120, 61-70 <in Russian>.

Kotelnikov, A.M., and V.F. Rylkov. 1986. Ecological role of forest in mountainous parts of the Baikal-Amur Railway zone and problems of forest exploitation and protection (a case study in Udokan forest resource exploitation area). In: Abstracts of Reports Presented at the All-Union Conference on Ecological Role of Mountain Forests, 19-21 August 1986, 127-129 <in Russian>.

Kruklis, M.V., and L.V. Milyutin. 1977. *Larix czekanowskii*. Nauka Publ. Moscow. 212pp. <in Russian>.
Panarin, I.I. 1966. Forests of Central Trans-Baikal Region. Proc. 2nd Conf. Forest Management. Chita. 321 pp. <in Russian>.
Parfenov, V.F. 1979. A complex in *Pinus sibirica* forest. Lesnaya promyshlennost. Moscow. 239 pp. <in Russian>.
Pobedinsky, A.V. 1962. Pine stand regeneration in Trans-Baikal Region. Lesnoe khozyaistvo (J. Forest Management) No.7, 19-22 <in Russian>.
Pobedinsky, A.V. 1991. Forestry assessment of the replacement of climax taiga forest species by secondary ones. Lesnoe khozyaistvo (J. Forest Management) No.11, 19-22 <in Russian>.
Polikarpov, N.P. 1966. Mountain forests of Siberia and scientific approaches to their management. Krasnoyarsk. 35 pp. <in Russian>.
Pravdin, L.F. 1963. Natural regeneration of pine and larch in Buryatia. Transactions Inst. For. Sib. Br. Russ. Acad. Sci. Vol. 57, 27-41. Krasnoyarsk, Moscow <in Russian>.
Rylkov, V.F. 1975. Peculiarities of conifer regenerationin cut areas in the mountain-taiga and forest-steppe zones of southern and Eastern Trans-Baikal Region. Scope of Scientific Papers of Moscow Forest Management Institute (MLTI), Vol.68, 33-46 <in Russian>.
Rylkov, V.F. 1991a. Justification of the amount of forest cultivation activities. Information Sheet, Chita Inter-Industry Science Center No.7. 4 pp. <in Russian>.
Rylkov, V.F. 1991b. A treatment of *Pinus sibirica* regrowth under birch canopy. Author Certificate No. 1639507 <in Russian>.
Rylkov, V.F., and V.B. Gerasimov. 1987. Man-enhanced larch stand regeneration using the method of *B.nana* restoration. Information Sheet of Chita Inter-Industry Science Center No.96. 4 pp. <in Russian>.
Rylkov, V.F., and V.B. Gerasimov. 1988. A technology for climax larch-*B.nana* forest type restoration. In: Problems of Forest Resource Exploitation in Trans-Baikal Region. Western Trans-Baikal Branch, USSR Geographers Society, Vol.122, 39-40. Chita <in Russian>.
Van Sitzy. 1987. A forest fire in Big Khingan. China N 9, 6-11 <in Russian>.
Volosevitch, I.V. 1968. The peculiarities of spruce regrowth and pole stand development in the north. In: Cutting and Regeneration of Forests in the North, 181-197. North-West Publ. House <in Russian>.
Yevdokimenko, M.D. 1977. Pyrological characteristic of mountain taiga forests of Baikal basin. In: Forest Protection and Regeneration in Trans-Baikal Region, 5-55. Krasnoyarsk <in Russian>.

J.G.Goldammer and V.V.Furyaev (eds.), Fire in Ecosystems of Boreal Eurasia, 227-238.

Pyrological Zoning: Principles, Methods, and Significance of the Role of the Geographical Factor in the Problem of Wildland Fires

M.A. Sofronov [1]

1. Introduction

Forest and other wildland fires must necessarily be controlled to prevent potential damage to human life, property and the environment. The greater the expenditures on fire control, the less the damage. Today it is commonly accepted by wildland fire managers that the best approach to funding is to try to minimize the sum of cost plus loss. Therefore, an effective way of arriving at an optimal solution to the problem is to consider the following conditions: (1) effective fire control, and (2) accurate and thorough assessment of fire damage. The control of wildland fires is rather complicated by the following properties of the phenomenon:

- The occurrence of wildland fires is very uneven in terms of time and space, so that firefighting resources have to maneuver between regions. To bring good results, the maneuvers should be performed in a timely and anticipatory manner; this is only possible on the basis of regional fire danger forecasts.

- If a wildland fire escapes control, it may spread across a large area. For this reason, the control of wildland fires must be free of mistakes and underestimates. Therefore, standard containment tables must inevitably be conservative, with over-estimated production costs and expenditure ratings, i.e. wasteful. Effective control of fires can only be achieved on the basis of sizing up by an experienced specialist, who has the ability to take into account both regionally dependent features of fire behaviour and the actual fire situation.

- In boreal forests, fires occur mainly in areas with poor road access, which often restricts the use of heavy fire-fighting equipment, making hand tools the main means of firefighting in such conditions. The limited productivity of hand crews often make it necessary to retreat to a convenient place to construct the control line, and to apply large-scale backfiring. Choice of the best tactics is much easier with the use of maps which display the character and state of the vegetative fuels for the area.

[1] V.N. Sukachev Institute for Forestry and Timber, Siberian Branch, Russian Academy of Sciences, 660036 Krasnoyarsk, Russian Federation

The influence and consequences of fire on natural communities result, first, from the impact of the qualitative and quantitative characteristics of the fire, and second, from the properties of the communities themselves. It seems reasonable to investigate and predict fire effects in ecosystems within the boundaries of natural regions, where burning pattern and the character of fire effects are similar across the region. We propose to use taxa of our pyrological zoning as such natural regions.

3. Principles of Pyrological Zoning

The term *pyrology* was introduced by Melekhov (1944). It is defined as a science which analyzes the behavior and effects of fire on vegetation. It includes forest pyrology (forest fire science). The concept of fire in the vegetation cover is less confined than that of *wildland fire* because it embraces plant communities not only of natural origin, but of agricultural origin as well (e.g., tree plantations). The term "zoning" is widely used in geographical sciences in Russia. Zoning means the partition of a territory into sections so that every region is characterized as homogeneous within a particular set of features appropriate to the objectives of the classification. Zoning may be performed through a hierarchy of levels, partitioned into sections of characteristic scale.

It is necessary to differentiate between economic and natural zoning. Economic zoning deals with jurisdictional and developmental units, but natural zoning usually has no relationship to jurisdictional demarcation. Zones of responsibility for fire agencies provide a clear example of economic zoning. Pyrological zoning is a natural one (Sofronov and Volokitina 1990).

There are two methodological approaches to natural zoning: one is typological and the other specific (Armand 1975). The typological approach is characterized by the selection of criteria (usually quantitative) on which to base the classification and determination of threshold levels. Zones falling into different threshold levels are segregated and delineated for the territory of concern. It is obvious that many of the zones determined with this method will be characterized identically by the criteria chosen. Climatic zoning may be regarded as an example of the typological approach.

If the specific approach is applied, the zones are established as unique objects of nature, objects of qualitative peculiarity. Specific zoning may be a very useful method for the description of poorly developed wildland regions (e.g., Siberia). The complexity and multi-component nature of the fuel properties of the vegetation cover make the specific method the only appropriate approach for pyrological zoning.

What is the aim of natural zoning? Exploration of vast expanses of wildland poorly developed by humans is often performed with the use of the so-called "key method." The essence of the method is that research results collected on a single object (stand, plot, etc.) are assumed as common for all similar objects. Nevertheless, such an assumption is only applicable within the boundaries of a particular natural region. The key method is also used for satellite image interpretation. Key objects must be located within the territories of their natural regions. It should be noted that the generalization of natural research data across a large territory of varying character must also be done within natural regions.

Pyrological zoning expresses fire response and fire regime characteristics of the vegetation cover of the globe. These characteristics are related to interactions between fire and vegetation, where plant biogeocoenoses are the object of combustion, and the prime condition for the existence of fires is their ability to spread across terrain from the point of origin. Therefore, the following conditions are considered in pyrological zoning: (1) the fuel properties of the vegetation, (2) fire start circumstances (that is ignition sources), and (3) fire spread conditions (that is spatial distribution of different patches of vegetation, and territory fragmentation - discontinuity of the fuel cover due to crossing by rivers, streams, roads and other barriers). A wildfire usually lasts for days, or sometimes for months, and fires may burn over large and diverse areas. Therefore, changes in vegetation character over time and territory fragmentation must also be taken into account in pyrological zoning. Both factors are mostly influenced by fires and humans.

Both the character of the spatial distribution of different patches of vegetation and the character of territory fragmentation are well reflected on satellite images in the patterns of vegetative cover. This structure is a result of climatic, geomorphologic and historic (including human-caused) conditions. We assume that similarity of vegetation cover is an indication of similarity in pyrological characteristics. Therefore, in the procedure of zoning we separate regions (or plots) with portions showing similar structure of vegetation cover, which are seen on satellite images as patches of a homogeneous character within the mosaic.

Every zoning is somewhat lacking in accuracy, so regional boundaries at any level are adequate for initial mapping. After the correction of pyrological characteristics of delineated portions of the territory their boundaries are to be revised.

Final fire sizes may vary from one hundredth of a hectare to hundreds of thousands of hectares, so we established a pyrological zoning system with 6 taxonomic levels. The lowest level is a homogeneous patch of vegetation - the pyrological plot. Higher incorporating levels are (in ascending order) pyrological: lot, wildland, terrain, district, and region. Zoning may be started at any level, or at several levels simultaneously. In the process of pyrological zoning of a territory, zones (or plots) are separated into areas with a homogeneous structure of vegetation cover, i.e., areas which show a uniform pattern on aerial images. According to the zoning level chosen, it is necessary to determine the right scale of images, the season of image taking, optical bands, etc. For example, zoning of a forested territory might be done with the use of black and white winter satellite images of the scale 1:2,000,000 to 1:10,000,000, though NASA's scanned infrared summer images with levels of grey depicted in contrast colours are the best for this purpose.

Unfortunately, in our country we have neither such scanned images nor a bank of cloud-free images. Under the circumstances, we have had to use vegetation and terrain maps along with available images (Sofronov 1979; Sofronov and Volokitina 1990). Using this approach, we have separated 49 pyrological regions within the forested and semi-forested zones of the territory of the former USSR (Fig.1). The next level - of pyrological districts - has thus far been applied only for Siberia. Such zoning units as the pyrological terrain and the pyrological plot have been used by Volokitina (1991) in her work of mapping vegetation fuels, where forest inventory plots supplemented with pyrological characteristics are used as pyrological plots. Furyayev (Furyayev and Kireyev 1979) investigates fire regimes of forest areas using such units as the natural district, terrain, and landscape, which are topographically separated by by researchers.

3. Characterizing Fire Danger

Pyrological characteristics of vegetation are, in fact, characteristics of its fire danger. The word "danger" here means possibility or probability of some damage. The most definite meaning implies possible damage from fires (Sofronov 1970).

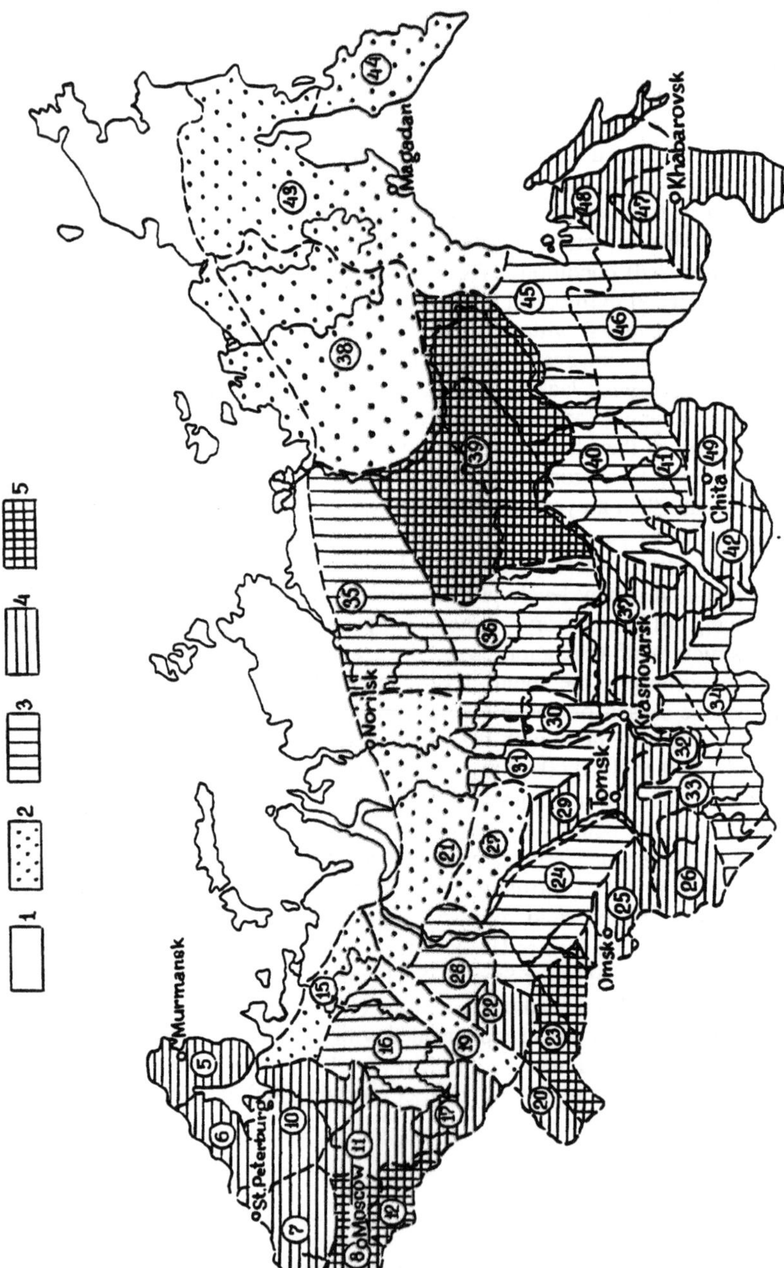

Fig.1. Fire Incidence Pattern on the Pyrological Region Level. Fire Incidence Levels: 1 - low, 2 - mean, 3 - high, 4 - very high, 5 - extreme.

It is necessary to differentiate between the terms "actual fire danger" and "natural fire danger." The actual fire danger of a particular territory results from the ignition risk (Kurbatski 1972) and the fire spread potential in the territory. One of its components is the probability of occurrence of an ignition source. This parameter depends mostly on the size of the area and the duration of the fire season (Tab.1). Fire agencies can use values of the parameter, falling into a realistic range (that is 0.1 -1.0), in activity planning. Assessment of a current actual fire danger (for a current day) will be usable for areas of 1,000,000 ha or larger (i.e. for the level of pyrological district). By taking into account precipitation variations across the territory of the district, the spatial distribution of different fuel complexes, and past years fire occurrence (which follows the distribution of ignition sources), it is possible to determine those portions of the territory where the predicted fires are most likely to occur. The assessment of current actual fire danger on this smaller scale is necessary for fire patrol and lookout scheduling.

Tab.1. System of actual fire danger rating and its application.

Time Coordinate	Space Coordinate		
	P - within plots (biogeocoenoses, forest inventory plots)	B - within blocks (forest inventory blocks, small natural-territorial complexes)	R - regional (large natural-territorial complexes, administrative regions, forest enterprises
C: Current (for the current time or date) a) predicted b) actual c) retrospective	C-P, SA-P: Impracticable due to very low probability of the emergence of an ignition source in a small area. (In the territory around an active fire it is possible to assess actual fire danger, because there the probability of the emergence of fire is equal to one.)	C-B: Predicted and actual - for scheduling operations of fire agencies, refinements	C-R: Predicted and actual - for scheduling activity of fire agencies
SA: Seasonal (of a fire season) a) predicted b) retrospective		SA-B. Predicted: for activity planning	SA-R: Predicted: for interregional resource dispatch. Retrospective: for the analysis of activity of fire agencies
MYA: Multi-year average (for a fire period, fire season, 10-day period)	MYA-P: Practicable for thoroughly detailed pre-suppression planning	MYA-B: For fire management planning	MYA-R. For large-scale fire management planning

Presuppression planning includes taking into account actual fire danger records for a long period of time (e.g., 10 years), or an average seasonal rate is calculated from multiple year statistics. Using data from a longer period makes it possible to reduce the lowest level of zoning to 1000 ha (i.e., the area of pyrological lot). Provided the attractability and accessibility of the area to visitors are taken into account, the assessment of real fire danger may be performed at the level of the pyrological plot.

The concept of natural fire danger relates to the nature of a fire-prone natural object, its properties and conditions. First of all, it indicates the ability of the vegetation to catch fire in case of contact with an ignition source. The probability of the emergence of an ignition source is not specifically estimated, but is conventionally regarded to be equal to 1 (100%). Therefore, a natural fire danger rating index can not be confined to a particular area in practice. Burnable areas, surrounding an active fire and not isolated from it by any barriers, certainly have an ignition source emergence probability equal to 1. As a result, the actual fire danger rate and the natural fire danger rate are equal in such situations.

There may be different types of fire danger rating: either integral or expressed with a set of component factors; either current or average for a period, either local (for a particular territory) or abstract (for forest types, for example).

In the next section of this paper we discuss methodology issues. There are basic differences between a methodological approach for characterizing pyrological plots as individual areas of homogeneous vegetation and for consideration of other taxa (like blocks or regions) where classification is based on the aggregation of vegetation patches with similar characteristics. Due to the limited length of this paper, we consider here only the second of these problems. The problem of characterizing of pyrological plots is discussed by Volokitina (this volume).

In those cases where climatic and human-caused conditions are sufficiently stable, past years' statistics of fire occurrence and fire effects can be used to predict the influence of fires on ecosystems. But mankind may now be about to face an increase in global temperatures. Vegetation would be expected to need tens of years or even centuries to adapt to the warming. But there is a real possibility of an increase in droughts and conflagrations in the boreal zone associated with the warming, resulting in serious changes in the character of the vegetation in a few years. Therefore, modelling of global climate change scenarios should include the prediction of possible fire danger changes in regions.

In any region, current fire danger is influenced not only by weather. Other factors are: vegetation character, seasonal alterations of vegetation, the spatial distribution of different vegetation types, presence of barriers to fire spread, presence and spatial distribution of ignition sources, and so on. It seems hardly possible to describe such a complex relationship in a mathematical model. Nevertheless, there is a different approach. Kurbatski (1963) proposed that within the boundaries of a relatively small area and during a short, annually recurrent, period (a phenological phase or a 10-day period, for example) all factors except for weather can be regarded as constant. Using past years' fire and weather records, an empirical function reflecting relationships of these factors can be derived easily for every such period. Weather conditions of a region can be assessed with a drought index (for example, Nesterov index or HI-1 ("humidity index-1") index), developed to reflect fire danger (Nesterov et al. 1968; Vonskiy and Shzdanko 1976). Its value represents the moisture content of the standard fuel, which is a cover of true mosses over a drained soil. We developed an improved index - HHI ("humidity & hygroscopicity index"), in which the hygroscopicity of mosses and forest debris is also taken into account (Sofronova and Volokitina 1990):

$$HHI_n=[HHI_{n-1}+(t+10°)_n\times(t-t_d-5°)_n]\times(K_r)_n$$

where:

t - is air temperature (°C) at 1-3 p.m.

t_d- dew point temperature (°C) at 1-3 p.m.

n - means that the factor is calculated for the current day

n-1 - the factor is for the previous day

K_r - precipitation coefficient dependent on 24-hour total of precipitation (R, mm) or the total duration of rain (T, hours).

It is possible to calculate the precipitation coefficient by two methods:

$$1)\quad K_r=\frac{1.8}{R+1};\qquad 2)\quad K_r=\frac{1.8}{1.3T+1}$$

1) Taking into account the sum of precipitation (R) for the last 24 hours. If the precipitation for the last 24-hour period is less than 0.5 mm, the precipitation coefficient is set to equal 1.

2) Taking into account the total duration of rain (T) for the last 24 hours. If the duration is less than 0.4 hours, then the precipitation coefficient is set to equal 1.

The 24-hour period for this pyrological index does not begin at midnight, but in the morning (typically at 9 a.m.), and ends the next morning. Calculation of HHI is carried out twice for every 24-hour period: initially at the beginning of the morning, using the weather forecast. It is then corrected at the end of the next morning, using real weather data.

To further characterize drought severity, we have established drought classes (DC), which are derived from the value of the drought index. The limits of these drought classes, expressed in terms of a fireweather drought index (Nesterov Index, HI-1, or HHI-1), are: 0 DC = drought index less than 60; I DC = 60-300; II DC = 301-1000; III DC = 1001-3000; IV DC = 3001-10,000; V DC = 10,001-30,000; VI DC = drought index higher than 30,000.

The severity of fire danger should be assessed not only by the probable number of fire starts, but spread and suppression conditions should also be taken into account. However, fire danger classes based on number of ignitions per area unit, do not allow direct comparisons of flammability of the vegetation between different regions or seasons.

In the U.S.A. current fire danger is being calculated with the use of the National Fire Danger Rating System, which is based on a mathematical model, in table form, with many factors being taken into account. In the procedure, all the vegetation of the region is regarded usually as one basic or composite fuel model; the ignition component is assessed in relative units, territory fragmentation is not considered, and so on. The resulting Fire Load Index (FLI), is not comparable between regions.

The comparability of ratings is necessary to determine fire danger across a large region during the planning of presuppression resource allocation for fire situation control. Comparison is attainable with the use of a non-relative rating. Such an absolute rating may be based on probable fire occurrence (probable number of active and starting fires for an area unit of 1 million hectares). On the basis of probable fire occurrence magnitude, a regional current day fire danger class can be derived.

We divided the range of observed active fire densities (number of fires burning per million hectares of land area) into five categories: I - 0.1 and less; II - 0.11-0.40; III - 0.41-1.60; IV - 1.61-6.40; V - more than 6.40 fires/10^6 ha. These five classes of probable active fire densities correspond to five classes of fire danger. According to our data, the active fire density typical of the third Drought Class (III DC) in each region and at every point in the season is doubled for IV DC, and quadrupled for V DC. The active fire density in II DC is approximately half that in III DC.

We have distinguished 8 typical relationships between the forest fire-related drought index and the probable fire density and have designed on this basis 8 typical regional fire danger scales (Tab.2). The boundaries of these scales in terms of probable active fire density in III DC are: N1--0.1 and less; N2--0.11-0.20; N3--0.21-0.40; N4--0.41-0.80; N5--0.81-1.60; N6--1.61-3.20; N7--3.21-6.40; N8--more than 6.40 fires/10^6 ha.

Tab.2. Regional Current Fire Weather Assessment

Standard Local Scales	Drought Classes (DC)						
	O	I	II	III	IV	V	VI
	Fire Weather Classes (FWC)						
N1	I	I	I	I	II	II	III
N2	I	I	I	II	II	III	III
N3	I	I	II	II	III	III	IV
N4	I	II	II	III	III	IV	IV
N5	I	II	III	III	IV	IV	V
N6	I	III	III	IV	IV	V	V
N7	I	III	IV	IV	V	V	V
N8	I	IV	IV	V	V	V	V

Note: Fire danger assessment: FWC I--low; FWC II--moderate; FWC III--high; FWC IV--very high; FWC V--extreme.

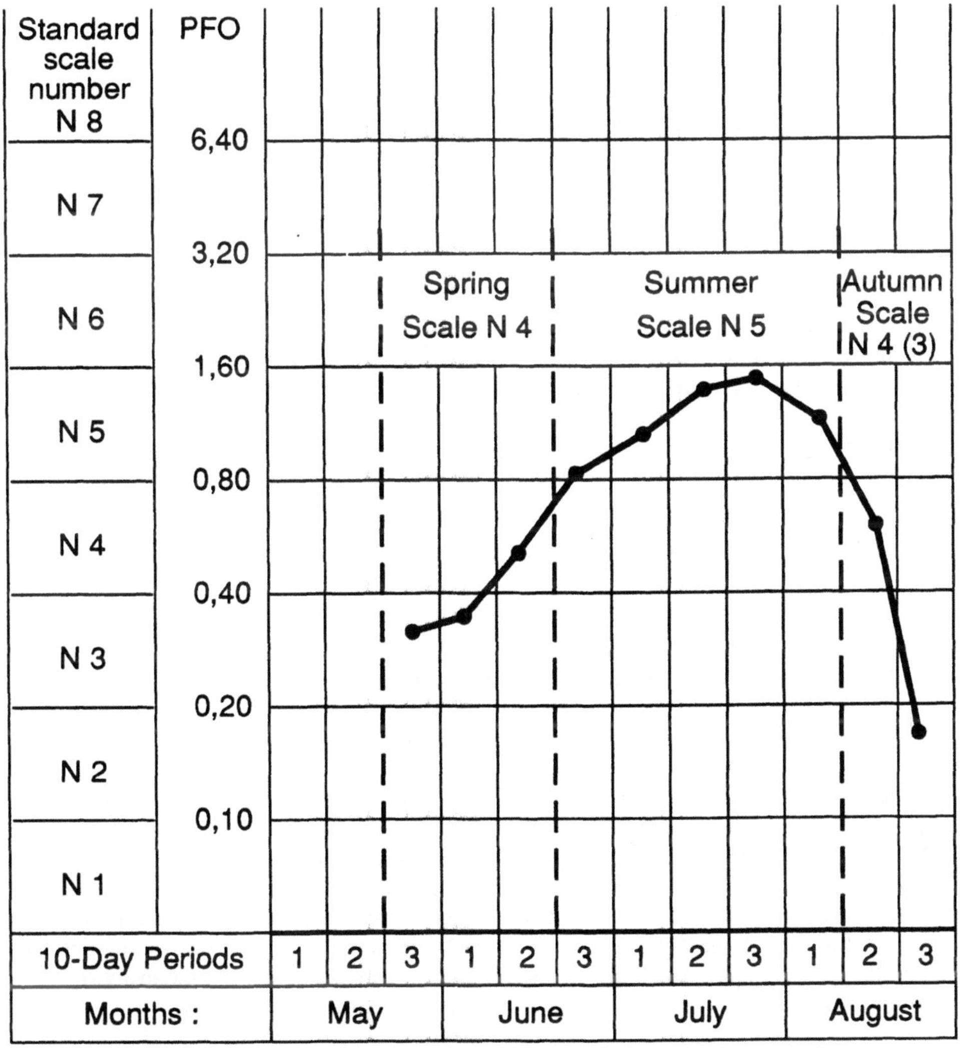

Fig.2. The dynamics of Probable Fire Occurrence (PFO) at Drought Class III during the fire season in a region of Priangarye. An example of dividing a fire season into fire periods and choosing a set of standard scales for the region. 10-day periods correspond to 1/3 of a month.

A method has been worked out to divide the fire season in every region into standard periods of one/third of a month (about ten days) and choose appropriate local scales for every period. It includes, with the use of multi-year statistics, determining the seasonal dynamics of fire density in DC III for the region considered. The graph (Fig 2.) and season are "automatically" divided by the limits of a typical scale into periods. The graph illustrates which scale must be used for each period of the season. The region considered should be 1 to 10 million hectares in area, which corresponds to one pyrological district. In practice, the fire danger class must be determined taking into account drought class and the local scale which corresponds to the region and the time within the season (Tab.2). The fire danger class shows the average (i.e., the most probable) active fire density, which was determined using many years of data. The actual active fire density on a given day may differ from the probable active fire density.

So, the division of the fire season into fire periods, with a fire danger ranking and probable fire occurrence (PFO) assigned to every period, should be regarded as a most important climatic characteristic of the pyrological district. This approach can also be applied to a section of the pyrological district with an area of 1 million hectares or more (for example, the responsibility zone of a local forest protection airbase). Pyrological regions, due to their heterogeneity, often cannot be characterized adequately without partitioning them into such smaller units.

One of the most important characteristics of pyrological regions and districts is their fire incidence. This may be regarded as the actual realization or expression of fire danger. Fire incidence of past periods may be used as a basis for the prediction of future fire incidence.

Usually, multiple year records of either the number of fires or the relative area burned are used to express fire incidence. We have proposed an additional method of fire incidence assessment - fire incidence index (I), which takes into account both number of fires (fire occurrence) and the area burned, and represents a sum total of the fire perimeters:

$$I_f=0.3(n\times S)^{0.5}$$

where

I_f is fire incidence index (km/100,000 ha per year)
n - fire occurrence (number of fires/100,000 ha per year)
S - relative area burned (ha/100,000 ha per year).

This equation is an empirical one; it takes into account the observed distribution of fires by size.

We took wildland fire statistics for 10 or more years from nine regions of Russia. For every region we sorted all fires in an ascending order by areas burned and divided the resultant array into 100 sections or ranks (r), then represented the area burned (s) of every fire as a fraction of the average burned area. We determined that different regions have nearly the same characteristics of the distribution of numbers of fires in the array of ranks, and the fires with areas burned, equal to the average, fall into the rank 88:

r	49	67	82	88	95	99	99.6	99.9
s	9.03	0.11	0.43	1.00	3.5	14	35	140

Provided the average area burned is known for a region of concern, it is easily possible to extrapolate the entire array of the distribution, to reckon a probable number of large fires, etc.

The following order of rating of fire incidence is proposed with the index: low (index value less than 0.7); mean (0.8-2.5); high (2.6-8.5); very high (8.6-26.0); extreme (more than 26.0).

When characterizing pyrological regions using the fire incidence index (Fig. 1), we were confined by using statistics from only a 13 year period (1965-1978), so a few years of high fire occurrence (1972, for example) were influential.

As a method of assessing natural fire danger, the concept of Natural Fire Danger Class (NFDC) is introduced. The NFDC is determined by the Drought Class (DC), at which the fuel cover becomes ignitable; for example: I NFDC at I DC, II NFDC at II DC, and so on.

Usually the territory of every region includes areas of vegetation with different NFDCs. To calculate an average NFDC, it is necessary to first convert qualitative degrees of NFDC into quantitative indices of a basic scale, calculate a weighted average index, then, using the basic scale, to determine an average NFDC.

It is proposed to determine the numeric indices of a basic scale as percentages of fire season days when the areas of relevant NFDCs can burn: I NFDC - 80 percent, II NFDC - 70, III NFDC - 50, IV NFDC - 25, and V NFDC - 3 percent. It is important to recognize that in many regions the distribution of vegetation areas by NFDC does not stay constant, but changes during the fire season. For such regions the calculation of weighted average NFDC must be first done for vegetation phenological phases (seasonal change stages), then for the entire season, with the number of days in every phenological phase being taken into account.

The pyrological fragmentation of a territory takes place with the presence of fire spread obstacles of two kinds: 1) linear obstacles, or barriers (such as rivers, streams, roads); 2) inflammable areas (dunes, rock or pebble outcrops, large sphagnum peats, plowed fields, etc.) and areas not ready to burn. The pyrological fragmentation of a territory into completely isolated areas may also occur (for example, dry islands amidst large sphagnum bogs).

It is proposed to assess pyrological fragmentation (F) by average distance between fire spread obstacles (km). Barriers of any width ought to be considered, because backfiring may be done from any linear fire spread obstacle. Overall pyrological fragmentation incorporates both a temporal fragmentation and a spatially constant one; the latter does not change seasonally.

Constant pyrological fragmentation of an individual territory may be determined with the use of a terrain map, on which two mutually perpendicular lines are drawn. On the lines all distances between fire spread obstacles are measured, then the average distance (the criterion of pyrological fragmentation) is calculated.

The following scale of degrees for assessment of pyrological fragmentation is proposed:

very high	$F < 0.5$ km
significant	F from 0.5 to 1.5 km
moderate	F from 1.6 to 5.0 km
low	F from 5.1 to 15 km
practically non-fragmented	$F > 15$ km

Temporal pyrological fragmentation may be assessed for all except the lowest taxonomic levels of zoning. Other useful characteristics merit consideration (e.g. probable character and intensity of burning, probable fire effects, the accessibility of an area), but we do not discuss them here.

4. Conclusions

In conclusion, it is believed that an information base such as that described here, which incorporates geographically based descriptions of pyrological zoning and fire danger characteristics within and across regions, would greatly help to predict, evaluate and respond to the development of fire situations.

References

Armand, D.A. 1975. Landscape science. Moscow, Mysl Publ., 288 pp.

Deeming, J.E., J.W. Lancaster, M.A. Fosberg, R.W. Furman, and M.J. Schroeder. 1972. The National Fire Danger Rating System. USDA For. Ser. Paper RM-84, 165 pp.

Furyayev, V.V. and D.M. Kireyev. 1979. Investigation of post-fire development of forests using landscape basis. Novosibirsk, Nauka Publishers. 160 pp. <in Russian>.

Kurbatski, N.P. 1963. Fire danger of forest and rating of it using local scales. In: Forest fires and fire control, pp.5-38. USSR Sci. Acad. Publishing Office, Moscow <in Russian>.

Kurbatski, N.P. 1972. The terminology of forest pyrology. In: Problems of Forest Pyrology, pp. 171-231. Sukachev Institute for Forest and Wood, Krasnoyarsk <in Russian>.

Melekhov, I.S. 1944. On theoretical basis of forest pyrology. Forestry and Technical Institute, Arkhangelsk, 19 pp <in Russian>.

Nesterov, V.G., M.V. Gritsenko, and T.A. Shabunina. 1968. The use of the dew-point temperature at the calculation of the fire occurrence index. Meteorologia and Hydrologia 9, 102-105 <in Russian>.

Sofronov, M.A. 1970. On assessment of forest fires damage. In: Voprosy lesnoi pyrologiyi, pp. 354-366. Sukachev Institute for Forest and Wood, Krasnoyarsk <in Russian>.

Sofronov, M.A. 1979. Pyrological zoning principles. In: Prediction of Forest Fires, pp. 108-122. Sukachev Institute for Forest and Wood, Krasnoyarsk <in Russian>.

Sofronov, M.A. and A.V. Volokitina. 1990. Pyrological zoning of taiga areas. Nauka, Novosibirsk, 204 pp. <in Russian>.

Volokitina, A.V. 1991. Forest fuel maps are necessary. Lesnoe Khozyaistvo, N.4, 14-16 <in Russian>.

Vonskiy, S.M. and V.A. Shzdanko. 1976. Principles of elaboration of meteorological fire danger indices in the forest. LenNIILH, Leningrad, 48 pp. <in Russian>.

J.G.Goldammer and V.V.Furyaev (eds.), Fire in Ecosystems of Boreal Eurasia, 239-252.

Forest Fuel Maps

A.V. Volokitina [1]

1. Introduction: Classification of Forest Fuels

The subject matter of this paper is a new area in the field of pyrology (wildland fire science) in Russia, the mapping of vegetation fuels (VF). Methods of composing VF maps of middle and large scales are briefly explained, along with the purpose of such maps. A classification of the basic groups of VF - the prime conductors of burning in surface fires - is presented.

The basic condition permitting the spread of forest fires is the continuity of forest fuels across the landscape. The composition, structure, loading and moisture content of forest fuels (FF) are spatially very diverse, and this diversity is variable with time. These features of the FF strongly influence fire danger, fire behavior, and fire effects. That means FF-maps of various scales are necessary to predict fire situations, monitor fire danger, and control large fires.

We started to develop methods for drawing FF maps in 1984 (Volokitina and Ryzhkova 1984). The basis for this work was a system of FF classification. There are three approaches to classification of FF. In the USA the FF are classified into two categories (live and dead), with the dead fuels divided into four classes by drying rates (1-hr timelag period [TL], 10-hr TL, 100-hr TL and 1000-hr TL) (Pyne 1984). In France FF are classified by their location in the forest (Trabaud 1977). In Russia the FF are classified by their contributions to fire incidence, spread and behavior.

The Russian forest fuel (FF) classification was initiated by Kurbatsky (1962, 1972). He classified FF into seven groups: 1) moss, lichens, fine litter; 2) duff, humus and turf layers of soils; 3) grass, seedlings, and low brush; 4) large wooden particles (dead branches, snags, limbwood, slash); 5) understory saplings and shrubs; 6) green foliage and branches of living trees; 7) stems of trees and branches thicker than 7 mm. The first group of FF plays the leading role in fire incidence and spread. We named this group of FF the "Prime Conductors of Burning" (PCB-fine fuels) and worked out its classification (Sofronov and Volokitina 1985). We have further subdivided this group into two subgroups; the first includes layers of living fuels (moss, lichens); the second subgroup includes dead fuels (litter, dead leaves and grasses; Tab.1). The first subgroup ("moss") covers 4 types: lichen, dry moss, moist moss, and bog-moss types. The second subgroup ("litter") is also classified into four fuel types: cured grass, loose litter, compact litter and non- conductor types of FF. The PCB types are characterized by specific vegetation and soil indicators (Tab.2).

---

[1] V.N. Sukachev Institute for Forestry and Timber, Siberian Branch, Russian Academy of Sciences, 660036 Krasnoyarsk, Russian Federation

Tab.1. Types of the Prime Conductors of Burning (PCB)

Subgroup "moss"	Lc	Dm	Mm	Bm (1)	Bm (2)
Subgroup "litter"	Cg ↔	Ll ↔	Cl ↔	c (1)	Nc (2)
Critical DSC	1	2	3	4	can't burn

Note: The critical class of drought for a PCB fuel class is the drought class (DSC), at which the PCB fuel class becomes flammable. Subgroup abbreviations are described in the text.

Tab.2. Types of "Prime Conductors of Burning" (PCB), characterized by specific vegetation and soil indicators

PCB Type	Vegetation and soil indicators
Moss Subgroup	
Lichen (Lc)	lichens in forest floor vegetation; dry soil
Dry moss (Dm)	cover of true mosses over drained soil
Moist moss (Mm)	true mosses with polytrichum and sphagnum; insufficiently drained soil
Bog-moss (Bm(1))	polytrichum over bog or nearly bog soil; patchy boggy areas that may dry out during drought
Bog-moss (Bm(2))	sphagnum and hypnum over nearly-bog or bog soil; areas of extensive bogs which do not dry out; usually do not burn
Litter Subgroup	
Cured grass (Cg)	cover mostly cured grass or sedge
Loose litter (Ll)	loose litter (fallen pine needles and birch leaves; cured herbs)
Compact litter (Cl)	compacted needles of fir, spruce, larch, or broadleaf trees; consolidated loose litter later in summer.
Non-conductor (Nc)	load of PCB fuel is below the critical level (0.05-0.20 kg m^{-2}); no flame propagation possible.
Nc(1)	smoldering fuels (litter, turf, duff) sustain ground fires only
Nc(2)	PCB fuel absent or scarce (dunes, rock or pebble outcrops, plowed fields, etc.); no fire possible.

Those PCB fuels that become flammable at one drought severity class, are attributed to the same PCB type, assuming the following standard environmental conditions: on a horizontal surface, under the canopy of a forest stand of middle canopy closure (0.5-0.7) with the foliage present. If actual drying conditions do not correspond to those described above, i.e. they are non-standard, proper corrections should be applied.

We determine a drought severity class (DSC) based on the Nesterov index (Nesterov et al. 1968). The Nesterov index (DI) is based on the product of air temperature and the difference between air temperature and dew-point temperature. Measurements are taken once in each 24-hour period, between 1300 and 1500 hours in the afternoon. DI is calculated by summing these values for previous days up through the current day, and is reset to zero any time there is rain exceeding 2.5 mm during the previous 24 hour period. Relationships between DSC and DI are: DSC = 1 at DI < 300; DSC = 2 at DI = 301-1,000; DSC = 3 at DI = 1,001-3,000; DSC = 4 at DI = 3,001-10,000; DSC = 5 at DI = 10,001-30,000; and DSC = 6 at DI > 30,000 (Sofronov and Volokitina 1990). At DSC=1, both lichen and cured grass PCB types become flammable, at DSC = 2 loose litter and dry moss PCB types become flammable when the conditions for drying are standard, etc. (Tab.1). Forest fuel characteristics of the first subgroup (moss, lichens) are practically constant for the whole fire season. Those of the second (litter) subgroup may change their rate of drying due to such seasonal factors as decomposition of dead fuels, growth of fresh biomass of grass in summer and its curing in autumn. So PCB types of the second subgroup can alter from one to another during a fire season, as indicated by arrows in Table 1.

Classification of PCB is based on the analysis and pooling of both our and published data on drying rates of the PCB in various regions of Russia (Sofronov and Volokitina 1985). At present we are the first to have experimentally studied the pyrological characteristics of some types of PCB, such as the rate of spread at zero wind speed (Vo), and unit energy release rate (I") as depending on the DI (see Tab.3). In our experiments we used the annular screen of Wright (1967) with a simple calorimeter for the investigation of the quantity and dynamics of heat emission during the combustion of forest floor vegetation.

There are two ways to describe pyrological characteristics of vegetation. In our opinion, pyrological characteristics are those that can be used for evaluating fire danger, and predicting fire behavior and fire impact. The first approach to description is based on typical characteristics of vegetation. For example, fuel models of the US National Fire Danger Rating System (Deeming et al. 1972) or the method of Kurbatsky (1954) for classifying the typical pyrological characteristics of forest types. But, diversity of the vegetation is very high, though the number of categories (and respectively, number of typical characteristics) used is limited. As a result, the internal pyrological heterogeneity of the delineated vegetation categories will be high, and it will be impossible to characterize a particular area accurately enough for mapping on both a middle scale (1:500,000-1:1,000,000) and a large scale (1:25,000-1:50,000) (Sofronov and Volokitina 1990).

The second approach is based on individual characteristics of forest plots. We are inclined toward this approach as being more appropriate for representing the heterogeneity of fuel conditions on the landscape. Descriptions of characteristics of individual forest plots includes evaluation of PCB types and of seasonal dynamics of the fuels, characteristics of other groups of FF, and patterns of moisture changes. These characteristics should be available on FF maps. Using this approach, we have developed methods for drawing FF maps on both middle and large scales (Sofronov and Volokitina 1990).

2. Middle-Scale Forest Fuel Maps

Two methods have been developed for drawing middle-scale FF maps. The first method we call "original" ("autonomous"). It is based on using satellite pictures to delineate "Natural-Territorial Complexes" (NTC). Several NTC's are integrated into

"Categories-Analogs" within natural regions, by images of vegetation in satellite pictures. Field studies are required on key plots to characterize NTC analogs, with the utilization of aerial pictures. Then usual operations for drawing FF-maps are erformed. This method requires huge financial expenses that we could not afford.

Tab.3. Pyrological characteristics of forest fuels - prime conductors of burning (FF-PCB)

Nesterov Index	Pyrological Characterisitics											
	Lichens		Dry Moss		Moist Moss		Loose Litter		Compact Litter		Cured Grass	
	I"	V_0	I"	V_0	I"	V_0	I"	V_0	I"	V_0	I"	V_0
100	68	0.20	--	--	--	--	--	--	--	--	--	--
200	80	0.25	--	--	--	--	--	--	--	--	--	--
300	105	0.32	--	--	--	--	--	--	--	--	--	0.30
400	121	0.40	--	--	--	--	--	--	--	--	--	0.33
500	136	0.46	80	0.14	--	--	--	--	--	--	--	0.35
700	140	0.54	100	0.16	--	--	57	0.10	--	--	--	0.38
1000	148	0.60	230	0.19	50	0.15	100	0.20	70	0.20	--	0.40
2000	156	0.64	370	0.25	190	0.22	132	0.30	210	0.23	--	0.40
3000	156	0.64	450	0.27	260	0.27	140	0.40	280	0.26	--	0.40
4000	156	0.64	480	0.29	290	0.30	146	0.44	310	0.29	--	0.40
5000	156	0.64	510	0.30	310	0.33	150	0.50	330	0.32	--	0.40

Notes:
1. Bog moss and non-conductors practically do not burn
2. I" - unit energy release rate, kJ m^{-2} s^{-1} (kW m^{-2})
3. Vo - rate of spread at zero wind, m/min.

The second method for drawing middle-scale FF maps we call the "conjugated" one. Here contours of existing maps of natural landscape characteristics, such as terrain, forest types, or stocking levels, are used as the basis for drawing the FF map. Pyrological characteristics are attached to these contours based on analysis of the legend of the base map and of other maps of the region, and also other data on the FF for the region. In cases of data shortage, field studies must be conducted. The general process for developing these maps is as follows:

I. Examination of materials on general natural characteristics of the territory is aimed at revealing the character of its heterogeneity and separate natural zones. It necessarily must be carried out with the application of usable existing natural zoning systems, for example, a forestry one, and especially, a pyrological (wildland fire) zoning.

II. Choosing of a background map. Preferable is a map of the most suitable scale, with the most detailed presentation of forest inventory plots (including detailed descriptions of the vegetation, especially forest floor vegetation). To develop a map of forest fuels, one requires the following additional information:

- general descriptions of the vegetation of the region of concern, and descriptions of the pattern of spatial distribution of the vegetation with regard to topography, especially for mountain terrain;

- the descriptions of the biogeocoenosis types, both forest and non-forest ones;

- data on the pyrological characteristics of forest types indigenous to the region;

- data on fire occurrence in the region.

III. Examination of satellite images. The aim is to outline burned and felled areas, farming lands, and areas of sapling-pole deciduous stands, which are not displayed on the background map. If aerial photographs are available, interpretation of them will help to more precisely determine the character of vegetation modification.

IV. The composition of a preliminary map:

- drawing of an outline map at a pre-determined scale;

- consideration of the legend of the background map and supplementary data materials, in order to develop the pyrological characteristics of forest inventory plot categories (here an assessment takes place of FF-PCB types and the dynamics of their seasonal change);

- designing of the fuel map legend;

- revision of the outline map and putting additional details on it on the basis of the information taken from satellite images;

- coloring of the pre-map according to pyrological characteristics in accordance with the legend.

V. The selection of key areas. Based on the examination of the pre-map and supplementary materials, key areas are selected for examination in the field (it is preferable to establish these areas in every natural zone). The key areas must meet the following conditions: be representative, accessible, and suitable for field work. Relevant aerial photographs must be obtained for the key areas.

VI. Field examination includes:

- the selection of observation travel routes and profiles;

- the pyrological description of sample points and profiles with the use of a special form. If stationary, fire danger buildup observations on some key areas (forest types) and fuel load measurements on some profiles are necessary; test plots are selected and supplied with meteorological observation points. Guidelines for the examination method are described in more detail by Volokitina and Sofronov (1979).

VII. Examination of the data taken in the field includes:

- checking whether the descriptions made on travel routes and the background map correspond with each other; if necessary, satellite images may be used to rectify the pre-map;

- revising the identification of PCB types (and their seasonal dynamics) indigenous to the present forest types and other area categories, based on route descriptions and results of test ignitions;

- revising the table of the relationships between PCB types and forest types;

- rectifying the pre-map.

VIII. Drawing of the fuel map and its legend. The original draft of the map must follow common standards of map-making.

Using the conjugated method, we have drafted FF maps for the Angaro-Yeniseisk region on the trapezia from 0-45 to 0-48 (the region between 55°-60°N and 85°-105°E). The 1:500,000 scale forest stock map was utilized as the base map. We have also drafted a 1:1,000,000 scale FF map for the northern part of the Lake Baikal region (Fig.1), using a vegetation map as the base map. Additionally, we used the information from a landscape map of the same scale and performed field studies. The map shows, first of all, the PCBs, their combinations and seasonal dynamics. For this purpose, the original was drawn with the use of both coloring and shading. In the paper only a shaded version of a fragment of the map is presented (Fig.1). Distributions of other FF-groups (trees and shrubs) are marked by special symbols on the map.

Here is a brief explanatory comment on the fragment. The least fire-prone PCB type, a non-flammable one with too low or none-fuel load, is common for the mountain tundra and bald hills on high mountainous terrain. On a more flat and lower terrain of the Ikat and Muya regions, thickets of dwarf creeping Siberian cedar (*Pinus pumila*) and boggy lands with sphagnum are common; therefore, a combination of loose litter and bog-moss PCB types is typical of those territories.

The moist-moss PCB type, of relatively low fire danger, is typical for a middle-mountainous terrain with stands composed of Siberian cedar (*Pinus sibirica*) and *Abies* spp., the forest type is "low brush and true mosses" (following the forest type classification of Sukachev [1972]); such terrains are situated along the seashore at the foot of the Barguzin mountain range, also in the southern middle-altitude part of the Ikat mountain range and in the Ulan-Burgasy mountain range.

The dry-moss PCB types are intermediate in fire danger. These, which get ready to burn under conditions of the second fire danger class, are typical of sparse larch stands ("low brush and true mosses" forest type) with the forest floor composed of brush and moss, situated on the spurs of the Southern Muya mountain range and in mid-altitude mountain valleys.

In spring and autumn the highest fire danger is in the cured grass PCB of the dead fuels PCB subgroup. It is typical of the vegetation of the closed Barguzin hollow and the Lower Angara hollow. On steppe-like terrains this PCB type usually alters from the live grass and cured grass PCB type in spring to the loose litter or even non-flammable PCB type in summer in response to seasonal drying. The loose litter PCB type is also typical of pine stands surrounding the Barguzin hollow; the compact litter PCB type is typical of the larch stands of this area.

The main function of middle-scale FF maps is to serve as the basis for drafting maps of current forest fire danger at various DSCs. These middle-scaled maps of current forest fire danger are drafted by the application of FF maps onto maps of fire occurence, useful for optimizing the net of aerial patrol routes. Figure 2 gives an example corresponding to DSC=3, but the real pattern can be more diverse because of variations of DSC across the region. In this illustrative example of the fuels' state at the given drought class, areas are categorized into three classes: flammable, non-flammable, and becoming flammable (intermediate). In the intermediate state, flammability of fuels can best be determined by the use of test ignitions of the forest floor.

3. Large-Scale Forest Fire Maps

Large-scale FF maps (Fig.3) assist greatly in suppression planning for large forest fires. These maps are drafted based on forest survey data (Volokitina 1988). The basis for this work are the uncolored plans of forest stands. The mapping is followed by a detailed pyrological description of each FF plot.

It is desirable for forests of the southern taiga to draft large-scale FF maps separately for spring and for summer. We have also developed a method for automatically mapping pyrological descriptions based on forest inventory descriptions (Volokitina et al. 1989). The pyrological description of a forest inventory plot includes information that make it possible to judge the ability of the plot to burn in different periods of the fire season and at different levels of drought, as well as the probable character of burning, the expected rate of spread, and probable fire effects. In practice, such information is a set of data characterizing the entire fuel complex (seasonal dynamics included) and its conditions of moistening and drying. The pyrological description comprises of the:

I. Location of the forest inventory plot, its area and designation (number);

II. Position on the landscape (aspect, slope angle);

III. Forest type, class of bonitet (site index);

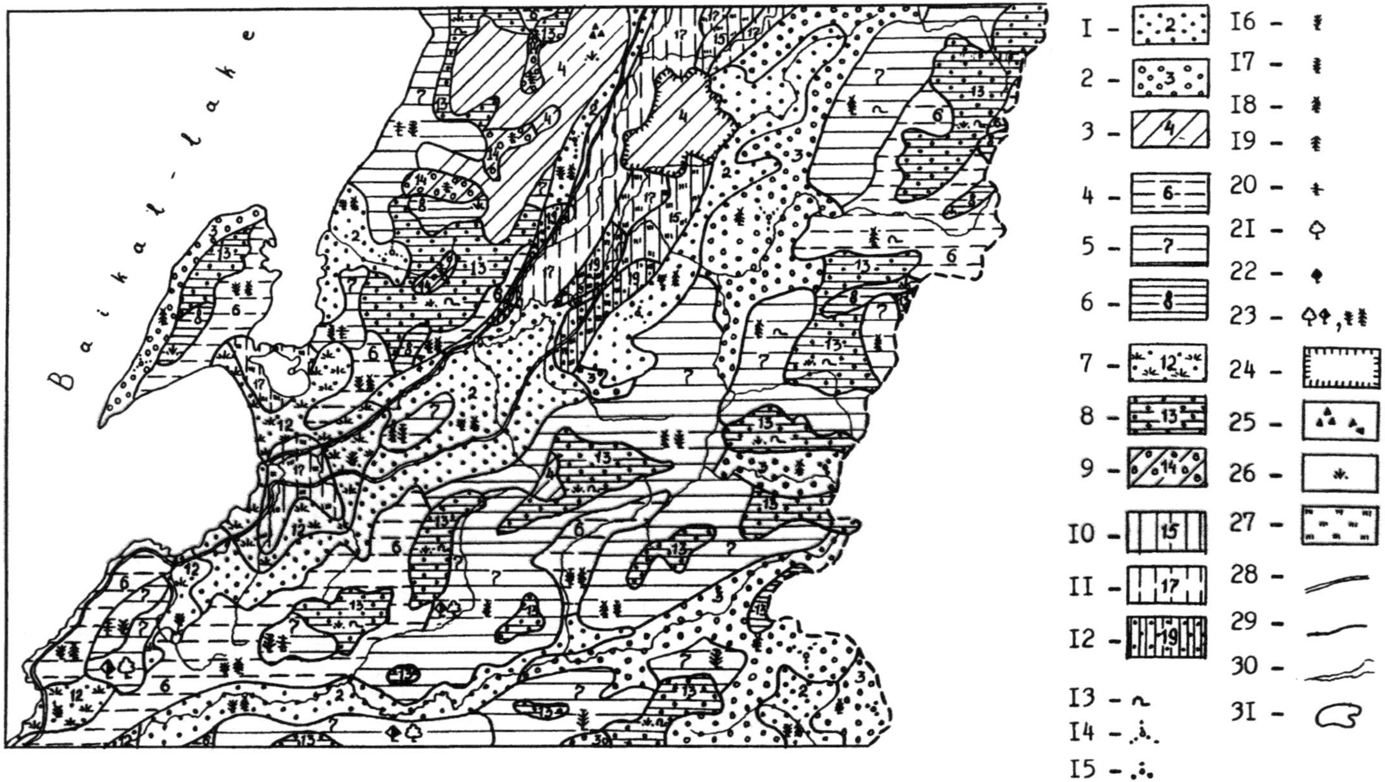

Fig.1. A fragment of a middle-scale map of forest fuels (for Lake Baikal Basin) (Scale 1:1 000 000). Types of prime conductors of burning (Types of PCB): 1 - loose litter, 2 - compact litter, 3 - non-conductors, 4 - dry moss, 5 - moist moss, 6 - bog moss. Combinations of PCB: 7 - loose litter and lichen, 8 - loose litter and bog moss, 9 - compact litter and non-conductors. Seasonal dynamics of PCB (spring/summer): 10 - dry grass/loose litter, 11 - dry grass/non-conductors, 12- loose litter/compact litter. Bushes: 13 - dwarf cedar-pine, 14 - rododendron, 15 - dwarf birch. Crown: 16 - pine, 17 - cedar-pine, 18 - larch, 19 - spruce, 20 - fir, 21- birch, 22- aspen, 23 - crowns of combination of species. N Open lands: 24 - ploughed lands, 25 - grassless lands, 26 - tundra, 27 - steppe. Other map symbols: 28 - roads, 29 - borders of forest plots, 30 - rivers, 31 - lakes.

Fig.2. A fragment of a middle-scale map of current forest fire danger (spring, DSC = 3) (Scale 1:1 000 000).
1 - flammable areas, 2 - areas about to become flammable, 3 - nonflammable areas (other notes: same as in Fig.1).

IV. Characteristics of the forest stand with respect to canopy layers, as follows:

- age (which determines the impact of surface fires);

- height (which affects the likelihood of a crown fire in conifer stands);

- species composition of every canopy layer, separately for evergreen and deciduous species (such separation is very important for estimating ground surface fuel shadowing by the canopy and fuel drying rates for summer and spring (when the canopy is foliage-free);

- canopy cover of the stand correlates with both solar light penetration (a very important factor for fuel drying) and wind velocity under the canopy (one of the most important factors in rate of flame spread). Forest stand cover (in combination with height) may be used to estimate such factors as foliage load of every tree species, maximum possible stocking of standing dead trees, etc.;

- dead tree stocking and the load of fallen wood;

- young trees are characterized in height and species separately for evergreen and deciduous species; the understory woody vegetation is characterized only in its canopy density; the young tree and the understory add to shadowing;

- overall shadowing (projected ground surface cover) from all layers of the forest canopy is calculated separately for spring and summer, expressed as proportion (decimal fraction) of the ground surface covered;

- present FF-PCB types are characterized separately for spring and summer.

Some of the values itemized above are to be found in tables or calculated by formulas on the basis of forest inventory descriptions. Accurate classification of forest inventory plots into FF-PSB types is the main problem because in current practice these are not included in the forest descriptions, although forest types are noted.

Determination of PCB types for forest inventory parcels (uniform parcels of the forest stand, the smallest mapped unit in Russian forest inventory) may be carried out in three ways:

I. Based on forest types: the forest types are characterized by PCB types indigenous to them, using the inventory description of the forest type; accuracy of this method is fairly low because the forest type descriptions lack information on dead fuels, and forest types themselves are sometimes difficult to identify correctly;

II. Based on forest types as well as determined locally through field examination. Forest type descriptions (for example, made at forest cruiser's field training) are used for PCB characterization of forest types instead of standard descriptions. This is a somewhat improved version of the method discussed above. Nevertheless, employment of both methods will lead to similar PCB descriptions of those forest parcels which, to a forest cruiser's mind, belong to the same forest type. These methods are applicable for preparing PCB fuel maps based on data from the most recent forest inventory campaign.

III. By direct description of PCB types of forest parcels, carried out in the process of aerial image description or forest cruising in the field. This method provides individual, more precise, PCB characteristics of forest parcels. Field identification of PCB types may be done using the basic PCB descriptions given in Sofronov and Volokitina (1990). Identification may face difficulties if the fuel on the observed spot shows attributes of two or three PCB types, or if characteristics are those especially particular to southern taiga forests with grass forest floor vegetation.

In case a FF-PCB type is not easy to identify, its identification is carried out by the following process:

I. Identification of the background type of PCB.

II. Examination of the stratification of the fuelbed (in order to detect a cured grass layer over the litter or moss).

III. Determination of the dynamics of dead fuel (leaf debris, curing of grass) accumulation and decomposition, rate of moss stem decomposition.

IV. Assessment of the FF-PCB subgroup for the period of the fire season.

V. Identification of FF-PCB types for different periods of the fire season, taking into account the discriminating features of every type and exceptional cases.

VI. Carrying out a series of test ignitions of the fuelbed at different levels of drought over the course of the fire season.

VII. Final determination of a FF-PCB type, with shadowing taken into account.

Field work and, to a greater degree, aerial image interpretation may be done more easily with the use of regional PCB type identification manuals. Each manual contains clearly defined, easy to recognize, categories with specific descriptions of seasonal changes in PCB types (spring-summer); the categories are designed to cover all vegetation diversity of the forested area for which the manual is composed.

We made such a manual for the Krasnoyarsk Priangarye region (Volokitina 1990). Twelve categories of forest parcels were defined. The main groupings of these are:

I. Deciduous forest stand with an admixture of short-needled coniferous species, with the forest floor composed mostly of sedge and grass (Ll-Cl);

II. Birch stand with an admixture of aspen and short-needled coniferous species, with the forest floor composed of a mixture of herb species (Cl-Nc);

III. Pine stand with an admixture of birch, with the forest floor PCB composed of true mosses (Dm-Dm);

IV. Old clearcut, or burned area, or area with sparse fuels (Cg-Ll);

V. Pine stand without any lichens or mosses over the litter (Ll-Ll).

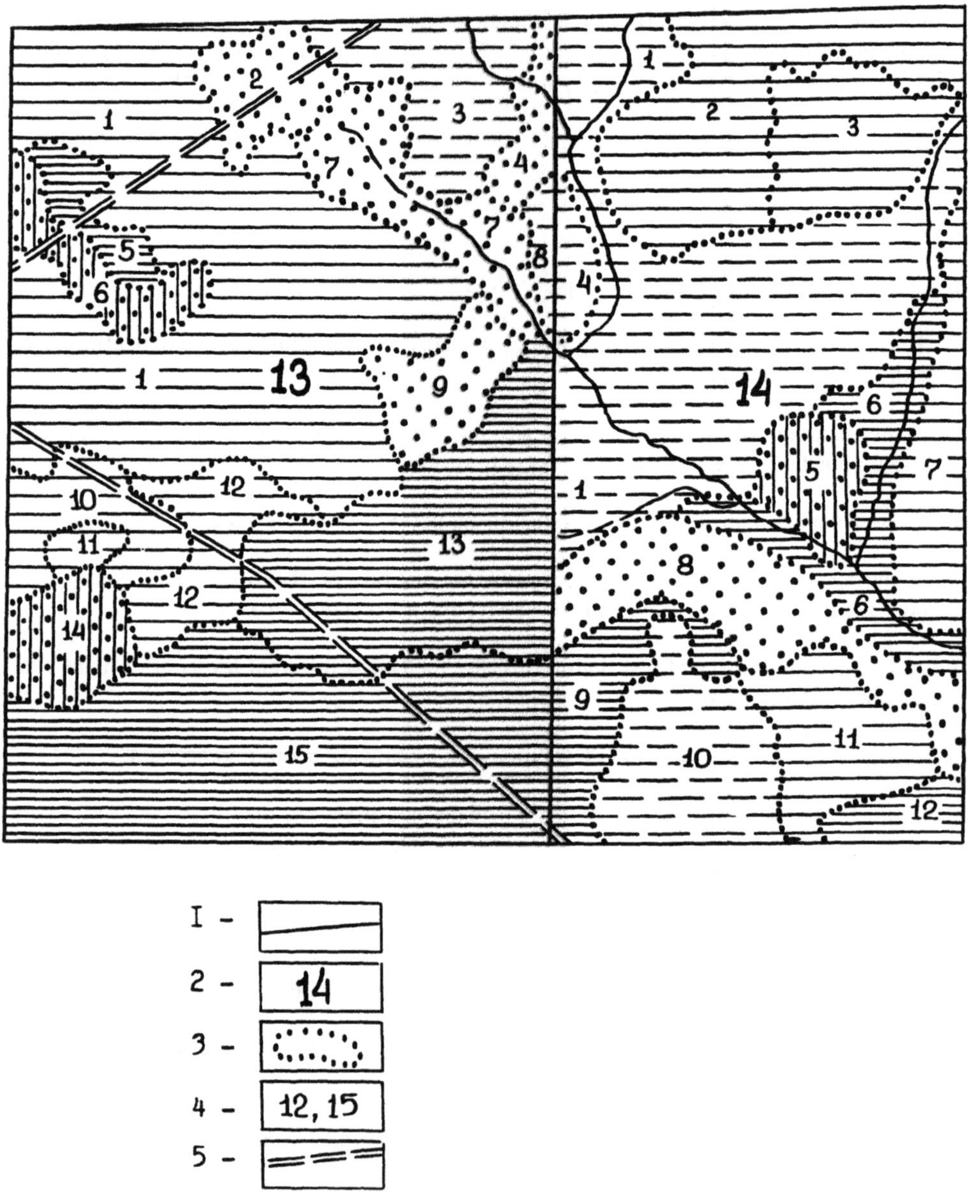

Fig.3. A fragment of a large-scale map of forest fuels (scale 1:25,000)

1 - borders of forest plots
2 - numbers of forest blocks
3 - borders of inventory blocks
4 - numbers of inventory blocks
5 - roads

Categorization of PCB types in the process of forest inventory would make it possible to work out more correct PCB fuel maps, which are necessary in forest fire control.

Colorless forest maps or forest parcel plots may be used as a background to develop PCB fuel maps, with the use of pyrological descriptions of forest parcels. This description is made by the computer processing of forest inventory data (Volokitina et al. 1989). Such a data package, including the geographic data base and corresponding descriptive data, can be useful for operational preparation of PCB fuel maps and fire danger maps in critical fire situations.

It is desirable for forests of southern taiga to draft large-scale FF maps separately for spring and summer. To achieve a greater accuracy of these maps it is desirable, in the course of forest survey, to note FF types. To assist foresters in obtaining this additional information as part of routine inventory surveys, we have developed the "Guide for Identification of Forest Fuels Types" as applied to Krasnoyarsk Priangarie (Volokitina 1990).

4. Conclusions

Vegetation fuel maps are useful for forecasting the seasonal development of fire danger and the behavior of individual fires. In this article we have briefly described the state of the classification and mapping of forest fuels in central Siberia. Our methods for drawing middle-scale and large-scale FF maps are described and illustrated by typical examples. Further improvement of FF-mapping methods is possible based on more detailed classifications of the FF and further studies of their characteristics.

Acknowledgements

This paper was prepared with the financial support of the Krasnoyarsk Regional Science Foundation.

References

Deeming, J.E., J.W. Lancaster, M.A. Fosberg, R.W. Furman, and M.J. Schroeder. 1972. The National Fire Danger Rating System. USDA For. Ser. Paper RM-84, 165 pp.

Kurbatsky, N.P. 1954. Methodological guidelines for experimental development of local fire danger rating scales. TsNIILKh, Leningrad, 33 pp. <in Russian>.

Kurbatsky, N.P. 1962. Techniques and tactics of fire suppression. Moscow, 154 pp. <in Russian>.

Kurbatsky, N.P. 1970. Studies of loading and characteristics of forest fuels. In: Problems of forest pyrology, pp. 5-58. Krasnoyarsk <in Russian>.

Nesterov, V.G., M.V. Gritsenko, and T.A. Shabunina. 1968. The use of dew point temperature in calculation of the fire occurrence index. Meteorologia and Hidrologia 9, 102-105 <in Russian>.

Pyne, S.J. 1984. Introduction to Wildland Fire. Fire Management in the United States. John Wiley, New York, 455 pp.

Sofronov, M.A. and A.V. Volokitina. 1985. The types of the prime conductors of burning in surface fires. Forest Journal 5, 12-17 <in Russian>.

Sofronov, M.A. and A.V. Volokitina. 1990. Pyrological regions of the taiga zone. Novosibirsk, 204 pp. <in Russian>.

Sukachev, V.N. 1972. The foundations of forest typology and biogeocoenology. Nauka, Leningrad, 420 pp. <in Russian>.

Trabaud, L. 1977. Fuel mapping helps forest firefighting in Southern France. Fire Manage. Notes 39, 14-17.

Volokitina, A.V. 1990. Principle for developing guidelines for identification of the types of prime conductors of burning (on the example of Krasnoyarsk Priangarie). Krasnoyarsk, Dep. in VINITY: 5352-W90, 31 pp. <in Russian>.

Volokitina, A.V. and M.A.Sofronov. 1979. Moistening, drying, and burning intensity of the forest floor. In: Modelling in forest fire control, pp. 45-85. Krasnoyarsk <in Russian>.

Volokitina, A.V. 1988. Methodological recommendations on drawing operative maps of forest fuels for Priangarie of Krasnoyarsk territory. Krasnoyarsk. ILID, 12 pp. <in Russian>.

Volokitina, A.V., T.M. Tartakovskaya, and E.G. Schevchuk. 1989. Creating database for rapid drawing of the maps of forest fuels (methodological recommendations). Krasnoyarsk, 20 pp. <in Russian>.

Wright, J.G. 1967. Forest-fire hazard research as developed and conducted at the Petawawa Forest Experiment Station. Forest Service, Department of the Interior, Ottowa, Canada, 40pp.

J.G.Goldammer and V.V.Furyaev (eds.), Fire in Ecosystems of Boreal Eurasia, 253-259.

Sectoral and Zonal Classes of Forest Cover in Siberia and Eurasia as a Basis of Clarifying Landscape Pyrological Characteristics

D.I. Nazimova [1]

1. Introduction

Because potentially large changes in plant cover may occur in response to climate change, we face the problem of predicting shifts in forest zones. To do this requires creating a uniform global ecosystem classification that is sufficiently informative concerning the major features of ecosystem structure and dynamics. Climate-vegetation correlations are the focus of this paper. Our research on climate-vegetation relationships for the territory of Russia, and particularly Siberia and the Far East, can provide a basis for modelling vegetation on a regional scale.

It is known that climate, biota and soil are the primary natural determinants of terrestrial ecosystems. It is reasonable to classify bioclimatic and lithologic-geomorphologic components of natural complexes separately. Therefore, we first investigated features and parameters related to climate and vegetation as the main system-forming factors of zonal categories of forest cover. Because forest ecosystems function in a landscape context, we use "zonal type of landscape" as the basic unit for any territory (ecoregion). This term, as perceived by geobotanists, landscape scientists and biogeographers (e.g. Sochava and Lavrenko 1980; Isachenko 1988), appeared to be the most appropriate for our purposes.

The zonal type of landscape (of ecosystem, or plant cover) is determined by the whole complex of bioclimatic characteristics, with an emphasis on the diagnostic role of its structure and physiognomy. However, recently, peculiarities of seasonal patterns and soil cover have often been used for determining zonal boundaries of ecosystems. This is caused by two reasons: disturbance of natural vegetation (plant cover) that makes it difficult to base classifications on current vegetation, and a good correspondence between zonal types of vegetation and soils. For example, the steppe zone (biome) can certainly be delineated by the distribution of chernozem soils, and the forest-steppe ecotone by the area of grey forest soil in combination with chernozem and other soils in intra-zonal sites. However, such a one-to-one correspondence is impossible in other situations for many reasons.

A zone, in our understanding, is a large geographic territory with a broadly homogeneous macroclimate. This means that each zone occupies some part of climatic space. As climatic space is continuous, it is divided between zonal units. The diagnostic features of forest zones include composition of plant types, some prevailing soil-forming processes, and one or more

---

[1] V.N. Sukachev Institute for Forestry and Timber, Siberian Branch, Russian Academy of Sciences, 660036 Krasnoyarsk, Russian Federation

typical major species of tree, dominant ectobiomorph, or group of species in lower vegetation layers. Description of diagnostic combinations of species is not yet complete, but it is in progress for some ecoregions of Siberia (Buks et al. 1976; Smagin et al. 1980; Nazimova et al. 1985; Anonymus 1990).

Zones are often named after one or more of the dominant climax species. For example, the east European taiga is classified as *Picea-Pinus sylvestris* (with intra-zonal pine types forming an edaphic climax on sandy soils). The west Siberian taiga is dominated by *Pinus sibirica*, *Picea obovata*, and *Abies sibirica* in the south of the taiga zone. Pine forests occur as intra-zonal patches on sandy soils and in bogs. But some regions in the Trans-Ural part are less humid than the eastern half of the Ob-Yenisei watershed. Here one can find many examples of post-fire succession with dominance of *Pinus sylvestris*, not restricted to sandy and boggy soils. Climate of this province is not optimal for dark-needled tree species. The highest class of fire danger and very frequent large wildfires are characteristic of this part of western Siberia (Valendik 1990; Sofronov and Volokitina 1990). Therefore, it is reasonable to distinguish a self-dependent pine zone in this and some other ecoregions of Siberia. The east-Siberian taiga is dominated by larch. More than 85 percent of this zone is dominated by two main species of *Larix*: *L. gmelinii* and *L. cajanderi* (Abaimov and Koropachinskiy 1984). Many features of this zone (subzonobiome according to Walter and Box [1976]) have been formed in response to wildfire. These examples show that not only latitudinal zonality but also longitudinal or sectoral and provincial changes influence the combination of forest-forming species, biotic composition, soil processes, and forest fire characteristics in boreal and adjacent zonobiomes.

2. Study Methods

Using various data bases and forest maps (e.g., Smagin et al. 1977; Anonymus 1973), we created a network of 130 polygons. Each region was characterised by 3-7 polygons (ecoprofiles). The characteristics of forest cover were generalized for a 10 by 40 km grid. We feel this is the minimum area for characterizing a zonal type of forest. Weather data were available for each polygon; our analyses incorporated data from more than 1,100 meteorological stations within the USSR. Most of the weather stations are located in valleys and characterize only this part of the ecoprofile. There are, of course, considerable local deviations of climatic parameters, for instance within a given mountain landscape. Experiments and calculations (Tchebakova 1981) in Siberia demonstrated that slope orientation and steepness can be so climatically significant that they cause different types of vegetation on southern and northern slopes. For these and other reasons, one should not expect one-to-one climate - vegetation linkages. Nevertheless, a well-defined picture of vegetation classes can be revealed by our approach.

In this approach, a climatic space filled with station data, or with data averaged for a given landscape characterized by specific zonal vegetation, is subdivided into homogeneous sections. Then, every section is characterized by an average border value. The degree of overlapping of classes can be specifically evaluated. We are satisfied with the informativeness of the resulting diagram, because not many stations appeared to be in zones and sectors other than their own, and every case could be explained. The resulting diagram indicates general patterns remarkably similar to those visible on a geographical map. It should be mentioned that geographical space is not the same as climatic space, and that, principally, axes in our diagram differ from those on a map.

According to our concept, the system-forming factors of zonal classes include regional biota, climate regime, and heat and water supply. For the purpose of constructing our ordination, we have used an index of continentality to represent climate regime, and a temperature sum to characterize heat supply. The index of continentality (here indicated by K) can be calculated in the simplest case as a function of annual temperature amplitude (A: mean July temperature minus mean January temperature [°C]) in the northern hemisphere and latitude (ϕ) (Conrad 1947):

$$K = [1.7\ A / \sin(\phi + 10°)] - 14$$

We have used the temperature sum (Growing Degree Days [GDD] 10) to characterize the heat resources of the climates. This parameter is accepted in the Russian literature and is calculated as the sum of all positive temperatures occurring in the period with daily temperature above 10°C. The GDD on the basis of 5°C, typically seen in Western literature, are always less than GDD 10 values, but strongly correlate with them (Tchebakova and Parfenova 1991).

Next the applicability of this combination of continentality and heat supply was tested using the concept of sectoral-zonal differentiation in plant cover and ecoregions (landscape zones) of Eurasia. We also tried to create a generalized scheme of life zones for the Eurasian continent using the Holdridge system (Holdridge 1967). This attempt was unrewarding because of regional specifics in biota and in relationships between species and climate. Unfortunately the ability to distinguish many features of structure and composition of forest types and formations was lost.

There is no index of humidity in this ordination, as the scheme is two-dimensional. It is well known that heat and water supply are the main factors of plant distribution on the earth. The distribution of zonal landscapes in climatic space of these parameters is shown by Isachenko (1988). These or similar parameters and indices (see Budyko 1974) have been used for belt complexes in mountains and zonal complexes of vegetation on plains (Buks et al. 1977; Polykarpov et al. 1986; Tchebakova and Parfenova 1991; Nazimova et al. 1990).

3. Results and Discussion

We have been convinced that this two-dimensional ordination approach works well on a regional level. It was rather applicable even for a large part of Siberia, though discrepancies did appear due to permafrost and other reasons.

It is known that ecosystems function within a landscape context, and many of their features are formed under the control of the environment. We consider wildfires as an important factor at global and subcontinental scales. There is no doubt about a close correlation between continentality of climate and nature of forest ecosystems, which is in a large part a function of fire regimes. On no other continent have fires such a distinct influence on the structure and distribution of forest types, formations, and succession.

Relationships between forests and climatic regimes of ecoregions are not only direct but also indirect. When speaking about continentality it is useful to take in account a whole range of factors limiting the distribution of dark-needled species as well as of small-leafed ones (*Betula* spp., *Populus* spp.). Spring drought, high insolation, superheating of crowns when the roots are in cold soil, permafrost, and many features of hydrothermic regime caused by permafrost in soils are some of the important stress factors.

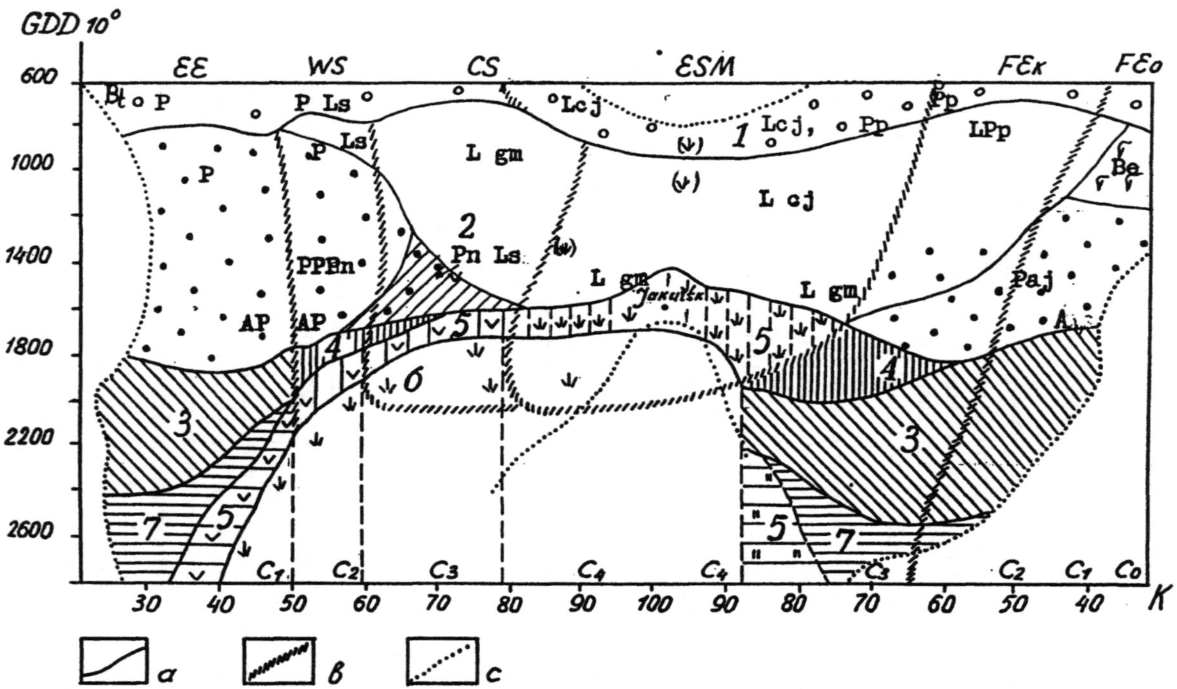

Fig.1. The climatic ordination of boreal and adjacent forests of Eurasia.

EE - East European; WS - West Siberian; CS - Central Siberian; ESM - East-Siberia-Mongolian; FEk - Far-Eastern continental; FEo - Far-Eastern oceanic.- C_0-C_4: Degree of continentality (classification after Conrad [1947]).- Boreal and subboreal forests and their subdivisions: 1 - open forest, woodland, creeping trees; 2 - boreal biomes (dark-needled, light-needled deciduous [or larch], small-leafed *Betula ermanii* biome; 3 - mixed conifer-deciduous (*Quercus, Tilia, Acer*) boreal-subboreal ecotone; 4 - small-leafed (*Betula pendula*), mixed with conifers microthermic forest; 5 - forest-steppe zonoecotones (broad-leafed in EE, small-leafed in WS, pine-birch in CS, larch-forest steppe and subtaiga in ES, broad-leafed in FEk sector); 6 - steppe; 7 - broad-leafed (nemoral) biomes. - Boundaries: a - biomes and zono-ecotones; b - sectors; c - limits of climate space according to data from 1200 meteostations. - Other abbreviations are explained in the text.

Among the main limiting factors in these ecosystems are recurrent wildfires. Neither *Picea obovata*, *Pinus sibirica* (and other pines), nor *Abies sibirica* are resistant to fire because of their characteristically thin bark. The rising generation suffers most. Young *Pinus sibirica* is also not resistant. It is more resistant to burning in mature stands without a second tree layer. We have found that dark-needled zonal complexes (Fig.1, dark points and indices P, PPPn, AP, Paj, A) rarely occur in eco-regions with an index of continentality more than 65-70. This threshold is similar for montane zones (belt complexes), though the dryness index DI is the major limit for dark-needled conifers (*Pinus sibirica, Picea obovata*). In the mountains of southern Siberia, DI 0.6-0.7 divides dark-needled (primarily *Pinus sibirica*) forest zones from light-needled (primarily larch or pine) ones. In a more continental climate this threshold occurs at DI values of 0.8-1.0 (e.g. in Transbaikal; Tchebakova and Parfenova 1991).

The southern border of dark-needled boreal forest is also influenced by fires, at least in continental climates of Siberia (Furyaev 1988). Not only fires but anthropogenic factors shift this border up in mountains and tablelands and to the north in western Siberia. At this boundary, K is equal to 50-65 and DI is about 0.8. There is no larch in this part of the ordination space, only *Betula pendula* and *Pinus sylvestris* dominate mature forest stands with herbaceous ground cover. There are no mosses among dominating synusia, as well as no typical taiga dwarf shrubs. This pine-birch sub-taiga is an example of a pyrogenic subclimax. Its distribution is correlated with rather high heat supply, with DI=0.8-1.0, and K=50-70.

Another area of light-needled sub-taiga without dark-needled conifers is represented in the Amur basin and consists of larch (*L.gmelinii*), *Betula pendula* and some other species of *Betula*, and locally, *Ulmus* and *Quercus mongolica*. The distribution of this subzone is rather restricted in Russia. It is close to forest-steppe (Fig.1,#5) and to mixed broad-leafed-coniferous zone (n3), and is located between these zones. The remaining subdivisions are not boreal vegetation zones. Some geobotanists call the next zone (# 3) sub-nemoral and nemoral (Sochava and Lavrenko 1956), others consider it cool-temperate. Landscape ecologists consider this zone sub-boreal, as well as the zones of broadleafed mesothermic forest (# 7), steppe (# 6) and the zono-ecotone between these two classes (# 5). All of these represent not the boreal but sub-boreal zonal classes of landscapes in humid, less continental, sectors.

In continental sectors of Siberia, the main border between boreal and sub-boreal landscape classes is the northern border of the forest-steppe zone. The biotemperature of 6°C is accepted as a thermic border - the common parameter of the boreal life zone (Holdridge 1967; Tuhkanen 1984). But local stations in eastern Siberia and Mongolia show steppe or forest-steppe vegetation even in the boreal area (note steppe symbols in extremely continental sector, Fig.1). Russian geobotanists and geographers (e.g., Sochava and Lavrenko 1956; Shumilova 1962) distinguish neither boreal nor sub-boreal forest- steppe. The other viewpoint (Hamet-Achti 1976) is that the boreal zone differs in continental and oceanic sectors and may contain different types of zonal vegetation. Some exact data from weather stations are likely to confirm this viewpoint.

The discussion of the northern border was not our task in this paper. Attention to the northern border of taiga, woodlands, and the tree line is given in current publications. Different sectors of Eurasia are characterized by the presence of different ecobiomorfs and tree and shrub species. These thresholds are continuous but rather definable. *Betula tortuosa* and *Alnus sinuata,* represented in oceanic climates, are replaced by *Picea obovata* in moderate continental and continental sectors, and then by larch in the severely continental and extremely continental sectors. Different species of larch dominate different eco-regions and replace one another with precise regularity not only in the forest-tundra ecotone but also in the taiga zone: *L. sibirica*, *L. gmelini*, *L. cajanderi*, *L.olgensis* (Fig.1).

There is some asymmetry in the location of biomes and zones in the right and left parts of the diagram, due to different paleogeography and climatic regimes of the western and eastern margins of the continent. The next example illustrates this. *Pinus pumila* (which forms a creeping "krummholz" forest, more sub-arctic than boreal) is the original dwarf-pine formation of montane monsoon-continental climate in the Far East. Its location on this diagram could be marked approximately by means of interpolation. Parameters of other zonal forest ecosystems in the Far East are likely to be specified. The unique combination of perhumid and superhumid climate with continental and extremely continental regime and rather warm growing season is the main reason that this region supports a great diversity of tree species and high overall biodiversity. In spite of summer humidity, the spring and autumn are really the stress periods in forest life. Forest fires are frequent and large, as well as being more destructive in mixed coniferous and broad-leafed stands (Valendik 1990). Post-fire succession leads to a wide distribution of species resistant to burning. The resistance of *Quercus mongolica* to fire plays a significant role in its relationships with man and with coniferous species, primarily *Pinus koraiensis*.

It is impossible to characterize all the sectors and zones in a brief paper. We hope the approach presented here is rather informative as the first attempt to represent all the major zones, biomes, and formations on the axes of continentality and heat supply. Although the basic scheme is two-dimensional, it may be regarded as multi-dimensional taking into account the diversity of relationships and patterns that it is able to illustrate.

4. Conclusions

We consider this empirical ordination as a useful conceptual model of Eurasian boreal forest cover. It is the first level of modelling where realistic geography of the earth (Eurasian continent) is involved. It is represented as an invariant system where continuous climatic space is filled by some biological (and maybe some physical) characteristics. In such a way considerable data on fires and typical fire danger situations can be plotted in this ordination space and analyzed.

The virtue of this graphic model is of some heuristic value as it discovers some things which are not able to be visualized on typical maps. It concerns general contours of the boreal forest climatic zone: the boreal zone is narrowest in the center of this diagram (while on a map it is widest in the middle due to mountain relief). The breadth of the boreal zone declines also toward the margins of the continent. Secondly, the border between forest-steppe and main light-needled taiga is typical. The zone of forest-steppe in Eurasia seems to be two or more principally different zono-ecotones with different levels of heat, water supply and continentality. Thirdly, the continentality index is rather informative for characterizing the boundary between the two major coniferous biomes of Eurasia: evergreen pine-dark-needled conifer, characterized by a humid continental regime, and deciduous-light-needled conifer (larch), which is crio-semihumid and extremely continental in most of the area. The central Siberian sector or at least a part of it (with K 62-70), appears to be a kind of zonoecotone. Here and in other ecotone zones forest fires and post-fire succession patterns are especially widespread. Abundance of birch and pine along the boundaries of the dark-needled biome has been caused by wildfires more than by other events.

References

Abaimov, A.P., and I.J. Koropachinskiy. 1984. Larch: *Larix gmelinii* and *Larix cajanderi*. Novosibirsk, Nauka. 121 pp. Novosibirsk, Nauka <in Russian>.

Anonymous. 1966-1970. Reference book of climate of the USSR. vol. 1-4, Issues 6-8, 17-29, 33. Hydrometeoizdat, Leningrad. (In Russian)

Anonymus. 1973. Atlas of Forests of the USSR. Moscow <in Russian>.

Anonymus. 1990. Vegetation of the USSR. 1990. Map for High Schools at 1:4,000,000 scale. Moscow, GUGK <in Russian>..

Budyko, M.I. 1974. Climate change. Hydrometeoizdat, Leningrad, 280 pp. <in Russian>.

Buks, I.I. et al. 1977. The correlation ecological-phytocoenotic map of 1:7,500,000 scale. Institute of Geography of Siberia. Irkutsk <in Russian>.

Conrad, V. 1947. Usual formulas of continentality and their limits of validity. Transactions of American Geophysical Union 27 (5): 663-664.

Furyaev, V.V. 1988. Analyses or fire consequences for evaluation of forest-forming processes. Lesovedenije No.1, 59-66 <in Russian>.

Hamet-Achti, L. 1976. The biotic divisions of the boreal zone. Geobotanical cartography, pp. 51-58. Moscow <in Russian>.

Holdridge, L.R. 1967. Life zone ecology. Tropical Science Center, San Jose.

Isachenko, A.G. 1988. Landscapes and the content of the world landscape map. Izvestia, VGO, Vol.120 (6), pp. 489-501.

Nazimova, D.I., N.I. Molokova, and K.K. Djanseitov. 1985. Altitudinal zonality and climate in the mountains of the Southern Siberia. Geography and Natural Resources 2, 53-58.

Nazimova D.I., N.P. Polikarpov, and N.M. Tchebakova. 1990. Climatic ordination of forest vegetation zones and altitudinal belts as a basis of general classification of forest cover in Siberia. In: Proc. Int. Symp. "Boreal Forests. Stability, Dynamics and Anthropogenic Inputs", pp. 49-61. The Forest State Committee Press. Archangelsk.

Polikarpov, N.P., N.M. Tchebakova, and D.I. Nazimova. 1986. Climate and mountain forests in Southern Siberia. Novosibirsk, Nauka. 225 pp.

Shumilova, L.V. Botanical geography of Siberia. 1962. Tomsk State University Press, Tomsk. 440 pp. <in Russian>.

Smagin, V.N. et al. 1977. Forest cover division of Siberia. Institute for Forest and Wood, Krasnoyarsk. 44 pp. <in Russian>.

Smagin, V.N., S.A. Iljinskaya, and D.I. Nazimova, et al. 1980. Forest types of the south Siberia mountains. Novosibirsk, Nauka. 280 pp. <in Russian>

Sochava V.B., and E.M. Lavrenko. 1956. Plant cover of the USSR. Explanation text to the geobotanical map. v., 2 pp <in Russian>.

Sofronov M.A., and A.V. Volokitina. 1990. Pyrological division of taiga zone. Novosibirsk, Nauka. 200 pp. <in Russian>.

Tchebakova, N.M. 1981. The evaluation of mountain forest potential productivity in the Western Sayan on the base of climatic parameters. In: Formation and productivity of stands, pp. 6-18. Novosibirsk, Nauka <in Russian>.

Tchebakova, N.M., and E.I. Parfenova. 1991. The role of radiational factors in plant cover differentiation into altitudinal belts in Baikal basin. Geography and Natural Resources 2, 75-82.

Tuhkanen, S.A. 1984. Climatic parameters and indices in plant geography. Acta Botanica Fennica 67. Uppsala, 110 pp.

Utkin, A.I. and N.V. Dylis. 1969. The biogeocoenotic classification of forests. Lesovedenije 3, 3-23.

Valendik, E.N. 1990. Large forest fires. Novosibirsk, Nauka. 205 pp. <in Russian>.

Walter, H., and Box E. 1976. Global classification of natural terrestrial ecosystems. Vegetatio 32 (2), 75-81.

J.G.Goldammer and V.V.Furyaev (eds.), Fire in Ecosystems of Boreal Eurasia, 260-270.

The Extreme Fire Season in the Central Taiga Forests of Yakutia

G.A. Ivanova [1]

1. Introduction

In Yakutia, as in Siberia as a whole, recurrent extreme fire seasons are common. These are characterized by long rainless periods, high air temperatures and low relative humidities. These seasons account for most large forest fires, which cover vast areas and have a global-scale influence on the biosphere. From the point of view of fire suppression, it is important to know the temporal and spatial patterns of large fires in Yakutia.

In the forests of Yakutia, large fires begin to burn as early as May. In dry years, however, their maximum number is observed only in the second half of June. They also continue to occur in July, and sometimes until the end of August. Drought periods promoting large fire development are caused by dry, warm air masses of cyclones coming from the south-west, and also by areas of high pressure formed in circumpolar regions. During such periods large fires (>200 ha) account for up to 60% of the total number of fires (Valendik et al. 1979).

In dry valleys, drought conditions cause both surface and deeper fuel layers to dry out intensively. Drought even affects forest sites which, under normal conditions, hamper fire spread, e.g. river flood plains, bogs, bog pools, and other overwetted sites. This leads to the destruction of natural fire breaks, and critical conditions resulting in large fire development (Kurbatsky 1962). The specific climate of the region in question, combined with the presence of permafrost, determines the pattern of large fire occurrence in the forests of Yakutia. The summer period is remarkable for the highest fire danger. The high rate of forest burning during this period is attributed to long rainless periods, when high flammability is attained by larch forests, which represent 98% of the total forest area (Shcherbakov et al. 1979).

In the forests of Central Yakutia, relatively active high-intensity fires were recorded for the Lena-Vilui interfluve in 1850, 1883, 1890, 1899, 1913, 1915, 1922, 1941, 1948, and 1952 and for the Lena-Amgin watershed in 1855, 1892, 1897, 1913, 1937, and 1952 (Utkin 1965). This high frequency of large fires prompted our decision to investigate the temporal and spatial dynamics of extreme fire seasons in the forests of Yakutia.

[1] V.N. Sukachev Institute for Forestry and Timber, Siberian Branch, Russian Academy of Sciences, 660036 Krasnoyarsk, Russian Federation

2. Study Area and Methods

To analyze the conditions associated with extreme fire season development data on large forest fires (exceeding 200 ha in area) for the 1947-1992 period and on the total number of fires for 1947 to 1970 were gathered. In addition, weather data (air temperature, relative humidity, and precipitation) for a 10-year period from 12 weather stations were analyzed. We also attempted an extreme fire season reconstruction for a longer period through the use of dendrochronological analysis and forest fire dating. Samples for dendrochronological analysis were taken in the Lena Plateau (Olekma Forest Management Area) during the 1988-1990 period. A description of sample sites is given in Table 1.

Tab.1. Description of sample sites

Region	Site Location	Forest type	Stand composition and (age)	Mean height (m)	Mean diameter (cm)
Lena Plateau: 60°30'N 120°05'E	Ridge; Lena flood plain	Pine-mixed-herb-*Rododendron*	10P* (140) with larch	16	20
Lena Plateau: 60°10'N 120°00'E	South-facing hill	Pine-mixed-herb	10P (230) with larch	18	36
Lena Plateau: 60°45'N 120°20'E	Plain	Pine-mixed-herb-green moss	5P(230) 5L (200) 5P(100) 5L (100)	18 12	40 16
Lena Plateau: 60°00'N 120°00'E	Plain	Pine-*Arctostaphylos uva-ursi* (dead surface cover)	10P (350) 10P (120)	20 16	44 24

Note: All stands experienced several fires. * P = Pine, L = Larch

Increment cores were taken from trees at a height of 1.3 m. Tree cross-sections with fire scars were analyzed to obtain forest fire dates. More than 100 increment cores and 20 fire-scarred tree cross-sections were used in this study. Tree-ring widths were measured with the help of a stratified structure autoanalyzer connected to a DVK-3M computer. The exact date of each tree ring was established by cross-dating (Shiatov 1986; Fritts 1976).

Mathematical analysis of the obtained data was accomplished by computer analysis, using a program specifically developed for constructing tree-ring chronologies by the University of Arizona in the United States. To remove individual tree-ring series variability from the age trend, methods of standardization and indexing were applied (Fritts 1976; Vaganov and Terskov 1977; Shiatov 1986). For each tree the tree-ring series was standardized through polynomial smoothing. Calculated increment indices were obtained by dividing the absolute annual increment value by the mean increment norm for the same year; the latter being obtained from a fitted age curve equation. This processing allowed the removal of both age factor and effect of site conditions. Further, general tree-ring chronologies were constructed for each site. Past fire dating was based on well-known methods (Melekhov 1948; Madany and Swetnam 1982) that include date correction through the comparison of dates from adjacent trees, and a cross-dating procedure using tree-ring series (chronologies) acquired for a given stand.

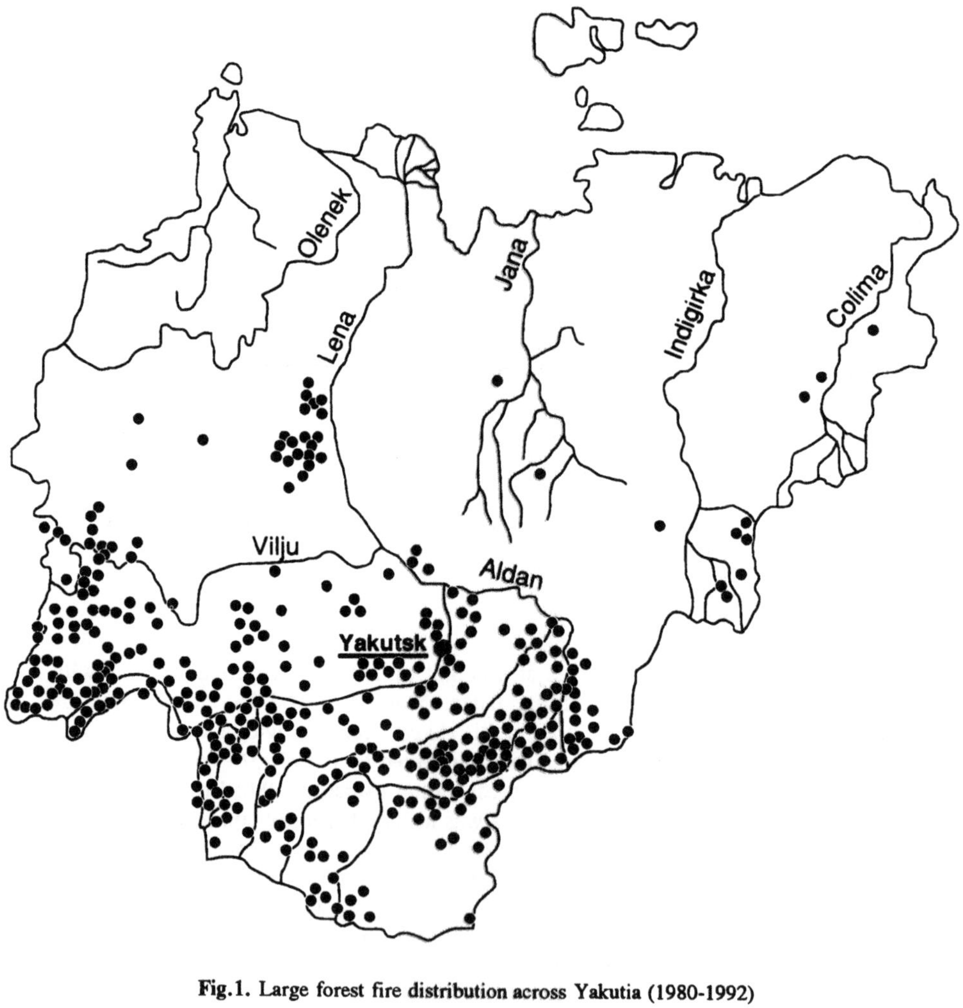

Fig.1. Large forest fire distribution across Yakutia (1980-1992)

3. Results and Discussion

Forest fires are a common event in Yakutia, whereas extreme fire seasons with large fire outbreaks are much more rare. Their occurrence pattern varies with climate, geography, vegetation patterns, and the level of management found in the region. Figure 1 shows large fire distribution in the area of interest over a 15-year period. They occur primarily in central and southwestern parts, where most of the area's forest resources are concentrated.

Peculiarities in climatic conditions are determined by circulation processes in the atmosphere which create an extremely continental climate over most of the Yakutian Republic. Spring is short and dry, and summer is largely dry. Precipitation amounts to 250-300 mm annually, with about 200 mm in the southwestern and central parts of the area. 70-75% of the total precipitation is recorded in summer, when cyclones prevail due to air mass transfer from the west. However, by the time these cyclones reach Yakutia, they usually have become highly occluded and therefore cause little precipitation.

Climatic peculiarities, which are reflected in wide daily and annual air temperature fluctuations, insolation rate and duration in summer, and considerable cloudiness combined with low precipitation amounts non-uniformly distributed across the area, induce drought periods of various duration, mostly in the first half of summer, and are responsible for high forest fire danger in the summer (Shcherbakov et al. 1979; Pozdnyakov 1983). The predominance of light conifer species in the area also contributes to high forest fire danger. These are represented mostly by larch stands, which account for 92% of the total forest area, while pine stands make up only 6%. In most larch stands, surface cover is dominated by *Vaccinium vitis-idaea* (Utkin 1965; Pozdnyakov 1963, 1983). Forest fires in Central Yakutia are known to occur as early as May, right after snow melt. During this period, fires occur mainly in higher-elevation forest stands where the living surface cover is represented by grasses and low shrubs that compose a loose fuel layer. Although the maximum number of fires is recorded in June, large forest fires are most frequent in July.

Figure 2 presents the general forest fire dynamics for the period 1947-1970 and, separately, large fire dynamics for 1947-1992. Fire distribution is of a cyclic nature. As many as 125 large fires have occurred in some years. Large fire outbreaks were found to occur once a decade for 2-3 consecutive years. Maximum large fire numbers were recorded for 1947-1948, 1954-1956, 1959-1960, 1962-1964, 1967, 1969-1970, 1973, 1985-1986, and 1990-1991. From 1974 through 1983 the number of annual large fires did not exceed 11. Over the past two decades, large forest fires decreased considerably in number, which was apparently due to certain improvements in the forest protection policy.

We have analyzed conditions associated with extreme fire season occurrence. Climatic diagrams for some years with large fire outbreaks, constructed according to Walter (1968), are shown in Figure 3. To determine drought period duration, precipitation is given on a 1:2 scale (i.e. 20 mm precipitation corresponds to 10°C air temperature). For comparison, a climatic diagram of a year with no large fires is presented, when the total number of fires was only 30.

In Yakutia, droughts are observed almost every year, but they differ in precipitation-free period duration. According to Smirnov (1958), drought occurs every 5 years, whereas catastrophic drought events occur only once a decade. In years with extreme fire seasons, drought was observed to last from June to August, but occurred most often in July. Long and severe drought was recorded in 1948. It prevailed across the area over the three summer months. 550 fires burned during this period, 125 of which were large fires. The longer and the more severe the drought, the greater the number of large fires.

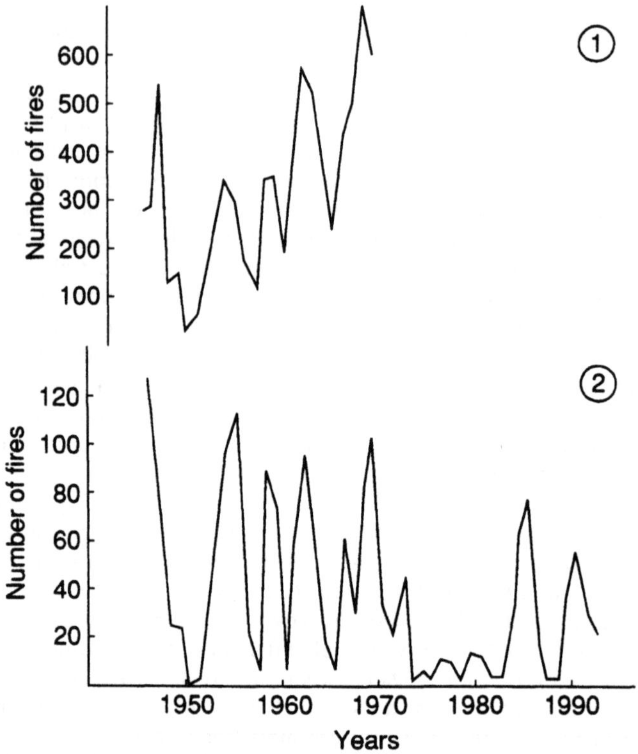

Fig.2. Forest fire dynamics in Central Yakutia: 1947-1970 total number of fires (1) and the number of large fires for 1947-1992 (2)

Analysis of precipitation-free periods (precipitation less than 2.5 mm was ignored since it does not reduce forest fire danger anyway) in years with extreme fire seasons showed that they can be as long as 40-50 days (Fig.4). However, a 10-12 day dry period in spring and a 30-40 day drought in summer is sufficient for large fires to start burning (Valendik et al. 1979).

Extreme fire seasons in the forests of Yakutia are characterized by a 50-60% precipitation deficiency, low (not more than 30%) relative humidity, 30-38°C air temperature, with wind speed being 12-15 m s^{-1}. During such seasons, high flammability is rapidly attained even by stands growing in moist sites, where fires do not normally burn.

Extreme fire seasons were reconstructed using dendrochronological methods combined with fire dating from fire scars; forest fire statistics were also used. To accomplish this four master tree-ring chronologies were developed for *Pinus sylvestris* to cover 150 to 350 years (Fig.5). These are described in Table 2. Then, using methods for comparing fire dates with the dates from adjacent trees and a dendro-series-based cross-dating procedure, dates of past fires in a given stand were obtained. Resulting fire chronologies for some stands are shown in Figure 6. To build a larch stand fire chronology, Zabelin's data on larch forests of southwestern Yakutia were used (Shcherbakov et al. 1979).

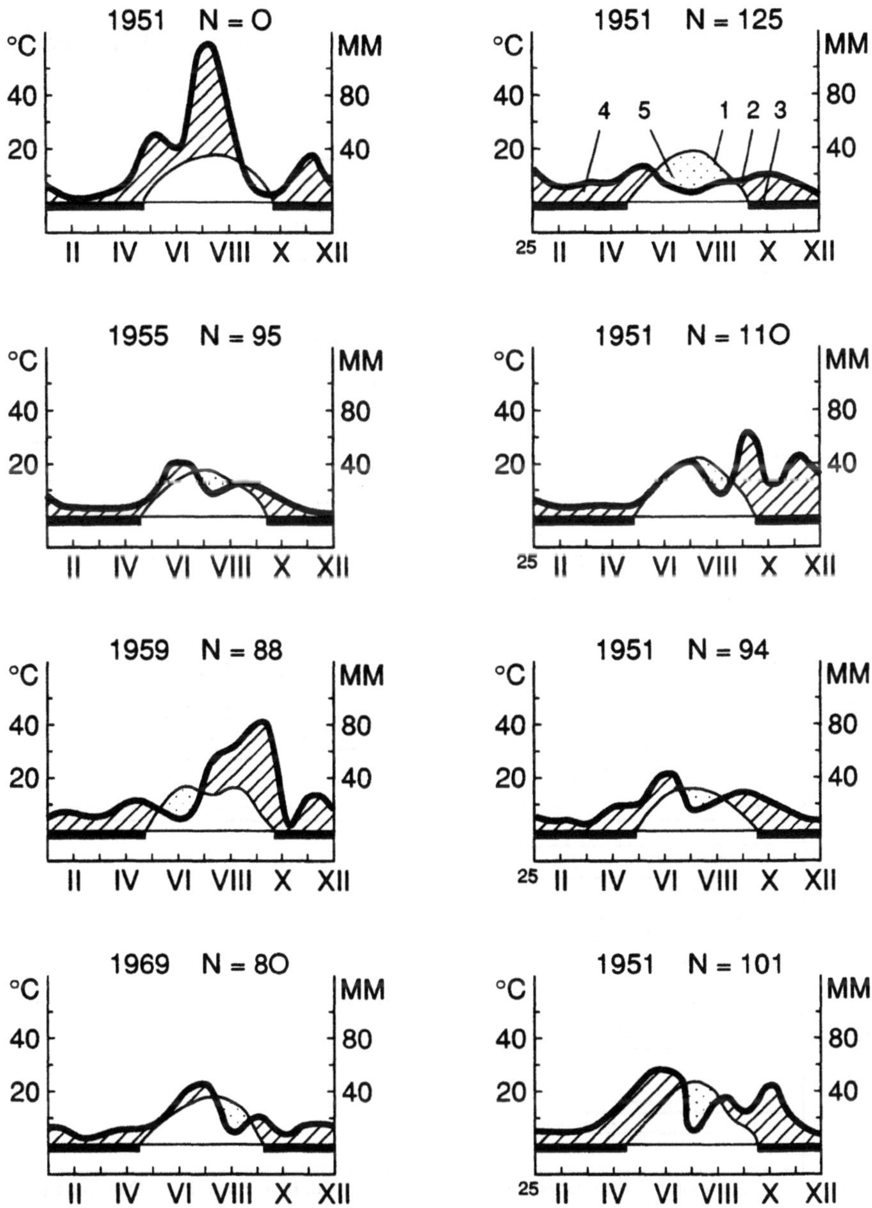

Fig.3. Gossen-Walter climate diagrams (Yakutsk weather station):

1 - Average monthly air temperature (°C)
2 - Precipitation (mm)
3 - Periods with mean monthly air temperature lower than 0°C
4 - Periods with high relative humidity
5 - Drought period
N - Number of large forest fires

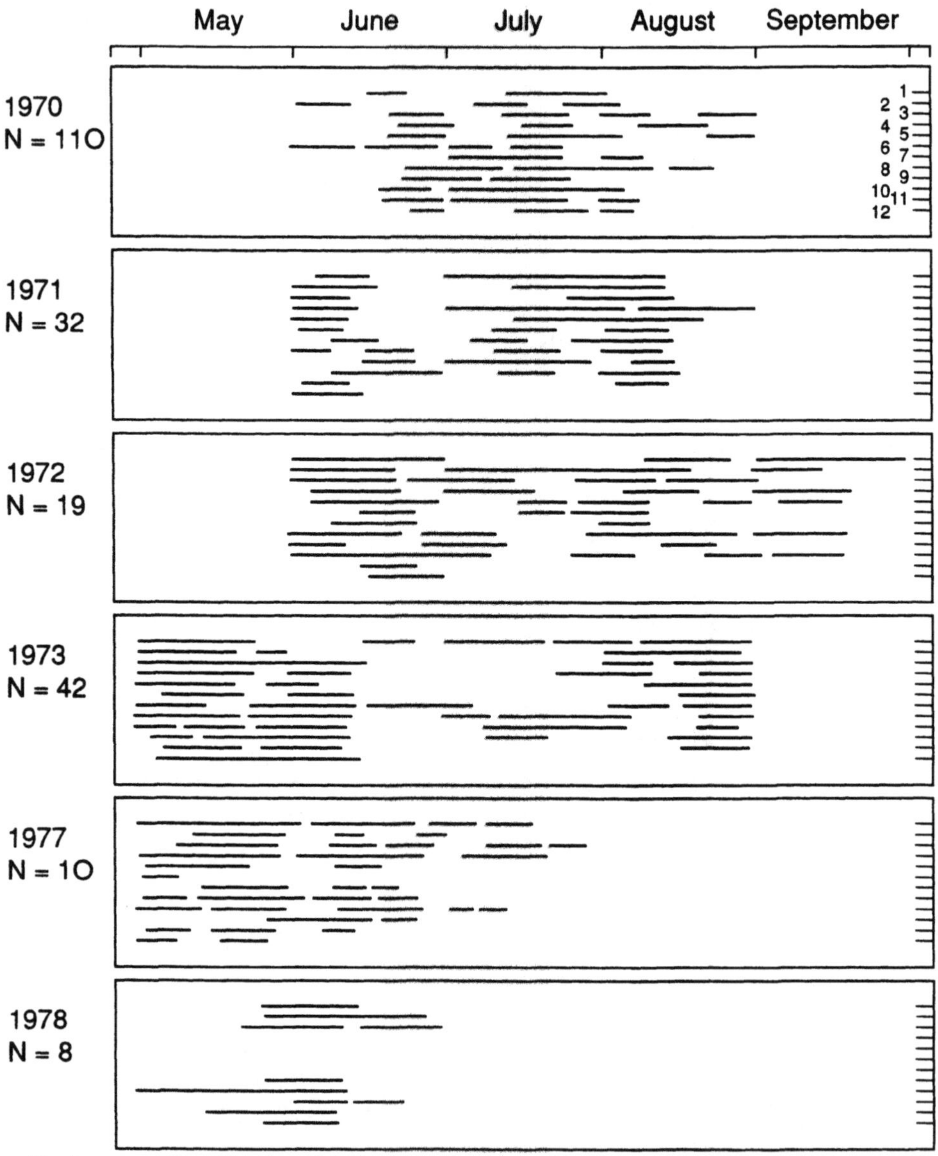

Fig.4. Duration of precipitation-free periods in years with large forest fires. Data were provided by the following weather stations:

1 -	Verkhoyansk	7 -	Isit
2 -	Higansk	8 -	Yakutsk
3 -	Viluysk	9 -	Ust-Maia
4 -	Oiyako	10 -	Vitim
5 -	Suntar	11 -	Aldan
6 -	Olekminsk	12 -	Chulman

N - Number of large forest fires

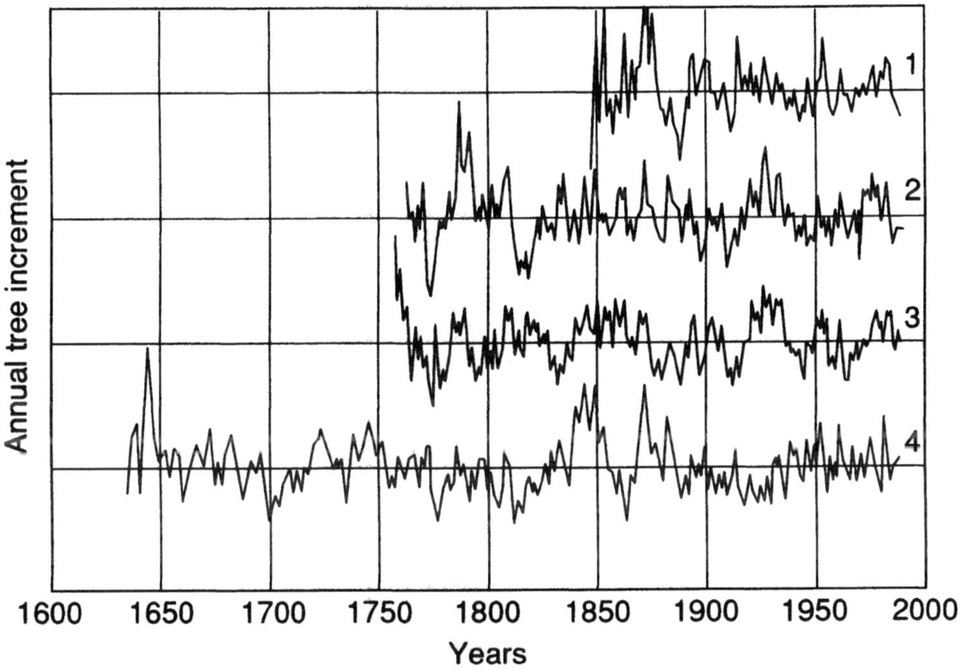

Fig.5. Dynamics of annual tree increment index. Detailed description of the four master tree-ring chronologies are given in Table 2.
1 - pine-mixed-herb-*rododendron* forest type
2 - pine-mixed-herb forest type
3 - pine-mixed-herb-green moss forest type
4 - pine-*Arctostophylos uva-ursi* forest type (dead surface cover)

Tab.2. Statistical characteristics of master tree-ring chronologies

Forest type	Species	Number of samples	Time period	Sensitivity coefficient	Synchronism coefficient (%)	Correlation coefficient
Pine-mixed-herb *Rododendron* forest	Pine	29	1847-1988	0.14	82	0.89
Pine-mixed-herb forest	Pine	22	1757-1988	0.12	70	0.78
Pine-mixed-herb green moss forest	Pine	10	1764-1990	0.13	61	0.67
Pine-*Arctostaphylos uva-ursi* (dead surface cover)	Pine	29	1635-1990	0.11	67	0.68

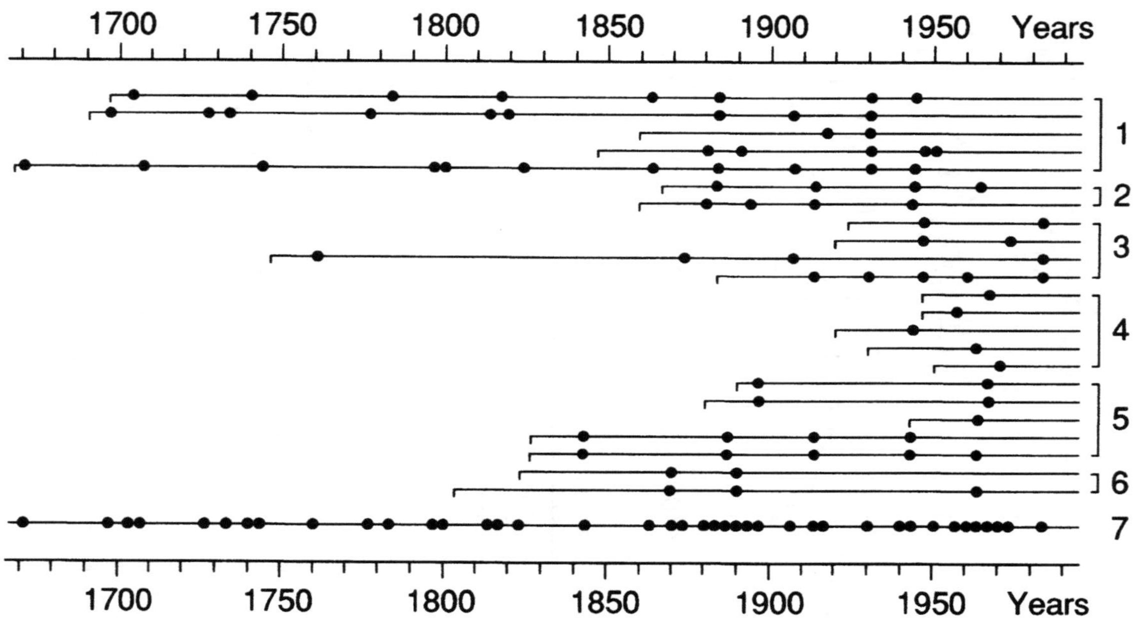

Fig.6. Fire chronologies for:

1 -	pine-*Arctostophylos uva-ursi* forest type;	2 -	pine-mixed-herb forest type;
3 -	pine-mixed-herb-green moss forest type;	4 -	larch-*Vaccinium vitis idaea*-moss forest type;
5 -	larch-*Vaccinium vitis idaea* forest type;	6 -	larch-alnus-*Vaccinium vitis idaea* forest type;
7 -	Master (general) fire chronology	● -	indicates fire dates from fire scars)

For some pine forest types, more than 10 fires were identified, with their maximum being in the pine-*arctostaphylos uva-ursi* forest type. Forest fire recurrence (rotation) was found to vary with forest type:

* 12 to 30 years in pine-mixed-herb forest (mixed-herb means that surface cover is composed of *Thalictrum simplex, Geranium transbaicalicum, Fragaria orientalis, Filipendula palmata, Cypripedium guttatum*, and others)

* 6 to 49 years in pine-*Arctostaphylos uva-ursi* stands (dead surface cover)

* 11 to 72 years in pine-mixed-herb-green moss forest (green mosses include *Pleurozium schreberi, Helacomium splendens, Dicranum scoparium* and others)

* 20 to 77 in larch-*Vaccinium vitis-idaea* and larch-*Vaccinium vitis-idaea*-green moss stands.

Our data confirm the conclusions of Isaev and Utkin (1963) that, in larch stands of Central Yakutia growing in dry-to-moist sites, the fire rotation period does not exceed 40-50 years, and in dry sites no longer than 20-25 years. During extreme fire seasons fuels are known to dry out intensively in overwetted sites, which are inflammable in normal years. To reconstruct extreme fire seasons it is thus reasonable to use data on fire periodicity obtained for forest types occupying moist sites.

Analysis of weather information and fire periodicity data has revealed that Central Yakutia experienced extreme fire seasons in 1670, 1697, 1705, 1708, 1734, 1759, 1777, 1785-87, 1811, 1818, 1821, 1846, 1862, 1870, 1872, 1881-83, 1887-1890, 1893, 1897, 1905, 1912-13, 1930, 1942-43, 1947-48, 1954-56, 1962-63, 1967, 1969-73, 1985-86, and 1990-91. From 1670 through 1992 some 50 extreme fire seasons occurred. Thus they repeated actually once a decade and covered 2-3, and sometimes 4, consecutive years.

4. Conclusions

The research studies of extreme fire seasons carried out in forests of Central Yakutia have thus revealed no distinct trend in their occurrence. These seasons occur each decade. Large fire outbreaks characteristic of extreme fire seasons can occur over several years in succession. Extreme fire seasons are remarkable for long (40-50 day) rainless periods, precipitation being 40-50% of the norm, relative air humidity as low as 30%, 30-38°C air temperature, and 12-15 m s^{-1} wind speed.

References

Fritts, N.S. 1976. Tree rings and climate. London, New-York, San-Francisco. 565 pp.

Isaev, A.S., and A.I. Utkin. 1963. Surface fires in larch forests of Eastern Siberia and the role of tree trunk pests in post fire stand regeneration. In: Forest Pest Protection in Siberia (Zashita lesov Sibiri ot nasekomykh vrediteley), 118-183. USSR Acad. Sci. Publ. Moscow <in Russian>.

Kurbatsky, N.P. 1962. Forest fire suppression methods and tactics. Goslesbumizdat. Moscow. 154 pp. <in Russian>.

Madany, M.N., and T.W. Swetnam. 1982. Comparison of two approaches for determining fire dates from tree scars. Forest Sci. 28, 856-861.

Melekhov, I.S. 1948. Influence of fires on forest. Goslestechizdat. Moscow-Leningrad. 126 pp. <in Russian>.

Pozdnyakov, L.K. 1963. Hydroclimatic regime in forests of Central Yakutia. USSR Acad. Sci. Publ. Moscow. 146 pp. <in Russian>.

Pozdnyakov, L.K. 1983. Forest on permafrost. Nauka. Novosibirsk. 96 pp. <in Russian>.

Shcherbakov, I.P., O.F. Zabelin, and B.A. Karpel. 1979. Forest fires in Yakutia and their influence on forest parameters. Nauka. Novosibirsk. 224 pp. <in Russian>.

Shiatov, S.G. 1986. Dendrochronology of the tree line in the Urals. Nauka. 136 pp. <in Russian>.

Smirnov, V.A. 1958. On estimation of drought rate. In: Droughts in the USSR. Their Occurrence, Periodicity, and Influence on Agricultural Crop Yield (Zasukhi v SSSR. Ikh proiskhozdenie, povtoryaemost i vlianie na urozhai), 135-144. Gidrometeoizdat. Leningrad <in Russian>.

Utkin, A.I. 1965. Forests of Central Yakutia. Nauka. Moscow 208 pp. <in Russian>.

Vaganov, E.A., and I.A. Terskov. 1977. Tree increment analysis based on tree ring structure. Nauka. Novosibirsk. 94 pp. <in Russian>.

Valendik, E.N., P.M. Matveyev, and M.A. Sofronov. 1979. Large forest fires. Nauka. Moscow <in Russian>.

Walter, G. 1968. Vegetation of the Earth Vol.1. Progress Publ. Moscow. 551 pp. <in Russian>.

J.G.Goldammer and V.V.Furyaev (eds.), Fire in Ecosystems of Boreal Eurasia, 271-276.

Forest Fire Spread as a Probabilistic Modelling Problem

O.Yu. Vorob'ev [1]

1. Introduction

Furyaev and Kurbatsky (1972) introduced the concept of probability of forest fires spreading across fire prevention strips. The present paper explores more general new mathematical theories on the process of geometrical forest fire spread, including a summation of a finite set of functions (Vorob'ev 1991a, b, 1992, 1993), a theory of mean sets (Vorob'ev 1975, 1984), and a theory of random spread (Bazhenov and Vorob'ev 1976, 1978; Vorob'ev 1973, 1975, 1976a, b, 1984; Vorob'ev and Valendik 1978. Vorob'ev and Dorrer 1974). These theories were intended to solve applied problems of probabilistic modelling of the geometrical shape of forest fires (Valendik et al. 1977; Vorob'ev and Valendik 1978; Konev et al. 1974; Kurbatsky and Telitzyn 1976; Sofronov and Volokitina 1983). Statistical models of fire characteristics were considered by Gorovaya and Korovin (1980); Dorrer (1978, 1986); Korovin (1984); and Furyaev and Baranov (1971).

2. Deterministic Spread

In deterministic theories describing situations with a weakly expressed turbulence (width of combustion zone and width of fire front are small in comparison with a characteristic size of a process) the fire front can be considered as a surface where spasmodic changes take place. The velocity of a moving of this surface is some physical and chemical constant, usually named a normal velocity of spread, *un*. Determination of flame front location requires solving a purely kinematic problem: to find a surface every point of which moves with velocity

$$u + unn$$

where n is a normal to a flame front (Grishin 1981, 1992; Dorrer 1979; Zeldovich et al. 1980; Konev et al. 1974; Kuznetzov and Sabelnikov 1986; Kumagai 1979; Kurbatsky and Telitzyn 1976; Ludanov 1973; Williams 1964). The simplest case follows Huygens' principle from geometrical optics (Zeldovich et al. 1980). In this approach we can fix some points on a flame surface and observe their displacement in space along trajectories of shafts

---

[1] Computer Center of the Russian Academy of Sciences, Siberian Branch, 660036 Krasnoyarsk, Russian Federation

of light. A parameter of this purely kinematic model of flame spread is a normal velocity of a front in which the total influence of all physical and chemical factors is described in the model. Note that an application of deterministic models for description of random forest fire spread is correct as used in the interpretation of kinematic construction velocities.[2]

3. Statistical Spread

Investigations of the mechanisms of forest fire spread describe turbulence statistically (Grishin 1981, 1992). Flame parts and non-flame parts form a large-scaled cellular flame front for which a width of fire zone and a heat-width are co-measurable with a forest fire size. That is why a flame front of a forest fire is not a surface with spasmodic changes of velocity, density, concentration, temperature and so on. Cellular flames are observed only near an inflammability limit, a forecasting and an explanation of which are a profound problem in a combustion theory, (Zeldovich et al. 1980; Kuznetzov and Sabelnikov 1986; Kumagai 1979; Frank-Kamenetsky 1987; Williams 1964). Following Mandelbrot (1977) we can consider that a flame front is distributed on a set of points, a topological bound, which can be considered as a space fractal with a dimension between 2 and 3. Basic difficulties are the determination of a shape and a position of the flame front and its physical description. Existing mechanistic theories of flame spread of forest fires are based on assumptions concerning statistical turbulence. Less mechanistic theories taking a full spectrum of turbulent sizes into consideration, are based on pure statistical descriptions of a large-scaled turbulence flame in a forest fire zone.

4. Probabilistic Finite Set Spread

In the probabilistic finite set theory of spread processes (Vorob'ev 1984, 1991a, b, 1992, 1993; Vorob'ev and Valendik 1978) intended for a description of situations non bounded by any fluctuations of pressure, density, concentration and so on; the front of a flame spread is defined as some set of points on which the changing of process characteristics takes place and the probabilistic distribution of random set mechanisms of moving flame front is a characteristic defined by physical and chemical constants called probabilities of a set spread:

$$px(y) = \frac{P(v(x,y) > r(x,y))}{P(v(x,y) > r(x,y) - dr)},$$

$$px(y) = \frac{P(r(x,y) <= u(x,y) < r(x,y)+dr)}{P(u(x,y) < r(x,y)+dr)},$$

---

[2] Supported by Krasnoyarsk regional science foundation (1993, 1F0061) as mean velocities of unknown probabilistic distributions which have no interpretation within a deterministic approach. An application of these models is more correct if the size of random fluctuations is small.

where u(x,y), v(x,y) are momentary velocities of maximal and minimal set spread from point x to point y from o(x), o(x) is a locality of point x, r(x,y) is a distance between points x, y, and dr is a distance between nearest points.

The essence of a probabilistic finite set approach is the notion of maximal and minimal velocities of a set spread. These velocities characterize a joint scope of a local spread and describe a spread generated by non-connected mechanisms apart from everything else. Thus, if a pure kinematic model of a deterministic spread is based on Huygens' principle then a pure statistical model of a set spread is based on the probabilistic set analogue of Huygens' principle, which adds a set description of point origin and its random nature. The joint set spread is described as a consequence of random finite sets { Kt, t=0,1,... } connected by a recurrent dependence

$$K_{t+1} = \bigcup_{K_t} S_x$$

which is analogous to Huygens' principle from geometrical optics and also has statistical properties allowing probabilistic distributions of initial data and geometrical characteristics to be described correctly. The only parameters of a pure geometrical model of a random set spread are probabilities of spread from point x during unit time: probabilities of covering px(y) = P(y belongs to Sx) via which this model takes all factors into consideration. In this situation the solution of a probabilistic problem (using theories of mean sets and sums of finite set functions) gives the description of a mean set front of flame; an estimation of set deviations from a mean front; an estimation of probabilities of hitting and covering separate forest parts; and describes the whole spectrum of random fluctuations of forest fire spread. Mechanisms of a set spread of flame from point x each of which is described by one of random finite sets Sx(I), I=1,...,m, plays an essential role here. The joint set spread from point x is described by a random set

$$S_x = \bigcup_{I=1}^{m} S_x(I)$$

which is obtained from a set-operation union of random sets, according to separate mechanisms of a flame spread of a forest fire. The result is a non-connected set which is a probabilistic model of a large-scaled cellular flame front of forest fire. Naturally from the definition of probabilities of spread, they may be represented by distribution functions of velocities:

$$p_x(y) = 1 - \frac{F_u(r(x,y))}{F_u(r(x,y)+dr)}, \qquad \frac{1-F_v(r(x,y))}{1-F_v(r(x,y)-dr}$$

It is possible to calculate a distribution of the random finite set Sx of a local spread from x via a distribution of velocities of a set spread:

$$P(Sx = A) = \prod_{A} \frac{1-Fv(r(x,y))}{1-Fv(r(x,y)-dr)} \prod_{AC} \frac{Fu(r(x,y))}{Fu(r(x,y)+dr)}$$

Thus probabilistic finite set theory is a pure statistical description taking the whole spectrum of fluctuations of a spreading front into consideration. This allows us to understand the probabilistic nature of a set spread. For analytical investigation of a probabilistic finite set spread the mathematical tools are worked out. These tools include, firstly, the theory of set-summation (Vorob'ev 1991a, b, 1992, 1993), allowing calculation of distributions of random finite sets; secondly, the theory of an average measure simulation (Vorob'ev 1973, 1975, 1976a, b, 1984; Vorob'ev and Valendik 1978) in which the notion of mean sets as a base of statistical estimations of a mean set spread is introduced.

5. Probabilistic Finite Set Identification

In order to link an abstract concept of set spread to real fire spread we must quantify estimations of the dependence of model parameters on natural factors. One can select either of two extreme approaches. First, is the conservative approach which supposes a consecutive investigation of physical and chemical variables of fire spread (for different types of fuels, topography and weather) and its influence on model parameters, for example, on a normal velocity of a moving flame front (Grishin 1981, 1992; Dorrer 1979, Konev et al. 1974; Kurbatsky and Telitzyn 1976). Second, is an operational approach: the estimation of all model parameters for a data base on spread of actual fires which then use these estimations to calculate further spread (Bazhenov and Vorob'ev 1978, Vorob'ev 1984, Vorob'ev and Valendik 1978). The first approach to the identification of a random spread is based on a known dependence between probabilities of covering and velocities of a set spread. The first approach uses results connecting velocities and physical and chemical factors accompanying a forest fire. Thus in this paper, the definition of velocities of a set spread makes possible the calculation of distributions of the random finite set Sx via forest fire physics. The second approach is based on two groups of initial physical and geometrical informations: (1) about real dynamics of a fire spread in the form of a map of consecutive fire sets; (2) about a space placement and physical properties of fuels in form of a map of parts of territory which have homogeneous physical and pyrological properties. Algorithms of identification of probabilistic distribution of the random set Sx uses an obvious analogy between the set Huygens' principle and the operation of addition of sets by Minkowsky. It allows us to use Minkowsky's set-operations on fire sets for estimations of probabilities of covering: P(Sx contains y) for every y from O(x).

6. Comparison With Existing Models of a Forest Fire

As a short comparison of the probabilistic finite set model of fire spread with existing models we consider Grishin's (1981, 1992) forest fire model based on modern theory of partial differential equations. His model is universal. It describes and considers the random nature of forest fire spread. We can also indicate partial kinematic models of fire spread (Dorrer 1979; Konev et al. 1974; Kurbatsky and Telitzyn 1976) and imitative models (Gorovaya and Korovin 1980) in which influence of all factors is taken into consideration via normal velocities. These models can take account of random fluctuations but can't describe non-connected spread. Thus we can conclude that probabilistic finite set models of fire spread are less mechanistic than the above models insofar as they describe the whole spectrum of random geometrical fluctuations in a combustion zone of a fire.

References

Bazhenov, V.V. and O.Yu Vorob'ev. 1976. Peculiarities of statistical modelling of random spread processes. Izvestiya SO AN SSSR No.13, 78-83 <in Russian>.

Bazhenov, V.V., and O.Yu. Vorob'ev. 1978. Identification of random spread processes. Izvestiya SO AN SSSR, No.8, I.2, 95-100 <in Russian>.

Dorrer, G.A. 1978. Estimation of statistical characteristics of forest fire contours. Physics of Combustion and Explosion. No.1, 39-46 <in Russian>.

Dorrer, G.A. 1979. Mathematical models of forest fire dynamics. Moscow: Forest Industry. 161 pp. <in Russian>.

Dorrer, G.A. 1986. Estimation of parameters of dynamic systems from its attainable fields. Automatics and Telemechanics No.1, 39-46 <in Russian>.

Frank-Kamenetsky, D.A. 1987. Diffusion and heat transfer in chemical kinetics. Moscow: Nauka. 492 pp. <in Russian>.

Furyaev, V.V., and N.M. Baranov. 1971. About precision of accounting of quantity of soil fuels. In: Questions of forest pyrology, 164-170. Institute of Forest and Wood SO AN SSSR, Krasnoyarsk <in Russian>.

Furyaev, V.V., and N.P. Kurbatsky. 1972. Effectiveness of protection strips in pine woods. In: Questions of forest pyrology, 140-152. Institute of Forest and Wood SO AN SSSR, Krasnoyarsk <in Russian>.

Gorovaya, E.N., and G.N. Korovin. 1980. Imitative model of forest fire. In: Economical and mathematical modelling of forest measures, 32-42. Leningrad <in Russian>.

Grishin, A.M. 1981. Mathematical models of forest fires. Tomsk: State University. 277 pp. <in Russian>

Grishin, A.M. 1992. Mathematical modelling of forest fires and new methods of its control. Novosibirsk: Nauka. 407 pp. <in Russian>.

Konev, E.V., A.I. Sukhinin, and E.K. Kisilyakhov. 1974. About combustion of soil cover in pine wood. In: Questions of forest pyrology, p.41-50. Institute of Forest and Wood SO AN SSSR, Krasnoyarsk <in Russian>

Korovin, G.N. 1984. Analysis and modelling of statistical structure of combustion field of forest. Leningrad: LenNIILKH <in Russian>.

Kumagai, S. 1979. Combustion. Chemistry, Moscow. 256 pp. <in Russian>.

Kurbatsky, N.P., and G.P. Telitzyn. 1976. Modern theory of forest fire spread. Modern investigations of typology and pyrology of forest. Arkhangelsk <in Russian>

Kuznetzov, V.R., and V.A. Sabelnikov. 1986. Turbulence and combustion. Moscow: Nauka. 288 pp. <in Russian>.

Ludanov, V.V. 1973. About geometrical shape of forest fire. Forest Economy No.3/1973, 48-50 <in Russian>.

Mandelbrot, B.B. 1977. Fractals and turbulence: attractors and dispersion. Lecture Note in Math. V. 615, 83-93. New York: Springer-Verlag.

Sofronov, M.A., and A.V. Volokitina. 1983. Formation of operative maps of forest fuels. In: Operative control of forest protection, 49-52. Institute of Forest and Wood SO AN SSSR, Krasnoyarsk <in Russian>.

Valendik, E.N., O.Yu. Vorob'ev, and A.M. Matavev. 1977. Forecasting of forest fire contours by computer. In: Questions of forest pyrology, 52-66. Institute of Forest and Wood SO AN SSSR, Krasnoyarsk <in Russian>.

Vorob'ev, O.Yu. 1973. Mathematical description of random spread processes and its control. Izvestiya SO AN SSSR. No.13, I.3, 145-152 <in Russian>.

Vorob'ev, O.Yu. 1975. Definition of probabilities of fire spread and estimation of dynamics of mean forest fire contours. In: Questions of forest pyrology, 453-67. Institute of Forest and Wood SO AN SSSR, Krasnoyarsk <in Russian>.

Vorob'ev, O.Yu. 1976a. Models of some distributed probabilistic processes. Izvestiya SO AN SSSR. No.3, I.1, 105-113 <in Russian>.

Vorob'ev, O.Yu. 1976b. Methods of modelling of random spread processes. Izvestiya SO AN SSSR. No.8, I.2, 20-94 <in Russian>.

Vorob'ev, O.Yu. 1984. Average measure modelling. Moscow: Nauka. 133 pp. <in Russian>.

Vorob'ev, O.Yu. 1991a. USSR DAN. V.318, No.4, 785-788 <in Russian>.

Vorob'ev, O.Yu. 1991b. Soviet Math. Dokl. 43, No.3, 747-751.

Vorob'ev, O.Yu. 1992. Doclady RAN. V. 326. No.4, 583-588 <in Russian>.

Vorob'ev, O.Yu. 1993. Set-summation. Novosibirsk: Nauka. 137 pp. <in Russian>.

Vorob'ev, O.Yu, and E.N. Valendik. 1978. Probabilistic set modelling of forest fire spread. Novosibirsk: Nauka. 151 pp. <in Russian>.

Vorob'ev, O.Yu, and G.A. Dorrer. 1974. Probabilistic model of forest fire spread. In: Questions of forest pyrology, 118-135. Institute of Forest and Wood SO AN SSSR, Krasnoyarsk <in Russian>.

Williams, F.A. 1964. Combustion theory. London: Addison-Wesley Publishing Company. 666 pp.

Zeldovich, Ya.B., G.I. Barenblatt, V.b. Librovich, and G.M. Makhviladze. 1980. Mathematical theory of combustion and explosion. Moscow: Nauka. 479 pp. <in Russian>.

J.G.Goldammer and V.V.Furyaev (eds.), Fire in Ecosystems of Boreal Eurasia, 277-284.

Information Technology for Forest Fire Danger Rating Evaluation

A.I. Sukhinin [1]

1. Introduction

An intensive forest resource management policy that is currently followed in Siberia necessitates a high level of forest fire protection. In Russia, fires occur annually on 6% of the total forest area, while in the West European countries and in Canada they burn only a hundredth percent of the area.

During the last decade, mass forest fires induced by droughts were observed every year in some regions of Siberia and the Far East. These fires tend to cover hundreds of thousands of hectares of forest area and often cause irreversible economic losses. In Canada, fire damage was appraised to be $100 per hectare. This sum also applies to our forests since they are similar in composition and in technical properties to Canadian forests.

The present situation is recognized to have resulted from the forest protection system being far behind forest management progress. Furthermore, no well-developed scientific technology currently exists for forest fire danger evaluation and prediction which could provide reliable data on the probability of forest fire occurrence, on potential fire behavior, and on fire damage to biogeocoenoses.

Prediction of conditions favorable for fire occurrence and development is of crucial importance in making decisions on optimal fire fighting resource distribution and in effective planning of forest fire prevention measures. The knowledge of potential fire intensity in any given forest environment is required to make most effective use of fire suppression resources and to choose appropriate fire fighting tactics.

2. The Problem Acuteness and State-of-the-art in Research

Forest fuel moisture content is known to be the ultimate control of the probability of forest fire occurrence and growth. This parameter varies with weather conditions.

In our country, the system of forest fire danger rating evaluation is based upon Nesterov's complex meteorological index which takes into account the relative humidity, air temperature and precipitation, and indirectly reflects fuel moisture content. The input information for this index is provided by weather stations.

---

[1] V.N. Sukachev Institute for Forestry and Timber, Siberian Branch, Russian Academy of Sciences, 660036 Krasnoyarsk, Russian Federation

Unfortunately, a small radius of weather measurement extrapolations and a lack of weather stations hinder the determination of spatial distribution patterns of both precipitation and fuel moisture content as a basis for constructing a fire danger map. As a result, dispatchers are limited in their choice of optimal fire detection flight routes and no scientifically justified system of pre-suppression resource distribution can be developed.

We propose to evaluate forest fire danger on the basis of the ground cover infrared (IR) and/or microwave (MW) radiation recorded by space-borne sensors which are directly related to fuel moisture content (Sukhinin 1991a,b).

A previously proposed method of current forest fire danger evaluation, using aircraft measurements of the forest surface microwave radiation, appears to be almost similar to the newly proposed approach (Valendik and Sukhinin 1980, 1986). This method involves the acquisition of microwave radiation data along the flight route using a smoothing procedure for the portions of records corresponding to the forest area which are characterized by abnormally low brightness temperature and the determination of current forest fire danger rating from the average brightness temperature values and its mean-square deviation. However, the above method has the following limitations:

- Acquired data cannot apply to the territory beyond the flight route coverage. For this reason, a fire danger map of a given area cannot be constructed.

- Smoothing of the record portions indicating abnormal brightness temperature does not allow one to account for natural water pools and wetland sites. Thus no fully reliable information on a real overall picture can be obtained and consequently, natural fire breaks are inevitably disregarded in forest fire fighting decision-making. Furthermore, brightness temperature values obtained on a single measurement basis do not describe fuel drying history and dynamics and thereby do not permit one to predict forest fire danger.

- The aim of the present project is to increase the accuracy of the weather-related determination of current and potential fire danger rating spatial patterns.

This can only be achieved through the acquisition of forest area images in IR and MW bands on a regular basis followed by image processing. For the two-dimensional forest area recording the IR spectral band (8-14 micrometers) is used because the maximum of radiation from natural objects falls within this band. The MW band (0.3-3.3 cm) is used since it takes atmospheric windows into account, best represents the close correlation between brightness temperature and available fuel moisture content, and permits all-weather survey. Additionally, the above bands of the electromagnetic spectrum allows one to evaluate forest fire danger rates under permanent smoke cover.

The data processing involves the spatial overlaying of images obtained in a number of successive satellite survey cycles, the calculation of both MW and IR brightness temperatures for each pixel of an individual image, and finally, the estimation of the current forest fire danger rating using the following equation:

$$FDR=\sum_{j=1}^{n}\frac{Tir_j(x,y)*Tmv_j(x,y)}{Tir_j(x,y)-Tmv_j(x,y)} \tag{1}$$

where j = 1,2,...n is a satellite survey cycle; n is the number of survey cycles, starting from the cycle that corresponds to the assumed boundary conditions; x,y are the coordinates of the area under analysis (image pixel); Tir_j and Tmv_j are infrared and microwave brightness temperatures, respectively. Rainfall intensity and distribution pattern can be determined through the comparison of successive images obtained in IR and MW bands based on temperature differences (D) recorded between two successive survey cycles.

$$Tir_j(x,y) - Tir_{(j-1)}(x,y) = Dir_j(x,y)$$
$$Tmv_j(x,y) - Tmv_{(j-1)}(x,y) = Dmv_j(x,y) \qquad (2)$$

If the temperature differences Dir_j and Dmv_j are less than assumed threshold values, the forest areas satisfying this condition are assigned to a zero fire danger class. In this case, the sum of fire danger values is set to zero and the calculation by Equation 1 starts again for the next survey cycle, with Equation 2 being considered.

3. The Theoretical Concept

The theoretical concept behind the proposed method is that fuel temperature (Tf) regulates the fuel drying rate (Lykov 1968), therefore, moisture content (W) appears to be more closely correlated with Tf than with air temperature (Ta), since the latter does not provide information on the radiation heat exchange between forest fuel and the environment. However, forest fuel radiation coefficient emissivity is 4 to 10% less than 1; that is to say that Tir is 7 to 16% lower than Tf; consequently, Tf can be measured by the IR spectral band (8-14 micrometers) with the above accuracy. Both the underlying surface radiation maximum and the atmospheric window fall within this spectral band. The integration of the temperature curve (Equation 1) over time permits the calculation of the current fuel moisture content. That is why we propose to add fuel temperature values obtained at consecutive moments of time in order to account for fuel temperature history.

Thus there is a possibility to predict fuel moisture content when predicting fuel temperature. MW brightness temperature turns out to correlate closely with forest fuel moisture content (Valendik et al. 1987). For the MW spectral band, the radiation coefficient appears to vary in a wider range (from 0.4 to 1). The conditions under which the brightest of forest fuel radiation is formed and the desire for all-weather measurements necessitated us to choose 0.3 to 3 cm wavelengths of the microwave band.

4. Application of the Method

The proposed method can only be employed practically provided that a regional centre for satellite data reception and processing (RCRP) is established in Krasnoyarsk coordinated by the Institute of Forest. The processed information is then transferred to the Krasnoyarsk Forest Fire Protection Air Base.

Experience in the execution of operational forest fire protection tasks gained in foreign countries so far shows that it is by the use of high resolution NOAA information that the highest effectiveness is achieved. The NOAA satellites are equipped with special scanning devices designed to contribute to the solution of ecological monitoring problems and have much to offer for the solution of the following forest fire monitoring tasks:

- Early detection of fires; estimation of their intensities.
- Operational determination of the current forest fuel flammability level based on computer-dependent synthesis of information from the near IR and the far IR spectral bands.
- Large fire dynamics mapping; estimation of energy characteristics of large fire fronts and convection columns.
- Spatial distribution of smoke zone intensity under multiple-fire conditions.
- Burned area inventory and analysis of the post-fire state.

NOAA satellite information is very promising for many other aspects including crop, pasture, and forest monitoring, hydrometeorology, geology, and the Russian Far North industrial development.

5. Technology of Forest Fire Danger Rating Evaluation

At such a centre, images are processed by a 1:2,500,000 scale or larger obtained from a NOAA satellite in the 10-12 micrometers wavelength range (AVHRR scanner, Channels 4,5) over the Krasnoyarsk Territory. The survey must be conducted only when the sun is in its highest position over a given area. As only thermal spectral band information is now available for RCRP, current fire danger can be determined, though at a lower accuracy level, by the following equation:

$$FDR=\sum_{j=1}^{n} Tir_j(x,y) \tag{3}$$

Fire danger rating estimation begins with the onset of snow melting in the southern taiga region. To evaluate fire danger, every image is converted into a digital radiation temperature map and the temperature values of those image pixels having the same geographical coordinates are added. Based on calculation results, maps are constructed of the daily fire danger rating in relation to weather conditions. To do this, two zones are delineated: the snow cover zone for $Tf < 273$ K and the snow-free zone for $Tf > 273$ K. The resulting fire danger map is painted and shaded by fire danger class. In this case, any given range of fire danger values corresponds to a certain fire danger class. In case the fire danger rate exceeds an assumed threshold value, for example, when it corresponds to the third fire danger class, additional rainfall analysis is needed based on Equation 2.

The resulting fire danger map is transferred daily via a communication channel to the dispatching units of forest protection air bases and is used to make decisions on optimal detection flight routes. This map must be improved using the results of weather situation analysis and fuel map information. When a part of the area, over which satellite survey is conducted, is covered by clouds, it is deemed reasonable to additionally analyze the subsequent images acquired on the same day by overlaying them in order to exclude cloud cover effects.

However, no cloud cover problems are confronted when anticyclone (high pressure) prevails over the area being surveyed. Anticyclone promotes most dangerous fire weather conditions.

At some time in the near future, when space-borne MW sensors are available, one will be able to estimate fire danger using the more informative Equation 1.

6. The Essence of the Project

The major goal of the project is to increase the accuracy of operational evaluation of spatial fire danger rating patterns. This will lead to labour-saving technology due to fire detection and containment planning optimization. High effectiveness of fire detection planning is in turn possible with forest cover remote sensing approaches which involve the acquisition of IR or MW data from space-borne sensors on a regular basis. Using this data, digital maps are constructed of forest cover radiation and/or brightness temperatures. The maps are then overlayed (compared) and fire danger value is calculated for each map pixel. This value is expressed as a sum of product-to-difference ratios of brightness and radiation temperature measurements obtained in each satellite survey cycle.

The above procedure is dependent on the information provided by forest fuel maps. These should be landscape-based maps that use typological data on forest stands, relief and hydrographical characteristics of a given area, and reflect drainage and ground water level regimes. Basic classification principles and methods for constructing operational maps of forest fuel flammability patterns are discussed in a number of works.

The aim of the project is to scientifically justify the proposed method, to obtain the correlations and empirical parameters needed, and to test the method.

7. The Project Stages

Stage 1 - 1994

- Installation of the HRPT station for receiving NOAA AVHRR imageries in Krasnoyarsk.

- The development of methods for constructing 1:500,000 landscape-based fuel maps for the territory covered by 46-48 sensing swaths taking into account forest types and hydrological and relief features.

- The development of a mathematical model describing fuel moisture content dynamics that will incorporate ground cover radiation and brightness temperatures as the main variables. Determination of model applicability limits. Estimation of the contribution of the parameters associated with thermal and biological fuel properties, their spatial arrangements, as well as with forest canopy and atmospheric screening effects.

- The development of application software to be used in designing a GIS for fire danger evaluation and prediction. The GIS will include forest area space image processing along with the analysis of the associated weather and fuel parameters.

- The development of in situ investigation methods, field testing of the proposed methods, and the establishment of the correlations and empirical parameters needed.

Stage 2 - 1995-96

- Testing the information technology for fire danger rating evaluation in Krasnoyarsk Forest Fire Protection Air Base using ground and satellite data.

  Expected product: A fundamentally new technology for forest fire danger evaluation based on forest ecosystem remote sensing which will provide for improved fire protection planning.

8. Comparison of the Results with the World Standard

In world-wide practice, fire danger evaluation is dependent on information provided by weather and storm detection stations. The network of these stations is sufficient to achieve a high fire danger calculation accuracy. However, the need is recognized for the development of a Geographical Information System (GIS) of fire danger evaluation technology where new parameters are introduced that characterize forest fuel ignitability (flammability), potential fire spread rate, current fire load, and fire season severity (Burgan and Rothermel 1984).

Methods of remote sensing are employed primarily to measure soil moisture content for agriculture, but they have relatively little applicability to forest fire danger estimation (Engman 1991).

9. Current Progress

- Basic classification principles and methods are now available to build operational fuel flammability maps (Valendik 1990).

- The relationship between microwave radiation coefficients and fuel moisture content has been investigated for major fuel types. The procedure to estimate ground water levels and methods of available fuel flammability estimation from airborne sensor data have been developed (Valendik et al. 1980).

- A semi-empirical model for weather-related dynamics of available fuel moisture content has been developed (Sukhinin 1975).

- Basic methods for the development of the GIS of forest fire danger evaluation are now available (Valendik and Sukhinin 1991).

- The main characteristics have been established of forest canopy screening effects on incident and back-scattered IR and microwave radiation (Sukhinin 1991a,b).

- An airborne information complex of infrared and microwave sensors has been designed to detect and map forest fires, to estimate their intensities, and to evaluate forest fuel moisture content (Valendik et al. 1983).

The idea behind our approach to forest fire danger evaluation has been proved valid by the results of preliminary field radiation and brightness temperature measurements conducted for major fuels available for burning (needle litter, dead grass, lichen, and moss layers). The close correlation identified between forest fuel moisture content and the new fire danger value permitted us to establish the critical fire danger value indicative of the conditions under which a fire occurs (starts).

10. Application of Scientific Results

The proposed information-dependent technology is potentially useful in measuring moisture content of agricultural and other natural and artificial objects (targets). It can be successfully applied for the determination of vegetation indices, particularly of water stress. It can also be relied upon to determine the level of risk when deciding on rent and insurance rates.

Acknowledgements

I wish to thank Academician Alexander Isaev, International Forestry Institute, RAS, Dr. Robert Murphy, NASA Headquarters, Dr. Michael Fosberg, US Department of Agriculture, Forest Service, Dr. Brian Stocks, Canadian Forest Service, Dr. Eric Kasischke, Duke University for their helpful reviews of the manuscript and important comments. This work was supported by Krasnoyarsk Science Foundation, Grant 1F0044.

References

Burgan, R. E. and R. C. Rothermel. 1984. Fire behavior prediction and fuel modeling system. USDA Forest Service. Gen. Tech. Report INT-167, 126 pp.

Engman, E. T. 1991. Applications of microwave remote sensing of soil moisture for water resources and agriculture. J. Remote Sens. Environ. 35, 213-226.

Lykov, A. V. 1968. Theory of drying. Energia, Moscow, 470 pp. <in Russian>.

Sukhinin, A. I. 1975. An experimental study of combustion spread over needles. Essay of thesis. Krasnoyarsk <in Russian>.

Sukhinin, A. I. 1991a. A method of forest fire danger evaluation. Certificate No 1648505, A 62 C 3/02, BI No.18 <in Russian>.

Sukhinin, A. I. 1991b. Probability of forest fire detection using methods of remote sensing. In: Scope "Forest Fire Suppression" (Lesnye pozhary i bor'ba s nimi),56-69.VNIIPOMLESHOZ, Krasnoyarsk <in Russian>.

US Forest Service. 1983. The 1978 national fire danger rating system. USDA Forest Service Intermountain For. and Range Exp. Station, Ogden, Utah, 144 pp.

Valendik, E. N. 1990. Major forest fire suppression. Nauka, Novosibirsk 192 pp. <in Russian>.

Valendik, E. N., E. K. Kisilyakhov, and A. I. Sukhinin. 1980. Taiga forest fire danger evaluation from thermal radiation data. Investigation of the Earth from Space (Issledovanie Zemli iz kosmosa) No.3, 14-19 <in Russian>.

Valendik, E. N., G. A. Dorrer, A. I. Sukhinin, V. V. Bagenov, and E. K. Kisilyakhov. 1983. A remote sensing system for forest fire control and operational fire spread rate prediction. In: Scope "Satellite-Based Methods of the Environmental Studies in Siberia and the Far East" (Kosmicheskie metody izuchenia prirodnoi sredy Sibiri i Dalnego Vostoka), 136-155. Nauka, Novosibirsk <in Russian>.

Valendik, E. N., and A.I. Sukhinin et al. 1986. A method for current forest fire danger evaluation. Certificate No. 1247020, A 62 C 3/02, BI No 28 <in Russian>.
Valendik, E. N., A. I. Sukhinin, E. K. Kisilyakhov, and B.A. Khrebtov. 1987. Forest fire monitoring. In: Scope "Investigations of forests using aerospace methods" (Issldovanie lesov aerokosmicheskimi metodami), 118-135. Nauka, Novosibirsk <in Russian>.
Valendik, E. N. and A. I. Sukhinin. 1991. Aero-space system for forest fire monitoring. In: "Forest Monitoring" (A.S.Isaev, ed.). Nauka <in Russian>.

J.G.Goldammer and V.V.Furyaev (eds.), Fire in Ecosystems of Boreal Eurasia, 285-302.

Mathematical Modelling of Forest Fires

A.M. Grishin [1]

1. Main Assumption and Physical Model of Medium and a Forest Fire

A burning forest is considered to be a polyphase, multi-layer, porous three-dimensional heterogeneous medium, consisting of dry organic substance (volume fraction φ_1), dispersed water (liquid medium water, φ_2), solid pyrolysis product (char, φ_3), stationary char combustion product (ashes φ_4), gas phase (φ_5), dispersed ash particles (φ_6), soot particles (φ_7), and drops of rain (φ_8). Forest fuel elements (thin twigs, conifer needles, foliage) all have the same temperature, while the gaseous and dispersed phases have another. The solid elements (trunks, branches and twigs) sway in the wind. Flexural vibration of these elements, i.e. aero elasticity of the medium, acts only on the resistance value and on the coefficients of heat-and-mass transfer of forest fuel elements with gas phase, the medium being nearly rigid (almost unchangeable in the wind gusts); the thermal energy, resulting from free and forced convection and radiation in the fire front, heats,dries and then decomposes forest fuels. The result are gaseous combustible products, inert gaseous pyrolysis products, and solid combustible pyrolysis products (char). Then the gaseous and solid pyrolysis products are ignited and burnt out, and the process is repeated in the above mentioned order (see Fig.1).

During a forest fire the photon free path length is assumed to be appreciably smaller than typical dimensions of forest phytocenosis $\bar{r}$ and $\bar{h}$,where is $\bar{r}$ an effective macropore diameter (an average tree spacing), $\bar{h}$ is an average height of a forest fuel layer. All assumptions are based on the physical model of the phenomenon and agree with well-known experimental results (Grishin 1981, 1992).

2. Basic Combined Equations

Using an arbitrary Cartesian coordinate system, fixed to the underlying surface, tensor symbolism, and taking into account the assumptions mentioned above, we get combined equations as follows (Grishin 1992):

---

[1] Physical Mechanics Department, Tomsk State University, 634050 Tomsk, Russian Federation

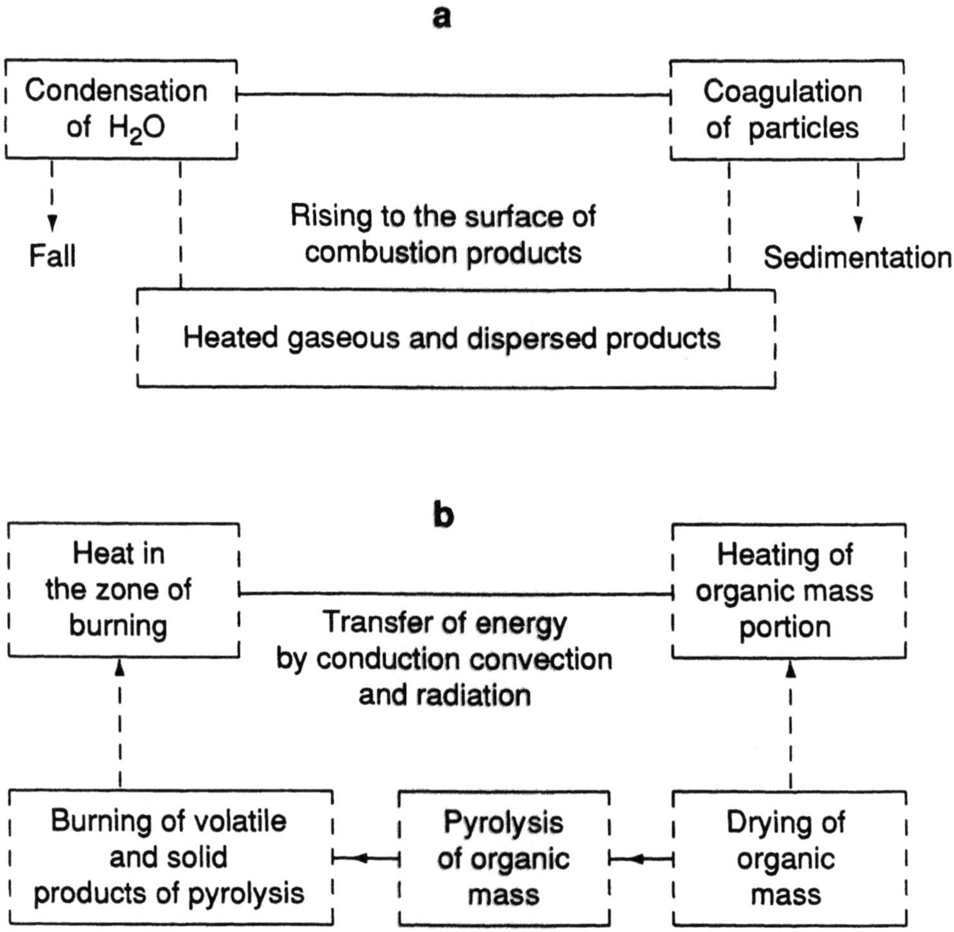

Fig.1. A diagram of the burning process of forest fires

$$\frac{\partial\rho}{\partial t}+\frac{\partial\rho V_j}{\partial x_j} = Q\ , \quad j=1,2,3 \tag{1}$$

$$\rho\frac{dV_i}{dt} = -\varphi\frac{\partial P}{\partial x_i}+\rho F_i+\frac{\partial\tau_{ij}}{\partial x_i}-QV_i-\rho SC_d\Big[\ V_i|V|+\frac{1}{2}\sum_{j=1}^{3}\overline{(V_j'-V_{js}')^2}\cos\alpha_i+$$
$$+\sum_{j=1}^{3}\overline{(v_i'-v_{is}')(v_j'-v_{js}')}\ \cos\alpha_j\Big]\ ,\ i,j=1,2,3,\varphi=\sum_{i=5}^{8}\varphi_i,$$
$$\vec{F}=\vec{g}+(\Omega\times\vec{r})\times\Omega+2\vec{V}\times\Omega \tag{2}$$

$$\frac{dT}{dt}\sum_{i=5}^{8}\rho_i\varphi_iC_{pi}=\sum_{i=5}^{8}\varphi_i\frac{dP}{dt}+\rho F_jv_j+\frac{\partial}{\partial x_j}(\lambda_{\text{э}\Phi}\frac{\partial T}{\partial x_j})+\frac{dT}{dx_j}(\rho\sum_{\alpha=1}^{N}D_{\text{э}\Phi\alpha}C_{p\alpha}\frac{\partial C_\alpha}{\partial x_j})+$$
$$+(\eta+\eta^{(S)})\ [cU_R-B(T)]+\alpha_v(T_s-T)+q_5R_5+q_{3W}R_{3W}^{(S)}+q_2(R_{8-}^{(S)}-R_{8+}^{(S)}) \tag{3}$$

$$\rho\frac{dC_\alpha}{dt}=R_\alpha+\frac{\partial}{\partial x_j}(\rho D_{\text{э}\Phi\alpha}\frac{\partial C_\alpha}{\partial x_j})-C_\alpha Q\ ,\ \alpha=1,2,\ldots,N \tag{4}$$

$$\sum_{i=1}^{4}\rho_i\varphi_iC_{pi}\frac{\partial T_S}{\partial t}=\frac{\partial}{\partial x_j}(\lambda_{sj}\frac{\partial T_S}{\partial x_j})+\eta_s[cU_R-B(T_s)]+$$
$$+q_{1S}R_{1S}+q_{2S}R_{2S}+q_{3W}R_{3W}+\alpha_v(T-T_S) \tag{5}$$

$$\rho_1\frac{\partial\varphi_1}{\partial t}=-R_{1S},\ \rho_2\frac{\partial\varphi_2}{\partial t}=R_{2S},\ \rho_3\frac{\partial\varphi_3}{\partial t}=\alpha_cR_{1S}-\frac{M_c}{M_1}R_{3W}-\alpha_4R_{3W}-R_6^{(s)} \tag{6}$$

$$\rho_4\frac{\partial\varphi_4}{\partial t}=\alpha_4R_{3W}-R_7^{(S)},\ \sum_{i=1}^{8}\varphi_i=1,\ P=\rho_5RT\sum_{\alpha=1}^{N-3}\frac{C_\alpha}{M_\alpha},$$
$$Q=(1-\alpha_c-\alpha_6)R_{1S}+R_{2S}+\frac{M_c}{M_1}R_3+R_6^{(S)}+R_7^{(S)} \tag{7}$$

Here t is the time; $\vec{r}$ is a radius vector (position vector) of an arbitrary point; x_j - Cartesian coordinates; η , η^S , η_S are integral absorption factors of gas, dispersed particles and solid phase; B is the Planck function; U_R is the integral radiation density; v_i , $|\vec{v}|$ are components and modulus of averaged gas and dispersed particle velocity; v_i' , v_{is}' are fluctuating components of flow velocity and fuel element flexural vibrations; ρ_i is the ith phase true density; $R_{8-}^{(s)}$, $R_{8+}^{(s)}$ are mass rates of steam condensation and free water heat of evaporation in the gas-dispersed phase; q_{1s} , q_{2s} , q_{3W} are the pyrolysis heat, bound water heat of evaporation and char heat of combustion respectively; q_2 is free water latent heat of evaporation (condensation); P is the gas pressure; τ_{ij} are tensor components of sheer stress; C_α is the mass fraction of α - component in the gas-dispersed medium; N is the number of the components in the gas-dispersed medium; R_α is the mass rate of formation of α -component, resulting from f.f. pyrolysis, water evaporation, heterogeneous and homogeneous chemical reactions;

$$\rho = \sum_{i=5}^{8} \rho_i \varphi_i \quad \text{gas-dispersed mixture density;}$$

D_α, D_T are the molecular and turbulent diffusion coefficients; λ , λ_T are molecular and turbulent thermal conductivities of the gaseous phase; $D_{N\text{эф}}$, $D_{N-1\text{эф}}$ are effective coefficients of smoke particle diffusion, where $N-1$ associates with soot particles, and N - with ashes; $D_{\alpha\text{эф}} = D_\alpha + D_T$ is the effective coefficient of α -component diffusion; α_v is the inter phase heat exchange coefficient; R_{1s} are mass rates of formation (disappearance) of the solid phase components; $\vec{q}_R$, $\vec{q}_{Rs}$ are intensity vectors of radiant heat flow in the gas-dispersive phase and porous medium; q_{Rj} , q_{Rjs} are components of vectors $\vec{q}_R$, $\vec{q}_{RS}$; $\cos\alpha_i = v_i / v$ are direction cosines of mean (average) velocity vector in the gas-dispersive phase; q_k are thermal effects of chemical reactions; α_c is the coke number (char yield) of the f.f.; R is the universal gas constant; M_α is molecular mass of gas phase component α ; $\vec{g}$ is gravity acceleration; Ω is the angular velocity of the earth's rotation; $\lambda_{\text{эф}} = \lambda + \lambda_T$, $\mu_{\text{эф}} = \mu + \mu_T$ are effective coefficients of thermal conductivity and gas viscosity; C_d is the drag coefficient of the vegetation; s - specific f.f. surface in the given forest layer; F_j are vector components of $\vec{F}$; i , j adopt values 1,2,3; k =1,2,3 where 1 associates with pyrolysis reactions, 2 - water evaporation rate, 3 - char heterogeneous burning rate.

It is necessary to choose turbulent and radiation heat transfer coefficients for a combined equation constant. It is also necessary to have structural, thermophysical and thermokinetics data. Turbulence model selection for the first generation mathematical model of forest fires has already been discussed in detail (Grishin et al. 1984, 1985, 1994). As to the first approximation, one can choose equilibrium approximation of the $K-\varepsilon$ turbulent model (Grishin et al. 1984, 1985, 1994; Patankar 1984) (modified theory by Prandtl). Perfect account of dissipation and turbulent energy generation in the fire front is more accurate if the exact turbulent $K-\varepsilon$ model is used (Patankar 1984). It is evident that f.f. element vibrations and chemical reactions in the forest fire front greatly influence the turbulent energy generation and dissipation processes. However, to model turbulence using all these effects is a complex independent task, deviating from the theme of this work.

A forest fire model of the first generation was based on the assumption of "grey" medium (Grishin 1981, 1992), which is not performed under real conditions, because the f.f. elements (green leaves, conifer needles) reflect radiation in the neighboring infra-red region (740-1200 nm) more accurately than in the visible region. Therefore, it is advisable to use a group diffusion model for energy transfer by radiation (Chetverushkin 1985). The equation for radiation transfer with regard to spectral properties of f.f. and dispersed phase particles is as follows:

$$div\, \vec{q}_{Rn\Sigma} + \eta_{n\Sigma} c_n U_{Rn} = Q_{Rn}, \quad n=1,2,\ldots,\mu$$

$$\vec{q}_{Rn\Sigma} = \frac{1}{3} \int_{\nu_n}^{\nu_{n+1}} \frac{c_\nu \nabla U_\nu}{K_{\nu\Sigma}} d\nu, \quad \eta_{n\Sigma} c_n U_{Rn} = \int_{\nu_n}^{\nu_{n+1}} \eta_\nu c_\nu U_{R\nu} d\nu \tag{9}$$

where $\vec{q}_{R\nu\Sigma}$, $q_{Rn\Sigma}$ is a general spectral radiation flow in the poly-phase porous dispersive medium and its group approximation; $U_{R\nu}$ is the radiation density in the polyphase medium; c_ν , c_n is the spectral velocity of light and its group approximation; $\eta_{n\Sigma}$ is a group approximation for the integrated coefficient of absorption for the medium ; $K_{\nu\Sigma}$, $K_{n\Sigma}$ is the spectral coefficient of radiation extinction and its group approximation:

$$Q_{Rn} = \int_{\nu_n}^{\nu_{n+1}} Q_{R\nu} d\nu, \quad Q_{R\nu} = 4\pi \left[(\eta_\nu + \eta_\nu^{(S)}) B_\nu(T) + \eta_{\nu S} B_\nu(T_s) \right] \tag{10}$$

B_ν -Planck's spectral function; η_ν , $\eta_\nu^{(S)}$, $\eta_{\nu S}$ - spectral coefficients of absorption for gas, dispersed particles and solid phase respectively.

3. Boundary and Initial Conditions

Boundary and initial conditions are necessary in order to solve the problem of forest fire origin and its propagation. In the general three-dimensional case, a fire area is a parallelepiped, in which the fire front is propagating. Let the first and the second faces of this parallelepiped (Γ_1 , Γ_2) be perpendicular to the axis x which is aligned with the wind; the lower and the upper faces (Γ_0 and Γ_3) are perpendicular to the axis z , directed vertically upward; the sidefaces (Γ_4 and Γ_5) are perpendicular to the axis y . The origin is at the origin of combustion. Then on the boundaries Γ_1 , Γ_3 , Γ_4 , Γ_5 the medium state variables concide with non-disturbed values

$$\vec{V}|_{\Gamma_i} = V_e, \quad T|_{\Gamma_i} = T_e, \quad T_S|_{\Gamma_i} = T_{Se} \tag{11}$$

$$C_\alpha|_{\Gamma_i} = C_{\alpha e}, \quad \varphi_i|_{\Gamma_i} = \varphi_{ie}, \quad i=1,3,4,5$$

where index e associates with the non-disturbed medium parameters. It is necessary to use the so called "soft" boundary conditions on the Γ_2 boundary.

$$\frac{\partial T_S}{\partial x}\Big|_{\Gamma_2} = \frac{\partial T}{\partial x}\Big|_{\Gamma_2} = 0, \quad \frac{\partial C_\alpha}{\partial x}\Big|_{\Gamma_2} = 0, \quad \frac{\partial \vec{u}}{\partial x}\Big|_{\Gamma_2} = 0 \tag{12}$$

The boundary conditions on the underlying surface are as follows:

$$\frac{\partial T}{\partial z}\Big|_{\Gamma_0}=0,\quad \frac{\partial T_s}{\partial z}\Big|_{\Gamma_0}=0,\quad \frac{\partial C_\alpha}{\partial z}\Big|_{\Gamma_0}=0$$
$$V\big|_{\Gamma_0}=V\big|_{\Gamma_0}=0,\quad \rho V\big|_{\Gamma_0}=f_0 \tag{13}$$

If a forest fire propagates (spreads) from an original ignition area, the initial conditions are as follows:

$$T_s\big|_{t=0}=T\big|_{t=0}=\begin{cases}T_{\Gamma H}, & M\in G_\Gamma\\ T_e, & M\notin G_\Gamma\end{cases}\qquad C_\alpha=\begin{cases}C_{\alpha\Gamma H}, & M\in G_\Gamma\\ C_{\alpha e}, & M\notin G_\Gamma\end{cases} \tag{14}$$

$$\varphi_i\big|_{t=0}=\begin{cases}\varphi_{i\Gamma H}, & M\in G_\Gamma\\ \varphi_{ie}, & M\notin G_\Gamma\end{cases}\qquad \vec{u}\big|_{t=0}=\begin{cases}\vec{u}_{\Gamma H}, & M\in G_\Gamma\\ \vec{u}_{\Gamma e}, & M\notin G_\Gamma\end{cases} \tag{15}$$

where G_Γ is a space region, corresponding to the ignition area; M is any point in the fire area; index ΓH shows the medium condition parameter of the ignition zone.

If forest fuels of tree crowns ignite, results are the following boundary conditions:

$$q_{Rn}\big|_{\Gamma_i}=-\frac{cU_n}{2}\Big|_{\Gamma_i},\quad q_{Rn}\big|_{\Gamma_0}=\begin{cases}E_n\sigma T_{\Gamma H}^4, & M\in G_\Gamma,\\ E_n\sigma T_e^4, & M\notin G_\Gamma\end{cases} \tag{16}$$

where q_{Rn} is the radiation flux density in the direction of the inner normal of any face Γ_i ,E (unter n) is the coefficient of radiation of emission, σ is the Bolthman-Stefan-coefficient, $i=1,2,\ldots,5$; Γ_0 is a surface of the soil cover.

4. Results of Numerical Solution of Problem on Aerodynamics of Forest Fires

By means of the methods stated above, a series of calculations for various values of speed of wind, temperature at the front of the fire, rate of blow-in of heated products of combustion and the width of the fire front $2x_\Gamma$ were carried out. The problem was solved by the Patankar-Spalding method (Grishin et al. 1994).

Figure 2 shows a field of velocities and isotherms for a steady flow in the ground layer of atmosphere with a logarithmic profile of the wind speed, injection velocity w_Γ = 2.56 m/s, T_Γ = 1200K, T_∞ = 300K, x_Γ = 3m. It is seen that a large vortex which induces the inflow of air to the fire front is formed on the leeward side. Near the fire front, from the windward side the lines of flow bend and the values of velocity over the fire front are about 1.5 times greater than the speed of wind in the undisturbed flow, i.e. the speed of wind at the fire front increases. The field of velocities agrees in qualitative respect with the experimental data given in (Grishin 1981).

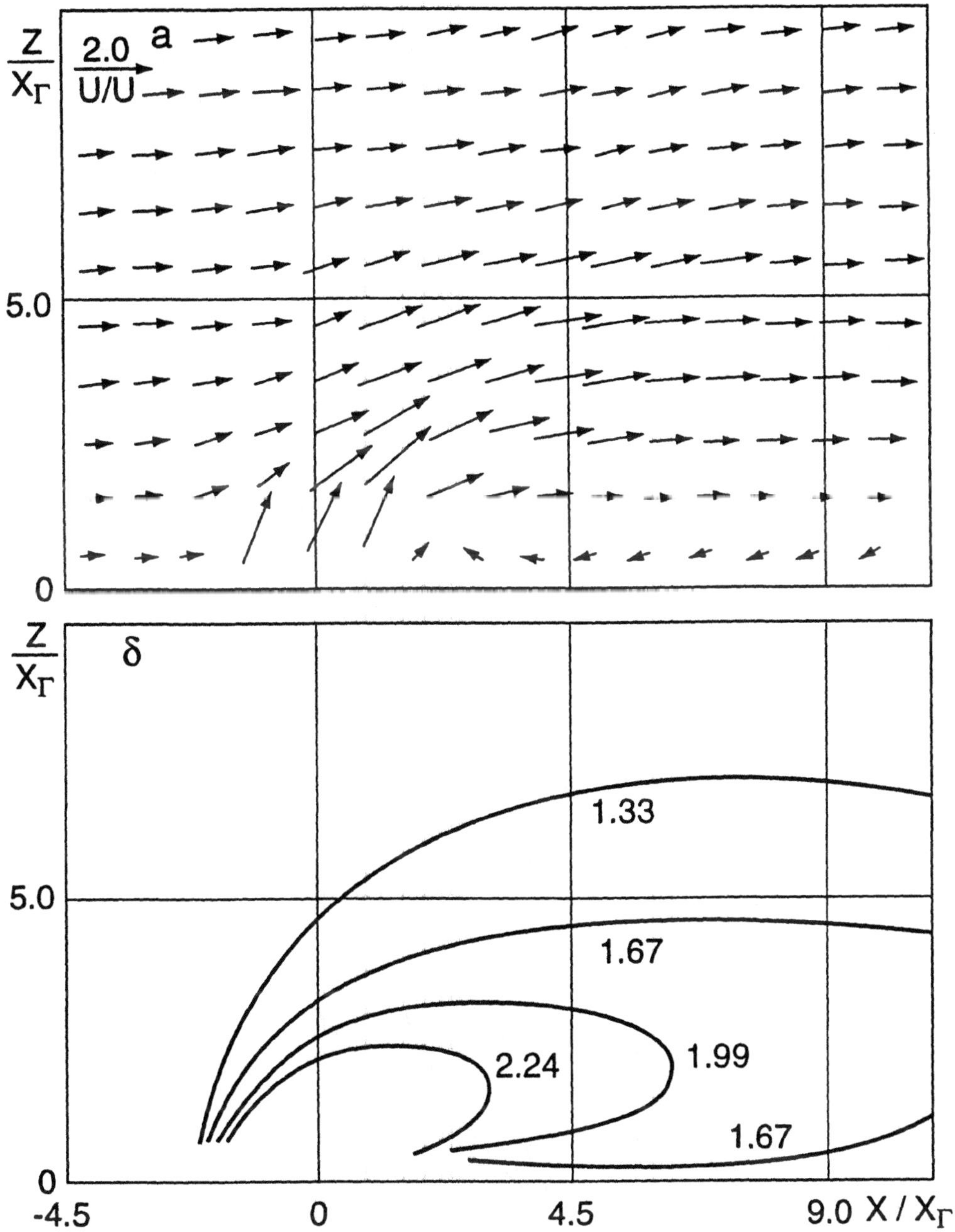

Fig.2. Velocity field (a) and isotherms over the burning zone in flow of the "plumage" type

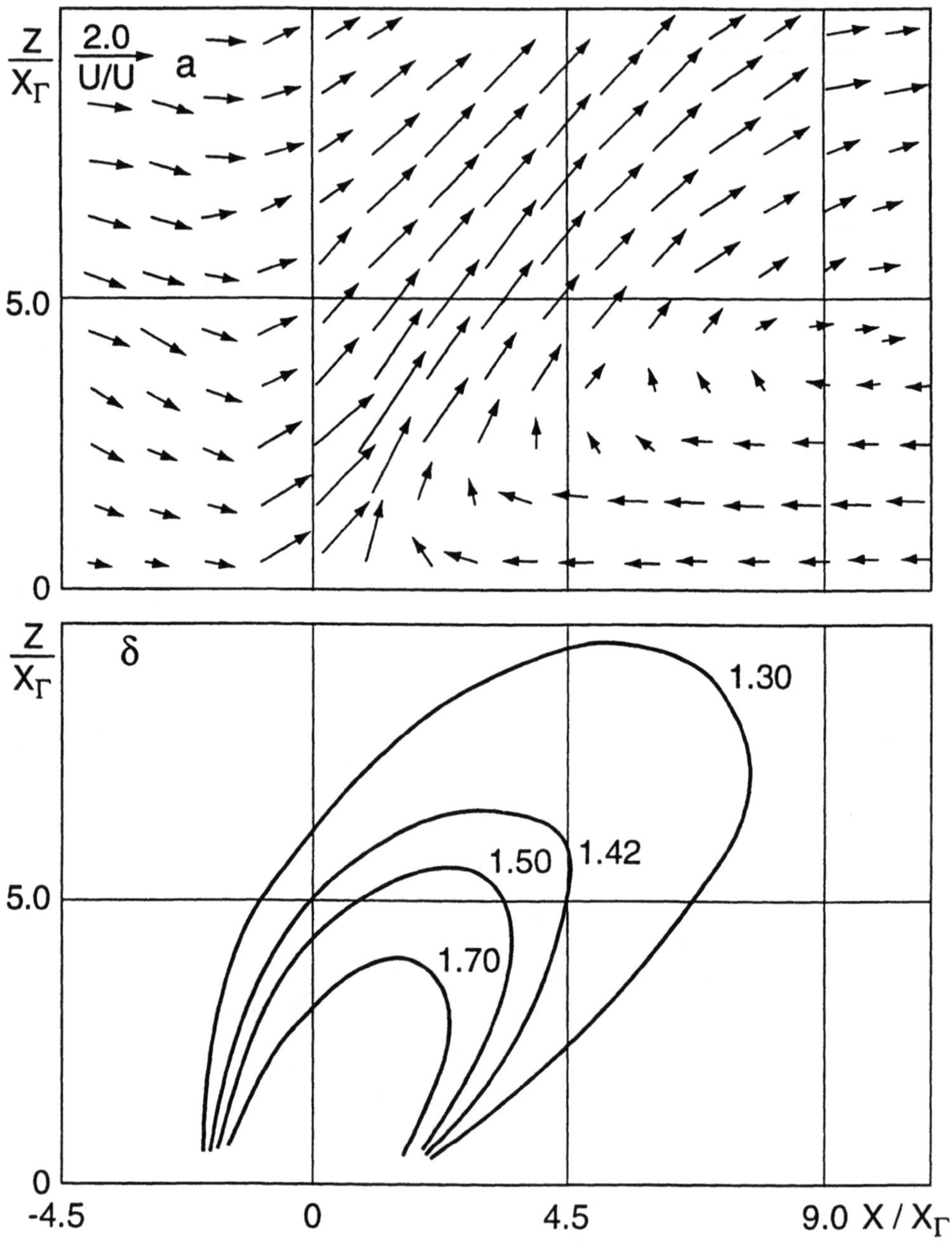

Fig.3. Velocity field (a) and isotherms over the burning zone for the inclined convective column

Analysis of isotherms show that the thermal plume is almost vertical. Therefore, in this case, one can speak of a flow - a type of convective column. (Hereafter we shall understand a convective column flow to be one with a vortex on the leeward side of the fire front.)

Another type of a flow in the fire zone arises with increased wind speed. In Figure 3 the field of velocities and isotherms with $U|_{z=h}$ =5.43m/s, W_r =18.1m/s, T_r =1200K, x_r=3m, T_∞ =300K is represented. It is seen that in this case far away from the fire front, the streamlines and isotherms are almost parallel to the underlying surface. According to Patankar (1984) and Grishin and Fomin (1985) such a regime of flow is called "plumage". This type of flow is characterized by the fact that the vortex on the leeward side of the front is absent and a boundary layer flow is realized near the underlying surface. As calculations show, the heat flow on the underlying surface changes sign. This fact as well as the picture of the lines of flow and isotherms prove that the front of a forest fire can be considered to be a peculiar thermal screen (Kutateladze and Leontiev 1976). This conclusion makes it possible to develop an analytic theory of forest fires as within the limits of the theory of thermal screens analytic formulae (Kutateladze and Leontiev 1976), allowing evaluation of characteristics of heat and mass exchange between the front and the environment.

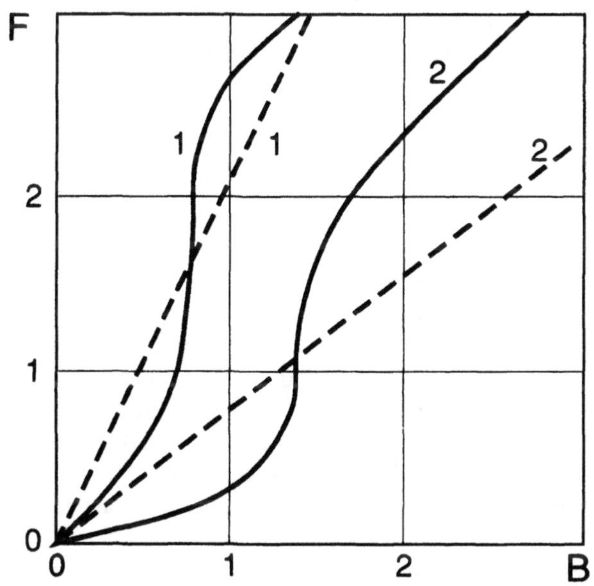

Fig.4. FLOW conditions in the boundary layer of the atmosphere at the time of forest fire (F>f(B)) are the conditions of plume ($B=\frac{W_\Gamma}{U_\infty}(i-\frac{T_\infty}{T_\Gamma})$).

As a result of the analysis of numerous mathematical experiments, aided by dimensional analysis, it is found that the type of flow in the ground layer of the atmosphere over the fire front is determined by the Froude number $Fr=U_{\infty}^2/gl$, where $l=x_r$ is the semi-width of the fire front, and by parameters of a complex type:

$$B=W_r/U_{\infty}[1-\frac{T_{\infty}}{T_r}] .$$

In the plane of these parameters one may determine the curve $Fr=f(B)$, above which, points correspond to the flow of the plumage type. From the analysis of the curves presented in Figure 4 it follows that the Byram criterion satisfactorily predicts types of flow in the near-ground layer of atmosphere for small moments of time as compared to the time of the fire and the Gostintsev criterion which predicts types of flow for moments of time, comparable to the time of restructuring the flow in the near-ground layer of the atmosphere.

The obtained results are true in a qualitative sense, not only for low intensity fires but for crown fires as well; as the plume rising in a strong wind in the case of crown fires is analogous to the plume rising in the ground layer of atmosphere from low intensity fires.

5. Dynamics of Crown Fire Front Formation

Mathematical modelling of one-dimensional front propagation of a crown forest fire was carried out by the use of a resonant circuit based on the iteration-interpolation method (Grishin et al. 1981). In Figure 5, graphs illustrating the dynamics of the front of a crown forest fire entering the steady regime of spreading with $U_{*\infty}$ =8m/s, ρ_c =0.2 kg/m³ ; W =66.66%; α_c =0.1; ν_r =0.8; β =0.08; h_2 =5m, μ_τ =1kg/(ms) are given. The curves

$$\bar{\varphi}_1=\varphi_1/\varphi_{1H},\ \bar{\varphi}_2=\varphi_2/\varphi_{2H},\ \bar{\varphi}_3=\varphi_3/\varphi_{3m},\ \varphi_{3m}=\alpha_c\rho_1\varphi_{1H}/\rho_3$$

are a maximum volume fraction of char; $\bar{T}=T/T_e$ is the non-dimensional temperature; $\bar{U}=U/U_{*\infty}$ is the non-dimensional velocity; $U_{*\infty}$ is the equilibrium speed of wind in the forest canopy (Grishin et al. 1984), which is discussed by Grishin (1992) in detail; oxygen mass fraction c_1 and combustible gas c_2 in the forest canopy at various moments of time: t =0.6; 0.8; 1.0; 1.2; 1.4; 1.6; 2.0; 3.2; 4.8; 9.86 s (curves 1-10 respectively) are presented. The fire front was initiated by a source of heat release with a volumetric rate of $Q=6\cdot10^5\ g/(m^3s)$. It is seen that in some time (t >0.8s) ignition of the forest canopy takes place and the fire front is formed which in the end with t >4.8 enters a steady regime of spreading characterized by the fact that all the profiles of parameters of the medium state retain their forms and shift downwind with the constant velocity $\omega=0.625U_{*\infty}$. It is interesting that the oxygen concentration as a result of oxidization of combustible products of pyrolysis drops almost to zero while $c_2\neq0$. Hence, in the combustion zone of gaseous products of pyrolysis, a lack of oxidizer occurs and diffusion controlled combustion is realized. Conditions for extinction of the steady fire front are described in detail in papers by Grishin (1981) and Byram (1966) therefore they will not be discussed here further.

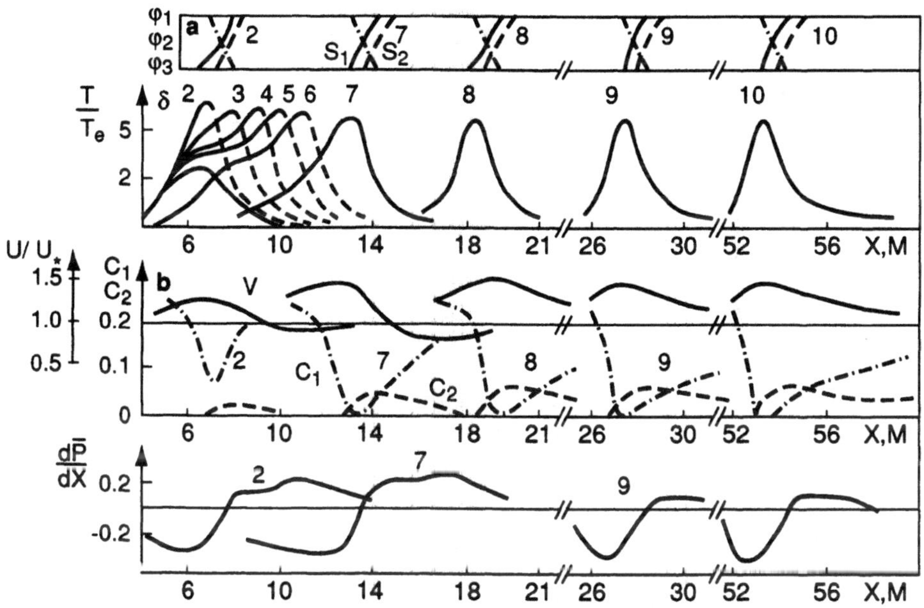

Fig.5. Volume fraction fields φ_1, φ_2, φ_3, fields of dimensionless temperatures T/T_e and speed U/U_*, oxygen C_1 and gaseous combustible mass fractions C_2 at different moments of time from ignition till steady propogation conditions.

The stationary front structure (in coordinates fixed to the fire front) of a crown fire is of interest. Volume fractions of phases (see Fig.6a), temperature, concentration and the speed of wind in the forest canopy (see Fig.6b), non-dimensional heat and mass flows

$\bar{q}^{(-)} = q_T^{(-)} / \rho_{5\infty} C_{p5\infty} T_\infty U_{*\infty}$ and $\bar{J}_\alpha^{(-)} = J_\alpha^{(-)} / \rho_{5\infty} U_{*\infty}$ (see Fig.6c) as well as

$\frac{d\bar{p}}{d\xi} = \frac{dp}{dx} h_3 / \rho_{5\infty} U_{*\infty}^2$ and $\bar{\tau}^{(-)} = \tau / \rho_{5\infty} U_{*\infty}^2$ (see Fig 6d).

In Figure 6e profiles of temperature and wind velocity, as well as a mass fraction of oxidizer and combustibles over the canopy of a forest at some values of ξ (see Fig.6b) are presented. It is easy to see that in the forest canopy the minimum c_1 (oxygen mass fraction) coincides with the maximum of T (temperature). Our attention is drawn to the fact that the curve $c_1(\xi)$ is asymmetrical in relation to $\xi = 0$, the value $c_1(\xi)$ rapidly decreases with the growth of ξ with $\xi < 0$ $c_1(0) \sim 0$, and then a relatively weak growth $c_1(\xi)$ downward is observed, which is explained by consumption of oxygen on oxidization of combustible products of pyrolysis. The maximum $c_2(\xi)$ is observed in the front zone where pyrolysis takes place. This is interesting that the volume fraction of condensed products of pyrolysis (char) almost does not change in the back part of the fire front with $\bar{\xi} < 0$.

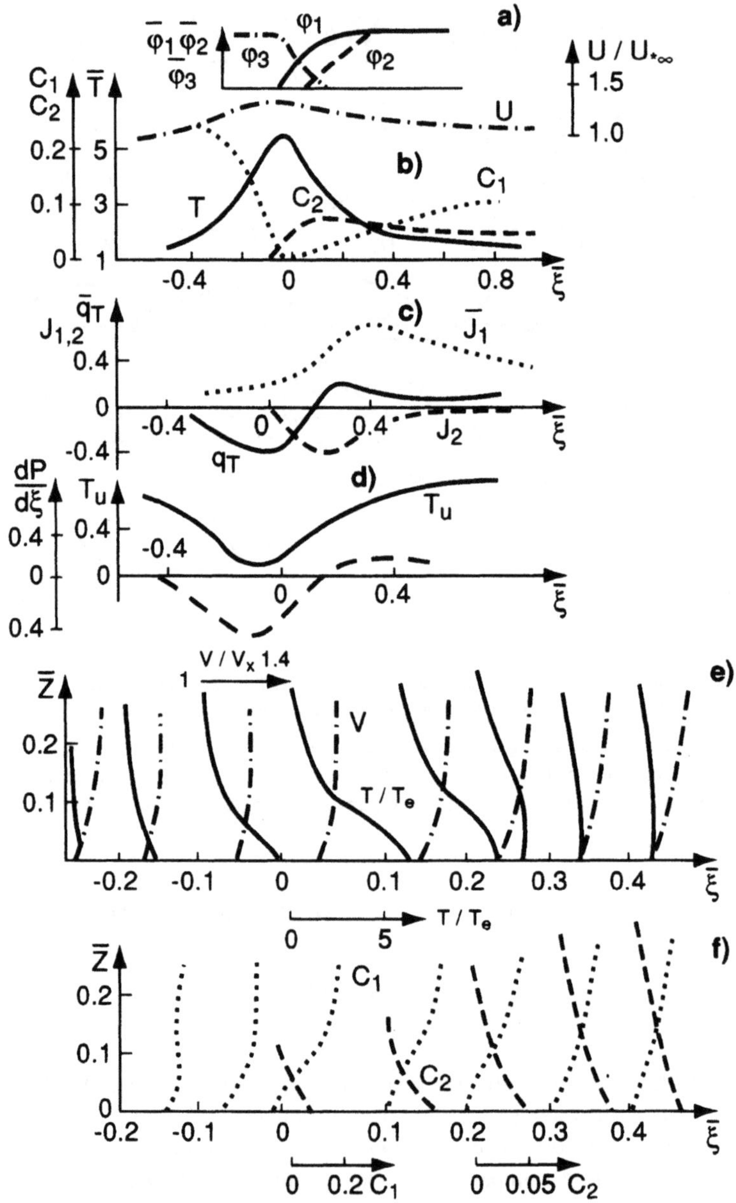

Fig.6. Structure of combustion wave when $U_{*\infty} = 8$ m/s, W = 66.6 %, $\rho_0 = 0.2$ kg/m, h_3 $h_3 = 5$ m, where $\bar{\varphi}_1 = \varphi_1 / \varphi_{1H}$, $\bar{\varphi}_2 = \varphi_2 / \varphi_{2H}$, $\bar{\varphi}_3 = \varphi_3 / \varphi_{3H}$, $\bar{T} = T / T_{*\infty}$, $\bar{U} = U / U_{*\infty}$, $\bar{Z}$ and $\bar{\xi}$ - dimensionless functions and independent variables.

It is explained by the fact that char does not have time to react as the combustion time of volatile products of pyrolysis is considerably less than the combustion time of solid products. As a result of calculations, it is found that with $U_{*\infty} \geq 5$ m/s burn-out of char does not exceed 4 % from the initial mass, while the growth of $U_{*\infty}$ decreases. In other words a crown fire spreads due to heat release as a result of the combustion of gaseous combustible products of pyrolysis, which agrees with the experimental data. It is interesting that the turbulent heat flow q_T (see Fig.6c) changes sign but in the main, the function $q_T(\bar{\xi})$ qualitatively reflects the behavior of $T(\bar{\xi})$, as $q_T<0$ with $T>T_\infty$ does in the back part of the front with $\bar{\xi}<0$. Only with $\bar{\xi}>0$, a change of the sign q_Σ conditioned by the fact that heated gaseous products of combustion in the ground layer of the atmosphere spreads faster than in the forest canopy as a result of which the heating up of canopy fuels with $\bar{\xi}>0$ takes place from above, through the ground boundary layer. This result allows us to consider the front of a crown fire as a peculiar heat screen which has already been spoken of by Grishin (1992), where the problem of heat and mass exchange of the front of a crown fire modelled by a surface source of mass and energy was considered. In the given case the result is confirmed within the limits of a more accurate model of the front of a forest fire. Dependence of intensities of diffusive flows $I_1(\bar{\xi})$ and $I_2(\bar{\xi})$ correspond to the profiles of concentrations $c_1(\bar{\xi})$ and $c_2(\bar{\xi})$ and differ from the profile of $q_T(\bar{\xi})$, i.e. strictly speaking, processes of heat exchange and mass exchange are not similar.

In the canopy of a forest near the fire front, the flow velocity increases (see Fig.6,b) which also agrees with the experimental data of Section 1.8 where the increase of the wind velocity in the fire front is pointed out. At the same time, the sheer stress and pressure gradient (see Fig.6c) have minimized at the fire front, i.e. they decrease with the growth of $\bar{\xi}$ and then rise again. Such behavior of $\bar{\tau}^{(-)}$ and $d\bar{p}/d\bar{\xi}$ agrees with the behavior of the flow velocity in the near-ground layer of the atmosphere and in the forest canopy.

6. Mathematical Modelling of Forest Undergrowth Ignition in the Explosion of the Tunguska Meteorite

From the analysis of the data (Grishin 1981, 1992; Grishin et al. 1984, 1994) it follows that the Tunguska explosion, according to its characteristics, resembles an atmospheric nuclear explosion (Grishin and Perminov 1992). Its heat effect on forest combustible materials can be modelled using the results of the papers of Dimitriev (1965), Gostintsev et al. (1989) and Korobeinikov et al. (1991):

$$q_{Rn} = \frac{P_m t_R \sin\varphi}{4\pi R^2} \begin{cases} t/t'_m & , t<t'_m \\ \exp[-k(t/t_m-1)] , & t>t'_m \end{cases} \tag{17}$$

where t_m is the time of maximum of heat release from the radiation source; R is the distance from the center of the radiation source to the forest canopy (Fig.7); t_R is the atmospheric transmission coefficient; P_m is the power of the light pulse at the moment of time t_m; φ is the angle of incidence of the radiant flux on the vegetation cover; k is an empirical factor; q_{Rn} is the intensity of normal heat flux normal to the surface.

It is supposed that a forest can be modelled as a two-temperature multi-phase, porous, reacting medium consisting of dry organic substance, water in liquid state, solid products of pyrolysis, ash, and a gas phase. For a description of the transfer of energy by radiation, a diffusion approximation is used (Chetverushkin 1985).

For reasons of simplicity, let us suppose that the horizontal component of the wind in the forest canopy affects the ignition process minimally. This assumption is true as the radiant heat flux is considerably greater than the convective one. For the same reason we consider the ignition process to be quasi-one-dimensional, i.e. we suppose that all the parameters depend on the time and the vertical coordinate z .

Thermo-dynamic, thermo-physical and structural characteristics correspond to a forest canopy chosen from (5) and they are equal numerically: $E_1/R=9400K$, $K_1=3.36 \cdot 10^4 s^{-1}$,

$q_1=0$; $E_2/R=6000K$, $K_2=3 \cdot 10^6 s^{-1}$, $q_2=6 \cdot 10^7 J/kg$; $E_3/R=10^4 K$, $K_3=10^3 s^{-1}$,

$q_3=1.2 \cdot 10^7 J/kg$; $E_5/R=11500K$, $K_5=3 \cdot 10^{13}$, $q_5=10^7 J/kg$;

$c_{p1}=2000 J/(kgK)$, $c_{p2}=4180 J/(kgK)$, $c_{p3}=900 J/(kgK)$,

$c_{p4}=10^3 J/(kgK)$, $c_{p5}=10^3 J/(kgK)$, $s_\sigma=10^3 m^{-1}$, $sc_d=0.01$, $h=10m$,

$\alpha=0.06$, $\rho_4\varphi_4=0.08 kg/m^3$, $\rho_1=500 kg/m^3$, $\rho_2=10^3 kg/m^3$, $\rho_3=200 kg/m^3$,

$\rho_{5n}=1.16 kg/m^3$, $P_e=10^5 N/m^2$, $\mu_c=1 kg/(m \cdot s)$, $T_e=300K$, $c_{1e}=0.23$,

$c_{2e}=0$, $c_{3e}=0$, $k=0.375m^{-1}$.

For a numerical integration of the system of equations (2)-(9), with initial and boundary conditions (11)-(13), a system of differences obtained on the basis of the iteration-interpolation method was used (Grishin and Bertsun 1981).

As. a result of numerical integration, the field of temperatures, mass fractions of components, and volume fractions of phases at various moments of time are obtained. In the figure, a distribution of relative temperatures and mass fractions of oxygen and combustible gas at the upper boundary of the forest canopy, in time for various distances from the explosion epicenter are shown. The explosion energy was 25 Mt. The analysis of these graphs show, the following succession of events takes place. At the beginning the solid phase is heated under the effect of light radiation. Evaporation of moisture and heating of dry material take place, with further decomposition into char and volatile products of pyrolysis.

The analysis of the curves (Grishin and Perminov 1992) makes it possible to state that two ignition regimes-normal and degenerate take place (Zeldovich et al. 1980). The latter occurs near the explosion epicenter and is characterized by increased values of temperature of the solid phase (up to 4000K) (curve 1), by nigh burn-out of solid and gaseous products of pyrolysis, as well as by a considerable consumption of oxygen. Essentially in this case, forced combustion under the action of a powerful radiant flux takes place. The heat released by oxidation does not cause essential influence on the temperature of the gas phase in the forest canopy, here due to heat exchange with the solid phase, the temperature of the gas phase grows and may be higher than the adiabatic combustion temperature (curve 1). The concentration of oxygen through the course of time decreases and that of combustible gas at first grows and then sharply decreases at the occurrence of burn-out. The maximum temperature of the gas phase and minimum oxygen concentrations occur at the upper boundary of the forest canopy (Fig.7). With increasing distance from the explosion epicenter up to r = 7.5 km, the temperature at the upper boundary of the forest canopy decreases (curve 2) as a result of a decrease of the intensity of the radiant heat flux. Therefore as gaseous products of pyrolysis oxidize, less oxygen is consumed (curve 2). Here the temperature of the gas phase with time (curve 2) has two points of inflexion. The so-called normal regime of ignition [17] which is characterized by relatively small burn-out of combustible material obtains.

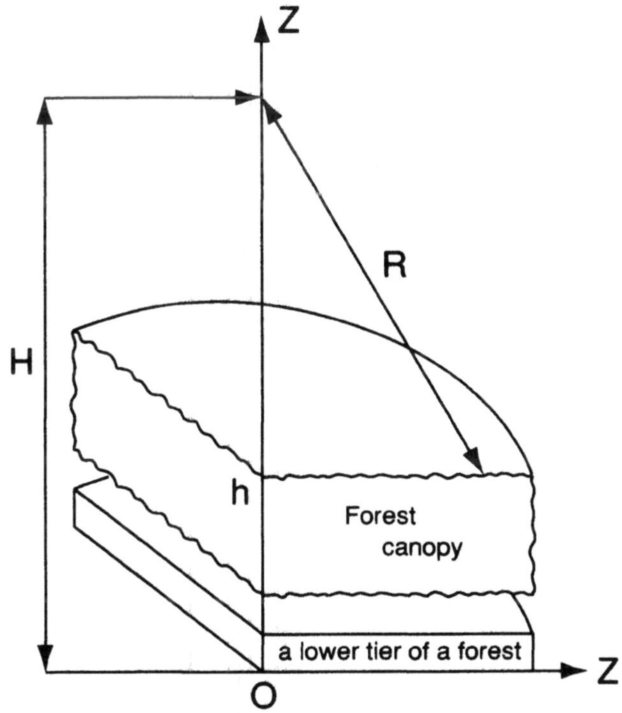

Fig.7. Statement of the problem of ignition of phytocenosis by the explosion of the Tunguska celestial body. The scheme of location of large tracts of forest and the system of coordinates.

As the criterion of ignition in this case, the condition:

$$\left.\frac{\partial^2 T}{\partial t^2}\right|_{\substack{r=h \\ t=t_*}} = 0 \tag{18}$$

is used. The numerical calculations show that with an increase of explosive energy up to 30Mt, the ignition radius increases up to 10 km and with W =20Mt- r_* =7km. Thus one can draw a conclusion that with an energy release of the order $2{\cdot}5{\cdot}10^{17}$ J , corresponding to the explosion of the Tunguska meteorite, crowns of trees ignite at the distance of the order 8 km, which agrees with the known data (Korobeinikov et al. 1991). On the basis of the analysis of the results obtained (see Figure 8) it follows that there exist two regimes of gas phase ignition at the explosion of the Tunguska meteorite: degenerate (high temperature) regime of ignition near the explosion epicenter, and normal (low temperature) conditioned by oxidation of gaseous products of pyrolysis at temperatures lower than the temperature of combustion.

Comparison of the results obtained according to one temperature and two-temperature models give practically equal dimensions of the ignition zones (r_* =7.5km and r_* =8km respectively) for the same energy of explosion. However, if one introduces the notion of tree burning as a degree of pyrolysis of their FCM (~20%), from the initial volume fraction, the numerical calculations according to the two-temperature model show that at the distance of about 16 km warm-up, evaporation of moisture and pyrolysis of FCM without ignition take place, which agrees with the data of Korobeinikov et al. (1991). Thus a two-temperature model in contrast to a one-temperature model allows for a study of such phenomenon as tree burning.

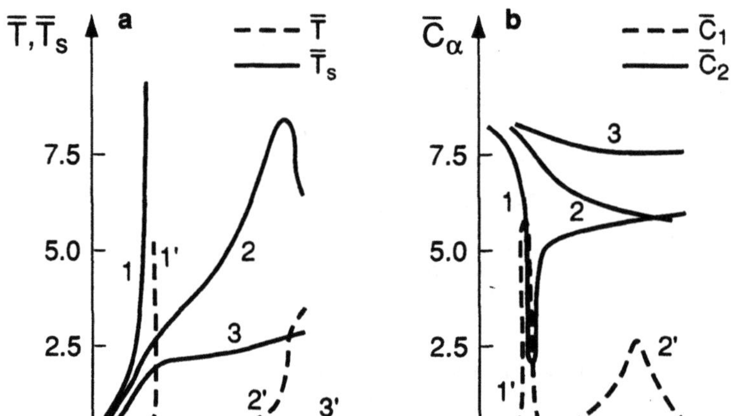

Fig.8. Dependences of relative temperatures (a) and mass fractions of oxygen and combustible gas (b) at the upper boundary of the forest canopy on time for various distances from the epicentre of explosion

$\bar{T}_s = T_s / T_H$, $\bar{T} = T_s / T_H$

7. Limiting Conditions

In sections 1 and 4 conditions under which the process of spreading a forest fire stops are investigated experimentally and theoretically. In accordance with the terminology accepted in the theory of combustion (Grishin 1981, 1992), these conditions are called limiting. Particularly, crown forest fires combustion stops under the following conditions:

$$\rho_c<\rho_*, \quad W>W_*, \quad U_\infty<\underline{U}_{*\infty}, \quad U_\infty>\overline{U}_{*\infty}, \quad \alpha_v>\alpha_{v*}$$

where ρ_* is the critical density of the FCM layer, the index $*$ is assigned to critical values of FCM moisture content W, equilibrium speed of wind U_∞ in the canopy layer and the coefficient of volume heat and mass exchange of an FCM element with the medium α_v, and the bar below and above limits of spread of a forest fire according to the velocity of wind.

It should be mentioned that for surface forest fires the velocity of wind is absent since they spread even in the complete absence of wind.

Let us point out that only the main limiting conditions are enumerated above. Similar to the classic theory of combustion (Zeldovich et al. 1980) in the theory of forest fires, there exists so-called combined limiting conditions, when FCM combustion stops due to the actions of several unfavorable factors for fire spread. For a quantitative determination of the critical values ρ, W, U_∞, α_v and the method of small perturbations and mathematical experiments are used and the checking of theoretically found limiting conditions of spread was realized as a result of semi-natural experiments.

As mathematical and physical experiments showed in carrying out at least one of the conditions (20), a combustion break-down takes place and the spreading of a forest fire stops. Knowing the limiting conditions of fire spread one can reproduce them in the neighborhood of the fire front, in such a way as to ensure its extinguishment.

8. On Comparison of Theoretical Results with Experimental Data

According to the above presented results of theoretical investigations the front of a forest fire consists of the zone of neating, of drying FCM, FCM pyrolysis, as well as of the combustion zone of gaseous products and the after-combustion of solid products of pyrolysis. These results agree with experimental data (Grishin and Plukhin 1985).

Mathematical modelling gives only averaged values by Renolds method (but not real) values of state parameters of the medium and the rate of spread of crown forest fires. In the paper of Grishin and Plukhin (1985), a generalized curve of dependence of the velocity of spreading a crown fire on the velocity of wind, which is obtained on the basis of treatment of literature data (see Grishin [1981] as well) on fires of various types, including crown fires is given. It should be mentioned that there is few experimental data on crown forest fires.

Particularly at the undisturbed speed of wind at the height of 10 m, $U_\infty=18$ m/s, $\rho_c=0.2$ kg/m, $W=66.6\%$, the calculated value $\omega_n=5.8$ m/s agrees with the experimental value $\omega_n=6$ m/s [17]. With $U_\infty=27$ m/s the calculated value $\omega_n=7.2$ m/s and the experimental value is 6-8 m/s (Konev 1977).

Thus the mathematical model used in the given section makes it possible to predict the velocity of the spread of a crown forest fire rather accurately. On the basis of comparison of averaged by Renolds' method, experimental profiles of temperature for crown forest fire (see Sec.1.8) with the theoretical curves presented above, one can state that with taking into account the accuracy of assigning initial data these results agree well.

In conclusion, we have got an analytical formula for the velocity of crown forest fire propagation (Grishin 1992), according to which ω increases in direct proportion to a wind velocity. As mentioned in the review (Zeldovich et al. 1980), the result is in agreement with the data obtained by several researchers, as quoted by Zeldovich et al. (1980).

References

Byram G.M. 1966. Scaling laws for modelling mass fires. Pyrodynamics, No.4, 271-284.

Chetverushkin B.N. 1985. Mathematical modelling of radiating gas dynamics problems. Moscow, Nauka, 304 pp. <in Russian>.

Dmitriev P.S. 1965. Effect of nuclear explosion. M. Voenizdat <in Russian>.

Gostintsev Yu.A., and L.A. Sukhanov. 1977. Aerodynamics of medium at large fires. Linear fire. USSR Academy of Sciences, Institute of Chemical Physics. Chernogolovka, 51 pp. <in Russian>.

Grishin A.M., and V.V. Plukhin. 1985. Experimental investigation of the crown forest fire front structure. The Physics of Combustion and Explosion, No.1, 21 <in Russian>.

Gostintsev Yu.I., G.M. Makhviladze G.M., and V.V. Novozhilov. 1989. Initial stage of development of a large fire initiated by radiation. Chemical Physics of processes of combustion and explosion. Chernogolovka: ICPh AS USSR <in Russian>.

Grishin A. M. 1981. Mathematical modelling of forest fires. The Tomsk University Press, Tomsk, 277 pp. <in Russian>.

Grishin A.M. 1992. Mathematical modelling of forest fires and new firefighting methods, Nauka, Moscow, 407 pp. <in Russian>.

Grishin A.M., and V.A. Perminov. 1992. Mathematical modelling of ignition of a forest cupola from the Tunguska meteorite with taking into account two temperatures of the medium. Siberian J. Mathematics 6, 107-111 <in Russian>.

Grishin A.M., V.N. Bertsun, and V.I. Zinchenko. 1981. Iterative interpolation method and its applications. The Tomsk University Press, Tomsk, 160 pp. <in Russian>.

Grishin A.M., A.D. Gruzin, and E.E. Gruzina. 1984. Aerodynamics and heat/mass exchange of the forest fire front with the lower atmosphere. J. Appl. Mathematics and Technical Physics 6, 91-96 <in Russian>.

Grishin A.M., and A.A. Fomin. 1985. Numerical investigation of heat/mass transfer in the ground atmospherelayer over a great forest fire place. In: Problems of dynamics of viscous liquid, 108-113. Novosibirsk: ITAM Siberian Branch of the USSR Academy of Sciences <in Russian>.

Grishin A.M., A.D. Gruzin, and V.G. Zverev. 1994. Mathematical theory of the crown forest fires. In:Thermophysics of forest fires, 38-75. Novosibirsk, Institute of Thermal Physics, Russian Academy of Sciences, Siberian Branch <in Russian>.

Konev E.V. 1977. Analysis of spreading forest fires and burnings down. Thermophysics of forest fires. Novosibirsk: ITPh SB AS USSR, 239 pp. <in Russian>.

Korobeinikov V.P., N.N. Chushkin, and L.V. Shurshalov. 1991. Complex modelling of flight and explosion in the meteor body atmosphere. Astronomical Vestnik 25 (3), 327-343 <in Russian>.

Kutateladze S.S., and A.I. Leontiev. 1976. Thermo-and mass exchange and friction in a turbulent boundary layer. M. Energy, 226 pp. <in Russian>.

Patankar, S. 1984. Numerical methods of solving heat exchange and liquid dynamics problems. Moscow, Energoatomizdat, 152 pp. <in Russian>.

Weber R.O. 1991. Modelling fire spread through fuel beds. Prog. Energ. Combust. Sci. 17, 62-82.

Zeldovich Ya.B., G.I. Barenblat, V.B. Zitrovich, and G.M. Makhviladze. 1980. A mathematical theory of combustion and explosion. Moscow, Nauka, 478 pp. <in Russian>.

J.G.Goldammer and V.V.Furyaev (eds.), Fire in Ecosystems of Boreal Eurasia, 303-313.

Mathematical Modelling and Optimization of Forest Fire Localization Processes

G.A. Dorrer and S.V. Ushanov [1]

1. Introduction

We can now see the intensive development of computer control systems for the Russian forest fire protection service. The system exists at the federal level in the Central Base for Aerial Fire Protection *Avialesookhrana* and implementation of control systems at the regional aviation bases of forest fire protection has begun. But most complex and demanding are the low levels of forest fire control - the levels of the operative departments and fire crew leaders in suppression operations on large fires (Valendik 1990).

Using modern equipment for forest fire detection and mapping, and modern methods of predicting propagation and development, the modern methods of mathematical modelling and control theory make possible the use of well-founded forest fire suppression actions and new tactical modes (Parlar and Vicson 1982). An important role in this work must be played by expert systems intended to collect and codify knowledge in specific subject fields (as a rule, obtained from the experts) and then use the knowledge in ways which simulate human decision making. In combination with the traditional mathematical models they permit the creation of knowledge bases for the national forest protection system (Kourtz 1987; Dorrer et al. 1991).

Tactics for extinguishing large forest fires are determined by the character of forest conditions, topography, peculiarity of fire behavior and propagation. Limitations of fire fighting forces and means may, to some degree, be compensated for by mobility.

In forest fire suppression several tactical stages can be identified. First is the process of forest fire localization. As this stage of forest fire fighting is important and labour-consuming it is of interest to attempt to optimize forest fire crews' productivity. The methodical basis for decision making in this field is the theory of localization processes developed over the last few years. Localization of a forest fire is a process of direct suppression at the edge of a fire or by surrounding it by an impenetrable fire barrier. In the conditions of real fire fighting, choice of the method of localization depends on many non-formalized factors. But it is useful to investigate some idealized cases which allows for a mathematical modelling that can be used for planning purposes.

Another problem of this type is connected with human safety in forest fire suppression - the problem of eluding contact between fire and moving crew (Dorrer 1990). Both of the problems mentioned are addressed in the theory of dynamic games (Krasowsky 1970).

---

[1] Krasnoyarsk State Technological Academy, 660036 Krasnoyarsk, Russian Federation

The work presented here is devoted to stating the problem of forest fire localization in the most common form as a problem of optimal control. In previous investigations (Albini et al. 1978; Ushanov 1990a; Dorrer 1992) the problem of localization by line construction when the crew moves with a constant speed along the fire edge, and the optimal localization control problem were studied (Ushanov 1990b; Dorrer 1992). A common peculiarity of works of Albini et al. (1978), Ushanov (1990a,b) and Dorrer (1992) is the polar coordinate system used for describing the localization control problems. Using the polar coordinate system sets some limitations, the greatest is the request for a singularity of fire edge descriptions. That means any radius drawn from the fire origin point must intersect the line of fire edge only once (Fig.1). For this reason, results obtained in the works mentioned cannot be used for describing the localization of large fires with complex and reentrant edge configurations.

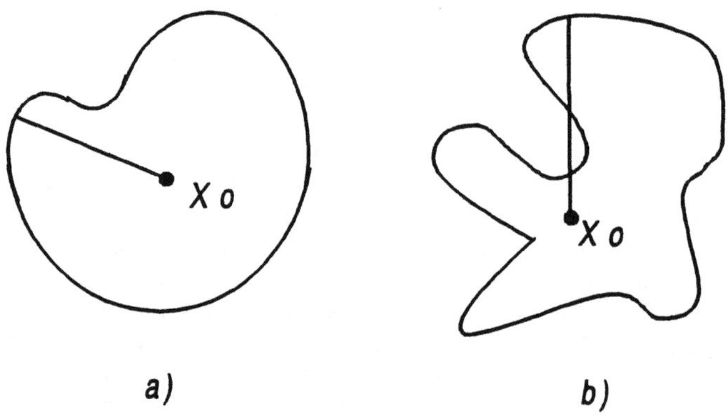

Fig.1. Examples of forest fire edge description in polar coordinate system.

a) Single-valued description b) Multiple-valued description

The mathematical theory suggested in the paper presented is free of this limitation and describes the most common mathematical model of localization control.

2. Mathematical Model of Forest Fire Localization

Let us consider the system when the object is control of a forest fire and the means of control are the fire crew and technical material. The process of forest fire propagation and localization is considered with the Decarte coordinate system

$$x = (x_1, x_2), \quad x \in D$$

The coordinates x_1, x_2 are considered as map (plan) cartographic coordinates of the forest territory observed - the set D.

2.1 Criterion of Localization Process Optimality

The efficiency of the control system is determined by the functional

$$J = \int_{T_0}^{T} F(t,x,x')dt, \tag{1}$$

where

To, T - are start and finish timees of the localization process.

J defines values of the generalized "expenditures" on fire suppression. In the general case the integrand

$$F = Fo + So + Lo, \tag{3}$$

where

$Fo = C_1(t,x)$ - is an expense that is proportional to localization time;

$So = a \cdot C_2(t,x) \cdot (x_2'(xo_1 - x_1) + x_1'(x_2 - xo_2))$ - is an expense that is proportional to the square of the fire area;

$Lo = C_3(t,x)\sqrt{(x_1'^2 + x_2'^2)}$ - is an expense that is proportional to the length of the fire barrier;

$x(t) = (x_1(t), x_2(t))$ - are the coordinates of the moving end of the fire barrier at time t;

$xO = (xo_1, xo_2)$ - are the coordinates of the fire origin.

Let α be a coefficient that determines the direction of localization barrier construction motion relative xo. If $\alpha = 1$ then localization direction is clockwise, and if $\alpha = -1$ the localization direction is counter clockwise.

2.2 System of Constraints on the Localization Problem

1) Limitation on the velocity of fire barrier construction:

$$x' = u, \quad \|u\| \le U(t), \tag{3}$$

where

$u(t) = (u_1(t),\ u_2(t))$ -

is vector of the control influence (velocities of the barrier construction along the coordinates x_1 and x_2);

$\|u\| = \sqrt{(u_1^2 + u_2^2)}$ - is length of the velocity vector;

$U(t)$ - maximum possible value of localization velocity.

2) Limitation imposed by the forest fire propagation process.

Let $G(t,h)$ be the set of points that includes the fire area and all points whose distance from the fire is not greater than h. If $h = 0$ then $G(t,h)$ is the set of points enclosed by the fire edge.

Then

$$x(t) \notin G(t,h), \tag{4}$$

where

$G(t,h) = \{\ \bar{x}(t) : F_2(t,\bar{x}(t),h) \le 0\ \}$;

$F_2(t,\bar{x}(t),h)$ - is a function that determines the fire edge. Then the set of the points permissible for barrier construction $X(t)$ is

$$X(t) = \{\ x(t) : F_2(t,x(t),h) \ge 0\ \} \tag{5}$$

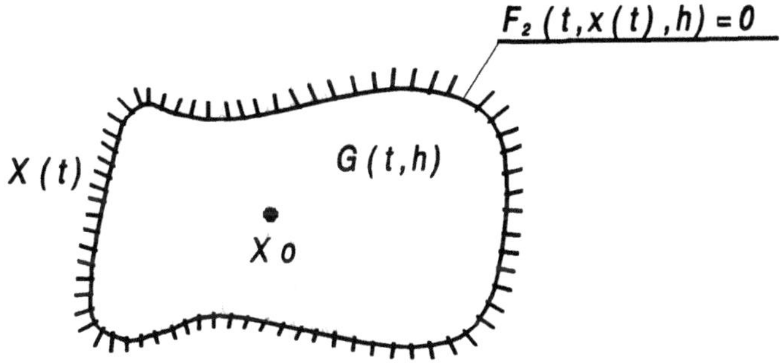

Fig.2. Permissible area for limitations (4)-(5)

X(t) - set of permissible points for the barrier creation;
G(t,h) - set of points that includes the fire area and those points whose distance from fire is not greater than ***h***;
- boundary of the permissible area for limitations (4)-(5) $(F_2(T,x(t),h) = 0)$.

The limitations (4), (5) have the clear sense: the fire barrier must lie outside the set of fire propagation and the distance from the barrier to the fire must be no less than ***h*** (Fig.2). The mathematical models and algorithms for ***G(t,h)*** and $(F_2(T,x(t),h) = 0)$ are given by Dorrer (1979, 1992).

3) Constraint on direction of burning movement

$$F_1(t,x,x') = \alpha(x_2'(xo_1 - x_1) + x_1'(x_2 - xo_2)) \geq 0 \tag{6}$$

Constraint (6) establishes localization movements as clockwise ($\alpha = 1$) or counter clockwise ($\alpha = -1$) relative to the localization center and eliminates multi-valued solutions (Fig.3).

4) Initial and final conditions of the localization processes

For closing the problem (1)-(6) it is necessary to determine initial and final conditions. Let us consider the most important cases.

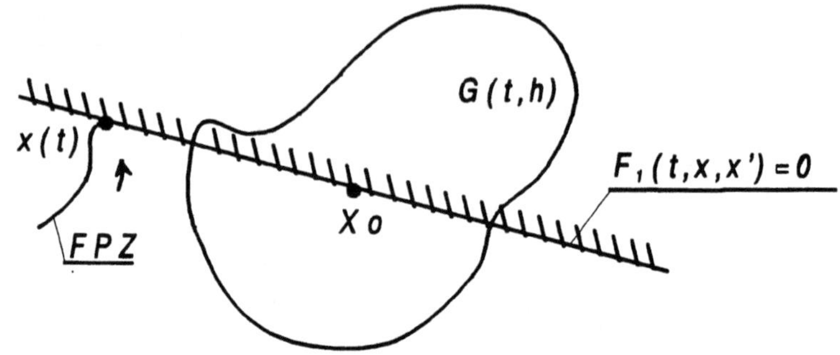

Fig.3. Permissible area for limitation (6)
X(t) - point on barrier at time ***t***;
G(t,h) - set of points includes the fire area and the points whose distancea from the fire are less than ***h***;
Xo - center of the spreading process;
- boundary of permissible barrier construction under constraint (6);
→ - localization direction.

(1) The initial and final points are determined. In this case we are given

$$x(To), \quad x(T).$$

(2) The localization center on which the localization process must start and finish is given. In this case are known

$$x(To) = x(T).$$

(3) It is necessary to determine the initial point coordinates. While moving from this point the optimality of the localization process under criterion (1) is to be achieved. In this case it is necessary to determine

$$x(To) = x(T).$$

(4) Find the functional (1) minimum amongst the localization curves $x(t)$ whose ends are determined by lines $R_1(t)$ and $R_2(t)$. An example of such a problem is constructing a localization barrier between two natural impediments. In this case

$$x(To) = R_1(To);$$
$$x(T) = R_2(T).$$

At the all cases considered time of localization finishing this not fixed.

2.3 Necessary Conditions of Optimality

The Lagrange function of the problem (1)-(6) is

$$\mathcal{L} = \int_{T_0}^{T} L(t,x,x',u)dt + l(To,x(To),T,x(T)), \tag{7}$$

where lagrangian

$$L(t,x,x'u) = F(t,x,x') + P_1(x_1' - u_1) + P_2(x_2' - u_2) - P_3 \cdot F_1(t,x,x') - P_4 \cdot F_2(t,x,h);$$

functional $l(To,x(To),T,x(T))$ is determed by the initial and final conditions.

For boundary condition cases 1 and 2

$$l(To,x(To),T,x(T)) = \lambda_1 \cdot x_1(To) + \lambda_2 \cdot x_2(To) + \lambda_3 \cdot x_1(T) + \lambda_4 \cdot x_2(T).$$

For the boundary conditions of case 3

$$l(To,x(To),T,x(T)) = \lambda_1 \cdot (x_1(To) - x_1(T)) + \lambda_2 \cdot (x_2(To) - x_2(T)).$$

For the boundary conditions of case 4

$$\begin{aligned} l(To,x(To),T,x(T)) = {} & \lambda_1 \cdot (x_1(To) - R_{11}(To)) + \lambda_2 \cdot (x_2(To) - R_{12}(To)) + \\ & + \lambda_3 \cdot (x_1(T) - R_{21}(T)) + \lambda_4 \cdot (x_2(T) - R_{22}(T)). \end{aligned}$$

The necessary conditions on the Lagrange functions (7) are determed by the following expressions

(1) Condition of stationarity on x :

$$L_{x_i} = \frac{d(L_{x_i'})}{dt}, \quad i = 1,2 \quad , \tag{8}$$

where

$$\begin{aligned} L_{x_i} &= F_{x_i}(t,x,x') - P_3 \cdot F_{1x_i}(t,x,x') - P_4 \cdot F_{2x_i}(t,x,h); \\ L_{x_1'} &= F_{x_1'}(t,x,x') + P_1 - P_3 \cdot F_{1x_1'}(t,x,x'); \\ L_{x_2'} &= F_{x_2'}(t,x,x') + P_2 - P_3 \cdot F_{1x_2'}(t,x,x'). \end{aligned}$$

(2) The transversality conditions on x :

$$L_{x_1'}(To,x(To),x(To)',u(To)) = l_{x_1}(To); \tag{9}$$

$$L_{x_2'}(To,x(To),x(To)',u(To)) = l_{x_2}(To); \tag{10}$$

$$L_{x_1'}(T,x(T),x(T)',u(T)) = -l_{x_1}(T); \tag{11}$$

$$L_{x_2'}(T,x(T),x(T)',u(T)) = -l_{x_2}(T); \tag{12}$$

where

$$l_{x_1}(To) = \lambda_1;$$
$$l_{x_2}(To) = \lambda_2;$$

$l_{x_1}(T) = \lambda_3$ - for boundary condition cases 1,2,4;

$l_{x_1}(T) = -\lambda_1$ - for the boundary conditions of case 3;

$l_{x_2}(T) = \lambda_4$ - for boundary condition cases 1,2;

$l_{x_2}(T) = -\lambda_2$ - for the boundary conditions of case 3;

$l_{x_2}(T) = \lambda_4$ - for the boundary condition of case 4.

(3) The optimality condition on u - the maximum principle

$$\max_{u \in U} H(t,\hat{x}(t),u,P(t)) = H(t,\hat{x}(t),\hat{u}(t),P(t)), \tag{13}$$

where

$H = -F + L_{x_1'} \cdot x_1' + L_{x_2'} \cdot x_2'$ is Pontryagin's function.

(4) Condition of stationarity on T

$$H(T,x(T),u(T),P(T)) = l_T(To,x(To),T,x(T)) \, , \tag{14}$$

where:

for the cases 1,2 of the boundary conditions

$$l_T(To,x(To),T,x(T)) = \lambda_3 \cdot x_1(T)' + \lambda_4 \cdot x_2(T)' \quad ;$$

for case 3 of the boundary conditions

$$l_T(To,x(To),T,x(T)) = -\lambda_1 \cdot x_1(T)' - \lambda_2 \cdot x_2(T)' \quad ;$$

for case 4 of the boundary conditions

$$l_T(To,x(To),T,x(T)) = \lambda_3 \cdot (x_1(T)' - R_{21}(T)') + \lambda_4 \cdot (x_2(t)' - R_{22}(T)') \quad .$$

(5) Condition of stationarity on To

$$H(To,x(To),u(To),P(To)) = -l_{To}(To,x(To),T,x(T)) \tag{15}$$

where

$$l_{To}(To,x(To),T,x(T)) = \lambda_1 \cdot (x_1(To)' - R_{11}(To)') + \lambda_2 \cdot (x_2(To)' - R_{12}(To)') \quad .$$

The condition (15) applies only to case 4 of the boundary conditions.

(6) The conditions of additional non-hardness

$$P_3 \cdot F_1(t,x,x') = 0 \quad , \tag{16}$$

$$P_4 \cdot F_2(t,x,h) = 0 \quad . \tag{17}$$

(7) The conditions of non-negativity

$$P_3 \geq 0 \quad , \tag{18}$$

$$P_4 \geq 0 \quad . \tag{19}$$

The simplest form of the optimality condition is the maximum speed problem (when $C_1=1$, $C_2 = C_3 = 0$).

3. Discussion

The optimal localization control problem stated above and the necessary conditions of optimality obtained are bases for developing numerical methods and software designs for the next problems.

a. Optimization of forest fire localization processes. The process of fire propagation is described by mathematical models that are given in works by Dorrer (1979, 1992). Solution of this problem is provided by the next iteration process:
   - On the first iteration calculation of $G(t,h)$ is realized without accounting for the localization process. determine the fire barrier configuration $x(t)$ by solving problem (1)-(6).
   - On the next iteration calculate $G(t,h)$ accounting for the fire barrier constructed on the previous iteration. Normal fire rate of spread on the barrier is equal to zero.

b. Fire localization by piecewise linear barriers. In solving this problem it is expedient to use the condition of conjugation of the extremals for the maximum velocity problem ($F=1$).

c. Estimation of the expert solution in the training systems. In this case it is necessary to examine the conditions of conjugation of the extremals. The results of such examination allows to determine the optimality of an expert solution. If the expert solution is not optimal, Problems 1 (or 2) may be solved, and the optimal solution compared to the expert one.

d. Verification of the validity of practical recommendations on forest fire suppression. This approach is similar to Problem 2.

e. Determination of the disposition of localization centers. When solving this problem the appropriate determinants of optimality must be taken into account and the iteration procedures of the Problem 1 applied.

f. Determining the minimum crew rate U that provides forest fire localization at given time. When solving this problem it is necessary to use the iteration procedures of Problem 1, the equations of functional (1) sensitive to the parameters U and T, and constraint (6).

g. Distribution on the forest territory of forces and means for fire suppression. When solving this problem it is advisable to use the simplified models of fire propagation process (Anderson 1989).

The mathematical models of localization control considered above may be a basis for software packages for forest fire control systems design. This package must include the following subsystems:

a. Cartographic database, that includes information on topography, forest fuel, infrastructure of district;

b. Database on ground and aviation capabilities for transporting forest fire control forces;

c. Database on technical equipment for forest fire suppression;

d. Package of the computer programs that implement the mathematical models of forest fire propagation under different forest and meteorological conditions, and

e. Package of computer programs that embody the mathematical models of localization control (Dorrer 1979).

References

Albini F.A., G.N. Korovin and E.N. Gorovaya. 1978. Mathematical analysis of forest fire suppression. USDA Forest Service Res. Pap. INT-207, 19 pp.

Anderson D.N. 1989. A mathematical model for fire containment. Can.J.For.Res. 19, 998-1003.

Dorrer G.A. 1979. Mathematical Models of forest fires. Moscow: Lesnaya promishlennost, 161 pp. <in Russian>.

Dorrer G.A. 1990. The problem of eluding contact with forest fire. In: Physical hydrodynamics of reactionable mediums, 73-76. Novosibirsk: Nauka <in Russian>.

Dorrer G.A. 1992. Modelling of forest fire spreading and suppression on basis of Hamilton mechanics methods. AMSE Transaction "Scientific Siberian", Series A, Vol.3, Forest Fires, 38-56.

Dorrer G.A., S.V. Ushanov and S.P. Yakimov. 1991. Expert system of forest fire danger estimation: methodology and design principles. In: Forest fires and suppressions, 159-172. Krasnoyarsk : VNIIPOMLeskhos <in Russian>.

Kourtz P. 1987. Expert system dispatch of forest fire control resources. AI Applications in Natural Resource Manage, Vol.1, 1-8.

Krasowsky N.N. 1970. The game problems on meeting of motion. Moscow: Nauka, 320 pp. <in Russian>.

Parlar M. and R.G. Vicson. 1982. Optimal forest fire control: an extension of Park's Model. Forest Science 28, N 2, 345-355.

Ushanov S.V. 1990a. Optimal localization control of flat process of propagation. In: Control of producing and technical systems, 145-153. Krasnoyarsk: KPI <in Russian>.

Ushanov S.V. 1990b. Optimization of forest fires localization. In: Control of producing and technical systems, 148-153. Krasnoyarsk: KPI <in Russian>.

Valendik E.N. 1990. Large forest fires suppression. Novosibirsk: Nauka, 193 pp. <in Russian>.

J.G.Goldammer and V.V.Furyaev (eds.), Fire in Ecosystems of Boreal Eurasia, 314-325.

A Mathematical Model of Spread of High-Intensity Forest Fires

H.P. Telitsyn [1]

1. Introduction

Fire impact on forests is as variable as are forests and fires, but the impact of high-intensity fires on forests is only negative: considerable or full elimination of surface organics and the replacement of stands. Boreal forests are certainly the most favorable environment for high-intensity fires; history of each large fire in these forests contains at least two or three days of extremely fast spread and tall flames at the fire head. Known for their high speed, high-intensity forest fires are, as a rule, large fires. They emit huge clouds of smoke covering vast areas and, by their smoke emissions, increase lightning activity and retard the fall of rains and dew, and strengthen the stability of the anticyclone, in which they usually occur. Forest fires are identified herein as high-intensity fires if they achieve, at the least, the following values of characteristics of burning, or even of one of them:

* Byram's intensity of fire head, MW m^{-1}	> 3
* flame height above the fuel layer, m	> 3
* rate of spread, m s^{-1}	> 0.2
* width of flame zone, m	> 5
* burned fuel loading, kg m^{-2}	> 0.5

These mimimum parameters for identification of high-intensity forest fires are rather vague because fire damage depends not only on these, but also on many other conditions such as residence time, thickness of tree bark, etc.

The above limits of fire characteristics are specified here as the minimum to identify forest fires as high-intensity fires also because at these and higher ones the fire spread mechanism most probably converts from the diffusion mode to the kinetic one (gaseous phase burning, blow up) which is the characteristic mode of spread of high-intensity fires. It is important to forecast this critical point of forest fire history, and the version of the mathematical model of forest fires presented herein may be a step toward achieving this purpose.

The basic feature of this version, by which it can be distinguished from similar models of Van Wagner (1967, 1968), Rothermel (1972), and of other authors, is its physical basis. Firstly, this model is derived based on the assumption that the penetration depth of radiative

---

[1] Research Institute of Computer Technologies, Khabarovsk Technical University, Khabarovsk 680035, Russian Federation

flux into the fuel bed is limited by the thickness of one thin exposed surface particle of the fuel bed, since all particles of forest fuels are not transparent to heat radiation. The second distinction is that, following Thomas (1965), the flame radiation flux in this model is considered to be composed of two fluxes - inward and outward fluxes, and the contribution of each flux in the fire spread rate is taken into account separately. These features make it possible to apply the model to thick fuel beds and high-intensity forest fires.

2. History of the Study

Almost half a century ago Fons (1946) presented probably the first mathematical model of forest fire spread along layers of forest fuels assuming that ignition of particles of the fuel bed is performed by heat transfer from flame gases enveloping the nearest particles of fuel. Later Byram et al. (1966) developed another version of the model based on the assumption that fire spreads by repeating and permanent contacts of fuel particles with the flame shifted by wind.

Notations:

c	specific heat of fuel
d	bulk density of fuel bed
d_{is}	density of material of fuel particles
e	base of Naperian logarithm
E_o	outward emissivity of the "flame-fuel" system
E_i	inward emissivity of the "flame-fuel" system
k	attenuation coefficient
L	flame length
l	mean free path of radiation in fuel bed
M	moisture content of fuel
R	rate of fire spread
R_h	rate of spread of high-intensity fire
R_i	the part of fire spread rate maintained by inward heat flux in the "flame-fuel" system, also rate of spread of backfires against strong winds, also rate of fire spread down steep slopes and vertical columns
R_o	the part of spread rate maintained by outward heat flux in the "flame-fuel" system
s	thickness (depth) of fuel bed
s_{is}	isothermal thickness of fuel particle, thickness of isothermal (thin) particle
T_f	absolute temperature of flames
t_i	ignition temperature of fuel
t_o	ambient temperature of fuel
H	flame height
U	wind speed
V	buoancy velocity of flames
Y	flame angle to fuel bed
W	fuel loading
W_{is}	isothermal loading of fuel bed

Then Emmons (1963) described a physical model of intensive fire spread along multi-story fuel beds as a process of radiative and convective heat transfer inside and outside the fuel bed from the flame zone to the fuel. This physical model became the basis for further mathematical modeling of forest fire spread performed by Thomas (1965, 1971), Van Wagner (1967, 1968), some Russian researchers (Telitsyn 1965, 1973a; Konev and Sukhinin 1977; Grishin 1992; and others), and Rothermel (1972). These models are described in short in an excellent review by Weber (1990), so we see no need to reproduce such a said review in this report, except in some cases of comparison of our version with other models.

3. Flame Spread in Thin Fuel Beds

This version of a mathematical model of forest fire spread was published by Telitsyn (1965) and, independently, by Van Wagner (1967). This version was assumed to be fit for thin fuel beds, but the limits of this fitness was not defined then; it became known afterwards in the results of further studies. The physical basis for this model was the assumption that fuel particles are ignited by flame radiative flux from flames above the fuel bed (outward radiative flux) on the fuel surface. The heat balance equation according to the Stefan-Boltzmann law for this case was given as follows:

$$5.67 \cdot 10^{-8} E_o (T_f^4 - T_o^4) FL = ds[c(t_i - t_o) + 2.6 \cdot 10^4 M] R_o \tag{1}$$

where:

- **c** - specific heat of fuel, J/kg·°C;
- **d** - bulk density of fuel bed, kg m^{-3};
- **Eo** - outward emissivity of the "flame-fuel" system;
- **F** - factor of flame inclination to the fuel bed (flame angle factor);
- **L** - flame length above the surface of the fuel bed, m;
- **M** - moisture content of fuel (kiln dried), %;
- R_o - rate of fire spread, m s^{-1}.
- **s** - thickness (depth of burning) of fuel bed, m;
- T_f - absolute temperature of flames, °K;
- T_o - absolute ambient temperature of fuel surface, °K;
- t_i - ignition temperature of fuel, °C;
- t_o - ambient temperature of fuel, °C;

The left part of this equation represents the forward radiative heat flux from the upper part of the flame (which towers above the fuel bed), according to the Stefan-Boltzmann law, and the right part expresses the entalpy of ignition of the fuel bed area of a given thickness, density, specific heat, temperature, and moisture content, ignited during a unit of time.

In the course of further development of the above equation the following parameters were assumed constant according to published data (Rothermel 1972; Davis et al. 1959; Telitsyn and Sosnovshenko 1970; Demidov and Saushev 1975) and to our relevant measurements:

T_f = 1200 °K, T_o = 293 °K, c = 1.4 KJ kg^{-1}.°C, t_i = 300 °C, t_o = 20°C.

The equation for the flame angle factor F was derived based on the technique of radiation algebra (for brevity, this derivation is not shown here, since it is published in Telitsyn [1965]) and is given by:

$$F=\frac{1+CosY}{2}, \tag{2}$$

where **Y** - flame angle to the fuel bed.

When the burning is spreading down a vertical column of fuel, then **Y** = 180°, so **F** = 0. If the burning is spreading along a horizontal fuel bed in still air, **Y** = 90° and **F** = 0.5. In the horizontal fuel bed this angle depends only on wind speed, and can be estimated by geometric adding of vectors of speeds of wind and flame (buoyancy) flows, and is given by Telitsyn (1965):

$$CosY=\frac{U}{\sqrt{U^2+V^2}} \tag{3}$$

where **U** - wind speed (m s^{-1}) and **V** - buoancy velocity of flames (m s^{-1}). The latter was measured by a fan anemometer at the upper level of flames of model (laboratory) fires, and, after some theoretical considerations, resulted in the following formula:

$$V=2.5\sqrt{H} \tag{4}$$

where **V** is in m s^{-1}, and **H** is the vertical height of flame (m).

Theory of burning states that the flame height of cellulosic fuels depends rather on the percentage of volatiles than on the caloric value of fuels, so the well-known Byram's formula for flame height (Davis et al. 1959) based on caloric value of fuels was modified into one based on the volatile percentage studied by Telitsyn and Sosnovshenko (1970):

$$H=7\sqrt{WR} \tag{5}$$

where **W** - fuel loading (kiln dry weight; kg m^{-2}), and **R** - rate of spread (m s^{-1}), and 7 the coefficient for the volatiles' percentage of 65-75 %; for a higher percentage it is 8.

All variables of Equation (1) are measurable on wildland or laboratory model fires exempt the flame emissivity which is known to be unmeasurable and calculated by Stefan-Boltzmann law on measurable variables of this law or otherwise. To evaluate the emissivity of flames of burning vegetation, the series of wildland and laboratory fire experiments were conducted when measurements of flame height, flame angle, the rate of spread and characteristics of fuel beds were performed.

The values of other parameters of this equation accepted to be constant are cited above. Based on measurements of flame angles, widths and heights, rates of spread and fuel bed parameters the flame emissivity was calculated and related to the width of the flaming zone according to the conventional law for flame emissivity (Demidov and Saushev 1975):

$$E_o = 1 - e^{-kD} \tag{6}$$

where **e** - base of Naperian logarithm, **k** - attenuation coefficient 0.1-0.3 (Thomas 1965; Demidov and Saushev 1975) depending on carbon content in flames, and **D** is the width of the flaming zone, m.

The analysis showed calculated emissivities to be much higher than those resulting from Equation (6) for the relevant widths of flames; and the thicker the fuel beds were the greater the over-estimations. Only very thin fuel beds with loading values within 0.1 -0.2 kg m^{-2} were values calculated close to those observed. This study resulted in the following conclusions:

* the outward radiative flux preheats and ignites surface particles only, since forest fuel particles are not transparent for this flux its influence cannot be extended to sub-surface particles of the fuel bed;
* heat transfer by conduction into the depth of the fuel bed is limited by the thickness of one particle, however thin it is, since contacts between particles are negligible and the heat conduction transfer for a short period into a thick particle is limited by the isothermal thickness of the material of this particle;
* so Equation (1) is applicable only to fuel beds composed of one row of the isothermal or thinner particles (one sheet of the isothermal or less than the isothermal thickness).

The isothermal thickness of the majority of forest fuels is, from various references, 0.4 mm (Lyons and Weber 1991) and 2.5 mm (Thomas 1965, 1971). So the isothermal fuel loading of the fuel bed, for which Equation (1) is assumed to be valid, is as follows:

$$W_{is} = s_{is} d_{is} \tag{7}$$

where $\mathbf{W_{is}}$ - isothermal fuel loading, kg m^{-2};
$\mathbf{s_{is}}$ - isothermal thickness of a fuel particle, or the thickness of thinner (isothermal) particles, m;
$\mathbf{d_{is}}$ - density of fuel particle, kg m^{-3}.

Also:

$$W_{is}=ld \tag{7a}$$

where l is the mean free path of radiation into the fuel bed, whose path depends on the surface-volume ratio of fuel particles (Thomas 1965, 1971), so the isothermal loading of a fuel bed depends on this ratio as well.

The thickness of leaves, of grasses and foliage of hardwoods is 0.1 mm, so their surface-volume ratio (or the specific surface) is 20,000 m^{-1}, and the mean free path of radiation in the fuel bed of leaves and grass is given by:

$$l=0.0002\frac{d_{is}}{d} \tag{7b}$$

Since the thickness of light forest fuels is 0.1-0.3 mm, and the density of particles is 200-400 kg m^{-3} (Telitsyn and Sosnovshenko 1970), the isothermal fuel loading for fuel beds composed of thin particles varies from 0.02 to 0.12 kg m^{-2}, and for those composed of thick particles 0.07-1.0 kg m^{-2}.

To analyze the flame spread mechanism along thick fuel beds whose fuel loading exceeds the isothermal value, it is necessary to take into account that sub-surface particles are never exposed to the outward heat radiative flux, and become exposed to internal heat flux (mainly radiative, and, though convection and conduction heat transfer also take part here, for the sake of simplification, their influence is assumed in this report to be covered by heat radiation) only inside the flaming zone where surface particles are already burnt.

This internal heat flux governs the downspread of burning along vertical fuel columns, where the flame angle factor **F** = 0, since the outward heat flux is excluded from the spread mechanism. So the vertical downspread of burning is a very simple and convenient case in analyzing the mechanism of spread of forest fires, since many complex factors do not take part in the process.

4. Downspread Burning of Vertical Fuel Layers

In this case the vertical spread of flame is ensured by the inward heat flux emitted from the interior of the flame onto the surface of particles which it covers. This process has been thoroughly studied for burning liquid hydrocarbons in open reservoirs (Blinov and Khudiakov 1961).

Similar conditions are available at every forest fire in thick fuel beds, where the burning penetrates from the surface of the fuel bed deep down to the horizon of mineral soil, as well as in backfiring against strong winds.

It is noteworthy that, in this case, the influence of wind is excluded, except some increase of flame temperature which is higher by some 20°C for moderate winds (Demidov and Saushev 1975) as related to that in calm air.

A series of laboratory experiments were conducted with vertical burning of fuel columns. Dried branches of pine and oak (foliage was preserved) with measured fuel characteristics (moisture content, density, weight, etc.) were arranged along vertical fire-resistant poles and, after measurement of the column, its upper part was ignited and flame size and downspread rate were measured. The heat balance equation derived for this downspread burning, is as follows:

$$5.67\cdot10^{-8}E_i(T_f^4-T_o^4)=d[c(t_i-t_o)+2.6\cdot10^4M]R_i \tag{8}$$

where E_i - the inward emissivity of the "flame-fuel" system, in which the internal space of flames emits the heat flux onto the burning surface of fuel particles covered by the flame.

Additional observations were conducted on wildland backfires in thick beds of dry grass during strong head winds, when the flame angle factor **F** was negligible.

Based on these laboratory and field observations and Equation (6), the inward emissivity E_i was calculated, and then, based on Equation (6), the average value of the attenuation coefficient was estimated. The following formula resulted from these studies:

$$E_i=1-e^{-0.16(L+0.4)} \tag{9}$$

where **0.4**, most probably, accounts for the heat flux from burning embers and heat transfer by convection.

For outward emissivity, by analogy with Equations (6) and (9), the following formula was drafted:

$$E_o=1-e^{0.16D} \tag{10}$$

where **L** is the flame length, m; and **D** is the width of the flame zone, m.

It is noteworthy for further understanding, that for very thin flames like those in low-intensity fires, where the width of the flames is negligible, outward emissivity is also negligible, and in this case, the inward emissivity is close to the constant value of 0.1 which accounts for heat flux from glowing embers and, probably, for some convection heat exchange. Inward heat flux is always available in forest fires even of minimum intensity.

So:

for low-intensity fires	$E_i = 0.1 - 0.3$; $E_o = 0 - 0.2$
for high-intensity fires	$E_i = 0.7 - 1.0$; $E_o = 0.7 - 1.0$
for fires of moderate intensity	$E_i = 0.3 - 0.7$; $E_o = 0.2 - 0.7$

5. Fire Spread in Thick Beds of Forest Fuels

Surface particles of thick fuel beds, being exposed to the outward heat flux before they contact the flame, are the first to be ignited, and so it is they which determine the rate of fire spread. The lower (sub-surface) particles are never exposed to the outward heat flux, so they are ignited only by the inward heat flux when they are covered by flame.
The role of these lower particles in the process of fire spread influences flame size and emissivity, and, through these parameters, fire spread rate. The inward heat exchange between the internal flame space and lower particles of the fuel bed is not influenced by slope and wind direction and speed, except as indicated above with the increase of flame temperature on some 10-20°C by winds of up to 10 m s^{-1}.
Theoretical considerations (Telitsyn 1973b) showed that the joint influence of outward and inward heat exchange on fire spread rate can be expressed by the sum of relevant contributions of each heat flux to the process of fire spread:

$$\mathbf{R} = \mathbf{R}_o + \mathbf{R}_i, \tag{11}$$

where $\mathbf{R}_o$ is the contribution of outward heat flux to fire spread rate $\mathbf{R}$ and is estimated by Equation (1), and $\mathbf{R}_i$ is that of inward heat flux and is estimated by Equation (8). These estimations were performed taking into account all above said constants and variables, and finally resulted in the following formulas:

$$R_o = \frac{3E_oFL}{W_{is}(16+M)} \tag{12}$$

$$R_i = \frac{3E_i}{d(16+M)} \tag{13}$$

Fires of low intensity, where $\mathbf{E}_o$ and, hence, $\mathbf{R}_o$ are negligible, are governed by $\mathbf{R}_i$ only. These are backfires propagating against strong head winds and down steep slopes. Rate of spread $\mathbf{R}_i$ is also the rate of penetration of burning into the depth of the fuel bed from its surface.

Substitution of Equations (12, 13) in (11), and also the above constants and formulas, gives the formula for rate of spread of fire front in ratio to backfire rate of spread:

$$R = R_i \left(1 + \frac{R_o}{R_i}\right) = R_i \left(1 + \frac{FLE_o}{l\,E_i}\right) \quad \text{m s}^{-1}, \tag{14}$$

where $\mathbf{l}$ is the mean free path of radiation in the fuel bed.

The ratio $FLE_0/l\ E_i$, when the fuel bed is horizontal is a certain function of speed and direction of wind, so:

$$R = R_i\ [1 + f\ (U)\ \mathrm{Cos}A], \tag{14a}$$

where **f** means "function", and **A** - the angle between direction of fire spread and that of wind.

For grass fires the relationship between rates of spread of head fire and backfire, as field observations showed, is:

$$R = R_i\ (1 + U) \quad m\ s^{-1}, \tag{14b}$$

This formula is convenient for fire reconnaissance. It is enough to measure the rate of spread of a backfire and the speed of wind to calculate the rate of spread of a fire head with this formula. The same relationship is valid for widths of flaming zones of backfire and fire head, so it is possible to estimate the width of the flaming zone of a fire head, having made relevant observations of width of a burning strip in the backfire.

Since the residence time is the ratio of depth of fuel bed to the vertical rate of its burning $\mathbf{R_i}$, and since this ratio is also of width of the burning strip to the rate of spread, it can be easily measured. Therefore the observer on the fire head can estimate this ratio and evaluate the $\mathbf{R_i}$, which is also the rate of backfire spread, should it be planned by the fire boss.

It is also easy to estimate fuel loading by measuring the height of flame and rate of spread in any part of the fire perimeter, and by the subsequent calculations by formula (5) to predict the height of flames of a planned backfire, or that of a fire head, if it is impossible to observe it directly, and the rate of spread calculated by the above method. The latter action is important to select tactics of fire suppression.

The basic distinction of this model of forest fire spread, described by the system of Equations (11...14) and the above stated accompanying formulas from those of Van Wagner (1967, 1968), Thomas (1965, 1971), Rothermel (1972), and others is that, in this model, the outward heat flux is related only to the surface particles of fuel beds (composed of more than one particle in depth) and is not considered to be equally distributed through the whole thickness of the fuel bed as it is assumed in the models of the above said authors. The model described here seems to be more realistic, and the experimental verification of its simplified corollaries showed their acceptable conformity to the observed data (Telitsyn 1973b; Kurbatsky and Telitsyn 1977).

6. Application of the Model to High-Intensity Forest Fires

Flames of high-intensity forest fires, according to the above identification parameters, are so large that their emissivities in all directions are close or equal to 1.0. The buoyancy velocity of flame gases at the flame top is close to or exceeds wind speed (see Equation 4), and the same is with their kinetic energy. So flame angle is close to 45°, and, for this angle, the flame angle factor $\mathbf{F} = 0.7$.

After substitution into equations (11...13), values $\mathbf{E} = 1$ and $\mathbf{F} = 0.7$ give the rate of spread of high-intensity fire $\mathbf{R_h}$ in the following semi-empirical formula:

$$R_h = \frac{3}{16 + M} \left(\frac{1}{d} + \frac{0.7\,L}{W_{is}} \right) \quad m\ s^{-1}, \tag{15}$$

$[W > 0.5\ kg\ m^{-2};\ U \approx V]$

In application to crown fires, say, to those in spruce forests, where characteristics of fuel (green foliage) are rather constant (moisture content 120 %, bulk density of layer of crowns 0.2...0.3 kg m^{-3}, particle thickness 0.0004 mm, particles density 350 kg m^{-3}, volatiles percentage 78 %, fuel loading 2.5...3.0 kg m^{-2} and the isothermal fuel loading about 0.12 kg m^{-2}) Equation (15) after substitution of these data and Equation (5) in it, and after solving the resulting quadratic equation for $\mathbf{R_h}$ by method of successive approximations, has resulted in the following empirical formula:

$$\mathbf{R_h = 0.6\ W}, \quad m\ s^{-1}, \tag{16}$$

$[W > 0.5\ kg\ m^{-2};\ U \approx V]$

According to this, when fuel loading in spruce forests is, say 3 kg m^{-2}, crown fires will spread downwind at the rate of about 0.2 m s^{-1}, this value being quite realistic for these forests.

This value is also the rate of spread of burning of the mixture of carbon monoxide with air at the ambient temperature 20°C (Demidov and Saushev 1975).

Obviously, a cloud of mixture of carbon monoxide moves, or rather, is permanently reproduced by powerful radiative heat flux from huge flames to the fuel bed (up to 130 kW m^{-1}) ahead of a high-intensity fire. If it is really so, then high-intensity fire spreads by kinetic, and not diffusion mechanism, or at least by the intermediate regime of burning along the cloud of carbon monoxide mixed with air.

This supposition is confirmed by the observed facts of reinforcement of air shock waves from explosions of hose-type charges of explosives in front of crown fires (Grishin 1992), giving evidence of available clouds of volatiles in front of high-intensity forest fires. Higher concentrations of carbon monoxide were also observed by Anufrieva et al. (1973) in front of fires of moderate intensity within a distance of 1-2 m from flames.

It is known that the higher the temperature of this mixture the quicker it burns (Demidov and Saushev 1975). Since temperature of a mixture of volatiles with air in forest fires is close to the ambient temperature (for the percentage of volatiles in this mixture is low), the spread rate of burning along this cloud, which is the same rate of spread of high-intensity fires, seems to be strongly dependent on the ambient temperature. Indeed, crown fires are not known to occur in cold weather, and most severe crown fires propagate when the weather is hot.

Observations of crown fires, which occurred on 16 October 1976 in the Khabarovsk province, known as the most severe fires in the Province during the whole history of Forest Protection Service here, showed that flames of such fires are blue, or bluish for observers located in front of such fires, and the same flames are yellow when observed from behind. This is evidence that high concentrations of carbon monoxide are available in front of high-intensity forest fires, a gas which is transparent mainly for the blue part of the light spectrum. So, predictions of probability of crown fire occurrence can be based on ambient air temperatures, but additional studies of characteristics and formation of carbon monoxide clouds in front of the flame zone should be conducted in various weather and fuels conditions.

7. Conclusions

This study was conducted for many years and included not only theoretical considerations but many field and laboratory observations. The author has also collected a considerable base of experimental information on the topic from Russian and foreign sources, and this model was verified by it. Yet nevertheless, the author is still dissatisfied with the fact that this theory which is well expressing the current state and correlation of parameters of fire spread processes, is in itself hardly fit for the prediction of fire behavior from initial parameters of fuel and other factors of the fire environment, but only through its semi-empirical corollaries. So this model does not make any pretensions to its completeness, but is rather of an informative essence, and an invitation to further studies.

Acknowledgements

This work was supported by the Far Eastern Research Forestry Institute up to 1992, and afterwards, by the Khabarovsk Technical University, and, under its umbrella, by the Institute on Computer Technologies.

The author is greatly indebted to the Director of this University Dr.V.K. Bulgakov and to the Project Leader of the Institute of Computer Technology Dr.A.I. Karpov for several helpful suggestions.

Conducting this study, the author took advantage of great assistance from Dr.P.H. Thomas, Dr.C.E. Van Wagner and Dr.J.M. Byram, and expresses his gratitude to all of them.

References

Anufrieva, G.P., M.D. Bodansky, S.M. Vonsky, and H.P. Telitsyn. 1973. K Raschetu Temperatury i Gazovogo Sostava Vozdukha pri Lesnykh Pozharakh" <Concerning calculations of temperature and gas mixture compositions in front of forest fires. Report to the All-Union Conference on Forest Fires, Krasnoyarsk 1971>, 120-125. Gorenie i Pozhary v lesu. Krasnoyarsk. Institute of Forest and Wood, USSR Acad. Sci. <in Russian>.

Blinov, V.I., and G.N. Khudiakov. 1961. Diffusionnoe Gorenie Zhidkostei <Diffusive burning of liquids>. Moscow, 208 pp. <in Russian>

Byram, G.M., H.B. Clements, M.E. Bishop, and R.M. Nelson. 1966. Final Report, Project Fire Model: an Experimental Study of Model Fires. US For. Serv. Southeast For. Exp. Sta., Southern Forest Fire Lab., Macon, Georgia, 36 pp.

Davis, K.P., G.M. Byram, and W.R. Krumm. 1959. Forest fire: Control and use. McGraw Hill, New York, 584 pp.

Demidov, P.G., and V.S. Saushev. 1975. Gorenie i Svoistva Goriuchikh Veshestv <Burning and characteristics of fuels>. Moscow, 278 pp. <in Russian>.

Emmons, H.W. 1963. Fire in Forest. Fire Res. Abs. Rev. 5, 163.

Fons, L.W. 1946. Analysis of fire spread in light forest fuels. J. Agr. Res.72 (13), 93-121.

Grishin, A.M. 1992. Matematicheskoe Modelirovanie Lesnykh Pozharov <Mathematical modeling of forest fires>. Novosibirsk, 408 pp. <in Russian>.

Konev, E.V., and A.I. Sukhinin. 1977. The analysis of flame spread through forest fuel. Combust. and Flame 28, 217-223.

Kurbatsky, N.P., and H.P. Telitsyn. 1977. Teoretichesky i Experimentalny Analiz Radiatsionnogo mekhanizma Rasprostranmia Lesnykh Pozharov <Theoretical and experimental studies of forest fire spread mechanism>. In: Sb. "Kharakteristika Protsessov Gorenia v Lesu", 4-26. Krasnoyarsk. Inst. of Forest and Wood, USSR Acad. Sci. <in Russian>.

Lyons, P.R., and R.O. Weber. 1991. Single-Strand tests of wildfire potential. A paper presented at the 11th Conference on Fire and Forest Meteorology, April 16-19, 1991, Missoula, MT (USA), 7 pp.

Rothermel, R.C. 1972. A mathematical model for fire spread predictions in wildland fuels. USDA For. Serv. Res. Paper INT-115. Ogden, 40 pp.

Telitsyn, H.P. 1965. Zavisimost Skorosti Rasprostranenia Lesnykh Pozharov ot Uslovii Pogody". Sbornik Trudov Dalniilkh. Vyp. 7 <Dependability of forest fire spread rate on weather factors. In: Collection of publications of Far Eastern Forestry Res. Institute, Vol.7>, 391-405. Khabarovsk <in Russian>.

Telitsyn, H.P. 1973a. Flame radiation as a mechanism of fire spread in forests". In: Heat Transfer in Flames, Vol. 2. (N.H. Afgan and J.M. Beer, eds.), 441-449. John Wiley, New York, Toronto, London.

Telitsyn, H.P. 1973b. O rasprostranenii gorenia v lesu. Gorenie i Pozhary v lesu <Concerning fire spread in forests. Report to the All-Union Conference on Forest Fires, Krasnoyarsk, 1971>, 164-176. Krasnoyarsk. Institute of Forest and Wood, USSR Acad. Sci. <in Russian>.

Telitsyn, H.P., and A.P. Sosnovshenko. 1970. Kharacteristiki lesnykh goriuchikh materialov i ikh sviaz s osobennostiami gorenia. Sbornik Trudov Dalniilkh. Vyp. 10 <Characteristics of burning of forest fuels". In collection of publications of Far Eastern Res. Inst., Vol. 10>, 248-252. Moscow <in Russian>.

Thomas, P.H. 1965. The contribution of flame radiation to fire spread in forests. Joint Fire Res. Organization. Fire Res. Note 594. Fire Res. Sta. Boreham Wood. U.K., 19 pp.

Thomas, P.H. 1971. Rates of Spread of Some Wind-Driven Fires. J. For. 44, 155-175.

Van Wagner, C.E. 1967. Calculation on forest fire spread by flame radiation. Can. Dept. For. Rep. No. 1185, 28 pp.

Van Wagner, C.E. 1968. Fire behavior mechanisms in a Red Pine plantation: Field and laboratory evidence. For. Branch Dept. Pub. No. 1229. Ottawa, 30 pp.

Weber R.O. 1990. Modelling fire spread through fuel beds. Prog. Energy Combust. Sci. 17, 67-82.

J.G.Goldammer and V.V.Furyaev (eds.), Fire in Ecosystems of Boreal Eurasia, 326-330.

Burned Forest Area Type Classification

I.S. Melekhov [1]

1. Introduction

Burned-forest area type classification is an integral part of dynamic forest-type classification. Fire influences on forest structure and post-fire stand transformations are directly related to fire parameters and stand characteristics. Burned areas themselves are remarkable for their complex and dynamic nature. Therefore, they are considered to be a complex object from the viewpoints of both fundamental study and practical exploitation. Even when roughly describing burned areas at an elementary level, one encounters quite a number of problems. However, present forest classification and inventory should involve thorough burned- and cut-area studies incorporating both fundamental and practical aspects.

2. Burned Area Classification

The aim of this paper is to specify some concepts and classification systems with respect to fire effects on vegetation. Apart from concepts widely accepted in forest pyrology and related disciplines, there exist some that need additional explanation. The latter include the concept of "burned-area type" and its hierarchical relationships. Basic burned area classification categories (burned forest, partially burned forest) can be determined using two approaches:

1) Burned area classification by vegetation cover type and dynamics;

2) Burned area classification based on the character of fire influence on the stand, particularly on the tree layer, including overstory and sub-canopy tree strata.

Both approaches are valid and are interrelated. Here, priority is of critical importance. Vegetation cover (i.e., surface cover), although related to fire type, is not normally a direct descriptor of fire activity except on fresh burns. The tree layer, or what remains of it after fire, provides both direct and indirect indicators of fire activity. The structure of the tree layer can preserve fire evidence for many years (even up to several hundred years). Changes of environmental and surface cover conditions resulting from fire-caused stand trans-

---

[1] Department of Forestry, Moscow University, 141001 Mytishchy, Moscow Region, Russian Federation

formations should also be considered. It is thus reasonable that the classification of burned areas and partially burned forests be based on fire-induced changes in the major stand component--the tree layer--with forest types being taken into consideration. Once categories of burned and partially burned forest areas are described on this basis, this system can be supplemented by burned-area classification based on vegetation cover patterns (including surface cover), colonization by natural regeneration and regrowth, and phytocenosis and biogeocenosis development dynamics (i.e., taking wildlife into account). Many years ago a burned-area classification by post-fire tree layer habitus was accepted that was underlain by the principle of characterizing the burned area based on the severity of fire influence on the stand, primarily on the tree layer as the major stand component. This classification includes three burned forest groups subdivided into smaller categories (Melekhov 1944):

Group I. Burned forest where the tree layer or its remains are absent as a result of complete or almost complete destruction by fire

Group II. Burned forest where the tree layer is present but has lost its viability

- a) burned forest dominated by downed wood
- b) burned forest dominated by snags

Group III. Burned forest with a living tree layer

- a) small number (less than 10%) of living trees in the overstory while subordinate tree layers are completely destroyed
- b) a considerable number (more than 10%) of live trees in the overstory (subordinate tree layers are completely destroyed)
- c) partial destruction of the subordinate tree layer only

This classification has been widely used by specialists in their investigations of burns and post-fire forest transformations in various regions of Russia (e.g., Kurbatsky 1964; Furyaev 1973; Voinov 1976; Savchenko 1982). In this paper, I do not discuss a number of interesting research studies of fire consequences currently being conducted by forest fire scientists (e.g., Sheshukov and Matveyev at the Siberian Technological Institute, Laboratory of Forest Fire Research of the V.N. Sukachev, Institute of Forest, and others in Siberia and the Far East).

The above classification has proven to be useful up to the present. However, we do not stick to strict terminology and use. Along with the principal "burned forest group" term, some other terms, such as "burned-area type," are often used. Although there is no exact uniform definition as yet of burned-area type, there is a cut-area classification where cut-area type, as a classification unit, is a well-defined term. This classification includes, among others, the type of areas cut after prescribed burning. For this reason, burned area-type classification terms can be specified using cut-area type classification.

Komarova (1992) states that the principal classification unit (burned-area type) should be analogous to cut-area type (according to Melekhov), since burned areas, like cuts, result in externally-induced transformations of vegetation and site conditions that represent the initial stage of the forest regeneration process. Komarova states "I believe that attempts to develop fundamental principles of burned area classification by type, based on cut-area type classification, are reasonable in all respects. In the analogy with cut-area type, as it is interpreted by Melekhov, burned-area type can be defined as the parts of a burned area similar in site conditions, surface cover, and microclimatic and microbiological regimes causing similar site conditions and forest regeneration trends." On the basis of this definition, Komarova has described and included into corresponding groups many burned-area types for the Skihote-Alin Region. According to Komarova (1992), "a burned-area type classification unit related to the overstory and subordinate tree layer rate-of-destruction is more reasonably defined as "burned-area kind." This is an interesting idea, especially as forest pyrologists use the analogous term "fire kind", and it is logical to relate fire-caused destruction to the kind of burned area (burned forest, partially burned forest.)

However, problems still remain as to hierarchy and terminology. Within the same burned-area kind (e.g., in burned forest where the tree layer was destroyed), burned areas of different types can be formed. This problem remains even in cases when burned areas of the same type are formed. In other words, the term "kind" has a wider meaning here than "type." This approach does not satisfy classification requirements. Thus, when speaking of burned forests which are identified, depending on the severity of destruction of the tree layer and other stand components, it is reasonable to use the term "burned- forest group" (Groups I to III) or possibly "burned-forest class." Unlike Komarova's interpretation, however, burned-forest class should not be considered as the upper limit of cut-area type classification. The relationship of burned-forest groups (classes) to burned-area types can be described as shown in Figure 1. Burned-area types are primarily categories of burned forests where the tree layer is completely consumed by fire (Group I) and of those where the tree layer (or its remains) is present but has lost its viability due to fire (Group II). As for burned forest with living trees (Group III), it should be treated, strictly speaking, not as burned areas but as partially burned forest (i.e., as a forest type). As an exception, burned forests with a small number of live trees remaining after fire can comprise an additional group (IIIa); in these forests, burned areas of various types can sometimes be formed, which is not to say that forest type can be ignored (in Figure 1, this relationship is indicated by dashed arrows to show conditionality). In the classification, each burned-area type is accompanied by the number assigned to the burned forest depending on its tree layer destruction rate. Some examples are as follows:

a) *Ribes* type (IIa): *Ribes*-dominated burned forest with a downed tree layer.

b) *Calamagrostis* type (IIb): *Calamagrostis*-dominated burned forest with snags.

c) *Hamaenerium angustifolium* type (1): *Hamaenerium angustifolium*-dominated burned forest with no tree layer.

d) *Calamagrostis* type (IIIa): *Calamagrostis*-dominated burned forest with a smaller number (less than 10%) of live individuals in the tree layer.

When conducting a forest inventory, burned-area type and fire date should be added to data on each numbered elementary forest inventory unit in question (as in the case with cut-area type classification). Research studies should also cover other parameters. As forest types of fire origin are a special area of interest of forest fire scientists, they are not discussed in this paper. It should be stressed, however, that all pyrological parameters are to be considered when classifying burned-area types.

Some terms must now be explained. The term "burned forest" has a general meaning as well as a more specific meaning. Loosely interpreted, this term covers all burned sites, ranging from completely burned areas (complete or near-complete consumption of all above-ground fuels) to very slightly burned forests. The term can be used more specifically to refer to burned forest with a downed tree layer or with a tree layer comprised of snags only, as well as those with a small number of live individuals in the tree layer. In the case of burned forest with downed tree layer the term "burned area" is often used by many scientists; however, the term is typically associated with bare areas, characterized by the complete absence of pre-fire trees. This term "burned area" is also used when describing the forest regeneration process. Classes (groups) cover the total range of fire severity and, therefore, use the same general definition of burned forest. For burned-forest areas, however, we divide them into "completely burned" and "partially burned" forest areas. Completely burned-forest areas get the narrower definition of "burned forest." Partially burned forests are not considered "burned area" under this definition. However, as is obvious from the diagram (Fig.1), burned forests of Group II (and partially Group III) are presumed to be formed in burned forest areas. This is especially true where a tendency to "pure burn" exists (i.e., Group I burned forest formation; Fig.1)

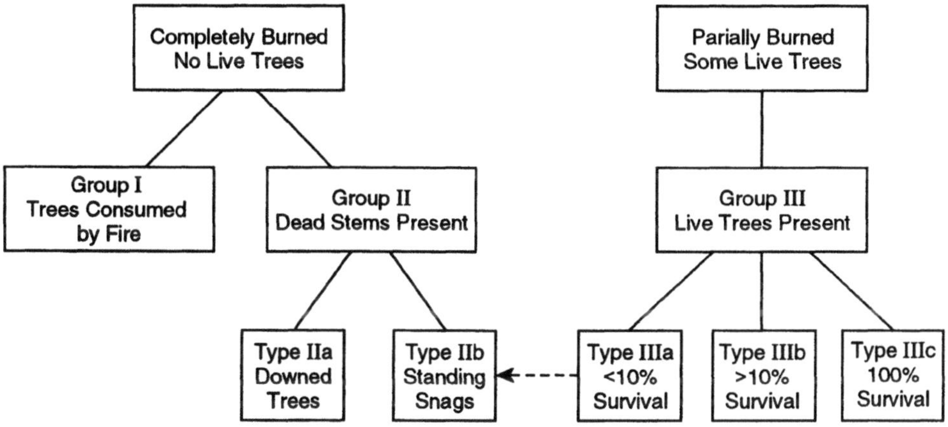

Fig.1. Burned forest area classification

3. Conclusions

It is hoped that the discussion presented in this paper will stimulate the study of specific burned-area types at a regional level. Like cut-area type classification, burned-area type classification, being a separate scientific problem, can be considered as an integral part of dynamic forest typology. Burned-area type is then the elementary inventory unit of the burned-area classification. It is indicated by changes in surface cover and included in a higher burned forest category, which in turn is determined based on the level of tree layer destruction. Burned-area type, as well as cut-area type, is a geographical phenomenon strongly related to vegetation pattern and fire parameters. Typological burned-area peculiarities should be taken into account when dealing with larger landscape units. For research studies at the landscape level, it is crucial to find optimal combinations of differentiation and integration criteria to underly classification configuration. Burned forest areas are an important object for landscape approach and have particular dynamics. Therefore, investigations at either a local or large landscape level should necessarily involve burned area monitoring (including aero-space monitoring).

References

Furyaev, V.V. 1973. Fire effect investigation and use of controlled burning. Combustion and Forest Fires (Gorenie i pozhary v lesu). Krasnoyarsk <in Russian>.

Komarova, T.A. 1992. Post-fire forest successions in the southern part of Sikhote-Alin Region. Vladivostok. 224 pp. <in Russian>.

Kurbatsky, N.P. 1964. Forest fire problem. In: Forest Fire Occurrence (Vosniknovenie lesnikh pozharov). Nauka. Moscow <in Russian>.

Melekhov, I.S. 1944. On the theory of forest pyrology. Arkhangelsk <in Russian>.

Savchenko, G.A. 1982. Investigation of Krimean pine fire resistance aimed at fire resistant forest formation in the Krimea. Abstract of thesis. Moscow <in Russian>.

Voinov, G.S. 1976. Prediction of tree mortality in a stand subject to surface fires (state-of-the-art in forest typology and pyrology). Institute of Forest and Forest Chemistry Archives. Arkhangelsk <in Russian>.

J.G.Goldammer and V.V.Furyaev (eds.), Fire in Ecosystems of Boreal Eurasia, 331-336.

Fires and Soil Formation

V.N. Gorbachev and E.P. Popova [1]

1. Introduction

The formation of a vegetation cover of a taiga zone is generally accompanied by fires of different intensity and impact. Kurbatsky (1964) has shown that beginning from the holocene forest fires have been occurring regularly, capable of spreading rapidly over large territories. It is estimated that at present lightning is the cause of only 8-9 % of all fires (Sofronov and Vakurov 1981); the remainder is caused by human activities. Most foresters consider fires a powerful ecological factor of forest succession (Kurbatsky 1964; Kolesnikov et al. 1973; Buzykin 1975; Furyaev 1977). Fires change forest tree species composition, influence the age structure of forest stands, and disturb and change the species composition of ground vegetation cover (Novoseltseva 1975). To a considerable extent the territorial location of forests of different age classes and generations, and the sinuous and mosaic structure of the vegetation cover are associated with forest fires (Pobedinsky 1965).

Forest fires have considerable effects on soil formation. Soils as an integral part of forest biogeocenoses undergo a versatile effect from fires. Numerous studies have investigated the changes in the physical, chemical, and biological properties of soils under the influence of fire (e.g. Firsova 1960; Grishin 1973; Sapozhnikov 1973; Shchegolkova 1988; Baranov and Stefin 1989; Beatty 1989; Hayes and Seastedt 1989; Vega and Diaz-Fierros 1989; Tarabunina and Savvinov 1990; Weber 1990; Marion et al. 1991). However, there are only a few studies on Siberia, where fires occur most frequently (e.g., Gorbachev et al. 1982).

In this project the effects of forest fires on soil were studied along three directions:

- effect of fires on the formation of a micro-relief and diversity of features and properties of soils;
- pyrogenic changes of soil properties under the direct influence of fire;
- pyrogenic changes of soil formation processes as a result of changing the vegetation cover.

2. Effects of Fire on Microrelief

In the southern taiga of Siberia long-term frozen soils are widely spread. They are

[1] V.N. Sukachev Institute for Forestry and Timber, Siberian Branch, Russian Academy of Sciences, 660036 Krasnoyarsk, Russian Federation

characterized by deep freezing (up to 2-3 m), below zero temperatures in the profile during 5-6 months a year, and slow thawing in spring. Because of this, root systems of trees, dwarf shrubs and herbaceous vegetation lie in the warm upper layers of soil. According to Kulagina (1978), in pine forests of the Angara region 58-80 % of total roots are concentrated in the soil layer of 0-10 cm, and 75-92 % of roots in the layer from 0-20 cm. As roots are damaged by ground fires, the stability of trees, due to exposure to wind, decreases sharply. Wind may blow down individual trees as well as entire groups of trees. Such unequal distribution of individual windfalls and larger gaps is accounted for by the unequal strength of wind under the forest canopy, the position and damage of roots, and the different intensity of forest fires.

Falling trees create windfall pits on the soil surface. The dimensions of these windfall pits are from 280-295 x 85-115 cm, corresponding to a surface of ca. 2.5-3.5 m^2, with a depth of ca. 30-40 cm. The height of the soil clump on tree roots can be 130-230 cm. In the course of time, under the action of rain, and wind and its weight, the soil from roots begins to crumble, forming mounds. Owing to this the windfall pits are partly filled and their size decreases considerably. The dimensions of old windfall pits are from 85-135 x 30-65 cm, corresponding to a surface area of 0.3-0.9 m^2. In this manner a specific post-fire microrelief is formed, whose dominant form is a coupled combination of mounds and pits.

The microrelief formed as a result of post-fire windfall results in a redistribution of temperature and moisture regimes in surface soil layers. In the spring the mounds become free from snow earlier and are warmed up better, but in the summer they lose more moisture than the pits do. Differences in the soil temperatures between the microrelief elements can be found during the entire vegetation period, but they are more distinct in spring and in the first half of summer, sometimes becoming as high as 5-6° C (Tab.1).

Tab.1. Soil temperature at the depth of 5 cm in the micro-relief elements

Month	Micro Relief	Taiga solodic loamy soil				Pseudo-podsolic sandy soil				
		Observation Sites								
		1	2	3	4	5	6	7	8	9
May	Mound	4.5	4.5	4.2	4.6	-	-	-	-	-
	Pit	0.5	0.5	0.5	0.5	-	-	-	-	-
June	Mound	6.5	7.2	5.1	5.9	13.4	12.8	13.1	12.5	15.7
	Pit	1.9	2.2	1.6	2.3	8.7	8.7	7.8	6.2	10.0
July	Mound	8.0	10.0	8.2	9.5	18.2	18.0	18.5	17.1	18.6
	Pit	5.0	5.6	5.0	5.1	15.5	12.0	12.0	11.0	14.8
August	Mound	8.0	9.4	7.6	8.1	16.2	14.6	14.7	14.7	16.3
	Pit	6.2	6.8	5.3	5.7	13.2	12.5	12.2	11.9	13.0
September	Mound	7.5	7.1	7.0	7.9	11.0	10.5	10.5	10.0	11.0
	Pit	5.0	6.0	5.1	5.5	9.5	9.5	9.0	8.8	9.6

The displacement of soil mass at the site of tree windfall is accompanied by the disturbance of the structure of the soil profile - the soil layers lying below are often brought to the surface; the genetic influence of horizons, their composition, structure, and color change.

Additionally, the microrelief stipulates a specific hydrothermal regime, which is reflected on the activity of microbiological and biochemical processes and eventually on the chemical activity of soil. Thus, the post-fire windfall of trees strengthens spatial heterogeneity of soil properties.

Direct Effects of Fire on Soil Properties

During fires forest soils undergo various fire effects. This is due to, first of all, an uneven distribution of combustible material on the soil surface, unequal soil moisture regimes, varying reliefs, and the character of the vegetation. At the experimental burning of a pine forest (rose bay - cowberry pine forest type) at a weak and average fire intensity, a broad range of temperatures - from 55 to 250°C - was recorded on the soil surface. At the places of burning slash and fallen down trees the thermal effect of fire is the strongest. When fire intensity increases a more complete mineralization of vegetation residues occur on the soil surface. The combustion of the forest floor is accompanied by the decrease of its thickness, quantity of organic matter, and change of acidity (Tab.2).

Tab.2. Change of thickness, quantity and acidity of forest floor organic matter after a surface fire in the rose bay-cowberry (*Chamaenerion angustifolium* - *Vaccinium vitis-idaea*) pine forest

Fire Intensity	Thickness (mm)	Quantity (kg m^{-2})	pH Water
Very low	43	2.58	5.63
Low	35	2.55	5.96
Average	23	1.29	6.76
High	21	1.28	8.12
Very high	11	0.88	9.39
Control	49	3.23	5.34

The higher the fire intensity, the more completely the organic matter on the forest floor is consumed. After burning of trunks of downed trees considerable amounts of ash remain on the soil surface, reaching quantities of 3-9 kg m^2. Finally after ground fires the spatial variability of acidity of the forest floor increases, which is distinctly observed during the first two years and can still be seen four years after the fire.

Weight and height of the litter layer increases relatively fast within 4-5 years after the fire, but its fraction composition stabilizes slowly. This is connected, to a considerable extent, with the post-fire decline of pine, which results in the increasing fall of needles and small branches. Four years after the fire an increased content of needles is observed in the litter composition, 34% on the burned site versus 15% on the control sites. Combustion products of the litter layer result in changes in the reaction of soil solution. The investigations have shown that an ash dose of 3 kgxm^{-2} changes the pH of water extract of surface soil layers to a weak acid, the dose of 6 kgxm^{-2} to neutral, and the dose of 9 kgxm^{-2} to alkali range.

Ash elements remaining in the soil, increase the general level of oxidation reactions. The soils are characterized by a relatively low redox potential of 370-415 mV, which indicates the possibility of reduction processes.

After surface fires of average and strong intensities the redox potential increases up to 425-600 mV. This creates the prerequsites for the activation of oxidation processes.

The fires of a weak intensity do not particularly have a considerable effect on the number and composition of adsorbed cations. However, the average fire intensity is accompanied by considerable changes in the soil adsorption complex. The share of calcium and magnesium increases, and the share of exchangeable hydrogen and aluminum decreases sharply. At a high fire intensity with aloss of the humus layer, the cation exchange capacity decreases. In the case when organic compounds of soils do not burn down, the content of exchangeable bases in surface horizons may increase 2-3 times.

The nutrient regime of soils also undergoes considerable changes in the post-fire period. A distinct dependence was observed between the fire intensity and soil calcination on the one hand, and the increase of concentration of mobile nitrogen, phosphorus, and potassium compounds on the other hand. The most favourable conditions for the nutrition of microbes and phytocenosis are created at average and high (but without the loss of soil organic matter) fire intensities. Also, some indices of soil biological activity distinctly respond to pyrogenic effects. If the humus was burned down, a decrease of soil carbon dioxide production and a decrease of catalase and urease activities were observed.

Effects of Fire on Soil Formation

The direct effects of forest fire on the soil - heating, calcination and appearance of ash - do not last longer than ca. 3-5 years. Initially, after a fire, the leading role in changing ecological conditions of particular habitats is played by soil factors - the temperature and nutrient regimes, the reaction of the medium, and the redox potential, which change biological activity.

Previous vegetation is either destroyed by fire or it turns out to be, to some extent, inappropriate in the new (post-fire) ecological situation and is replaced by another one. The change of vegetation associations often has a stable character and sometimes takes more than one decade. More often in the taiga zone there occurs a change of green-moss forest types by herbaceous ones. This affects the directive and intensity of primary soil formation processes.

To elucidate the role of fires in this respect on taiga solodic loamy soils, a number of sample areas were selected, on which during the unequal post-fire period, 120-year-old stands of different types were formed, (1) a cowberry-green-moss pine forest which had not been affected by fire during its formation, (2) a wild rosemary-green-moss pine forest burned in 1902 and 1942, and (3) a herbaceous-cowberry pine forest burned in 1902, 1916, 1942, and 1956. In the latter case, surface fires with the interval of 14, 26, and 14 years had consumed the moss cover, thus promoting the development of herbs with leguminous plants in an area of prevailing green-moss forest types.

Unequal composition of ground vegetation at burns of different ages results in forming peculiar forest floors which differ in thickness, quantity, fractional composition, water holding capacity, and biological activity. Different thickness and unequal fractional compositions of the forest floor contribute to the creation at these burns of peculiar hydrothermal conditions. Thus by the end of May the soil of herbaceous-cowberry pine forest thaws out to the depth of 15-38 cm, in the wild rosemary-green-moss pine forest frozen ground occurs under a moss cover, in the cowberry-green-moss pine forest zero temperatures are at the depth of 10-25 cm.

The most favorable redox conditions are formed in the taiga solodic loamy soil under a herbaceous-cowberry pine forest, where the changes of potential for the period of May-August are within the interval of 405-437 mV. In the cowberry-green-moss wild rosemary pine forests, the redox potential of soils is somewhat lower - 344-422 mV. This is contributed to both by a slow thawing out of the soil thickness and the presence of an overcongealed upgrade-streamwater in the first half of the vegetation period.

The soils of burns are not equally supplied with organic matter, the range being 2.5 to 7.6%. To the same extent they differ in the content of exchangeable cations. With the increased post-fire period the amount of the adsorbed calcium decreases, and those of hydrogen and aluminum increases.

The most favorable conditions for nitrogen and potassium nutrition are formed in the herbaceous-cowberry pine forests, and for phosphorus one in the wild rosemary-green-moss and cowberry-green-moss sites. A more favorable temperature regime of the herbacoeus-cowberry pine forest soil and a better quantity of organic matter against the background of the acid reaction of the medium contribute to the activation of biological processes. The biological activity of soils under the herbaceous-cowberry pine forest is considerably higher than under green-moss stands. It concerns especially urease activity responsible for the decomposition of nitrogen-containing organic compounds at the final stages of their decomposition. Ground fires also greatly affect the soil biota.

3. Conclusions

The study of fire effects on forest soils has shown that fires are a powerful actively functioning ecological factor of modern soil formation. Any fire action is reflected on soil properties, with the degree determined by fire intensity. The most favourable conditions for plant nutrition are established in soils after a fire of average intensity. Changing of green-moss types of forest by herbs and the improvement of redox conditions and temperature regimes of soils promotes activation of a turf process which results in accumulating humus, biogenic concentration of calcium and magnesium in surface horizons, decreasing acidity, and increasing the biological activity of soils.

The post-fire windfall of trees and unequal quantity of ash on the soil surface increase spatial heterogenity of soil properties in the forest zone.

References

Baranov, N.M., and V.V. Stephin. 1989. Post-fire disturbance of antierosion stabiliy of mountain-forest soils. In: Ecological principles of protection of Siberian nature, p.48-51. Krasnoyarsk <in Russian>.

Beatty, S.W. 1989. Fire effects on soil heteregeneity beneath chamise and redshanks chaparral. Phys. Geogr. 10 (1), 44-52.

Buzykin, A.I. 1975. The effect of surface fires on pine forests of the Middle Angara region. In: Protection of forest resources of Siberia, p.141-153. Krasnoyarsk <in Russian>.

Firsova, V.P. 1960. On the problems of the effect of forest fires on soil. Problems of development of forest management in the Urals. Proceed. Institute of Biology of the UF of the USSR AS Vol.1, 24-36. Sverdlovsk <in Russian>.

Furyaev, V.V. 1977. Forest fires as an ecological factor of taiga formation. In: Problems of forest science of Siberia, p.136-147. Nauka, Moscow <in Russian>.

Gorbachev, V.N., V.K. Drimitrienko, E.P. Popova, and N.D. Sorokin. 1982. Soil-ecological studies in forest biogeocenoses. Nauka, Novosibirsk. 185 pp. <in Russian>.

Grishin, I.A. 1973. Effect of fires on soil properties under oak forest of the Amur region. Problems of Geography of the Far East. No.12, 162-175. Vladivostok <in Russian>.

Hayes, D., and T.R. Seastedt. 1989. Nitrogen dynamics of soil water in burned and unburned tallgrass prairie. Soil Biol. and Biochem. 21, 1003-1007.

Kolesnikov, B.P., N.S. Sannikova and N.S. Sannikov. 1973. Effect of surface fire on the structure of a stand and regeneration of forest species. In: Burning and fires in the forest, p.21-22. Krasnoyarsk <in Russian>.

Kulagina, M.A. 1978. Biological productivity of pine forests and a turnover of macroelements. In: Productivity of pine forests, p.90-161. Nauka, Moscow <in Russian>.

Kurbatsky, N.P. 1964. Initiation of forest fires. In: Problem of forest fires, p.5-60. Nauka, Moscow <in Russian>.

Marion J.M., J.M. Moreno and W.C. Oechel. 1991. Fire severity, ash deposition and clipping effects on soil nutrients in chaparral. Soil Sci. Soc. Amer. 55, 235-240.

Novoseltseva, I.F. 1975. Initial stages of reestablishment of herb-dwarf shrub layer in the pine forests of the Angara region after ground fires. In: Protection of forest resources of Siberia, p.179-186. Krasnoyarsk <in Russian>.

Pobedinsky, A.V. 1965. Pine forests of Middle Siberia and beyond the Baikal. Nauka, Moscow. 267 pp. <in Russian>.

Sapozhnikov, A.P. 1973. On the role of pyrogeneous processes in the formation of forest biogeocenoses. Theses of papers of the meeting "Results of scientific invest" on forest science and forest biogeocenology, p.32-34. Moscow <in Russian>.

Shchegolkova, N.M. 1988. The soils of burns and changing of their properties at reforestation in the Sikhote-Olinsky reserve. Author essay of cand. disser. Moscow. 21 pp. <in Russian>.

Sofronov, M.A., and A.D. Vakurov. 1981. Fire in the forest. Nauka, Novosibirsk. 128 pp. <in Russian>.

Tarabunina, V.G., and D.D. Savvinov. 1990. Effect of fires on frozen soils. Nauka, Novosibirsk. 120 pp. <in Russian>.

Vega, J.A., and F. Diaz-Fierros. 1989. Wildfire effects on soil erosion. Ecol. Mediter.13, 120-125.

Weber, M.I. 1990. Forest soil respiration after cutting and burning in immature aspen ecosystem. Forest Ecol. Manage. 31, 1-14.

J.G.Goldammer and V.V.Furyaev (eds.), Fire in Ecosystems of Boreal Eurasia, 337-349.

Soil Microbial Biomass: Determination and Reaction to Burning and Ash Fertilization

J. Pietikäinen and H. Fritze [1]

1. Introduction

Soil microbes and invertebrates are essential for nutrient cycling, and microbial biomass in soil determines the rate of organic matter turnover. Although soil microbial biomass carbon constitutes only a small percentage of soil organic carbon, e.g. 1.19% in coniferous forest soil (Martikainen and Palojärvi 1990), the contribution of soil microbes to nutrient mineralization is notable. Living trees require a constant supply of mineral nutrients which are formed in the decomposition process carried out by microbes. Consequently, microbes play a major role in silviculture and timber production. In boreal coniferous forests, litter tends to accumulate on the forest floor indicating slow microbial litter decomposition. The rate of decomposition is not limited by the amount of soil organic carbon, which is present in large amounts; in fact the quality of organic matter is more important to decomposers than the quantity. A significant proportion of the soil organic carbon is recalcitrant, and microbial biomass may be limited by the forms of organic carbon present rather than by the absolute amount (Wardle 1992).

Fire reduces the thick humus layer and liberates the nutrients bound to litter and organic layers. Fire also changes abiotic factors including soil moisture, temperature, aeration, light intensity, pH, amount of extractable nutrients in the soil solution, and the quality of soil organic matter. Forest fires were frequent before the modern fire suppression policy was pursued in Finland. Zackrisson (1977) studied the frequency of forest fires in northern Sweden within the boreal forest ecosystem and concluded from fire scars in living and dead trees that, before fire suppression started, forest fires had occurred at mean intervals of 80 years. This is probably valid also for Finland and Russian Karelia, as the climate, bedrock, and tree species of these areas are very similar. Forest fires influenced both the structure and composition of forest vegetation. Fire represented a natural rejuvenating factor in boreal forests and enabled a new start in vegetation succession.

In modern silviculture, wildfires in forests are not desirable, but the beneficial effects of fire are utilized in prescribed burning. After clear-cutting and timber harvesting, the remaining logging slash and the upper layer of the forest floor are burned under controlled conditions. The burned area is reforested by either sowing or planting new seedlings. It must be stated here that neither burning-over of woodland for cultivation nor periodic prescribed burning under living stands is used in Finland. When prescribed burning is used, it is always

[1] Finnish Forest Research Institute, Department of Forest Ecology, FIN-01301 Vantaa

carried out after clearcutting a forested site. Changes in environmental factors influence soil microbial biomass (MB), species composition and activity of MB. Quantification of MB is important in studying the effect of fire or other disturbances on soil. MB can be measured by direct or indirect methods. Direct methods include microscopy and culturing in the laboratory; indirect methods are based on measuring some known cellular component, e.g. biomass carbon or nitrogen, ATP, muramic acid, ergosterol, or are based on metabolic activity such as substrate-induced respiration (SIR). We will present here our preferred methods, which are used in our studies, reviewed in the following sections. More detailed information about these and other methods for measuring MB and their applications can be obtained from the excellent reviews by Tunlid and White (1992), Gray (1990), and Frankland et al. (1990).

2. Some Methods Used to Measure Microbial Biomass

Chloroform fumigation extraction method

The chloroform fumigation extraction method (CFEM) is based on the biocidal effect of $CHCl_3$ fumigation, followed by immediate extraction in 0.5 M K_2SO_4 and measurement of extracted nitrogen (Brookes et al. 1985) or carbon (Vance et al. 1987). CFEM has been shown to be applicable to acid forest soils with high organic matter content (Martikainen and Palojärvi 1990). Diaz-Raviña et al. (1992) have shown that CFEM can also be used for heated soils. A coefficient (k_{EC}) is needed in order to transform the amount of extracted C to microbial biomass carbon (C_{mic}). The k_{EC} value is a measure of the efficiency of the extraction procedure and can be determined by calibration against direct microscopic counting (Martikainen and Palojärvi 1990), SIR (Sparling et al. 1990) or by ^{14}C-labelling of soil (Sparling and West 1988). The k_{EC} value chosen is a major source of variation in CFEM. Ross (1990) suggests initial calibration of CFEM for individual soils. We have calculated our microbial C and microbial N values according to the linear regression model presented by Martikainen and Palojärvi (1990).

Substrate-induced respiration

SIR is a method originally described by Anderson and Domsch (1978) used to determine soil microbial biomass. It is based on the assumption that soil microbes are carbon limited. When sufficient amounts of glucose are added to soil samples, the initial rate of carbon dioxide production is related to the biomass present. Sieved soil samples are weighed into flasks, supplemented with glucose and the CO_2 produced is measured. Automated devices enable the handling and measurement of large numbers of samples (Heinemeyer et al. 1989). The weakness of SIR is that it does not take into account anaerobes, microbes in the resting stage, or non-glucose utilizers. Anderson and Domsch (1978) have defined a conversion factor for converting the measured CO_2 production to microbial biomass carbon. As incubation temperature affects respiration, a standard temperature of 22°C should be used. Calibration must be redone for other incubation temperatures and for different soil types.

Fungal biomass

We have estimated fungal biomass both by fungal hyphal length, and biochemically by ergosterol content. The length of fungal hyphae is determined by direct microscopic measurement. The sample is prepared on a membrane filter (Hanssen et al. 1974) which is mounted on a slide and placed under the microscope. The hyphal length is determined by counting the intersections of a graticulated eyepiece and the hyphae (Olson 1950). Counting is rather laborious and all samples must be handled by the same person in order to get comparable results. These limitations have led to the development of chemical methods for estimating fungal biomass. We have chosen to measure the ergosterol content of soil, although other compounds specific to fungi, e.g. chitin or certain fatty acids, can also be used.

Ergosterol is the predominant sterol in most fungi (Tunlid and White 1992). It is a membrane component and its amount in soil has been shown to correlate with fungal surface area in arable and grassland soils (West et al. 1987). The method for isolating ergosterol was developed by Grant and West (1986). In brief, ergosterol is extracted from soil with methanol, the extract treated with alkali, which makes other lipids than sterols hydrophilic, and thereafter sterols can be re-extracted with hexane, as described in detail, e.g. by Fritze and Bååth (1993). Ergosterol is analyzed by high-pressure liquid chromatography using UV detection, as ergosterol has a characteristic UV absorption maximum at 282 nm separating it from other sterols.

3. Microbial Responses to Burning

Reduction of microbial biomass in soil

The direct effect of burning on soil microflora is clearly detrimental as the heat destroys microorganisms during combustion. Generally, bacteria are more resistant to heat than fungi (Dunn et al. 1985). The number of microorganisms drops immediately after burning. The time needed for recovery of MB depends on the intensity of the fire, rainfall, climate and types of microorganisms involved. In the tropics and subtropics, the time needed for recovery may be only a few days or weeks (Meiklejohn 1955; Tiwari and Rai 1977; Theodorou and Bowen 1982) compared to temperate regions, where recovery may take years (Viro 1969).

Changes in soil chemistry

The indirect effect of burning on soil organisms is via the changes in environmental conditions, e.g. higher pH, temperature and cation concentration or lower moisture content. The decrease in soil acidity due to fire has been reported in numerous studies, e.g. Eneroth (1928), Austin and Baisinger (1955), and Dyrness et al. (1989). According to Viro (1974), the immediate decline in the acidity of the humus layer after prescribed burning was as much as 2.4 pH units in coniferous forest soils in Finland. Fire liberates mineral nutrients from plant tissue, debris and logging slash, which results in a high concentration of basic cations in ash. Although a considerable amount of plant nutrients may be lost to the atmosphere during combustion, the remaining nutrients are deposited in the ash layer. The high concentration of nutrients in soil after a fire is documented for many different ecosystems (Kivekäs 1939; Austin and Baisinger 1955; Smith and James 1978; Kutiel and Naveh 1987; Macadam 1987).

Short-term effects of a prescribed burning and a forest fire

We carried out an intensive three-year study on the short term effects of burning on humus layer microbiology in boreal coniferous forests in Finland (Pietikäinen and Fritze 1993). Two types of burning treatments where involved in our study: 1) a typical prescribed fire after clear-cutting, and 2) a forest fire simulation in a 90-year old Scots pine (*Pinus sylvestris*) - Norway spruce (*Picea abies*) stand. The prescribed burning (PB) was a more intensive treatment than the forest fire simulation (FF) and consequently it caused more severe environmental changes: the pH rose by two units and concentrations of calcium, magnesium, potassium, soluble carbon and nitrogen increased two to three fold. Similar but significantly smaller changes in soil chemistry after FF were detected. Both fire treatments resulted in reduced soil microbial biomass carbon (C_{mic}) and fungal hyphal length, determined by CFEM and the membrane-filter method, respectively (Figs.1,2). Neither C_{mic} nor fungal hyphal length showed any recovery towards the control values during the three-year study after the fire treatments.

Activity of soil MB was measured as CO_2 evolution in field moist soil, under laboratory conditions. The respiratory activity of soils from both treatments was reduced compared to that of untreated controls (Figs.1,2). This is consistent with Weber (1990), who found temporary declines in CO_2 evolution in situ after separate clear-cutting and surface burning treatments in aspen ecosystems. However, when we measured the rate of needle litter decomposition in the field by the litter bag method described by Pietikäinen and Fritze (1993), the results indicated the contrary: decomposition was more efficient during the first year of recovery in the PB plots than in the control (Fig.3). To explain this difference, we measured the respiration of soils adjusted to 60 % of water holding capacity (WHC), which caused the CO_2 evolution rate of PB soils to increase to approximately 12 times greater than the controls. This showed that the PB soils had a high capacity for decomposition, providing that soil moisture was sufficient. Decomposition of needle litter in PB soils had obviously taken place in the spring when soil was still moist from snow melt and after heavy rains. In the FF area, insufficient soil moisture was not a problem as the soils did not show any respiratory response to elevated moisture content. Needle litter mass loss was lower in the FF area compared to its control (Fig.3).

Long term effects of prescribed burning

We observed that the MB in coniferous forest soil could not recover from burning over three growing seasons. The question arose as to how much time would be needed for complete recovery of MB. This was answered by a time series study which covered prescribed burned sites, ranging in age from 0 to 45 years in southern Finland (Fritze et al. 1994a). C_{mic} (Fig.4) and fungal biomass, estimated by CFEM and ergosterol content, respectively, both showed an initial decline in the youngest plots, but increased within twelve years to the values usually found in a mature forest. Within the same period of time, the pH value of soil had fallen from an initial 6.0 to 4.6. The increase in C_{mic} expressed per gram of soil organic matter (SOM) means that the same amount of SOM can support an increasing number of microbes. This is probably made possible by the emergence of easily decomposable carbon in the form of herbaceous litter and roots.

Basal respiration was measured in soils adjusted to 60 % of WHC. The respiratory values were converted to metabolic quotient (qCO_2) which is defined as CO_2 evolved per unit of microbial biomass carbon in one hour and it describes the specific respiratory activity of MB (Insam and Haselwandter 1989). The youngest sites, representing 0 and 1 years since PB, showed the highest qCO_2 values; the sites ranging in age between 2 and 39 years after PB had a rather stable qCO_2, but there was a sharp decrease in qCO_2 in the oldest sites, 43 and 45 years after PB (Fig.4). Immediately after fire, MB is low but metabolically active and qCO_2 is high although decreases with time. This is consistent with the hypothesis that ecosystem succession is accompanied by a decrease in the metabolic quotient (Insam and Haselwandter 1989).

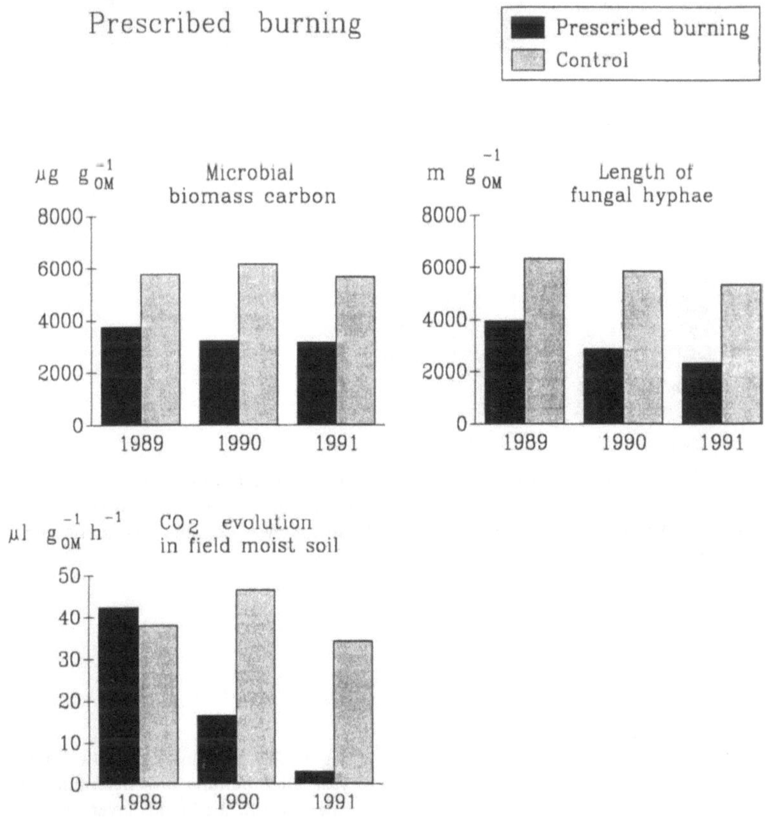

Fig.1. Microbial biomass carbon, length of fungal hyphae and basal respiration in field moist soil measured during three growing seasons after prescribed burning. Burning was carried out in May 1989, and, in the first summer after the fire all measurements were taken six times. In 1990, biomass carbon and basal respiration were measured four times, and in 1991 twice. Fungal hyphal length was measured twice in 1990 and once in 1991. The burned and control areas each consisted of five separate plots, of which the yearly means are presented.

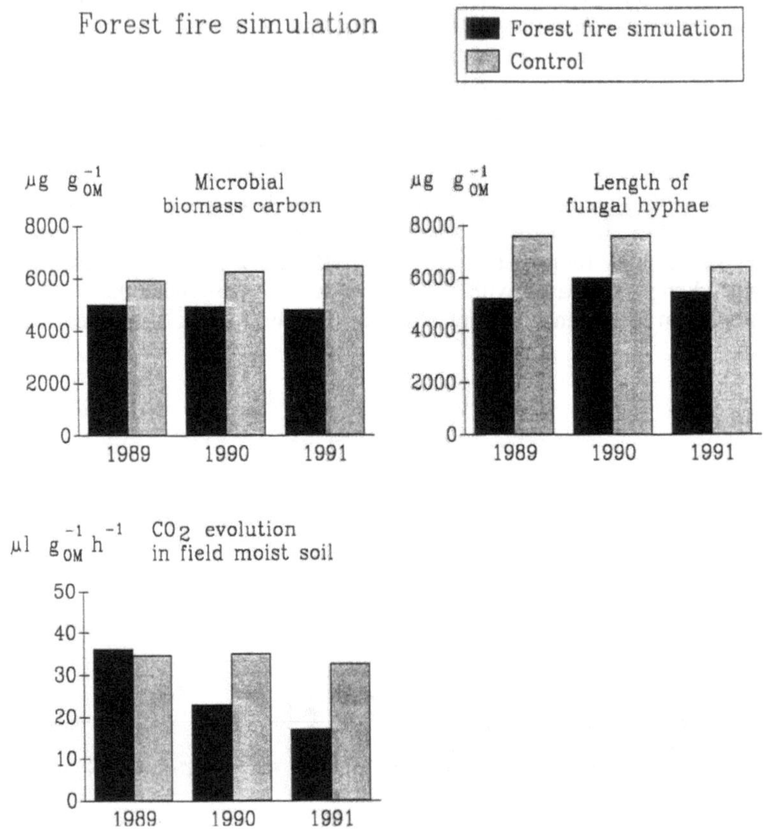

Fig.2. The forest fire simulation was carried out in a living Scots pine stand in June, 1989. The microbiological measurements were done only four times in the first year after burning compared to six times in the prescribed burned site, because of the later date of burning. Otherwise all the measurements were done identically and simultaneously with the prescribed burned site described in Figure 1.

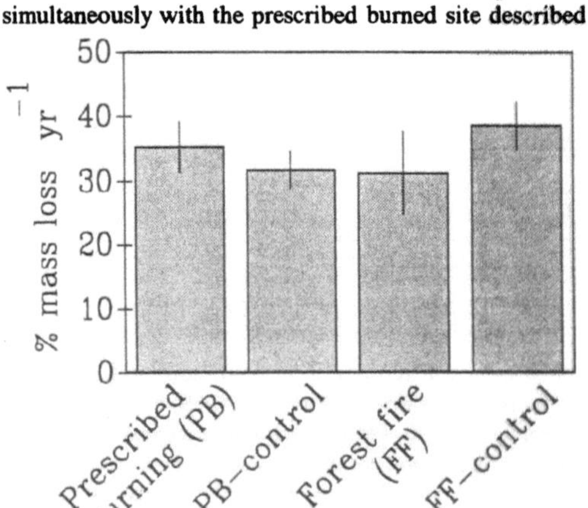

Fig.3. The mass loss of needle litter in the first year after burning was determined by burying litter bags in the humus layer. The mass loss is expressed as percentage of the original needle mass. Prescribed burning (PB) and forest fire simulation (FF) areas both had an untreated control nearby, labelled PB-control and FF-control, respectively. 100 litter bags were buried in each area. The bars indicate the standard error of the mean.

Effects of wood ash fertilization

The effect of PB on soil is the combination of: 1) clear-cutting; 2) heat treatment; and 3) ash deposition. The effects of PB and ash fertilization were compared in an experiment in a growing Scots pine (*Pinus sylvestris*) stand in central Finland (Fritze et al. 1994a). The effect of clear-cutting could be eliminated by carefully burning only the ground layer of the forest stand and not harming the living trees. Ash fertilization was done one week after the burning on separate plots at three levels (1000, 2500 and 5000 kg ha^{-1}). Soil sampling was done two years after the treatments. None of the ash treatments influenced the amount of C_{mic} determined by CFEM or the fungal biomass determined as ergosterol concentration of soil. Instead, burning lowered C_{mic} and fungal biomass to approximately half of the non-treated reference values.

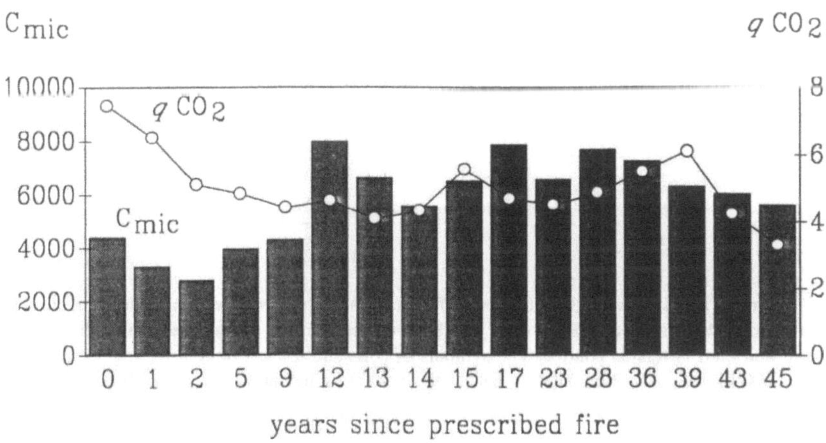

Fig.4. Soil microbial biomass C_{mic} ($\mu g\ g_{OM}^{-1}$) and its specific respiratory activity qCO_2 ($\mu l\ CO_2\ mg\ C_{mic}^{-1}\ h^{-1}$) in a series of individual soils representing time after fire from 0 to 45 years. The values presented are means of two samplings, which were done in June and August, 1991.

In addition, C_{mic} was measured by the SIR method, which takes into account only the metabolically active part of the population. Both CFEM and SIR gave almost identical results in untreated soils, but an increased application of ash resulted in increased C_{mic} measured by SIR, although C_{mic} measured by CFEM remained constant (Fig.5). However, the increasing trend in C_{mic} is statistically not significant. We claim that the favorable soil conditions (higher pH and base saturation ratio) in ash-fertilized plots shift a number of dormant microbes into a metabolically active state. Respiratory activity of soils adjusted to 60 % of WHC showed a similar increasing trend over the range of treatments from control to ash dose of 5000 kg ha^{-1}. Here the difference between the high and the low end was statistically significant. The corresponding respiratory activity in the PB soils did not differ from control soil indicating that burning did not stimulate the activity of soil microbes even in the presence of sufficient soil moisture.

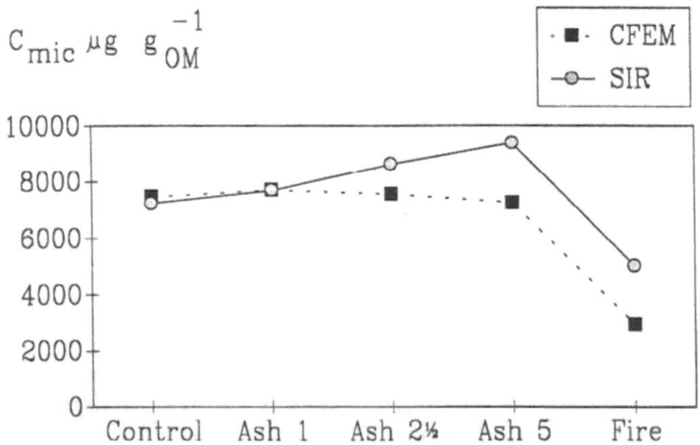

Fig.5. Soil microbial biomass determined by two different methods: chloroform fumigation extraction (CFEM) and substrate-induced respiration (SIR) in soils which have been burned, or fertilized with increasing amounts of wood ash, and their untreated control. The symbols of soils and the ash treatments are: Control: untreated soil, Ash 1: 1000 kg ha^{-1}, Ash 2½: 2500 kg ha^{-1}, Ash 5: 5000 kg ha^{-1} and Fire: prescribed burning.

Nitrification following burning

After PB , the majority of inorganic nitrogen exists as ammonium-ion (NH_4^+) because: 1) the heat of combustion chemically converts soil nitrogen into NH_4^+ (Dunn et al 1979); and 2) ammonification increases as a result of optimal conditions after fire. The liberated ammonium can be used by microbes in the nitrification process. This increase in nitrification after burning was observed as early as the 1930s (Kivekäs 1939; Fowells and Stephenson 1934). In a large study by Viro (1969), which included 92 PB sites, the NH_4^+ concentration of the humus layer was higher than in unburned controls for two years after fire, and then declined to values 70 % of the control. The nitrate content of the humus increased immediately after fire to values 150 % higher than the control. Three years after burning, the NO_3^- concentration started to fall, reaching the lowest values 13-19 years after burning (Viro 1969).

We studied nitrogen turnover in soils described by Fritze et al. (1993). A time series of soils from 0 to 45 years after burning was used. The measured NH_4^+ concentration in soils (expressed on an organic matter basis, OM) was highest in the summer of burning, declining to a stable level within five years. The NO_3^- concentration in soils varied less and showed no clear trend with respect to time elapsed since burning. We tested the nitrification and ammonification capacity of these soils under laboratory conditions (pot incubation for 4 weeks at +14°C). We observed an extremely strong nitrification capacity in the youngest soils which had been burned 1 or 2 years earlier. During incubation, the nitrate concentration of these soils, sampled in August, increased 1000 % and 600 %, respectively (Fig.6). In the field, the nitrate formed is either taken up by vegetation or leached. Leaching was confirmed by Viro (1969), who found increasing nitrate concentrations in all mineral soil layers (from 0 to 30 cm) after burning. This indicated that nitrification had occurred in the humus layer and nitrate had been leached out.

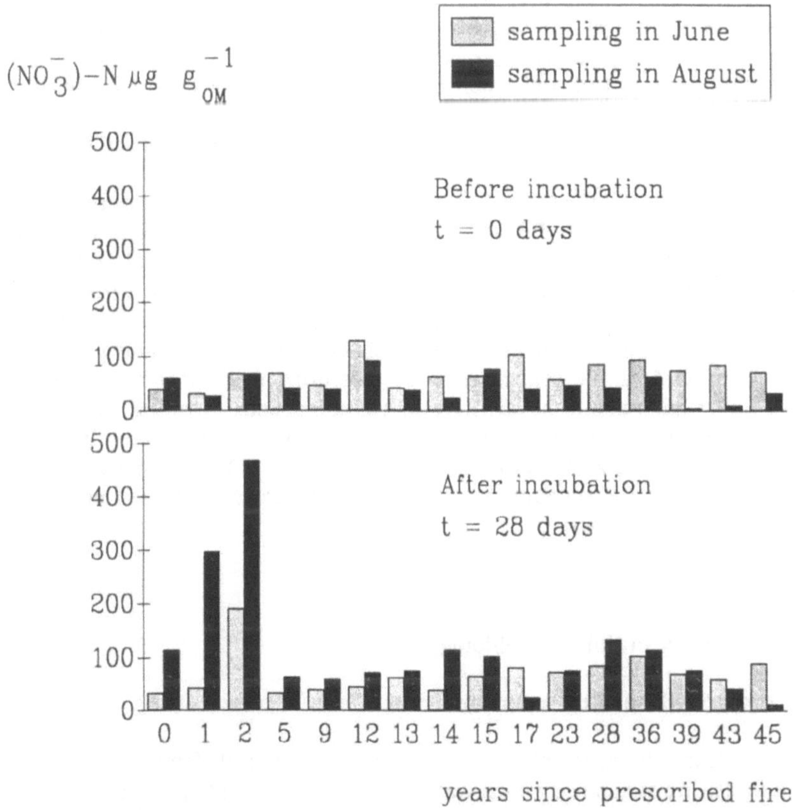

Fig.6. The nitrate nitrogen (NO_3^-) -N concentration in soil after and before incubation at +14°C for 4 weeks. The soils studied are the same as in Figure 4. The sites were sampled in 1991, first in June and then in August and the results of the two samplings are presented separately.

At very low ammonium concentration, nitrifying bacteria are blocked as many other organisms compete for the sparse substrate. In the soils burned 1 or 2 years previously, where nitrification was observed, the initial NH_4^+ concentration was from 289 to 735 μg g_{OM}^{-1}, whereas in the ash fertilization experiment, NH_4^+ concentrations varied between 22 and 32 μg g_{OM}^{-1} and consequently, no nitrate was observed before or after incubation in any of the soils studied (Fritze et al. 1994a).

Some reasons for the slow recovery of MB after fire

The maximum soil temperatures during PB have been shown to vary over 690-870°C in Finland (Vasander and Lindholm 1985). Although it is evident that the heat of fire kills a major part of soil MB, in most cases, sufficient numbers of propagules remain alive in mineral soil and can reinoculate the ash/humus layer. Figures 1 and 4 show that C_{mic} decreases immediately after fire, as expected, but it takes as long as five years before C_{mic} starts to recover. This slow recovery cannot be caused only by the heat damage; instead some environmental factors probably play a major role in the phenomenon.

Söderström (1979) has shown that soil fungal biomass and soil moisture content are positively correlated in the humus layer. The moisture content of the burned soil was less than half of that of the control soil, whereas the ash-fertilized soils and the control soil had equal moisture contents (Fritze et al. 1994a). This indicates that the recovery of fungal biomass to control levels is not expected before the soil moisture level becomes more favourable.

Another major factor influencing the well-being of soil microflora is the available substrate. In the ash-fertilized, PB and control soils, the concentrations of organic and soluble carbon were the same two years after the treatments (Fritze et al. 1994a) thus, the MB in the PB soil is not restricted by the absolute amount of soil organic carbon but instead, we believe, by the quality of SOM controlling the microbial population. Burning not only converts organic carbon into carbon dioxide but also modifies the remaining organic carbon compounds. The SOM can be characterized by measuring the near-infrared spectra (NIR) which gives information about the structure of the compounds (Socrates 1980) in the soil. We measured the NIR-spectra of the previously mentioned ash-fertilized, PB and control soils in order to test the correlation between the quality of SOM and the quantity of MB. Our first results from NIR spectroscopy show clear differences between control and fire treatments (Fritze et al. 1994b).

4. Our Future Work in Soil Microbiology

Effects of prescribed burning separated from effects of clear-cutting

The effects of clear-cutting, together with prescribed burning, differed drastically from the effects of a forest fire simulation which was carried out in a standing forest (Pietikäinen and Fritze 1993). A major reason for this difference may be the effect of clear-cutting, not the differences in fire intensity. Clear-cutting is the silvicultural practice where timber is removed from a large area. It causes three major changes in forest soil chemical and physical factors: 1) the annual litter fall ceases, but large amounts of logging slash is deposited on the forest floor; 2) the roots of the felled trees die and no more root exudates are secreted; and 3) the microclimate of the site changes as evaporation increases and the range of soil temperature becomes wider (Bååth 1980). Sundman et al. (1978) found increases in both biomass and activity of soil fauna and bacteria after clear-cutting in Finland, but, on the other hand, Bååth (1980) reported that clear-cutting resulted in the decrease of total and fluorescein diacetate-active fungal hyphae in central Sweden. Because most studies dealing with PB do not separate the effects of fire from the effects of clear-cutting, our future work will include a combined study of clear-cutting, PB and a growing forest stand in a randomized block experiment (see Pietikäinen and Fritze 1995).

Microbial community structure

Biomass measurement gives only a rough picture of the soil microbial community. In environmental microbiology, the determination of species composition is problematic as it requires culturing which is successful only for a minor proportion of soil microflora. At present, shortcut methods can be used to reveal the community structure of soil microbes. Phospholipid fatty acids (PLFA) present in all living cells can be used to study community composition in environmental samples (Tunlid and White 1992). Phospholipids extracted from soil form a specific signature of the community of microorganisms present. In the soils described by Fritze et al. (1994a) a different pattern of soil phospholipid fatty acids was observed in the fire-treated and ash-fertilized soils, reflecting a treatment-related change in community structure (see Bååt et al. 1995). PLFAs will be a future tool in our laboratory which might reveal more about individual species in soil.

Acknowledgements

We thank Taina Pennanen for providing the nitrification results. The translation into English was revised by Judith Hammond.

References

Anderson, J.P.E., and K.H. Domsch. 1978. A physiological method for the quantitative measurement of microbial biomass in soils. Soil Biol. Biochem. 10, 215-221.

Austin, R.C., and D.H. Baisinger. 1955. Some effects of burning on forest soils of Western Oregon and Washington. J. For. 53, 275-280.

Brookes, P.C., A. Landman, G. Pruden, and D.S. Jenkinson. 1985. Chloroform fumigation and the release of soil nitrogen: a rapid direct extraction method to measure microbial biomass nitrogen in soil. Soil Biol. Biochem. 17, 837-842.

Bååth, E. 1980. Soil fungal biomass after clear-cutting of a pine forest in Central Sweden. Soil Biol. Biochem. 12, 495-500.

Bååt, E., Å. Frostegård, T. Pennanen, and H. Fritze. 1995. Microbial community structure and pH response in relation to soil organic matter quality in wood-ash fertilized, clear-cut or burned coniferous forest soils. Soil Biol. Biochem. 27, 229-240.

Diaz-Raviña, M., A. Prieto, M.J. Acea, and T. Carballas. 1992. Fumigation-extraction method to estimate microbial biomass in heated soils. Soil Biol. Biochem. 24, 259-264.

Dunn, P.H., S.C. Barro, and M. Poth. 1985. Soil moisture affects survival of microorganisms in heated chaparral soil. Soil Biol. Biochem. 17, 143-148.

Dunn, P.H., L.F. DeBano, and G.E. Eberlein. 1979. Effects of burning on chaparral soils: II Soil microbes and nitrogen mineralization. Soil Sci. Soc. Am. J. 43, 509-514.

Dyrness, C.T., K. Van Cleve, and J.D. Levison. 1989. The effect of wildfire on soil chemistry in four forest types in interior Alaska. Can. J. For. Res. 19, 1389-1396.

Eneroth, O. 1928. Bidrag till kännedomen om hyggesbränningens inverkan på marken. Svenska skogsvårdsföreningens tidskrift 26, 685-758.

Fowells, H.A., and R.E. Stephenson. 1934. Effect of burning on forest soils. Soil Sci. 38, 175-181.

Frankland, J.C., J. Dighton, and L. Boddy. 1990. Methods for studying fungi in soil and forest litter. In: Methods in Microbiology, Vol. 22 (R. Grigorova and J.R. Norris, eds.) 343-404. Academic Press, London.

Fritze, H., and E. Bååth. 1993. Microfungal species composition and fungal biomass in a coniferous forest soil polluted by alkaline deposition. Microb. Ecol. 25, 83-92.

Fritze, H., T. Pennanen, and J. Pietikäinen. 1993. Recovery of soil microbial biomass and activity from prescribed burning. Can. J. For. Res. 23, 1286-1290.

Fritze, H., A. Smolander, T. Levula, V. Kitunen, and E. Mälkönen. 1994a. Wood ash fertilization and fire treatments in a Scots pine forest stand: effects on the organic layer, microbial biomass and microbial activity. Biol. Fert. Soils. 17, 57-63.

Fritze, H., P. Järvinen, and R. Hiukka. 1994b. Near-infrared characteristics of forest humus are correalated with soil respiration and microbial biomass in burnt soil. Biol. Fert. Soils. 18, 80-82.

Grant, W.G., and A.W. West. 1986. Measurement of ergosterol, diaminopimelic acid and glucosamine in soil: evaluation as indicators of microbial biomass. J. Mic. Meth. 6, 47-53.

Gray, T.R.G. 1990. Methods for studying the microbial ecology of soil. In: Methods in Microbiology, Vol. 22 (R. Grigorova and J.R. Norris, eds.), 309-342. Academic Press, London.

Hanssen, J.F., T.F. Thingstad, and J. Goksøyr. 1974. Evaluation of hyphal lengths and fungal biomass in soil by a membrane filter technique. Oikos 25, 102-107.

Heinemeyer, O., H. Insam, E.A. Kaiser, and G. Walenzik. 1989. Soil microbial biomass and respiration measurements: an automated technique on infra-red gas analysis. Plant Soil 116, 191-195.

Insam, H., and K. Haselwandter. 1989. Metabolic quotient of the soil microflora in relation to plant succession. Oecologia 79, 174-178.

Kivekäs, J. 1939. Kaskiviljelyksen vaikutus eräisiin maan ominaisuuksiin. Commun. Inst. Forest. Fenn. 27.2

Kutiel, P., and Z. Naveh. 1987. The effect of fire on nutrients in a pine forest soil. Plant Soil 104, 269-274.

Macadam, A.M. 1987. Effects of broadcast slash burning on fuels and soil chemical properties in the sub-boreal spruce zone of central British Columbia. Can. J. For. Res. 17, 1577-1584.

Martikainen, P.J., and A. Palojärvi. 1990. Evaluation of the fumigation-extraction method for the determination of microbial C and N in a range of forest soils. Soil Biol. Biochem. 22, 797-802.

Meiklejohn, J. 1955. The effect of bush burning on the microflora of a Kenya upland soil. J. Soil Sci. 6, 111-118.

Olson, F.C. 1950. Quantitative estimates of filamentous algae. Trans. Am. Microsc. Soc. 69, 272-279.

Pietikäinen, J., and H. Fritze. 1993. Microbial biomass and activity in the humus layer following burning: Short term effects of two different fires. Can. J. For. Res. 23, 1275-1285.

Pietikäinen, J., and H. Fritze. 1995. Clear-cutting and prescribed burning in coniferous forest: comparison of effects on soil fungal and total microbial biomass, respiration activity and nitrification. Soil Biol. Biochem. 27, 101-109.

Ross, D.J. 1990. Estimation of soil microbial C by a fumigation-extraction method: influence of seasons, soils and calibration with the fumigation-incubation procedure. Soil Biol. Biochem. 22, 295-300.

Smith, D.W., and T.D. James. 1978. Characteristics of prescribed burns and resultant short-term environmental changes in Populus tremuloides woodland in southern Ontario. Can. J. Bot. 56, 1782-1791.

Socrates, G. 1980. Infrared characteristic group frequencies. John Wiley & Sons, Chichester.

Sparling, G.P., C.W. Feltham, J. Reynolds, A.W. West, and P. Singleton. 1990. Estimations of soil microbial C by a fumigation-extraction method: use on soils of high organic matter content and a reassessment of the k_{ec}-factor. Soil Biol. Biochem. 22, 301-307.

Sparling, G.P., and A.W. West. 1988. A direct extraction method to estimate soil microbial C: calibration in situ using microbial respiration and ^{14}C labelled soils. Soil Biol. Biochem. 20, 337-343.

Sundman, V., V. Huhta, and S. Niemelä. 1978. Biological changes in northern spruce forest soil after clear-cutting. Soil Biol. Biochem. 10, 393-397.

Söderström, B.E. 1979. Seasonal fluctuations of active fungal biomass in horizons of a podzolized pine forest soil in Central Sweden. Soil Biol. Biochem. 11, 149-154.

Theodorou, C., and G.D. Bowen. 1982. Effects of a bushfire on the microbiology of a South Australian low open (dry sclerophyll) forest soil. Aust. For. Res. 12, 317-327.

Tiwari, V.K., and B. Rai. 1977. Effect of soil burning on microfungi. Plant Soil 47, 693-697.

Tunlid, A., and D.C. White. 1992. Biochemical analysis of biomass, community structure, nutritional status, and metabolic activity of microbial communities in soil. In: Soil Biochemistry, Vol. 7 (G. Stotzky and J.-M. Bollag, eds.), 229-262. Marcel Dekker, Inc., New York.

Vance, E.D., P.C. Brookes, and D.S. Jenkinson. 1987. An extraction method for measuring soil microbial biomass C. Soil Biol. Biochem. 19, 703-707.

Vasander, H., and T. Lindholm. 1985. Fire intensities and surface temperatures during prescribed burning. Silva Fenn. 19, 1-15.

Viro, P.J. 1969. Prescribed burning in forestry. Commun. Inst. Forest. Fenn. 67.7.

Viro, P.J. 1974. Effects of forest fire on soil. In: Fire and ecosystems (T.T. Kozlowski and C.E. Ahlgren, eds.), 7-45. Academic Press. New York.

Wardle, D.A. 1992. A comparative assessment of factors which influence microbial biomass carbon and nitrogen levels in soil. Biol. Rev. 67, 321-358.

Weber, M.G. 1990. Forest soil respiration after cutting and burning in immature aspen ecosystems. For. Ecol. Manage. 31, 1-14.

West, A.W, W.D. Grant, and G.P. Sparling. 1987. Use of ergosterol, diaminopimelic acid and glucosamine contents of soils to monitor changes in microbial populations. Soil Biol. Biochem. 19, 607-612.

Zackrisson, O. 1977. Influence of forest fires on the North Swedish boreal forest. Oikos 29, 22-32.

J.G.Goldammer and V.V.Furyaev (eds.), Fire in Ecosystems of Boreal Eurasia, 350-357.

Ecological Effects of Peat Fire on Forested Bog Ecosystems

T.T. Yefremova and S.P. Yefremov [1]

1. Introduction

Persistent outbreaks of forest fire within large territories of the taiga zone are now considered an important factor influencing forest floor dynamics, structure, formation and productivity of current forest ecosystems (Volokitina 1985; Melekhov 1948; Sannikov 1981, 1983; Tkachenko 1939; Furyayev 1988) and soil formation (Popova 1975; Sapozhnikov 1976).

Peatlands and bog forests are landscape units of the taiga bioclimatic zone, which are also subjected to fire. However, only few studies are available on the ecological and ontogenetic effects of fire on the state and productivity of peatland ecosystems. Data on bog fire traces were discussed mainly in terms of horizon boundary (Kozlovskaya et al. 1978; Tyuremnov 1956). The role of fire in the development of pine stands on oligotrophic peatlands has been mentioned by Pyavchenko (1978).

Fire-related processes in peatlands have been studied mainly in regard to damage and financial losses. Two types of peat fires have so far been considered: ground fire and underground fire (Volokitina 1985; Kurbatsky et al. 1957; Sofronov and Volokitina 1986).

The aim of the present study is to estimate exogenic influence of fire on peatland ecosystems: (1) to study fire effects on the characteristics of peat substrate as a root medium, and (2) to study the reaction of different root types of forested bog ecosystems to fire as a stress factor within the range of stand ontogenesis.

2. Material and Methods

The study has been carried out in the south taiga subzone of western Siberia on oligotrophic and mesotrophic peatlands, consisting of *Sphagnum*- and *Eriophorum*-peat. Six peatland areas were investigated in the territory of Tomsk region. Influencing each other, they form the bog macro-landscape which is a characteristic feature of the area between the Ob and Tom rivers. The detailed complex study covered 17 sites characterizing the ecological variation of Scots pine (*Pinus sylvestris* L.) communities of the hydromorphic series.

These groups of pine associations are also characterized by representatives of the genera *Carex, Eriophorum, Oxycoccus, Menyanthes, Comarum, Rubus, Vaccinium, Ledum,* Betula,

---

[1] V.N. Sukachev Institute for Forestry and Timber, Siberian Branch, Russian Academy of Sciences, 660036 Krasnoyarsk, Russian Federation

Chamaedaphne, Andromeda, Sphagnum, Aulacomnium, Tomenthypnum, Calliergon, Mnium, Helodium, Dicranum, Drepanocladus, Pleurozium, Ptilium and also a few occasional inclusions of other botanical families.

All of the bogs had been subjected to fire, and some of them were burnt several times: in 1712, 1738, 1833, 1867, 1918 and 1950. This information was obtained from the study of annual rings of fire-damaged trees (fire scars) and by the analysis of the age structure of the stands. Fire dates were also obtained by computing peat accumulation rates (Tab.1). In one of the sites, the depth of peat accumulated between 1984 and the last fire in 1959 was 50 mm. Thus, average annual peat accumulation during the last 34 years amounted to 1.47 mm. Radiocarbon dating provides accurate data for the bogs, even for previous conditions, and hence can be successfully used when studying fire history. It has been shown that 150-300 years ago, fire frequency was approximately two times less than nowadays. At present, bogs burn once every 30-50 years. This is a consequence of intensive cultivation activities in the area.

Tab.1. Fire dating of peat deposits within the root system zone

Depth of Post-fire Peat Bed Accumulation (mm)	Fire Age (years)		
	Calculated Age - According to the Rate of Peat Growth (mm yr^{-1})		Age - According to Annual Tree Rings
	^{14}C Age	Age - According to Current Records of Fires	
50	37	34	34
100	74	68	66
150	112	102	117
200	149	136	151
300	224	205	246
400	298	272	272

3. Results and Discussion

Peat fires do not cause the complete elimination of the substrate since there is superfluous water in water holding horizons. They do not affect the level of soil and ground waters and, moreover, do not change the alcalinity and the acidity of the ecotone either. Fires do not cause disastrous, irreversible changes to the ontogenesis of peatlands and succession dynamics of the major types of phytocenoses.

Occasional plant groups which are not characteristic of the area can occupy the burned site only for a short period of time. Thus, in the first decade following fire one could find a birch thicket with a willow bush on the burned site. But the population usually does not develop further and dies out due to the build-up of oligotrophic and oligo-mesotrophic peats, i.e. low amount of trophic substrate and high acidity of the medium.

Quasi-indigenous associations, particularly those with a versatile botanic species such as *Pinus sylvestris*, can gradually be restored in the conditions of a natural ecotone. The structural organization of peatland stands appears to be mostly affected by fires. Pine stands growing on burned-over peatlands, especially mature generations, demonstrate a limited range of ages, with relatively big trees of similar age groups and a higher volume prevailing over the trees of control groups. Productivity of pine stands on burned-over peatlands range from 126-178 m^3ha^{-1} compared to 52-97 m^3ha^{-1} in usual conditions. These data are from even-aged stands (120 yr) with DBH of 70 cm. The total aboveground biomass yield is determined by the burn depth of the peat layer, its thickness and distribution.

During the regeneration process, the growth of the first post-fire generation of pines, with their roots concentrated in the burned-over substrate, cumulates at the age of ca. 110-150 yr, their relative diameter usually ranging from 63-92 cm and the volume up to 218-300 m^3ha^{-1}. Second and third generations generally yield 2-3 times lower volumes as their root systems cannot get into burnt layers to employ oligotrophic peat.

An important effect of fire on forested bog ecosystems is observed during the redistribution of the function of surface relief. Positive elements generally occupied by woody plants are known to exhibit a certain dryness due to the drainage of mobile waters. Therefore, one can observe their stronger warming, compared to sites with tree groups situated in minor pits or subsoil where soil and ground water get to the surface or near to it. This factor greatly influences the future space and age structure of bog forests through initial regeneration phases of plant successions, including the distribution of shoots, seedlings and young trees.

A number of specific structural characteristics are demonstrated by the burned-over peat horizons, including a unique plasticity of peat. Like clay, it can be freely rolled into a homogeneous mass. During laboratory experiments it has been shown that the level of peat softening is determined by lignin content and the range of temperatures it was subjected to: 150-600°C (Rakovsky et al. 1959). The highest plasticity has been observed to result in 230-320°C. Four phases of fire effects on organic substance have been established depending on the complex of morphologic characteristics and the level of plasticity. The burned-over peat horizons have been denoted by a double index: "T" and additional "B" for describing the horizons of a humus profile with fire traces (Tab.2).

The phases of thermal peat destruction are characterized by additional index with digits. For sphagnum peat layers subject to lowest fire action (TB1) the following peculiarities were observed: charcoal either in the form of single pieces or thin layers (0.5-1.0 cm) of small coal particles in horizontally bound low-decomposed peat. This "fire line" is mostly uneven, of a tongue-like shape, but sometimes nearly horizontal to the surface.

Peat-producing plants bear black or dark-brown scorched spots. One can occasionally see a dark ash-grey coloring in the peat layers of slime soil consistency, with the major part of roots of pines and bog shrubs. There is also abundant whitish-colored coral ectotrophic mycorrhiza.

The thickness of the horizon is generally about 10 cm. Depending on the level of the microrelief differentiation and fire period the peat layer is situated at a depth of 5-30 cm.

Peat horizons with marked fire traces exhibit a great variety of morphologic characteristics. The peat layers show mixed coloring. Black is produced by the charcoal which is sometimes highly dispersed. White-ash grey is due to the ash, while mycorrhiza and birch remains are responsible for the whitish coloring. Red is the color of well-preserved but macerated wood. The brownish color is caused by the decay of peat humus. The main part of the TB2 layer consists of decomposed substrate of dark brown color. Large concentrations of ash result in the ash-grey color of a substrate. Here, peat is of silt consistency and exhibits high plasticity. There are many grain-like well-formed aggregates.

The horizon is often occupied by plant roots, its thickness ranging from 1-10 cm, but more often varying within 5-10 cm. It is situated 5-10 cm lower than the surface of the peat layer. The horizon is often intermediated by the layers providing morphological evidence of minor fire effects.

Tab.2. Data on humus characteristics of burnt-over forest peat soils.

Depth (cm)	Horizon Index	C:N	Cha/Cfa 1)	Humification Level (%) 2)	Bitumen Content (%) 3)	Peat Conservation 4)	Humus Quantity	
							Depth (cm)	t ha^{-1}
	Top Peat Soil of Subshrub-Sphagnum Pine Stands							
0-5	T1	70.8	0.9	23.3	9.6	1.4	-	-
5-20	T2	75.0	1.0	24.7	7.6	1.4	0-20	14.7
20-40	T3	76.8	1.0	23.7	5.7	0.9	20-40	15.8
40-45	T4B2	45.5	1.4	34.5	17.2	2.2	40-60	61.1
45-59	T5B3	56.2	1.7	34.6	12.1	1.7	60-80	86.2
59-66	T6B2	52.0	2.4	42.4	17.7	4.9	-	-
66-80	T7B3	60.3	2.0	42.0	15.5	2.7	-	-
	Transition Peat Soil of Subshrub-Sphagnum-Herbs Pine Stands							
0-5	T1	42.6	1.0	34.0	9.4	2.0	0-5	9.2
5-10	T2B2	31.8	1.4	39.8	12.9	2.5	5-25	118.2
10-20	T3B2	34.9	2.6	44.2	17.4	4.7	-	-
20-30	T4	35.7	3.2	47.2	13.5	5.4	-	-

Notes:
1) Ratio between the carbon content in humic acids to that in fulvic acids
2) Carbon of humus compounds (% of total carbon)
3) Carbon of alcohol-benzol extraction (% of total carbon)
4) Ratio of carbon in humus compounds to the total amount of hydrolized substances 1.0 and 80% HSO.
For B2, B3: see text above.

Peat layers of the third stage of fire destruction (TB3) are of a brown or red-brown color, often due to the inclusion of highly softened tree trunks. The peat is of a medium degree of decay and firmly bound. Single inclusions of coal can be observed. One can also see single spots of a slimy consistency, with a weakly developed grain structure. The remains of peat producing plants are highly macerated but still maintain their original morphological structure. When slightly pressed, the peat easily turns into a homogenous mass of low plasticity. TB3 peat layers are actively occupied by roots. They belong to different depths (20-80 cm) but are generally found below the layers of highly pronounced fire effects (TB2).

Peat layers of the fourth stage of fire destruction (TB4) are mainly of a dark brown color. Here, plant remains, as a rule, fail to retain their morphological structure. Those maintaining the structure are easily destroyed upon slight pressing, turning into a homogeneous mass. The peat is of a slimy consistency, high plasticity, and firmly bound. It can hardly be crumbled, yielding only large fragments. From time to time one could observe single spots of grain structure. The peat has a specific bituminous smell which is strongest at the bottom of the peat layer. The peat is not favored by roots. These are the most powerful layers, up to 60-80 cm. One can find them among the lowest layers of peat, but sometimes they are met at the depth of 40-50 cm. Thus, peat contains unique fire records, preserving fire traces for thousands of years. The historic memory of dry upland sites is comparatively brief, making up only two or three tree generations.

Peat fires significantly change the composition of peat organic matter. In peat fire from the top peat to the transitional peat layer, the level of its humification correspondingly rises from 23.0-24.7 up to 34.5-42.4 and from 34 up to 40.0-47.0 % (Tab.2). So, accumulated humus, taking a layer to equal 20 cm, should yield 61-86 t ha^{-1} in top peat and 118 t ha^{-1} in transitional peat soils. It is 3-6 times as high compared to analogous fire-free peat. The proportion of humic acids in humic matter greatly increases. The Cha:Cfa ratio increases from 0.9 for the peat of natural development to 2-3 for those in the regime of thermal destruction. Active processes of humus formation during burning greatly decrease the values for peat deposits: from 1 to 4.9 (for some horizons) and for transitional peats from 2 to 5.4.

The bitumen content (the product of alcohol-benzole extraction) significantly rises in burned horizons: 12.1-17.7 compared to 5.7-9.6. High plasticity of peat in burnt layers should be accounted for by the increase of bitumen content.

A lower C:N ratio in burnt-over layers can be due to the rise of biologic activity. However, the values for the ratio are fairly low: 45-60 for top peat layers, VS 77 for virgin peat, and 32-36 VS 43 for transitional peat soils. There is a controversy between the high C:N ratio in burned-over horizons and humus type, high content of humus and low content of turf conservation in them. Presumably, weak agreement between data should be accounted for by humus compound formation via chemical mechanisms of fast reaction rates.

The data on the influence of fires on physical and chemical characteristics of peat soils and peat deposits seem ambiguous. The discrepancy is mostly pronounced in peat compaction (Tab.3). The density of burnt layers equals 0.09-0.163 g cm^{-3}. This is 1.5-2.5 times higher than compared to burned-over peat layers. Peat compaction is also known to occur in the course of the natural development of bogs. Its value for oligotrophic and mezotrophic conditions is, however, much lower. Since total moisture is the derivative of density, it also decreases 2 or 3 times (from 1132-3413 to 546-1373 %). The data on porosity decrease to a lesser extent: 95-99 in comparison to 89-96 %. The ash content of burned peat layers is closely connected with fire intensity. In the horizons of intensive burning (TB2), the ash content (5-13 %) is much higher than in boundary peat layers with a natural accumulation of dead mass. In decay horizons the ash content is virtually constant along the whole of the profile (2-4 %).

Peat fires do not change the acidity and alcalinity of the medium. This is confirmed by little disturbance of the pH, hydraulic acidity and the level of base saturation. This phenomenon seems quite natural when considered in more detail. Organic acids of individual and humus origin are known to be the major agents of the acidity of peat soils. It follows from Table 2 and the data available in literature (Yefremova 1988; Rakovsky et al. 1959) that their content in burnt peat layers significantly increases. To average this amount of organic acids and to neutralize the medium one needs many more bases than the amount obtained during the burning of sphagnum peat of low ash content. In the burnt-over horizons no marked increase in the content of biogenous elements is observed. The exceptions are nitrogen and phosphorus, while their amounts increase insignificantly. However, the process is due to some activation of the biological state of peat upon thermal destruction and its enrichment with microbial biomass.

Tab.3. Physical and chemical characteristics of burned-over oligotrophic and oligo-mezotrophic peat deposits

Depth (cm)	Horizon Index	Volume Mass (g cm^{-3})	Ash content (%)	Hydrolitic Acidity (%)	Base Content (%)	Water pH	Total Water Capacity (%)	General Porosity (%)
	Subshrub-Sphagnum Pine Stands							
0-5	T1	0.029	5.7	78.3	19.2	3.9	3412	99
5-20	T2	0.040	7.9	102.3	12.1	3.5	2455	98
20-40	T3	0.040	2.9	106.7	11.5	3.3	2455	98
40-45	T4B2	0.095	5.3	83.9	14.2	3.4	993	94
45-59	T5B3	0.090	2.3	101.4	12.9	3.7	1052	95
59-66	T6B2	0.131	7.3	93.8	9.3	3.8	701	92
66-80	T7B3	0.097	3.2	104.2	9.5	3.8	971	94
	Sedge Fen - Sphagnum - Subshrub Pine Stands							
0-10	T1	0.053	5.1	115.5	13.6	3.5	1836	97
10-20	T2	0.064	7.0	111.9	12.9	3.7	1508	97
20-30	T3B1	0.123	4.5	108.6	14.5	4.1	751	92
30-40	T4B2	0.164	5.6	73.2	19.7	4.2	546	89
40-50	T5B4	0.143	3.2	88.9	12.4	4.2	636	90
50-60	T6B4	0.130	4.2	92.3	12.8	4.1	707	92
60-70	T7B4	0.130	3.0	99.6	12.5	4.2	707	92
70-80	T8B4	0.144	2.9	99.6	13.4	4.1	631	91
80-90	T9B4	0.137	3.0	86.9	13.7	4.4	667	91
90-110	T10B4	0.143	3.7	86.5	13.4	4.2	636	90
	Subshrub - Sphagnum - Herb Pine Stands							
0-5	T1	0.060	8.6	66.2	21.5	3.7	1613	97
5-10	T2B2	0.107	9.4	68.8	21.5	3.9	874	93
10-20	T3B2	0.161	6.6	81.6	15.4	4.1	561	90
20-30	T4	0.131	4.6	87.2	14.3	4.0	701	92

The influence of fire on the changes of substrate trophicity is most fully exhibited in the store of nitrogen and mineral compounds, which is indirect evidence for the significant increase of peat density (Tab.4). There, one could see a great difference between the store of nutrients in burnt layers compared to layers of post-fire genesis, e.g. they change within 2.7-7.5 times for nitrogen, 1.7-3.7 for phosphorus, 0-1.7 for potassium, 0-2.7 for iron, 1.7-3.5 for calcium, 0-4 for sodium, and 1.8-2.8 times for ash substances (these calculations are based on values given in Table 4 and additional data). Intensive ground fires produce the highest effect on the trophicity. This is due not only to substrate compaction but also to ash concentration because of local peat fire. Thus, bog fires caused by natural ecological processes are an important element of peat fire ontogenesis and under their influence productive forest ecosystems can successfully develop even on poor hydromorphic substrates. However, the soil-improving effect of peat fires gradually decreases due to a number of natural factors characteristic of the West Siberia lowland inducing the active formation of bogs and bog peat.

Tab.4. Comparative quantity of nitrogen and mineral elements in naturally formed peat and those in burned-over beds (kg ha^{-1})

Depth (cm)	Horizon Index	Depth of Bed	N	P_2O_5	K_2O	Fe_2O_3	CaO	Na_2O	Ash
	Subshrub - Sphagnum Pine Stands								
0-40	T1-T3	40	976	133	148	521	535	60	7922
40-80	T4B2-T7B3	40	4136	339	157	395	1215	52	16441
	Sedge Fen - Sphagnum - Subshrub Pine Stands								
0-20	T1-T2	20	1291	148	98	433	476	35	7202
20-40	T3B1-T4B2	20	3300	336	160	709	931	57	14678
	Subshrub - Sphagnum - Herb Pine Stands								
0-5	T1	5	369	87	36	189	138	18	2580
5-10	T2B2	5	840	171	48	289	235	27	5029

4. Conclusions

The four stages of thermal destruction of peat caused by different types of fire and their intensity have been studied. Fire factors completely change the composition of peat organic matter and strongly influence its compaction, thus producing the accumulation of ash, humus and biogenous elements in burnt peat layers. Fires activate forest development on peatlands by influencing soil conditions which have an improving effect on the productivity of stands in the first stage of forest regeneration.

References

Bunting, B.T., and J. Lundberg. 1987. The humus profile - concept, class and reality. Geoderma 40 (1-2), 17-36.

Furyayev, V.V. 1988. Analysis of the effect of fire for the estimation of forest formation. Forestry Science 1, 59-66.

Gundar, S.V. 1979. Ground fires in the Nizhni Amur basin, their prophylaxis and extinguishing. Avtoref. dis. cand. selsk. khoz. nauk. Krasnoyarsk. 24 pp. <in Russian>.

Klinga, K., R.N. Green, R.L. Trowbridge, and L.E. Lowe. 1981. Taxonomic classification of humus forms in ecosystems of British Columbia, First approximation. Prov. British Columbia, Min. of Forest, Land Manag. Report 8, 54 pp.

Kozlovskaya, L.S., V.M. Medvedeva, and N.I. Pyavchenko. 1978. The dynamics of organic matter development in the process of peat formation. Nauka, Leningrad. 172 pp. <in Russian>.

Kurbatsky, N.P., N.N. Krasavina, and V.A. Zhdanko. 1957. Forest ground fires and their prophylaxis. LenNIILKH, Leningrad. 32 pp. <in Russian>.

Kurbatsky, N.P. 1964. On the problem of forest fire. In: Forest fire development, p.5-60. Nauka, Moscow <in Russian>.

Melekhov, I.S. 1948. Influence of fires on forest. Goslestekhizdat, Moscow - Leningrad. 124 pp. <in Russian>.

Popova, E.P. 1975. The effect of ground fire on the characteristics of forest soils of Angara area. In: The protection of forest resources of Siberia, 166-168. Inst. Forest and Wood, SD AS USSR, Krasnoyarsk <in Russian>.

Pyavchenko, N.I. 1978. Biogeocenotic regularities of genesis of the bogs and the dynamics of their plant life. In: Genesis and dynamics of the bogs, 13-18. MGU. Moscow <in Russian>.

Rakovsky, V.Ye., F.L. Kagonovich and Ye.A. Novichkova. 1959. Chemical processes during fires. AS BSSR, Minsk. 208 pp. <in Russian>.

Sannikov, S.N. 1981. Forest fire as a factor of structure, development, renovation and evolution of biogeocenosis. Ecology 6, 23-33.

Sannikov, S.N. 1983. The theory of cyclic renovation of *Pinus vulgaris* due to fire-erosion processes. Ecology 1, 10-20.

Sapozhnikov, A.P. 1976. The influence of fire on the formation of forest soils. Ecology 1, 42-45.

Sofronov, M.A. and A.V. Volokitina. 1986. Fires in the bogged forests of West Siberia. In: Hydromorphic forest-bog ecosystems, 139-150. Inst. Forest and Wood, SD AS USSR, Krasnoyarsk <in Russian>.

Tkachenko, M.Ye. 1939. General forest management. Gostekhizdat, Leningrad. 729 pp. <in Russian>.

Tyuremnov, S.N. 1956. On the problem of boundary horizon. Devoted to the academician V.N. Sukachev on his 75 years anniversary, 572-580. AS USSR, Moscow - Leningrad <in Russian>.

Volokitina, A.V. 1985. Fire development of bogged forests of the South of West Siberia. In: Forest fires and their after-effect, 64-73. Inst. Forest and Wood, SD AS USSR, Krasnoyarsk <in Russian>.

Yefremova, T.T. 1988. Humus analysis of the soils of top peat in terms of microrelief. Izvestia SD AS USSR, Ser.Biol.N.I, 38-44 <in Russian>.

J.G.Goldammer and V.V.Furyaev (eds.), Fire in Ecosystems of Boreal Eurasia, 358-365.

Effects of Fire on the Regeneration of Larch Forests in the Lake Baikal Basin

R.M. Babintseva and Ye.V. Titova [1]

1. Introduction

Forest fires play the role of an ecological factor in the process of forming and functioning of forest ecosystems by determining the rates and direction of their stand development. The most fundamental pyrogenic influences are during the initial stages of forest regeneration.

In the forests of the Baikal basin fires are a characteristic phenomenon. Investigations of fire effects on the natural regeneration of *Larix dahurica* were conducted within the bounds of the light coniferous taiga high-altitude complex of forest types (HAC). Its ecological characteristics are: range of elevation 600-1500 m; the annual sum of temperatures above 10°C is 600-1400°C; total evaporation is 300-400 mm; total precipitation is 400-700 mm; and the moistening coefficient by Mezentsev is 0.7-1.3 (Polikarpov et al. 1978). Within these conditions there are wide-spread larch forests of rose bay (*Rhododendron dahurica*), green moss (mainly *Pleurozium schreberi* and *Hylocomium splendens*), dwarf shrub-moss (main shrubs: *Vaccinium vitis-idaea*, *Vaccinium myrtillus*, *Vaccinium uliginosum*), and wild rosemary (*Ledum palustre*) - cowberry (*V. vitis idaea*) groups of forest types. A specific ecological niche is occupied by yernik (*Betula rotundifolia*) - larch forests. These larch forests, with a low, dense understory of mixed shrubs such as willows and birches, are characteristic of the northern treeline.

The system of forest management techniques for this HAC allows industrial harvesting to be forbidden in other high altitudes of the Baikal basin. Therefore, the amount of larch regeneration under tree canopies in unlogged areas is used to estimate the potential for natural regeneration in clearcut areas.

Light coniferous taiga (larch) HAC, on the whole, is characterized by average indices of fire danger. According to the classification of Utkin and Isayev (1962), the potential fire susceptibility of rose bay, cowberry, wild rosemary, sphagnum, and yernik larch forests can be both average and high. However, fires occur in them much more rarely than in pine forests and the great majority are surface fires. Yevdokimenko (1985) notes a comparably low fire danger in the studied forests, associating it with climatic peculiarities of HAC and the dominance of such fire resistant species such as larch.

Larch does not possess less pyrofitness than pine. To a large degree, it is characteristic of those properties which, in Sannikov's (1981) opinion, confirm the pyrofitness of *Pinus*

[1] V.N. Sukachev Institute for Forestry and Timber, Siberian Branch, Russian Academy of Sciences, 660036 Krasnoyarsk, Russian Federation

spp. which successfully cope with the process of natural selection under the effects of fire. They are a heightened fire resistance of mature trees (a thick bark in the basal part and on root branches, depth of the roots, and an elevated open crown), with a rapid growth in young trees, an early and abundant seed productivity, and a good dispersal of seeds.

Periodically repeated ground fires leave their mark on natural patterns of regenerating stands causing stands with similar initial conditions of regeneration to become drastically different in the amount, composition, and age structure of young growth.

2. A Study on Post-Fire Regeneration

A study of regeneration under the shelterwood of larch forests in different areas of the Baikal basin permitted this process to be fully characterized for the whole of the light coniferous taiga HAC. The study followed the methods of Pobedinsky (1965).

In larch forests of the rose bay group of forest types the density of the conifer seedlings and saplings in the understory depend on the time since the last fire, and on the intensity of the last fire, while varying from 1,700 to 90,400 individuals (n) per ha. Larch dominates the young growth in 80% and cedar (*Pinus sibirica*) in 20% of these stands. The age range is 5 to 30 years. Wild rosemary-cowberry larch forests are characterized by smaller amounts of young growth (on the average 5,100 n ha^{-1}) predominantly of larch with a stable trace of cedar and pine, and, on contact with spruce forests, of fir. The maximal young growth of 30 years occurred in cedar. In green moss larch forests, we observed 100 to 9,800 n per ha of young growth. Young growth was mainly large up to the age of 50 years. Cedar and larch are represented in approximately equal amounts (cedar predominates in 45% of the test areas and larch in 45%). The presence of cedar in the overstory and in the young growth, and also the age structure of the latter, allows one to assume that green moss-larch forest types are comparatively seldom subject to fires.

Analogous characteristics have been noted for dwarf shrub-moss larch forests where cedar predominates in 75% of the test areas. Here, density of the young growth varies from 80 to 11,300 n ha^{-1}, and the age range is from 5 to 45 years.

Yernic-larch forests are characterized by a general predominance of larch in the young growth, with its number varying from 1,300 to 76,100 n ha^{-1}. The young growth age (5-20 years) is determined by the duration of the fireless period.

The analysis of quantitative characteristics of the regeneration process under the canopy of taiga larch forests confirms the conclusions of Utkin (1960), Cheremkhin (1961), and Pozdnyakov (1975) on the great variability in these indices, not only within groups but also within individual types of forest.

Fires which inhibit succession to climax species such as spruce, fir, and siberian pine, favor a continued dominance of larch. As Shanin and Falaleyev (1973) noted, only long fireless periods could lead a larch forest to be dominated by slow-growing cedar. Larch grows quicker and therefore more often manages to reach the second layer in a shorter period of time. The appearance of young growth of larch in forest types with a developed moss cover is mostly due to fire. By destroying the moss cushion, slash, and dead-wood part of the stand, fires favor soil mineralization, nutrient enrichment from ash, and increased soil temperatures. All of these factors create favorable conditions for seed germination, the rooting of seedlings, and the formation of a self-sown crop.

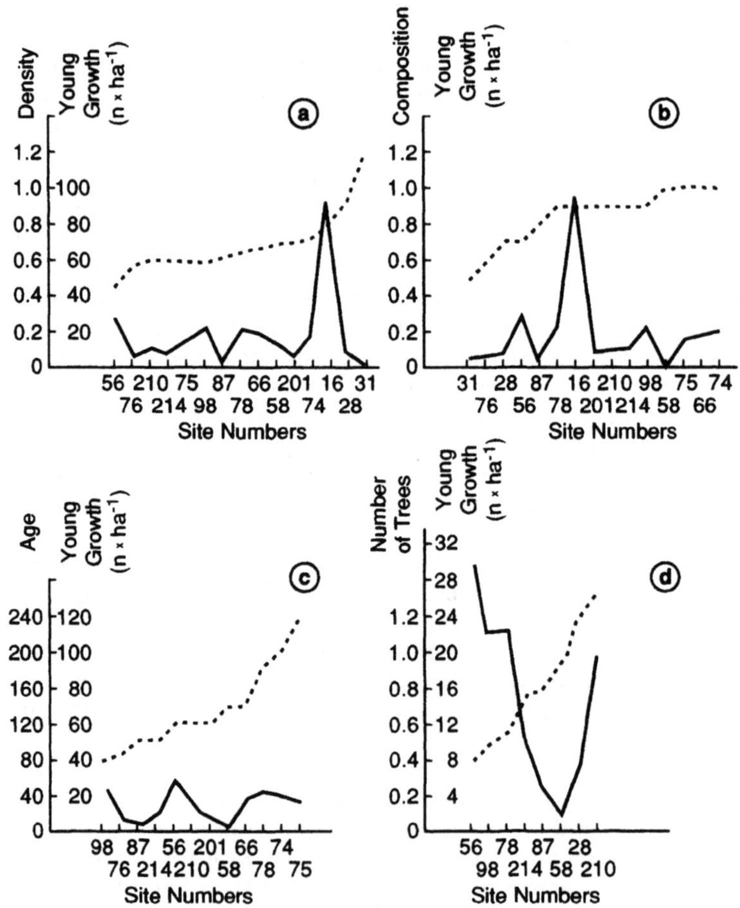

Fig.1. Density of young growth (solid line -- thousands of trees per ha) in rose bay larch forests relative to various overstory stand characteristics (dotted line): a - canopy cover (proportion of larch in total stand cross-sectional stem area, measured at breast height); b - composition (% of trees in stand that are larch); c - stand age (time since fire); d - stand density (thousands of stems per ha).

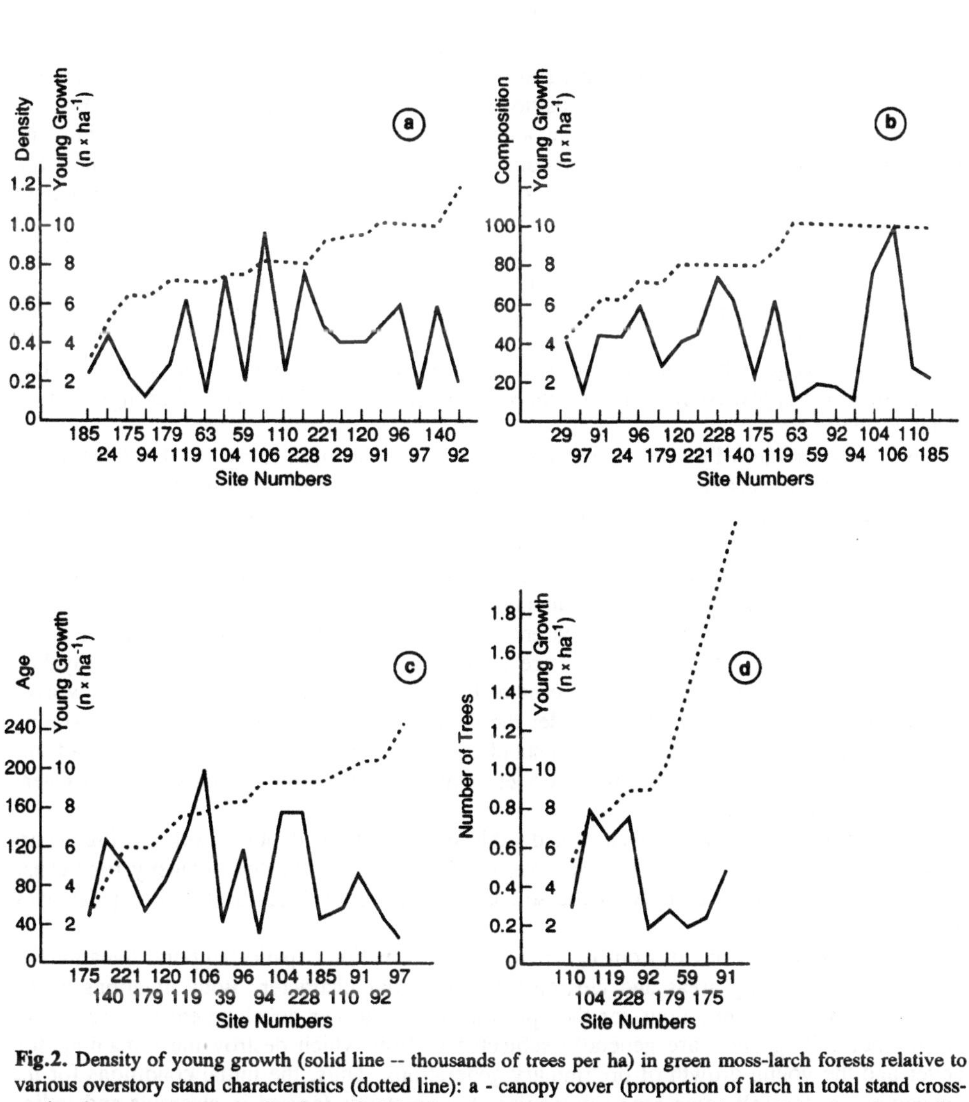

Fig.2. Density of young growth (solid line -- thousands of trees per ha) in green moss-larch forests relative to various overstory stand characteristics (dotted line): a - canopy cover (proportion of larch in total stand cross-sectional stem area, measured at breast height); b - composition (% of stand that is larch); c - stand age (time since fire); d - stand density (thousands of stems per ha).

A regulating role of a stand of trees in the process of regeneration is defined by Buzykin et al. (1985) as follows: "The structure and composition of a tree stand plays the role of a density regulator of young growth, as it determines the number of individuals and their life span which it "allows" under its canopy." However, in analysing quantitative indices of the regeneration process in taiga larch forests, their dependence on structural characteristics of the parent stand was not noted. This can be related back to the example of larch forests of rose bay and green moss types. As shown in Figures 1a and 2a, the density of young growth can differ considerably in stands of equal overstory density and be quite similar in stands of different densities.

The density of young growth does not depend on the composition of the stand either (i.e., on the share of the parent species) or on its density (Fig.1 b,d,2; Fig.2 b,d,2). As to the qualitative characteristics of the regeneration process, we also failed here to observe a relationship of the young growth composition and its height-age structure to the composition and density of the parent stand. For example, the density of young growth within one density of stands of practically the same composition can vary from 100 to 50,000 n ha^{-1}, the share of larch in young-growth composition from 10 to 100%, and its predominant height from 10 cm to 1 m.

The study of regeneration in the stands of different ages did not reveal the relation between the average age of the stands and the character of the regeneration process. As it is seen from Figures 1c(2) and 2c(2), in the framework of one age category of stands (middle-aged, early mature, mature, and declining) the density of young growth varies across a wide range, which is characteristic of larch forests of both green moss and rose bay groups of forest types. Analogous patterns were noted in larch forests of other high-altitude zones as well.

Thus, the character of a forest regeneration process in the larch forests of the Baikal basin is not directly connected with estimation indices of parent stands. Following Pobedinsky (1962) we explain the absence of such a connection by the effect of forest fires disturbing natural interrelations of the stand and young growth.

The next fire destroying the young growth simultaneously creates conditions for the appearance of the next generation. In the period without fires the larch young growth under the wood-canopy is short-lived. In a parent forest stand it can grow and develop for about 5 years; after that it begins to experience a marked inhibition and dries up. Pozdnyakov (1975) schematically presented the process of larch forest regeneration as follows: "Waves of regeneration periodically emerge under the wood-canopy and subside until the maximum in turn coincides with a change of forest stand conditions that favors the development of this new generation." The given scheme describes a natural process, but it seems to be consistent with fires in their periodicity.

In larch forests on frozen soils a considerable increase in the depth of soil thawing occurs only after severe fires which are hazardous for forest stands. But low-intensity fires cause a negligible increase of activity in the soil layer in the first 2 years after fire (Matveyev 1981).

The role of fires is obvious during young stand formation. Spatio-temporal dynamics of this process in felling areas and clearcuts has a similar character. Both situations create practically the same conditions for the appearance and further development of a self-sown crop, since felling areas are generally subjected to fires which destroy the remaining soil cover and the young growth of preliminary generations. Thus, the initial conditions for the appearance of the self-sown crop are similar. Young stands formed in clearcuts and felling areas have a bio-group character. The distribution of bio-groups throughout the area and their sizes correspond to the presence and sizes of burned spots, and in the fireless felling areas to mineralized plots. In the case of a fireless formation of young stands, reforestation of larch forests occurs without the changing of species and in an economically viable timeframe. After ground fires (especially after repeated ones) short- and long-duration changes occur.

Despite the fact that forest fires are an essential part of the natural environment and a consequence of the interaction of the electrical field of the atmosphere and biosphere, only 3% of them are caused by lightning worldwide (Kurbatsky 1964, 1975; Komarek 1966, 1973). The others occur because of man's activity. Taiga larch forests, which are often remote from the centers of population concentration and roads, are visited less often than forest-steppe ones and accordingly burn less often. However, all the stands we studied had traces of old and recent fires, and their age structure was a result of different characteristics of post-fire regeneration.

For instance, the analysis of tree distribution by age in yernic larch forests has shown that the regeneration process under these conditions was rapid (Fig.3). Inhabiting the territory by larch occurred during 1-2 age classes, which testifies to the participation of pyrogeneous factors in this process. Such forest stands are typically of single age classes (Bogdanov 1971).

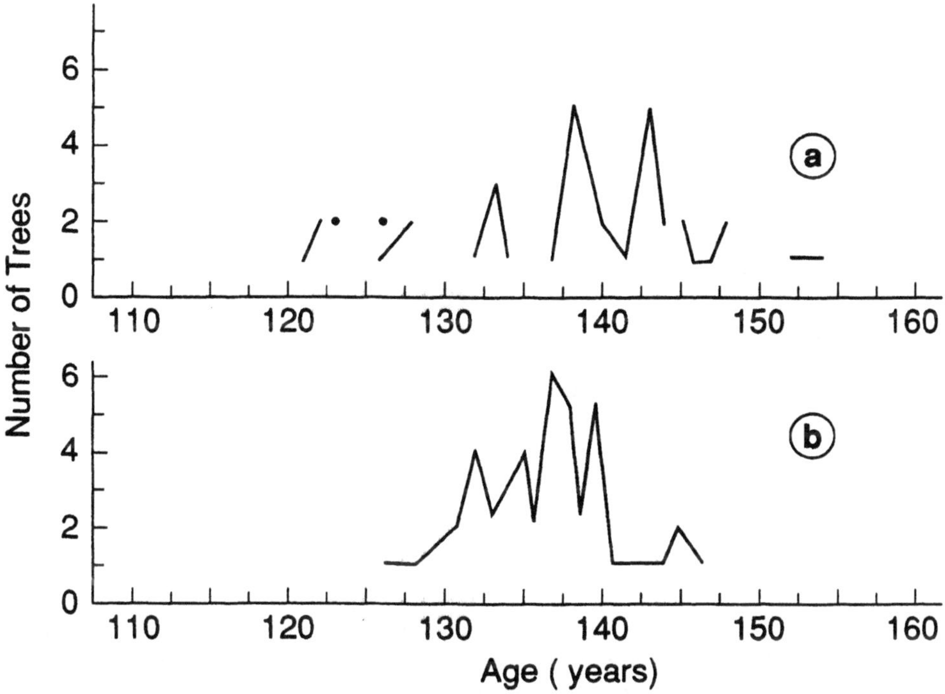

Fig.3. Distribution of trees by age in two typical sample plots, a) and b), from rose bay type larch forests.

As it has already been noted, green moss-larch forests burn much less frequently, which is also confirmed by the age structure of stands. The character of tree distribution in wild rosemary-cowberry-green moss gives grounds to assume a long and temporally irregular regeneration process (Fig.4). This is testified by the occurrence of three different forest-age groups (100-130, 140-160, 170-200 years) as well as a range of maximum ages.

At present, various technologies of prescribed burning, both in the felling areas and clearcuts and also under the tree canopy, which shorten the regeneration period, have been developed and tested. Koropachinsky (1958) has observed the development of a high density self-sown crop over a 6 to 7 year period. According to the data of other researchers, periods of seed regeneration of larch in burned plots are much shorter and occur from 2 to 4 years after a fire (Sannikov 1973; Bulygina 1981; Matveyev 1988; Komarova 1989). Stimulating regeneration by burning allows one not only to prolong the period favorable for the appearance of the self-sown crop, but also takes into account the cyclicity of larch seed production in this region.

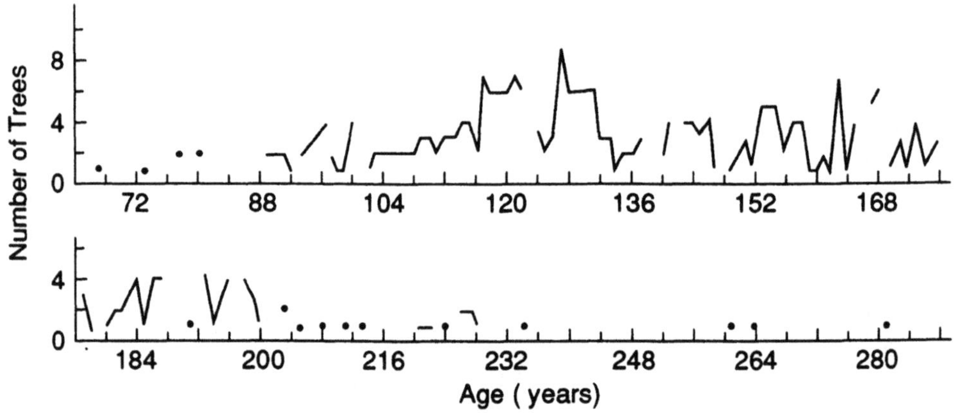

Fig.4. Distribution of trees by age in rosemary-cowberry green moss type larch forests

3. Conclusions

Thus, despite the fact that forest fires are the main factor determining the success and direction of the regeneration process of taiga larch forests, their effect cannot be estimated unequivocally. We believe one of the ways to solve this problem lies in a complex retrospective analysis with regard to successional stages and forest fire periodicity in the region of interest while taking into account all stand characteristics.

References

Bogdanov, V.M. 1971. Peculiarities of the age structure of larch-crops of South-West Yakutia. Investigations on forest management. Pskov, 34-56 <in Russian>.

Bulygina, P.K. 1981. Natural regeneration in pine forests of the Middle Angara region after ground fires. Forest Science, Forest Cultures and Soil-science. L. 18, 75-79 <in Russian>.

Buzykin, A.I., V.L. Cavricov, O.P. Sekretenko, and R.G. Khlebopros. 1985. The analysis of structure of forest cenoces. Nauka, Novosibirsk, 94 pp. <in Russian>.

Cheremkhin, S.S. 1961. Forests of the upper flow of the Vilyni. Works of the Institute of Biology of Yaf, SB, USSR AS. Yakutsk, p. 243-260 <in Russian>.

Komarek, E.V. 1966. Meteorological basis for fire ecology. Proc. Ann. Tall Timbers Fire Ecology Conf. 5, 85-125.

Komarek, E.V. 1973. Ancient fires. Proc. Ann. Tall Timbers Fire Ecology Conf. 12, 219-241.

Komarova, T.A. 1989. Seed regeneration of plants at the fresh slashes of Sikhote-Alin. Forest Science N 2, 51-60 <in Russian>.

Koropachinsky, I.Yu. 1958. The effect of fires on the age structure and peculiarities of regeneration of larch forests in the South of Tuva. Forest Journal. N 5, 43-48 <in Russian>.

Kurbatsky, N.P. 1964. Problem of forest fires. In: Fire outbreaks, pp. 5-60. Nauka, Moscow <in Russian>.

Kurbatsky, N.P. 1975. Natural and economic factors of increasing fire danger in forests. Problems of forest fire science, pp. 9-17. Krasnoyarsk <in Russian>.

Matveyev, P.M. 1981. Effect of fire action on successfulness of natural regeneration in larch forests. In: Larch: growing and cultivation, pp. 40-43. Krasnoyarsk <in Russian>.

Matveyev, P.M. 1988. Assistance in forest regeneration in the North taiga larch forests of Evenkiya. Problem of reforestation in the taiga zone of the USSR, pp. 147-149. Krasnoyarsk <in Russian>.

Pobedinsky, A.V. 1962. The study of forest regeneration processes. Krasnoyarsk Publ.

Pobedinsky, A.V. 1965. Peculiarities of forest regeneration of Eastern Siberia. In: Proceedings of scientific conference on the investigation of forests of Siberia and the Far East, pp. 144-154. Krasnoyarsk <in Russian>.

Polikarpov, N.P., R.M. Babintseva, Yu. S. Cherednikova, and L.M. Uskova, 1978. High-altitude ecological systems as the basis of organization of rational nature utilization in the Lake Baikal basin. Rational nature utilization and protection of the BAM environment, pp. 40-45. Irkutsk <in Russian>.

Pozdnyakov, L.K. 1975. *Larix dahurica*. Nauka, Moscow, 310 pp. <in Russian>.

Sannikov, S.N. 1973. Forest fires as an evolution-ecological factor of pine regeneration beyond the Urals. In. Burning and fires in the forest, pp. 112-118. Krasnoyarsk <in Russian>.

Sannikov, S.N. 1981. Forest fires as a factor of a transformation of the structure of regeneration and evolution of biogeocenoces. Ecology N 6, 23-33 <in Russian>.

Shanin, S.S and E.N. Falaleyev. 1960. Regularities of the age structure of coniferous forests of Siberia. Forest Management N IO, 20-21 <in Russian>.

Utkin, A.I. 1960. On natural regeneration of *Larix dahurica* in Central Yakytiya. Comm. Laboratories of Forest Science of the USSR AS.-M.2, 44-68 <in Russian>.

Utkin, A.I. and A.S. Isayev. 1962. Ground fires in larch forests of Eastern Siberia and their effect on the conditions of forest stands. In: Larch 29, 60-70. Krasnoyarsk <in Russian>.

Yevdokimenko, M.D. 1985. Fire danger of high-altitude vegetation complexes in the Lake Baikal basin. In: Forest fires and their consequences, pp. 46-55. Krasnoyarsk <in Russian>.

J.G.Goldammer and V.V.Furyaev (eds.), Fire in Ecosystems of Boreal Eurasia, 366-371.

Post-Fire Mortality and Regeneration of *Larix sibirica* and *Larix dahurica* in Conditions of Long-Term Permafrost

P.M. Matveev and V.A. Usoltzev [1]

1. Introduction

The focus of this investigation is on larch forests that occur on long-term permafrost. In spite of their comparatively small volume of wood they are of great importance, since they carry out an important stabilizing role in conditions where ecological equilibrium is easily disturbed. Powerful exogenous influences such as forest fires actively interfere with the life of the forest and may determine its future existence. Larch forests help to stabilize the ecological balance of the surrounding environment and provide a barrier impeding the expansion of the tundra to the south.

Knowledge of forest resistance on long-term permafrost to a damaging influence, and of the role of forest fires in the regeneration of taiga forests (Tikhomirov et. al. 1961; Viereck 1973; Belov 1973; Kryuchkov 1976; Romanovskii 1980; Aksamit and Irving 1984) is important in predicting post-fire mortality and regeneration of larch, the main forest-forming species of the taiga. Therefore we attempted to conduct integrated studies on the factors determining post-fire mortality and regeneration of larch with the aim of predicting them.

The investigations were made in the Krasnoyarsk and Yakutia regions, on the plains of the Central Siberia plateau, the Lena/Aldan plateau, the Central Yakutia plain and the Southern Koluma lowland (Gvozdetskii and Mikhailov 1978). We subdivided the study area into three permafrost zones: northern, central and southern. The thickness and temperature of the permafrost layers (Kudryavtsev et al. 1978), which are the criteria for defining geocryological regions, integrate the climatic, structural geological, geomorphological, hydrogeological, deep geothermal, and forest growth conditions. In the zone of continuous distribution of long-term permafrost layers, the forests grow mainly from the sixth to the ninth subzones of this classification (Kondrat'eva and Kudryavtsev 1977), with the frozen layer having a mean temperature of -1 to -9°C and a thickness of 100-500 m; forests do not extend north of the sixth subzone. Therefore we demarcated the northern and central permafrost zones along the middle of these subzones, i.e., along the line dividing the seventh and eighth permafrost subzones and characterized by a permafrost layer having a mean temperature of -5°C and a mean thickness of 300 m. The central and southern zones are divided by the boundary between the geocryological zones of continuous and patchy distribution of long-term permafrost layers (Kondrat'eva and Kudryavtsev 1977), which

[1] Siberian Technological Institute, 660036 Krasnoyarsk, Russian Federation

largely coincide with the boundary between the areas of *Larix sibirica* (Siberian larch) and *L. dahurica* (Dahurian larch) (Pozdnyakov 1975).

2. Materials and Methods

Due to these reasons the investigations were made in the zone of long-term permafrost, on lowland and plateau plains, in stands not previously damaged by fires. Minimum stand age was taken as 80 years; below this age the adverse effect of fires of any intensity is obvious.

Stand composition was determined based on counts of individual trees on 0.25 ha plots. Typical species compositions of the experimental areas of larch stands are given in Table 1.

Tab.1. Typical species compositions of the experimental areas of larch stands

Permafrost Zone	Number of Post-Fire Sample Plots		Typical Species Composition on 0.25 ha Plots *
	Mortality	Regeneration	
Northern	47	21	10 larch and 2-5% spruce; 10 larch and 2-5% birch.
Central	69	25	7 larch, 2 spruce, 2-5% *Pinus sibirica*;
Southern	78	30	8 larch, 2 spruce, 2-5% birch; 10 larch, 2-5% spruce, < 2% birch; 7 larch, 2 birch, 2-5% spruce, 2-5% *Pinus sibirica*; 8 larch, 2-5% pine, 2-5% birch

* Note: Description of stand composition according to Russian standards. Example: "10 larch" means that 100% volume (m³) of living trees are represented by larch only. A percent (%) number means that individuals of other species occur in the stand and account for up to this percentage of the total standing volume in a sample site.

As potential factors determining the post-fire mortality of trees we selected the permafrost zone (northern, central, southern), and within each zone four forest-type groups (from lichen to sphagnum), and within each forest-type group, the fire severity (low, medium, high), and within each fire severity, stands of different mean diameter. Mortality was counted as the percent of total standing volume. Post-fire regeneration was studied only for the first three factors.

Establishment and description of sample plots in the selected areas, stand mensuration, regeneration counts, and observations on features of fire spread and development and fire-edge parameters were done by conventional methods (Melekhov 1948; Sukachev and Zonn 1961; Kurbatskii 1964; Anuchin 1971). Fire severity was determined by the mean scorch height on the tree stems: up to 1.0 m = low; from 1.1 to 2.0 m = medium; and over 2.0 m = high. The number of years elapsed since fire was based on counts of annual rings on disks taken from the middle of fire-affected areas. From the sample plots we counted and examined trees of all species, but as different species react differently to the same fire we included the data only for larch mortality to ensure comparability in the prediction model. Post-fire regeneration was counted 4-8 years after fire.

Siberian larch (*L.sibirica*), Chekanovsky larch (*L.czekanovskii*), and Dahurian larches (*L.dahurica*, its western race, *L.gmelinii*, and its eastern race, *L.cajandiri*) grow throughout the investigated territory. The largest differences in the dimensions of trees and deflection of seed scales from the ripe cone axis occur between *L.sibirica* and *L.gmelinii* (Abaimov Koropachinsky 1984).

The purpose of this study is to investigate the possibility of post-fire mortality prediction and regeneration throughout the investigated territory. The objective is to determine the influence of the variables discussed above and of climatic differences on the levels of post-fire larch mortality and regeneration.

To investigate the presence or absence of such influences, we have compared tree mortality following fire in stands of moderate moisture forest growth conditions in the Baikit district with mortality in stands in the Central Yakutian area of the southern permafrost zone. These forest-districts were used for comparison because they have the most differences in climatic characteristics such as temperature and precipitation (Baikit district: annual precipitation 400-500 mm and average annual temperature -7°C; Central Yakutian district: annual precipitation 200 mm and average annual temperature -11°C).

3. Results

Comparison of the data on mean post-fire mortality in the stands of these districts (Tab.2) has shown that differences between them are insignificant, in spite of differences in climatic characteristics.

Tab.2. Mortality of trees in the stands of Baikit and Central-Yakutian forest-growth districts after fires of varying severity

Mean Diameter of Stand (cm)	Fire Severity	Larch Tree Mortality (Percent of pre-fire standing volume)	
		Baikit District	Central Yakutian District
14	low	13.7 ± 1.8	10.0 ± 2.0
16	medium	35.0 ± 3.6	30.0 ± 3.0
16	high	59.5 ± 4.7	66.3 ± 4.5

Furthermore our earlier investigations established that reliable differences in the amount of post-fire regeneration in the *L.sibirica* and *L.gmelinii* stands occur only during years of poor regeneration after a low-severity fire. After high-severity fires, which cause mortality of most larch trees, the regeneration is unsatisfactory in all larch species. During heavy seed crop years the small quantity of past year seeds preserved in Gmelin larch cones does not reliably increase its post-fire regeneration in comparison with other larch species.

So, larch stands of one forest type group, growing within the same permafrost zone but in different forest-growth districts, are characterized by a similar resistance to fires of varying severity and a capacity for regeneration regardless of species.

The following multiple regression model for predicting post-fire mortality was obtained from an analysis of experimental data:

$$\ln Y1 = -7.33 + 4.96 \ln X1 + 2.23 \ln X2 + 6.54 \ln X3 - 2.81 \ln X4 - 0.72 \ln^2 X1 + 0.49 \ln^2 X1(\ln X3) - 3.31 \ln X1(\ln X3) - 0.64 \ln X2(\ln X3) - 0.22 \ln X3 (\ln X4) + 0.54 \ln X1(\ln X3)(\ln X4) - 0.07 \ln^2 X1(\ln X3)(\ln X4),$$

$$R^2 = 0.938 \quad (1)$$

where Y1 is the percent of tree mortality by volume; X1 is the forest-type group with the coding: lichen = 10, green moss = 20, subshrub-moss = 30, and sphagnum = 40; X2 is the permafrost zone with the coding: southern = 10, central = 20, northern = 30; X3 is fire severity with the coding: low = 10, medium = 20, high = 30; X4 is the mean stand diameter class in cm.

We obtained the following regression model for predicting post-fire regeneration:

$$Y2 = 18.94 - 16.49 X1 + 2.85 X1^2 + 13.47 X3 - 21.33 X3^2 + 1.64 X3^3 - 16.82 X1 X3 + 32.05 X1 X3^2 - 15.60 X1^2 X3^2 + 2.21 X1^3 X3^2 + 13.34 X1^2 X3 - 2.44 X1^3 X3 - 2.57 X1 X3^3 + 1.23 X1^2 X3^3 - 0.17 X1^3 X3^3,$$

$$R^2 = 0.978 \quad (2)$$

where Y2 is the number of regenerations in 1000 stems per ha; and X1 and X3 are the same as in equation (1). Constants with variables apply for the southern permafrost zone. Analysis of the obtained data reveals some patterns of post-fire mortality and regeneration of larch. Firstly, there is a pronounced increase in percent mortality with decreasing mean stand diameter and severity of fire, irrespective of the other factors. The greater the moisture supply of the site, the greater the mortality and the smaller the differences in tree mortality percent in stands of different mean diameter.

These patterns persist in the transition from the southern permafrost zone to the northern. The increase in mortality from the southern to the northern permafrost zone, with other variables held constant, has the same explanation as with the increase in moisture supply: with increasing thickness of the permafrost layer, but with an unchanged moisture supply, the level of the frozen horizon rises, the root system stretches closer to the surface, and the proportion of roots in the total phytomass increases to 45-50% (Palumets 1988). All these factors reduce stand resistance to fire.

Regeneration patterns are closely interrelated with the complex of post-fire ecological processes, especially with the mortality of main canopy trees and with changes in soil and hydrological conditions. These processes are mediated in the models by the fire severity and forest-type factor. It was shown above that tree mortality increases from the southern to the northern permafrost zone. The reduction in the number of viable trees reduced the possibility of seeding the areas, but the patterns of increasing mortality and decreasing regeneration in this case cannot be regarded as mutually dependent. The process of reduction in regeneration, together with an increase in mortality, is further accentuated by the fact that with the frozen horizon close to the surface, especially in wet and very wet growth conditions, combustion of the heat-insulating layer of the ground cover and litter is accompanied by intensive soil heating, melting of ice particles in the soil, and partial or complete reduction or cessation of moisture consumption by the dead stand. This causes water logging, leading to the death of any advance growth, which survives only in drier places associated with various micro-elevations.

Tree mortality also increases within a permafrost zone from the lichen to the sphagnum group of forest types. And here too the complex of soil-hydrological conditions affects the association of mortality and regeneration. Maximum regeneration occurs in green-moss types, corresponding to optimum forest-growth conditions. Moving from these towards the sphagnum group, the process of progressive post-fire water-logging affects the increase in mortality and regeneration is sharply reduced. Moving from green-moss to the lichen forest type groups, the deterioration in regeneration is not associated with tree mortality. The reason for the decrease in regeneration here is the light soils whose upper horizons dry after fires and are impoverished in nutrients.

The deterioration of regeneration from the green-moss to the sphagnum and the lichen forest-type groups is apparently also caused by a reduction in stand productivity and mean height, and by the lower canopy, though the effect of the latter on regeneration is not statistically significant.

In analyzing the relationship between regeneration and fire severity, a shift in the regeneration peak becomes clear in the different forest type groups as fire severity increases. In the green-moss and subshrub-moss groups of forest types in the southern permafrost zone the regeneration peak occurs in stands after high-severity fires. These fires reduce stand competition for moisture and nutrients and destroy the herb-subshrub, moss, and litter layers, which prevent seedling emergence and development. In the lichen group of the same zone, the regeneration peak occurs in the stands after medium- and low-severity fires, which burn off the ground cover sufficiently for regeneration to occur and cause the mortality of 20-30 % of the trees.

Moving towards the northern permafrost zone, in the subshrub-moss forest type groups the regeneration peak shifts from high-severity to medium-severity fires. As was the case in the southern zone, with the great mineralization of the soil after severe fires, the level of the permafrost being closer to the surface, and the corresponding water-logging, an increase in post-fire regeneration is not induced.

In the northern permafrost zone, the effect of these factors is intensified and as a result, even in optimum growth conditions (green moss), the regeneration peak shifts from high- to medium-severity fires with a lower level of regeneration in all forest-type groups compared to the central and southern zones.

4. Conclusions

As a result of our work we can describe several main patterns of larch mortality after fire: increasing mortality with decreasing mean stand diameter; convergence of stands of different mean diameter in their inability to withstand fire with increasing fire severity within the same forest growth conditions, and with increasing site moisture supply with the same fire severity; and increasing percent mortality from the southern to the northern permafrost zone under otherwise equal conditions.

The regression models allow us to predict the final values of post-fire mortality and regeneration of *L.sibirica* and *L.dahurica* with the aim of predicting fire loss and prospects of larch forest regeneration dynamics on long-term permafrost sites.

References

Abaimov, A.P., and I.Yu. Kopachinsky. 1984. *Larix gmelinii* and *L.kayander*. Nauka Publ., 121 pp.

Aksamit, S.E., and F.D. Irving. 1984. Prescribed burning for lowland black spruce regeneration in northern Minnesota. Can. J. For. Res. 14, 107-113.

Anuchin, N.P. 1971. Forest Mensuration. Lesn. Prom. Moskow, 512 pp. <in Russian>.

Belov, S.V. 1973. Controllable fire in the forest - a means of regenerating the pine and larch stands of the taiga zone. In: burning and fires in the forest, 213-222. Krasnoyarsk <in Russian>.

Gvozdetskii, N.A., and N.I. Mikhailov. 1978. Physical geography of the USSR Mysl. Moscow <in Russian>.

Kondrat'eva, K.A., and V.A. Kudryavtsev. 1977. Map of the geocryological Regions of the USSR. Moscow University Press. Moscow <in Russian>.

Kryuchkov, V.V. 1976. The delicate subarctic. Nauka, Moscow <in Russian>.

Kudryavtsev, V.A., V.N. Dostovalov, N.N. Romanovskii et al. 1978. General permafrost science. Moscow University Press, Moscow <in Russian>.

Kurbatskii, N.P. 1964. Taiga fires - patterns of origin and development. Doctoral dissertation, Krasnoyarsk, Inst. Lesa Drev. Sib. Otd. Akad. Nauk SSSR <in Russian>.

Matveev, P.M., and A.P. Abaimov. 1985. The post-fire larch regeneration in different permafrost zones. In: Larch, 28-34. Krasnoyarsk <in Russian>.

Matveev, P.M. 1988. Especially of fire influence to regeneration of larches in northern taiga. In: Larch, 18-20. Krasnoyarsk <in Russian>.

Melekhov, I.S. 1948. Effect of fires on forest. Goslestekhizdat, Moscow <in Russian>.

Palumets, Ya.K. 1988. Distribution of fractions of the phytomass of Norway spruce as a function of age and climatic factors. Lesovedenie 2, 34-40.

Pozdnyakov, L.K. 1975. Dahurian larch. Nauka, Moscow <in Russian>.

Romanovskii, N.N. 1980. The gold of the Earth. Prosveshchenie, Moscow <in Russian>.

Sukachev, V.N., and S.V. Zonn. 1961. Methodical instructions for study of forest types. Akad. Nauk SSSR, Moscow <in Russian>.

Tikhomirov, V.N., I.YA Koropachinskii, and E.N. Falaleev. 1961. Larch forests of Siberia and the Far East. Goslesbumizdat, Moscow <in Russian>.

Viereck, L.A. 1973. Wildfire in the taiga of Alaska. Quat.Res. 3, 465-495.

J.G.Goldammer and V.V.Furyaev (eds.), Fire in Ecosystems of Boreal Eurasia, 372-386.

The Main Trends of Post-Fire Succession in Near-Tundra Forests of Central Siberia

A.P. Abaimov and M.A. Sofronov [1]

1. Introduction

The term "near-tundra forest" was first introduced into the Russian literature on forests by Andreyev (1954). Later, from 1959 onwards, a forest belt located along the southern tundra boundary was officially called near-tundra forest. These forests have an important environmental and stabilizing role. They are distributed from Norway to the Bering Straits and vary remarkably in vegetation pattern. However, the question concerning the southern boundary of their areal extent is still open, since no uniform fundamental criteria have been established for identifying near-tundra forests. Also, no scientifically justified methods now exist for managing the near-tundra forests of Siberia. This can be attributed to the lack of knowledge of these northernmost forest ecosystems.

It is noteworthy that, over so vast an area, the distribution of tundra and forest vegetation and permafrost soils vary considerably with climatic conditions. In the area where cold marine climate prevails (the coastal zones of the Barents Sea, and Hudson Bay and Labrador), the southern boundary of continuous permafrost, in fact coincides with the northern tree line. The transitional area is represented by a relatively narrow forest-tundra belt; permafrost soils are characteristic of the tundra zone, and the forests adjacent to the boundary are represented by typical northern taiga tree species supported by podzolic soils. In this case it is reasonable to consider near-tundra forests as a shelter belt in which forest management should be largely similar to that practiced in the taiga forest zone. There is every reason to call a relatively narrow taiga forest strip located in the north of the European part of Russia which acts as a climatic shelter belt, "forests bordering tundra". The term "climatic shelter belt" has already been used by some researchers for describing the functions of forests bordering the tundra zone (Tchertovsky 1983; Tchertovsky et al. 1987).

In the northern part of Siberia where continental and extreme continental climate prevails, the boundary of continuous permafrost is shifted, because of severe winters, to the south (down to 62°N in Central Siberia), while the timberline is shifted, because of warm summer, far to the north (up to 72°N). This results in the occurrence of a very special natural zone with woody vegetation being supported on permafrost and very cold soils instead of podzols. Within this zone, tundra ecosystems are quite common; in Western Siberia they are represented by raised bog-tundra vegetation communities (Shumilova 1962)

[1] V.N. Sukachev Institute for Forestry and Timber, Siberian Branch, Russian Academy of Sciences, 660036 Krasnoyarsk, Russian Federation

and in Central and Eastern Siberia by mountain tundra vegetation in watersheds extending as far south as 65-63°N.

Kolesnikov (1969) defined this zone as "a forest-shrub zone" and, Parmuzin (1979) suggested that it should be called "a tundra-forest zone". We believe that the term "near-tundra forest" is more accurate. In this area, forest formation is fundamentally different from that in the taiga zone (Sofronov and Abaimov 1991) and requires the thorough investigation and introduction of specific forest management policies. For this reason we suggest that the whole of the area of Siberian vegetation growing on continuous permafrost be referred to as near-tundra forest.

2. Study Area Description

The study area, which covers a vast territory of the northern part of Central Siberia, is bordered by the Yenisei river on the west and by the Lena river on the east. In this area, the northern boundary of near-tundra forest distribution (areal extent) corresponds to the northern timberline, reaching 70° to 72°N. The southern limit of near-tundra forest is between 65° and 67°N (Korotkov 1991). The area is notable for a considerable variety of environmental conditions which are controlled largely by mountain massifs (Putoran and Anabar plateaus).

The western part of the near-tundra forest area of Central Siberia has prevailing northwesterly winds and is characterized by relatively mild climate, which is somewhat similar to that in the catchment area of the Yukon river, Alaska (Viereck 1973). Forest cover is represented by *Larix sibirica*, *L. gmelinii*, *Picea obovata*, *Betula pubescens*, and, in some places, *Pinus sibirica*. Normally, birch forms secondary post-fire stands and dominates elfin wood distributed along the timberline in mountains. Spruce occurs mostly in mixed forest. Pure spruce stands are rare and occupy relatively small sites. In general, the whole of the study area is dominated by larch (Korotkov and Dzedzula 1969).

The central and eastern parts of the area are characterized by a cold, extremely continental climate, with mean annual air temperature ranging from -8° to -12.7°C. In the part of the area north of the polar circle, birch and spruce stands occur very rarely; the major tree species is *Larix gmelinii* which is replaced, on the left bank of the Lena river, by *L. cajanderi*. In mountains, the timberline ranges from 200 m a.s.l. in the north to 800 m a.s.l. in the south. Therefore, in the northern high mountains, forest makes up not more than 5-15% of the total area, while forest accounts for 70-80% of the area in the southern middle mountains and plains. Vertical (altitudinal) vegetation zonality, characteristic of other regions, is disturbed because of air temperature inversion induced by anticyclones in the winter (Parmuzin 1979). Summertime anticyclones are associated with severe drought and catastrophic wildfires.

Comprehensive analysis of literature data on near-tundra forests of Central Siberia (e.g., Tyulina 1937; Sochava 1956; Lukicheva 1963; Mironenko 1967, 1970; Korotkov and Dzedzula 1969; Vodopyanova 1975, 1976; Kuvaev 1975; Norin et al. 1986; Sofronov 1988; Abaimov et al. 1990), when combined with our recent unpublished observations, allows one to draw some general conclusions concerning specific dynamics of these forests.

3. Near-tundra Forest Classification

Many researchers who have studied zonal distribution of near-tundra forests state that these forests are characteristic of 1) the plain forest-tundra zone; 2) mountain systems of the north that support open woodlands; and 3) the northern part of the northern taiga subzone. Kolesnikov (1969), however, does not include these forests in the taiga zone. According to this researcher, "...outside the zone of taiga and mixed forests distributed on plain landscapes, closed-root-system light forests occur which are highly specific in botanical, geographical, and management respects..." We agree with Kolesnikov that this is a separate natural vegetation zone.

It is common practice, when describing subarctic woody vegetation, to classify forest by the degree of canopy closure (canopy density). Using this approach, highly, moderately, and little open woodlands (i.e. open forests of low, normal, and high canopy density) are distinguished. In some places, typical northern taiga forest occurs. Highly open woodlands represent the upper limit of the altitudinal vegetation zones. Fragmentary northern taiga forest occurs in flood plains, near streams, and on relatively warm, well-drained slopes. Most of the area, however, supports moderately and little open forests. The majority of authors (e.g., Tyulina 1937; Sochava 1956; Lukicheva 1963; Norin 1978, 1979; Parmuzin 1979; Demyanov 1988) are prone to define forest communities with canopy closure index (CCI) less than 0.1 as "highly open forest" and those having CCI of 0.1-0.3 as "moderately open forest". Also, some authors (Sambuk 1937; Dylis 1939; Kuvaev 1975) suggest that "moderately open forest" should refer to the forest communities whose CCI is 0.1-0.2 while others (Andreyev 1956; Mironenko 1970) state that this term applies to the communities having CCI of 0.2-0.3. By "little open forests" are meant the phytocoenoses with CCI of (0.3) 0.4-0.5 (e.g., Mironenko 1967, 1970; Norin 1979).

We agree that highly open forest should be treated as a separate vegetation category, since when CCI is less than 0.1 tree roots do not form a closed system, and thus the stand fails to be an environment edificator. In contrast, we see no qualitative differences between moderately and little open forests. In both cases tree roots form a closed system (structure) due to overlapping, and overstory is sparse. Variations in stand density and canopy closure are mostly due to factors influencing forest regeneration, such as fire behavior, time elapsed between a fire event and occurrence of a good seed production year, and spatial distribution of seed sources. Furthermore, the term "little open forest" implies low stand density. Under favorable conditions, however, very dense stands can develop having an extremely thin canopy; i.e. they are typical moderately open woodlands. We thus recommend use of a uniform term - "open boreal woodland" for both "moderately open" and "little open" forests.

In Central Siberia, the near-tundra forest area, which totals about 65 x 10^6 ha, is dominated by uneven-aged open larch woodland characterized by site-specific wood quality indices V-Vb (i.e. mean tree height is 7-12 m) and CCI 0.1 to 0.4. Thin stand canopy provides for the development of a brush layer represented by *Dushekia fruticosa* or *Betula nana* mixed with *Salix* spp. and typical tundra plants. Standing crop is 15-80 m^3 ha^{-1}. These forests are of low commercial value since small-diameter individuals account for 50-80% of the total number of trees, and large-diameter trees normally suffer from trunk rot. The stands develop slowly and are highly sensitive to stresses, since they grow under extremely unfavorable conditions. As fire is the main disturbance factor, stands of fire origin dominate the area.

4. Fire in Near-Tundra Forests

The high fire danger typical of near-tundra forests is due to the surface cover being dominated by green mosses and bunch lichens (*Cladonia* spp., *Zetraria* spp.) and the highly flammable *Ledum palustre*. Paludification level is relatively low. Sedge marshes, which can burn, are common, while raised sphagnum bogs that do not burn are very rare. Rate of fuel drying increases proceeding from the west to the east with increasing continentality, and from the South to the North with increasing daylength in summer and forest canopy growing thinner.

Natural fire danger, as manifested in vegetation readiness to burn (i.e. vegetation flammability level), is not the only factor determining fire occurrence. Another factor is the number of fire-starting agents present in the area. In the near-tundra forest zone of Central Siberia, population is very sparse and man-caused fires are thus a relatively rare event. Lightning fires account for 50 to 80% of all fires. Storm frequency varies considerably across the area. In the southern part of the Taimyr peninsula, behind the Putoran mountains, storms almost never occur. As for the mountains themselves, lightning strikes mostly rocky, vegetation-free peaks. But in hilly plains of southeastern Taimyr, lightning fires often occur.

The third important factor controlling the number of fires is the possibility for fires to spread, i.e. pyrological characteristics of the area. Pyrological non-uniformity is best manifested in high mountains, where narrow forest belts are surrounded by non-flammable mountain tundra and arctic barrens. The western part of the region in question is also considerably non-uniform from the pyrological viewpoint. In this area, crossed by many rivers, average annual precipitation is up to 800 mm. The southeastern part of the region (basins of Moiero, Kochechuma, and Olenka rivers), where average annual precipitation is 300 to 400 mm, fuels are more uniform and thus fires can spread here over large areas.

In the southern part of the region, the fire season starts in June and lasts until September, with peak fire danger occurring in July. Steady-state surface fires are common. As most trees are small, and their roots are located in the top part of the soil, they are highly susceptible to fire. This often results in complete destruction of stands by fire. Forest regeneration in burned areas depends on both seed source availability and seedbed conditions. Partial exposure of mineral soil provides for the most successful regeneration, because when all of the surface organic matter is removed by fire, the soil dries out quickly, especially on south-facing slopes.

Most of the forest area of Central Siberia is not covered by fire protection. Therefore actual fire activity can be assessed through the use of aero-space images and by aerial and ground observations. Data available to date suggest that fires are the major environmental factor, and their effects determining vegetation succession trends differ greatly from those in the taiga forest zone.

5. Regeneration, Soil Temperature Regime, and Fire

Patterns of open boreal woodland regeneration, which differ from those of taiga forests, confirm that woody vegetation on permafrost soils is a distinct vegetation zone. As the depth of summer soil thawing is shallow (0.2 to 0.7 m), the root layer is not thick. The result is that the amount of roots per unit area is much less than in typical taiga forest. A strong relationship is known to exist between the amount of physiologically active roots absorbing water and the biomass of needles and leaves which transpire water. Consequently, a small number of roots per unit area leads inevitably to a small amount of needles (or leaves), which is manifested in thin tree crowns and a low level of canopy closure. In open woodlands there is no competition for light, whereas in taiga forests this is one of the main factors influencing forest community development.

In taiga forests, high natural thinning of young stands is known to occur because stronger, more viable individuals overtop weak ones and deprive the latter of light. Since forests on permafrost have a thin canopy, there is enough light for both strong and weak individuals. Therefore, self thinning proceeds at a very low rate; the result is the occurrence of special communities, which seem to stop developing at the stage of pole stands, in burned areas initially colonized by a dense self-sown crop.

In the near-tundra forest zone, soil temperature regime is the main contributor to site conditions. Unfortunately, this parameter is still under-investigated. The few data available to date only allow one to get a very general understanding of the subject. Permafrost soil temperature regime is peculiar in that the soil is heated for a short period of time (not more than four months) and to a small absolute level (depth), with the temperature gradient being relatively high (Norin et al. 1986). For open larch woodlands of the northwestern Putoran plateau, the depth of soil thawing, in the middle of the growing season, appears to vary from 15 to 60 cm, and the average monthly soil temperature in July, measured at a depth of 20 cm, does not exceed +3° to +5°C (Norin et al. 1986). Our observational data obtained in the Lena lowland at the height of the growing season show that, under forest canopy, soil normally thaws to a depth of 30 to 40 cm, and only occasionally to 70 cm, while the temperature of the root-filled soil layer does not exceed +5°C (Stepanov 1985).

The above examples show that permafrost soils, at a depth as shallow as 20 cm, are considerably underheated during the growing season. As a result, root systems of trees, shrubs, and grasses develop very close to the soil surface, thereby increasing root competition (Dadykin 1952). This greatly hampers seed germination and seedling establishment in the first years of self-sown crop development. These negative effects are confirmed by the results of field experiments conducted in various parts of northern Siberia (Norin 1958; Vipper 1964; Pozdnyakov 1975) as well as by observational data showing low rates of natural regeneration under stand canopies (e.g., Tyulina 1937; Lukicheva 1963; Mironenko 1967; Tchugunova 1971; Vodopyanova 1976; Abaimov 1978).

However, according to a number of authors (Pozdnyakov 1975, 1986; Tchugunova 1979; Matveyev and Abaimov 1980; Stepanov 1985), forest regeneration in burned areas previously covered by larch is normally successful in the year following fire. This is attributed to two factors: a sudden decrease in root competition resulting from the killing of many competing individuals by fire, and changes in the soil temperature regime favouring regeneration (e.g., Matveyev 1984; Stepanov 1985; Tarabukina and Savvinov 1990).

Since a loosely packed surface layer composed of soil organic matter, mosses, and lichens is characterized by low heat conduction, it decreases the rate of soil heating by solar radiation. Because sphagnum turns out to be the best heat insulator, permafrost appears right below it in sphagnum sinusia. Heat insulation characteristics of the soil organic matter improve as the layer dries, which results in soil remaining cold in hot weather. The thicker the organic layer, the colder the soil and the shallower the depth of soil thawing. Heat-insulating organic matter is partially or completely consumed by fire. This results in increasing soil insolation and depth of thawing. Soil organic layer thickness and the depth of soil thawing were 7 and 59 cm, respectively, on 2-year old burns, compared with 12 and 47 cm on 7-year old burns, and 19 and 30 cm on an unburned control site. Increased depth of soil thawing in burned areas provides for the development of a deep root-filled soil layer. In 30-50 years, however, the pre-fire soil organic layer recovers, the depth of thawing decreases, and, as a result, the lower parts of root systems die off. In stands of normal density (200-500 individuals per hectare), the permafrost-induced loss of lower roots is compensated by the development of additional roots in top soil and organic layer. In stands of high density (1,000 to 5,000 individuals per hectare), additional roots fail to develop, which leads to partial tree mortality, while the crowns of the remaining trees grow thinner and a sudden decrease in their height increment is observed.

6. Post-Fire Succession

The results of ground investigations conducted in near-tundra forest of Central Siberia have revealed the following main trends of post-fire forest succession:

- Open larch woodland regeneration with a range from no stand replacement to normal, high, or low stand density.
- Forest regeneration through the replacement of pre-fire climax larch stands by birch-dominated elfin woodland of fire origin.
- Regeneration through the replacement of larch stands by secondary low bush tundra.
- Regeneration through the replacement of larch forest by *Carex-Eriophorum* communities.

Larix gmelinii and *L. cajanderi* stand regeneration normally occurs with no stand replacement. Variants of potential post-fire forest dynamics depend on many factors, most of which are still under-investigated. The typically succesful *L. gmelinii* and *L. cajanderi* regeneration in burns is largely due to the fact that, whatever the fire intensity and the area burned, a certain number of trees usually survive the fire and provide seeds for larch regeneration to follow. Even after a high-intensity surface fire, surviving trees are observed in micro-depressions, hollows and other wet sites, and those with roots only slightly damaged by fire appear to be viable enough to produce seed annually over several years following fire. In burns characterized by a good rate of regeneration, provided that some trees remained intact following fire, typical open woodlands of normal forest density usually develop.

Table 1 gives the description of stands that occurred after two or three fire events. Over the past 60-80 years, however, these stands did not experience fires. Usually, these are uneven-aged stands characterized by low rate of canopy closure and standing crop varying from 20 to 30 m^3 ha^{-1}. According to forest inventory data, 40 to 100 trees per hectare survived fires, thus providing for larch being the edificator.

As root competition is severe, uneven-aged stand density of 250-500 trees per hectare should be considered optimal for near-tundra forests of Central Siberia. The typical unevenaged nature of these stands is confirmed by the presence of both small (10-12 cm in diameter) and larger (14-24 cm in diameter) snags, which account for about 5% of the total number of trees (Tab.2). Moreover, it is noteworthy that dead trees (snags) can keep standing for decades. Relatively low density of regrowth under the canopy of these open forests of fire origin (see Tab.1) indicates that both overstory and understory vegetation are involved in competition. However, the amount of regrowth is typically sufficient to replace dead trees.

Fire-related partial mineral soil exposure and improvement of soil temperature regime, when combined with seed abundance and favourable seed bed conditions, can promote the development of extremely dense larch stands in burned areas. At the initial stages of development, their density can vary in the range from several tens of thousands to 1.5 million trees per hectare (Pozdnyakov 1975, 1986). Depending on the density of individuals at the initial regeneration stage, as well as on the initial soil conditions in the burned area and the rate of worsening of the soil temperature regime, the development of extremely dense stands can stagnate for an indefinite period of time at the stage of young or pole stands. In northern and northwestern Evenkia, 40-50 year-old larch stands of fire origin characterized by densities of 20,000 individuals per ha, 3-5 m tree height and 4-6 cm tree diameter are relatively common on fluvial terraces above flood-plains and in flat interfluves. Most trees have extremely thin crowns dominated by short branches and are remarkable for almost inappreciable increment in height and diameter. Individuals with dead crowns account for 25-40% of the total number of trees.

As moss cover and soil organic layer recover, the depth of soil thawing comes back to the norm, i.e. depth to permafrost decreases, while the root-filled soil layer becomes thinner. This hampers tree growth and leads to partial tree mortality in both extremely and highly (1500-5000 trees ha^{-1}) dense stands. As is obvious from Table 3, such stands, with ages from 90 to 200, have similar tree height and diameter, woodstock density, and standing crop. Standing dead trees (snags) occur, in almost equal amount, in all classes of trunk diameter (Tab.4). This suggests that tree mortality results from site conditions getting generally worse, but not from self thinning. Because of severe root competition, very little, if any, natural regeneration occurs under the canopy of extremely dense stands.

The structure of even-aged stands in open woodlands is different from that in the taiga zone. First, the range of tree diameter variation is 1.5 times greater in open boreal woodlands than in taiga (eg. 4-18 cm in taiga and 2-26 cm in open woodland), with average tree diameter being the same (in our example, 9 cm). Second, the number of trees of lower trunk diameter classes in open boreal woodlands is 8-10 times greater than in taiga forests, which results from a low rate of self thinning. Third, a 10-12-fold increase in the number of trees of high trunk diameter classes is observed in open woodlands as compared to the taiga zone. The latter phenomenon may be due to significant differences in site conditions under which individual trees grow, i.e. these are the so-called intracommunity differences in site conditions which are not observed in taiga.

Consider for example a 50 year-old pine stand located in the central Angara region. According to the data of Prokushkin (1982), the nutrient uptake area (rooting zone) for any given tree is eight times greater than its crown area. In a taiga forest stand whose CCI is 0.9, tree roots overlap each other to a great extent, thus creating a closed system. So, each microsite is used by several trees with the result that the differences in growing conditions between individuals are smoothed out. Absence of these differences is obvious from regular pattern of canopy of even-aged taiga stands. Conversely, shallow depth of the root-inhabited soil layer makes multiple root overlapping impossible in open woodlands. Therefore, each tree has its own rooting zone of certain size and quality. This results in irregular canopy pattern, with some large individuals towering over the majority of trees.

Tab.1. Description of uneven-aged, normal-density open larch forests of fire origin

	Sites					
	15-T	2-M	16-E	12-T	5-M	6-T
Geographical Location	Tembenchy River	Moiero River	Embenchime River	Tembenchy River	Moiero River	Tembenchy Lake
Topography (Relief, Aspect)	Lower part of a SW-facing slope	Lower part of a SW-facing slope	Flat hilltop in a valley	Middle part of a N-facing slope	Plateau-like hilltop	Second terrace
Stand						
Composition	10L *	10L	10L	10L	10L	10L
Age (years)	180 (160-278)	210 (107-397)	100 (94-209)	140 (110-264)	160 (138-265)	100 (85-240)
Tree diameter (cm)	15	14	13	10	11	8
Tree height (m)	12.0	10.8	11.9	8,3	9.8	6.2
Total stocking density (m^2 ha^{-1})	5.5	5.0	5.7	5.9	6.3	4.8
Canopy closure index (CCI)	0.3	0.2	0.3	0.3	0.3	0.3
Density of living/dead standing trees (N ha^{-1})	270/110	302/12	416/16	508/35	583/93	833/63
Standing crop living/ dead standing (m^3 ha^{-1})	32/2	32/1	25/2	23/1	33/2	18/2
Regrowth						
Density (N ha^{-1})	1,800	600	1,500	1,100	700	1,500
Abundance (%)	40	20	36	36	20	40
Lower vegetation layers Composition	*V. uligunosum*, green mosses (Briales)	Dwarf shubs, *V. uligunosum*, lichen green mosses	*L. palustre* *V. vitis-idaea* green mosses	*L. palustre* lichen green mosses	Dwarf shrubs *V. uligunosum* *V. vitis-idaea* lichens,mosses	*V. uligunosum* lichens

* Note: 10L indicates a well-stocked larch stand

Tab.2. Classification of living trees and dead standing trees (snags) by diameter (individuals/ha) in open larch forests of normal density

Diameter class (cm)	Site number												Total	Snags	
	15-T		2-M		16-E		12-T		5-M		6-T				
	live	snags	live	snags	live	snags	live	snags	live	snags	live	snags		Number	%
2	5	30	2	-	46	2	73	-	24	12	181	10	385	54	14
4	5	35	48	-	52	4	140	7	64	28	135	14	532	88	16
6	10	15	48	2	50	4	93	-	105	28	128	14	497	63	13
8	5	20	46	-	56	-	27	-	112	14	114*	10	404	44	11
10	50	5	28	2	34	-	27*	-	63*	7	90	10	316	24	8
12	35	-	14	4	28*	-	21	4	64*	-	48	-	190	8	4
14	70*	-	16*	-	28*	-	13	4	45	-	28	-	206	4	2
16	20*	-	12	2	28	-	27	-	36	2	19	-	146	4	3
18	20	-	14	-	28	2	20	-	26	-	5	-	115	2	2
20	15	-	24	2	22	2	20	-	17	2	5	-	109	6	5
22	15	-	18	-	22	2	7	4	7	-	10	-	85	6	7
24	10	5	14	-	8	-	33	-	5	-	10	-	85	5	6
26+	10	-	18	-	14	-	7	7	9	-	-	5	70	12	17

Note: * average diameter class for the site

Tab.3. Characteristics of even-aged, high-density open larch forests of fire origin

	Site				
	6-D	20-E	13-D	25-K	3-ST
Geographical Location	Dyupkun Lake	Kochechum River	Dyupkun Lake	Kochechum River	Kochechum River
Topography (Relief, Aspect)	Lower part of a NE-facing slope	Middle part of a SW-facing slope	SE-facing hollow slope	Middle part of a NE-facing slope	Middle part of a SE-facing slope
Stand					
Composition	10L*	10L	10L	10L	10L
Age	90 (80-97)	190 (178-207)	140 (120-162)	200 (190-214)	80 (63-87)
Tree diameter (cm)	10	10	11	10	7
Tree height (m)	10.3	8.8	11.9	9.5	8.2
Canopy closure index (CCI)	0.6	0.5	0.5	0.4	0.6
Total stocking density (m^2 ha^{-1})	11.9	12.5	14.5	12.6	13.7
Density of living/dead standing trees (N ha^{-1})	1,065/300	1,320/204	1,148/683	1,076/883	2,904/1,509
Standing crop of living/ dead standing (m^3 ha^{-1})	65/24	59/6	78/21	63/16	58/13
Regrowth					
Density (N ha^{-1})	500	600	300	0	300
Abundance (%)	16	16	12	-	13
Lower vegetation Layers Composition	*Ledum palustre* *V. vitis-idaea* lichens	Dwarf shrubs lichens green mosses	*Ledum palustre* *V. vitis-idaea* green mosses	low shrubs lichens green mosses	*L. palustre* *V. vitis-idaea* green mosses

* Note: 10L indicates a well-stocked larch stand

Tab.4. Classification of living trees and dead standing trees (snags) by diameter (individuals/ha) in open larch forests of high density

Diameter class (cm)	Site										Total	Snags	
	6-D		20-E		13-D		25-K		3-ST				
	live	snags	live	snags	live	snags	live	snags	live	snags		Number	%
2	11	-	68	20	39	22	42	133	267	244	846	421	48
4	183	44	144	40	155	150	50	242	771	644	2,363	1,120	47
6	205	39	272	44	155	117	117	183	778*	400	2,310	783	34
8	211	50	292	44	156	117	292	183	689*	155	2,188	549	25
10	139*	55	200 *	16	228 *	105	225*	67	400	22	1,457	265	18
12	139	6	132	16	183 *	100	125	58	67	-	826	180	22
14	94	50	72	12	83	44	125	-	-	-	480	106	22
16	44	28	40	4	72	11	67	17	-	44	327	104	32
18	22	11	36	-	44	17	33	-	-	-	113	28	17
20	17	17	28	4	11	-	-	-	-	-	77	21	27
22+	-	-	36	4	17	-	-	-	-	-	53	4	8

* average diameter class for the site

When forest regeneration in burned areas is unsuccessful, climax open forest can be replaced by post-fire highly open larch forest. Although *Larix gmelinii* can maintain live seeds in its cones for 2-3 years, poor regeneration typically occurs when the whole of the overstory is killed by fire and when the fire is preceded by a succession of years characterized by abundant seed production (Yegorov 1961; Karpel and Medvedeva 1977; Abaimov and Koropachinsky 1984). In such cases, seed amount and germination may be insufficient to provide for the development of a highly viable new generation having the role of edificator. In this regeneration scenario, the number of larch trees can vary from several individuals to several dozens per hectare. Rapid regrowth of understory species, such as *Alnus*, *Betula nana*, and willow, result in the establishment of well-developed (0.6-0.8 CCI) understory layer 10 years following fire. This layer hampers larch distribution, even when young larch trees developing in burned areas begin to produce abundant seeds. Our investigations conducted in Taimyr, northern Evenkia, and northwestern Yakutia show that *L. gmelinii* starts to produce abundant amounts of seed at the age of 10-15.

Fire origin of highly open larch forests is confirmed by the presence of standing dead trees (snags) of pre-fire origin and a considerable amount of downed wood showing the signs of fire, which can remain for 80-100 years or longer, due to low biological activity under permafrost conditions.

The results of our investigations, as well as data obtained by other researchers (Dzedzula 1969; Korotkov and Dzedzula 1969), show that in the western part of the region dominated by *L. sibirica* and *L. czekanowski*, post-fire forest regeneration occurs through species replacement. Pre-fire climax larch and mixed larch-spruce-birch stands are replaced by secondary birch stands and birch-dominated elfin woodlands. On low (up to 400 m a.s.l.) hilltops, along the southern boundary of tundra forest distribution in Evenkia and northwestern Yakutia, birch-dominated elfin woodland sometimes replaces *L. gmelinii*. According to I.P. Shcherbakov's (1982) data obtained in the upper reaches of Vilui river, postfire birch stands can exist for quite a long time, provided larch seed sources are absent, and conifers begin to colonize the area from the outside. Spruce and larch regrowth normally develops under the canopy of secondary birch stands of Yakutia and the western Putoran plateau. In 50-80 years birch stands are completely replaced by pure larch or mixed conifer-hardwood stands, since birch regenerates mostly by stump sprouting.

Secondary shrub tundra made up of *Alnus fruticosa*, *Salix*, and *Betula nana* occurs in place of stands completely destroyed by erratic high-intensity fires. The considerable amount of soil organic matter surviving such fires dries out and prevents the development of larch seedling root systems. Conversely, the brush layer develops successfully by stump sprouting. As shrubs are not suppressed by larch regrowth, they form communities having CCI of 0.8-0.9. Post-fire *Alnus* communities develop usually at timberline, i.e the near alpine tundra belt, where larch is not much of an edificator. Thus, in the course of post-fire regeneration, highly open larch forests and elfin woodlands are replaced by brush. Such brush communities can also be observed in taiga forest, but they are usually limited to north-facing slopes along rivers.

In flat, low-elevation watersheds, small areas of *Betula nana*-dominated tundra intermix with open larch woodlands. Sometimes they occupy high terraces above river flood plains and high level (raised) flood plains of the biggest streams. Canopy closure index of the brush layer varies with site conditions from 0.5 to 0.9. In some places, three or four *Salix* species can occur as an admixture in shrub communities. Such post-fire phytocoenoses do not cover much of the area and are relatively insignificant contributors to the vegetation cover of the region. However, to achieve a greater understanding of the postfire dynamics of near-tundra forests of the region, thorough investigation of both the mechanism of secondary shrub tundra development and its replacement over time by climax larch stands is required. Data obtained so far only show that these processes take decades and that larch tends to colonize shrub-covered areas gradually from the outside.

Carex-Eriophorum communities occur after steady-state surface fires, usually where the top soil layer is underlain by fossil ice lenses. When fire consumes the whole of the forest vegetation and the moss layer, which has the role of thermo-regulator, drastic changes occur in soil temperature regime and thermokarst is often promoted. Around thermokarst lakes, *Carex-Eriophorum* community development is favored by abundant stagnant water. These communities become common in dry years, when fires can occur in sites characterized by abundant stagnant water, as well as in river valleys, interfluves, and lightly dissected parts of the north-Siberian lowland. Because of the thick sod layer and soil waterlogging, decades are also needed for these forests to regenerate.

7. Conclusions

To conclude, in Central Siberian near-tundra forests growing on continuous permafrost soils the post-fire vegetation regeneration pattern is very specific. This pattern is remarkably different from that in the taiga zone in that competition for light is replaced by root competition for mineral nutrients and water availability from a shallow root-filled soil layer. Under the extreme climatic conditions of the cryolithic zone, forest fires become the major factor determining trends and dynamics of forest succession. The knowledge of post-fire forest succession mechanisms, being of great fundamental importance, is crucial in developing scientifically justified policies of subarctic forest management and protection. Also, this would be of much help in studying forest dynamics under environmental and anthropogenic stresses.

References

Abaimov, A.P. 1978. Some peculiarities of natural larch forest regeneration at the head of the Vilui river. In: Larch and its use, Issue 9, pp.38-44. Krasnoyarsk <in Russian>.

Abaimov, A.P. and I.Yu. Koropachinsky. 1984. *Larix gmelinii* and *Larix cajanderi*. Nauka Publ, Novosibirsk. 120 pp. <in Russian>.

Abaimov, A.P., A.I. Bondarev, and P.A. Tzvetkov. 1990. Brief description of larch forests of the northern part of eastern Evenkia. Proc. Int. Symposium on Boreal Forests: Their State and Dynamics and Anthropogenic Impact. Part 2., pp. 3-12. Moscow <in Russian>.

Andreyev, V.N. 1954. Larch promotion in tundra related to the protection (sheltering) role of artificial forest plantations in the north. Botanical Journal (Botanichesky Zhurnal) 39 (1), 28-47 <in Russian>.

Andreyev, V.N. 1956. Current forest distribution pattern in tundra. In: Vegetation of the Far North and its management, Issue 1, pp. 27-46. USSR Acad. Sci. Publ., Moscow-Leningrad <in Russian>.

Dadykin, V.P. 1952. Behaviouristic features of vegetation on cold soils. USSR Acad. Sci. Publ., 276 pp. <in Russian>.

Demyanov, V.A. 1988. On the terms "moderately open woodland" and "highly open woodland" used in tundra science. Botanical Journal (Botanichesky Zhurnal) 73 (9), 1313-1318 <in Russian>.

Dzedzula, A.A. 1969. Forestry and inventory peculiarities of larch forests in the Khantaika river catchment area. Abstract of Thesis. Siberian Technological Institute. Krasnoyarsk. 28 pp. <in Russian>.

Dylis, N.V. 1939. Open birch woodlands and treeless dwarf shrub formations in the taiga forests of the Pechera region as signs of the landscape of the glacial period. Botanical Journal (Botanichesky Zhurnal) 24 (4), 314-338 <in Russian>.

Karpel. B.A. and N.S. Medvedeva. 1977. Seed production of *Larix dahurica* in Yakutia. Nauka Publ. 117 pp. <in Russian>.

Kolesnikov, B.P. 1969. Forest management areas of the USSR: taiga forest zone and forest management systems in the context of long-term prediction. Inf. Sheet of Scientific Council for Complex Taiga Zone Management No 2, pp. 9-39. Irkutsk <in Russian>.

Korotkov, I.A. and A.A. Dzedzula. 1969. The forests of the Khantaika river basin. Siberian forest types, Issue 2, pp. 230-242. Krasnoyarsk <in Russian>.

Korotkov, I.A. 1991. Classification of near-tundra forests of Siberia by site conditions. In: Ecogeographical problems of boreal forest sustenance and artificial regeneration. Proc. All-Union Sci. Conf., pp. 303-307. Arkhangelsk <in Russian>.

Kuvaev, V.B. 1975. Description of the vegetation of the southern Putoran mountains. In: Natural landscape patterns of Putoran lakes, pp. 160-186. Nauka Publ., Novosibirsk <in Russian>.

Lukicheva, A.N. 1963. Vegetation of the northwestern Yakutia in relation to the geological structure of the region. Moscow-Leningrad. 168 pp. <in Russian>.

Matveyev, P.M., and A.P. Abaimov. 1980. On the role of fire in larch stands on permafrost soils. In: Forest fires and their effects, pp. 123-129. Krasnoyarsk <in Russian>.

Matveyev, P.M. 1984. Postfire soil thawing in larch stands distributed in the permafrost zone. In: Larch (problems of complex produce processing). Interuniversity Scope of Scientific Papers, pp. 36-41. Krasnoyarsk Polytechnical Institute, Krasnoyarsk <in Russian>.

Mironenko, O.N. 1967. Description of the forest cover of the northern Central Siberian highland (Kotui river basin). J. Forestry (Lesovedenie) No 5, 28-36 <in Russian>.

Mironenko, O.N. 1970. Vegetation at the Kotui river head (Northern Evenkia). Abstract of the Thesis. Krasnoyarsk. 31 pp. <in Russian>.

Norin, B.N. 1958. On the investigation of wood species regeneration from seed and sprouts in the forest-tundra zone. In: Vegetation of the USSR Far North and its management, Vol.3., pp. 154-244. Moscow-Leningrad <in Russian>.

Norin, B.N. 1978. Vegetation pattern of the central Ary-Mas forest block. In. Environmental conditions, flora, and vegetation of the northernmost forest area, pp. 133-162. Nauka Publ, Leningrad <in Russian>.

Norin, B.N. 1979. The structure of vegetation communities of the east-European forest-tundra zone. Nauka Publ., Leningrad. 200 pp. <in Russian>.

Norin, B.N. et al. 1986. Mountain phytocoenotic systems of the subarctic zone. Nauka Publ., Leningrad. 290 pp. <in Russian>.

Parmuzin, Yu.P. 1979. The tundra-forest zone of the USSR. Misc. Publ. Moscow. 295 pp. <in Russian>.

Pozdnyakov, L.K. 1975. *Larix dahurica*. Nauka Publ., Moscow. 309 pp. <in Russian>.

Pozdnyakov, L.K. 1986. Permafrost forest science. Nauka Publ., Novosibirsk. 190 pp. <in Russian>.

Prokushkin, S.G. 1982. Mineral nutrition of pine supported by cold soils. Nauka Publ., Novosibirsk. 202 pp. <in Russian>.

Sambuk, V.F. 1937. Brief description of the vegetation of the Taimyr peninsula. In: The Problems of the Arctic Zone, Vol1., pp. 127-154 <in Russian>.

Shcherbakov, I.P. 1982. Forests at the head and central parts of Vilui river. Scientific paper scope: Botanical Studies in the Cryolithic Zone, pp. 91-104. Yakutsk <in Russian>.

Shumilova, L.V. 1962. Botanical geography of Siberia. TGU Publ. Tomsk. 440 pp. <in Russian>.

Sochava, V.B. 1956. Dark coniferous forests. In: Vegetation cover of the USSR, Vol.1, pp. 133-256. USSR Acad. Sci. Publ., Moscow-Leningrad <in Russian>.

Sofronov, M.A., and A.P. Abaimov. 1991. Peculiarities of the regeneration of forests on cold soils. In: Proc. Conf. on Fundamental Concepts of Forest Formation Process, pp. 154-155. Krasnoyarsk <in Russian>.

Sofronov, M.A. 1988. Pyrological characteristic of the vegetation of the head portion of Turukhan river basin. In: Forest fires and their control, pp. 205-211. VNIILM Publ., Moscow <in Russian>.

Stepanov, G.M. 1985. Forest regeneration in burned areas in the northern taiga of Yakutia. Abstract of the Thesis. Krasnoyarsk. 17 pp. <in Russian>.

Tarabukina, V.G., and D.D. Savvinov. 1990. Fire impact on permafrost soils. Nauka Publ., Novosibirsk. 120 pp. <in Russian>.

Tchertovsky, V.G. 1983. Peculiarities of near-tundra spruce stand structure and growth. In: Problems of the management of near-tundra forests of the USSR Europen part, pp. 5-16. AILiLH Publ., Arkhangelsk <in Russian>.

Tchertovsky, V.G., B.A. Semenov et al. 1987. Tundra bordering forests. Agropromizdat Publ. Moscow. 168 pp. <in Russian>.

Tchugunova, R.V. 1971. *Larix dahurica* regeneration in the Zhigansky district of Yakutia. In: The study of vegetation and soils of the Northeastern USSR, pp. 76-82. Yakutsk <in Russian>.

Tchugunova, R.V.1979. Fire influence on regrowth and postfire regrowth development. In: Forest fires in Yakutia and their impact on vegetation patterns, pp. 158-181. Nauka Publ., Novosibirsk <in Russian>.

Tyulina, L.N. 1937. Khatanga region forest vegetation at the northern boundary of its areal extent. Arctic Institute Transactions (Geobotany) Vol.63, 83-180. Leningrad <in Russian>.

Viereck, L.A. 1973. Wildfire in the taiga of Alaska. Quat. Res. 3, 445-465.

Vipper, V.I. 1964. The influence of grass-brush layer on *Larix dahurica* regeneration in Central Yakutia. In: Artificial regeneration and improvement of the forest, pp. 122-132. USSR Acad. Sci. Publ., Novosibirsk <in Russian>.

Vodopyanova, N.S. 1975. Vegetation of the southwestern Putoran mountains. In: Putoran Lake Province, pp. 122-140. Nauka Publ., Novosibirsk <in Russian>.

Vodopyanova, N.S. 1976. Vegetation of the Putoran mountains. In: Flora of the Putoran, pp. 11-31. Nauka Publ., Novosibirsk <in Russian>.

Yegorov, O.V. 1961. Ecology and hunting of the Yakutian squirrel. USSR Acad. Sci. Publ., Moscow. 266 pp. <in Russian>.

J.G.Goldammer and V.V.Furyaev (eds.), Fire in Ecosystems of Boreal Eurasia, 387-392.

Fire Effects on Larch Forests of Central Evenkia

P.A. Tzvetkov [1]

1. Introduction

The forests of Central Evenkia are made up largely of *Larix gmelinii*, whereas *Larix sibirica* and *Larix czekanowskii* are distributed only in the western part of the area. In this area, forest fires are a great concern. Increasing resource exploitation and extremely poor forest fire protection in the north are the two factors accounting for the high rate of forest burning. As permafrost is continuous in the area, fires strongly disrupt the ecological equilibrium (balance) of northern forests.

This paper addresses only some of the wide variety of ecological consequences of fire for northern larch forests. Char formation on tree trunks, post-fire fuel load dynamics, depth of soil thawing for a variety of burns, and post-fire larch stand regeneration will be discussed. Since crown fires are very rare in the area, only surface fire effects will be considered.

The height and direction of char on tree trunks are known to be descriptors of major fire parameters. They indicate surface fire intensity and spread pattern and allow one to predict post-fire tree mortality. Literature review has shown that scientists disagree on relationships between char formation and tree diameter. Most opinions, disparate as they are, are based on visual observations. Actual data on either the absence or existence of a relationship between char characteristics and tree diameter are rare. These relationships have not yet been investigated for larch forests of the northern Krasnoyarsk Territory (Evenkia). In forest biogeocoenoses that occur in the permafrost zone, char formation can be expected to have some particular features.

2. Factors Influencing Char Height

In general, char height and direction are controlled by a set of factors that include fire intensity, wind, topography, stand morphology, and individual tree characteristics (species, diameter, bark structure, etc.). The influence of these factors, and their combinations, on char formation varies with forest location. In any given location, some of these factors can be assumed to be approximately constant. This assumption can be made because sample sites are established in burned areas having similar relief and site conditions. In this case char height thus depends on wind and tree diameter.

---

[1] V.N. Sukachev Institute of Forest, Siberian Branch, Russian Academy of Sciences, 660036 Krasnoyarsk, Russian Federation

It is known that char height on the leeward side of the trunk is greater than on the windward side. This can be attributed to vortices developing on the leeward side when the wind goes around the tree. This results in a zone of low air pressure that develops near the trunk surface. The zone extends up the trunk and is rapidly filled with gaseous combustion products. This leads to greater flame height on the leeward side in comparison with the windward side of the trunk. It is noteworthy that stronger vortices develop near larger diameter trees, which result in greater char height on their trunks. Therefore, the wind-related difference in char height can be expected to be greater for large diameter trees as compared to small diameter ones. In one of his papers, Amosov (1964) mentioned vertical vortices that caused the flames to go up the leeward side of the trunk to a considerable height. This regularly observed trend of char to be of greater height on the leeward side of the trunk is called char direction. This char formation pattern is observed for all tactical parts of a fire (the front, the flanks, and the rear).

The above considerations underlie the hypothesis that a direct correlation exists between tree diameter and difference in char height between the windward and the leeward side of the trunk. Absolute char height should thus correlate directly with tree diameter.

Observations conducted in burned areas in the larch-small shrub-moss-lichen forest type have revealed that the leeward side char height (H_{max}) is, on average, four times greater than that of the windward side.

3. Observations in *Larix gemilinii* Forests of Central Evenkia

We have established the form of the relationship between char parameters and tree diameter. Figure 1 presents char height difference as a function of trunk diameter. A linear equation was developed to satisfactorily fit the experimental data:

$$Y = 1.2 + 0.16\,X,$$

where Y is H_{max} - H_{min} (m), and X is tree diameter class (in 2 cm intervals).

The degree of the dependence of difference in char height on trunk diameter was judged by the correlation coefficient (r). The latter was found to be +0.875, with a standard error (s.e.) of ± 0.08. Reliability (confidence) index ($r/s.e._r$) is 11. The correlation coefficient value suggests the existence of a strong direct relationship between char height difference and tree diameter. Based on determination coefficients, the difference in char height was observed to be related primarily to tree diameter for 77% of sample sites. For the remaining 23% of the sample sites, other factors appeared to control char height differences. Difference in char height was established for *L.gmelinii* to increase with tree diameter, thereby confirming our hypothesis.

Figure 2 illustrates the dependence of the maximum char height on larch tree diameter. This relationship is somewhat different from that discussed above. The results of computer-aided regression analysis show that the relationship can best be described by an exponential equation:

$$Y = e^{(0.71 + 0.057X)},$$

where Y is the char height on the leeward side of the trunk (m), and X is the tree diameter class (cm).

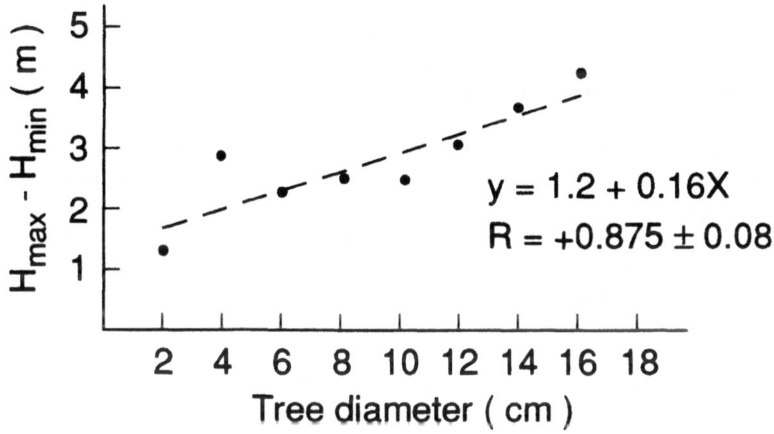

Fig.1. Dependence of char height difference on tree diameter. X-axis: Tree diameter (cm); Y-axis: H_{max} - H_{min} (m).

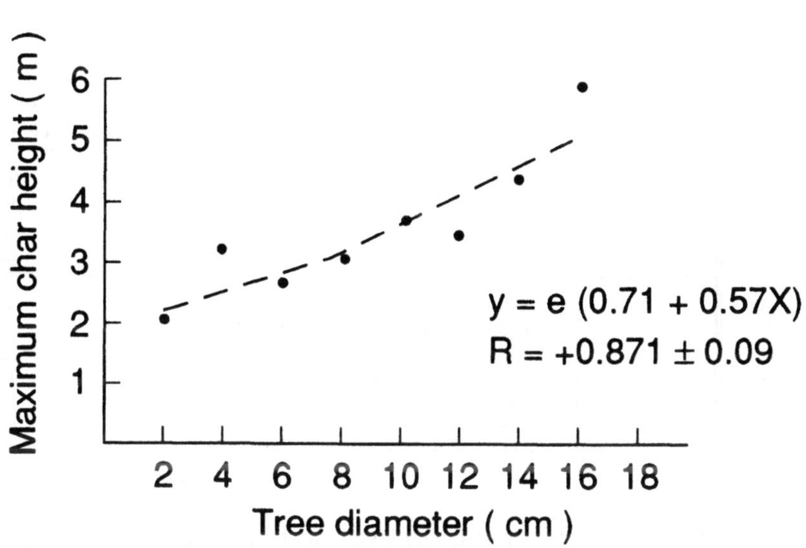

Fig.2. Maximum average char height related to tree diameter. X-axis: Tree diameter (cm); Y-axis: H_{max} (m).

The correlation coefficient between mean-maximum char height and larch diameter was +0.871, with a standard error of 0.09 and a reliability index of 10. Maximum char height is thus closely and directly related to tree diameter. This is confirmed by observational data from other regions (Amosov 1964; Tzvetkov 1976; Voinov and Sofronov 1976). The determination coefficient (r^2) of 0.76 also indicates that tree diameter accounts for 76% of the variability in char height.

In general, based on observational data from burned areas, surface fire causing char from 0.1 to 0.5 m high normally leads to complete stand mortality. This is due to a number of factors. Crowns of relatively short larch trees are scorched by convective heat from surface fires. More than half of the trees were observed to have yellow needles soon after a surface fire. Also, the thin bark of larch (less than 1 cm) cannot protect the cambium from thermal stress. Partial injury of the cambium leads to fire scar formation. In cases where the cambium is completely destroyed, the tree dies. More than 3/4 of the trees in our sample sites were found to have cambial injury on their stems. Finally, the root systems of more than 1/4 of the trees were observed to be injured by surface fires. This is because permafrost does not allow the roots to penetrate into deep soil layers. Typically, every tree turns out to have fire injury to trunk, crown or roots, each of which promotes tree death. To conclude, the resistance of *Larix gmelinii* to fires under permafrost conditions is much lower than in central and southern taiga areas, where permafrost is not a factor.

Fires induce considerable changes to the forest fuel load, composition, and structure. The major fuel parameters are presented in Table 1. The total fuel load decreased by 46% after surface fires. Over time, fuel load is restored primarily due to contributions by grass, low brush layers and litter. Fuel load increases to 86% of the total pre-fire load by 10 years after fire, and to 96% after 20 years. Burned sites have the potential to reburn 5-7 years after the initial fire.

Tab.1. Major fuel parameters in larch-low shrub-green moss* forest.

Site	Period since last fire (years)	Total fuel load (kg^{-1})	Mean thickness (cm)		Density (kg m^{-3})	
			moss-lichen layer	forest floor organic layer	moss-lichen layer	forest floor organic layer
Burned site	1	36,100	4.3±1.2	2.6±0.4	39.2	74.0
Duplicate	-	66,836	11.9±1.5	0.9±1.3	15.0	43.7
Burned site	10	57,564	2.6±0.5	3.6±1.1	44.5	122.2
Duplicate	-	66,692	6.9±0.9	7.6±1.2	24.1	64.0
Burned site	20	78,828	2.8±0.4	4.9±0.7	18.2	145.7
Duplicate	-	81,652	8.8±0.2	5.5±0.4	20.2	122.6

In unburned (duplicate) sites, the moss-lichen layer was found to be 3 times as deep as that in burned sites. This suggests that mosses, and particularly lichens, need a time period much greater then 20 years to recover. In contrast, the forest floor organic layer recovers much faster. For example, in a newly-burned area, the organic layer load was measured to be 3.8 times less than in the duplicate, whereas in 20 years they differed by only 10%. In burned areas, fuels are normally more compacted than in unburned areas.

Measurements of the depth of soil thawing have revealed that it varies widely with microrelief and with the densities of moss-lichen and grass layers. For example, depth of thawing is 1.5-3 times less in depressions than on hummocks. In sites covered by dense *Alnaster fruticosus* bush or *Ledum palustre*, the depth of thawing is 1.7-25 times less than that in open sites. A thick (20-25 cm) moss layer actually prevents any thawing, and permafrost is located right below the moss layer. The above difference is maintained for 20-30 years and is closely related to post-fire regeneration of surface vegetation cover.

Post-fire regeneration of *L. gmelinii* was studied in transects constructed across the centre of burned areas. One of these transects is presented in Figure 3. The transect is four kilometers long. As is obvious from our observations, the larch stand was almost completely killed by fire and stem insects. Dead trees can remain standing for decades since, under permafrost conditions, soil microfauna are not particularly active.

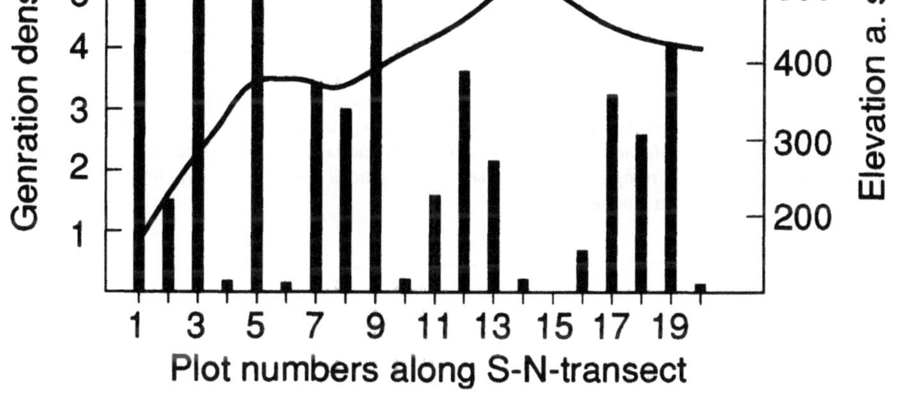

Fig.3. Density of larch regeneration along a South-North transect. X-axis: sample site numbers; Y-axis: regeneration density, thousands of individuals per hectare (left), and elevation a.s.l. (m) (right).

L.gmelinii was observed to regenerate much more slowly in the central part of the burned area than at its boundary. For example, at the boundary of the burn, the density of seedlings and regrowth is 30,000-50,000 individuals (n) per ha, and in some places, 70,000-90,000 n ha^{-1}, whereas in the center of the burned area it rarely exceeded 6,500 n ha^{-1} (only in three of all sample sites). Average regeneration density was 2,300$\pm$510 n ha^{-1}. On south-facing slopes (sites 1 to 5 and 8 to 14), under more favorable climatic conditions, forest regeneration was more successful (1,500-6,500 n ha^{-1}, with the average being 3,300$\pm$700 n ha^{-1}) in comparison with north-facing slopes, where the density of viable regrowth varied from 0 to 4,000 n ha^{-1}. Average density was 2,200$\pm$500 n ha^{-1}. Regrowth occurred on about 70% of the sample sites thereby indicating non-uniform forest regeneration in the burned area.

On the whole, no tree species replacement was observed in the forest regeneration process, because *L.gmelinii*, as a climax species, has no competitors in Central Evenkia, except on flat interfluves, where the short-term replacement of larch by *Betula pendula* is possible.

4. Conclusions

To sum up, in larch stands of Central Evenkia, surface fires usually result in the complete destruction of the overstory. Tree mortality is typically near 100 percent, even when the char height is only 0.1 to 0.4 m. Forest stand regeneration is normally quite successful in the area of interest. We attribute this success to particular biological features of *L. gmelinii* (e.g., an individual is capable of scattering its seeds for 3-4 years in succession), and to fire-caused mineral soil exposure and decreases in root competition. The ecological niche of *Larix gmelinii* is thus successfully maintained.

References

Amosov, G.A. 1964. On surface forest fire growth patterns. In: Vozniknovenie lesnikh pozharov <Forest fire occurrence>, 152-171. Nauka. Moscow <in Russian>.

Tzvetkov, P.A. 1976. On fire indication problem. In: Sovremennye issledovania tipologii i pirologii lesa <Current investigations into forest typology and pyrology>, 108-114. Arkhangelsk <in Russian>.

Voinov, G.S., and M.A. Sofronov. 1976. Predicting stand mortality after surface fire. In: Sovremennye issledovania tipologii i pirologii lesa <Current Investigations into Forest Typology and Pyrology>, 115-121. Arkhangelsk <in Russian>.

J.G.Goldammer and V.V.Furyaev (eds.), Fire in Ecosystems of Boreal Eurasia, 393-403.

Ecological Estimation of Forest Succession Patterns in Central Angara Region

F.I. Pleshikov and V.A. Ryzhkova [1]

1. Introduction

It is now widely recognized that the forests of the Angara region, as they look today, developed essentially under the influence of fires (Popov 1957, 1982; Krauklis 1975, 1985; Buzykin and Popova 1978). Fires are a common disturbance event and an important ecological factor in taiga forests. Over the last few decades, spatial patterns of forest communities, at different stages of post-fire succession, are undergoing complicated changes due to increasing forest resource exploitation. Intensive forest harvesting leads to a great increase of areas occupied by young stands. In most parts of the Angara region, fires and cutting cause the replacement of climax dark conifers by light coniferous and hardwood species. Relatively frequent surface fires destroy dark conifer regrowth and the subordinate (lower) wood layer composed of spruce, fir, and *Pinus sibirica*, thereby hamper their regeneration. Wide local variations in age structure, species composition, and productivity of southern taiga forests of the Angara region are attributed to the heterogenity of ecological site conditions and different patterns of fire occurrence and harvesting, which in turn determine specific microclimatic situations. Since regeneration processes after human-caused disturbances in fact obey, the same laws as natural vegetation succession, it is reasonable to study post-fire and post-cutting vegetation successions in the context of general vegetation cover dynamics.

The multi-variant nature of forest community dynamics introduces problems to the analysis and systematization of succession stages. In this regard, the importance of type classification of forest communities differing in age, species composition, structure, and succession stage should be stressed. There is a long-standing opinion among scientists (Kopp and Schwanecke 1972; Layser 1974; Chambless and Nixon 1975; Dyrenkov 1984; Manko 1989; Filrose et al. 1990) that the main control of successional development is ecotope (i.e. site type) which is in turn determined by constant landscape parameters. Therefore, dynamically different phytocoenoses should be classified using an ecological approach.

The objective of our investigation was to determine forest succession patterns for a variety of site conditions in order to predict succession rates and trends. We studied a 15x5 km sample site crossing the Chuna river valley which represents a considerable range of ecological conditions found in the study area.

---

[1] V.N. Sukachev Institute for Forestry and Timber, Siberian Branch, Russian Academy of Sciences, 660036 Krasnoyarsk, Russian Federation

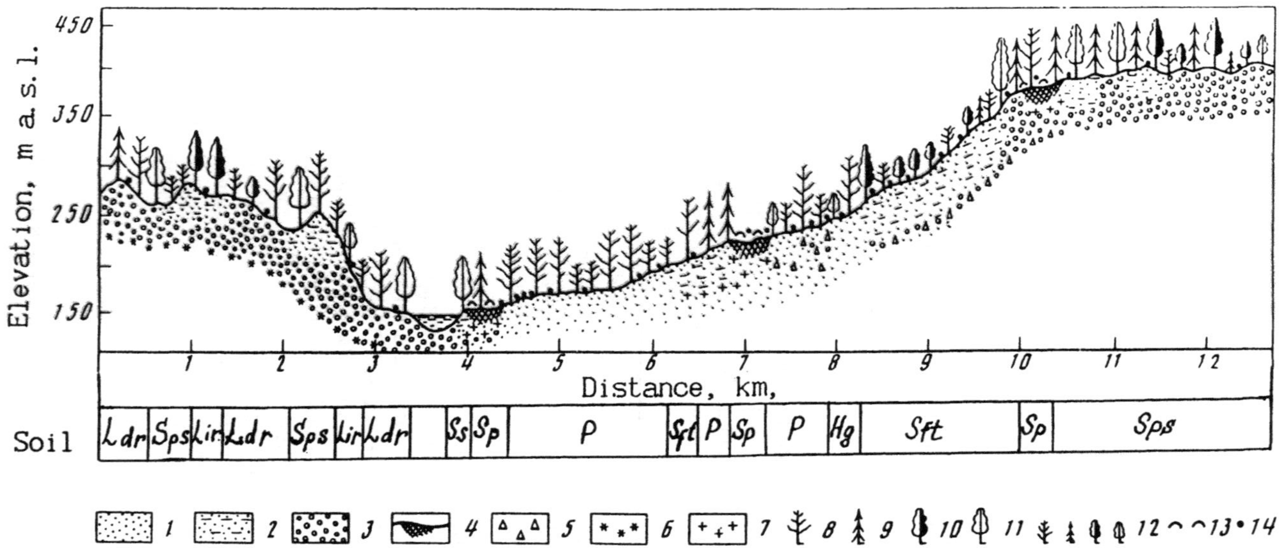

Fig.1. Topoecological transect of the sample sites. Soil-forming constituents and underlying bedrock: 1- sand; 2 - loamy sand; 3 -loam; 4 - peat; 5 - trapps; 6 - sandstone; 7 - permafrost. Vegetation: 8 - mature pine stands; 9 - mature dark coniferous forest; 10 - mature aspen forest; 11 - mature birch forest; 12 - young stands of the above species; 13 - bogs; 14 - sample plots. Soils: П - podzolic soil and shallow podzols; Пг - humic gley soil; Дп - soddy podzolic soil; Дпт - soddy forest soil of trapps; Дкм - leached deep rendzina; Дкс - leached immature rendzina; Пс - stratified soil of flood plains; Тм - shallow peat soil.

2. Material and Methods

One can get the idea of the main soil and vegetation cover features of the area in question, as well as its landscape structure when looking at the transect going through the centre of the sample site (Fig.1). Spatial soil and vegetation distribution reflect relatively well the heterogenity of geomorphological conditions and composition of the bedrock. Aerial imagery analysis and ground data interpretation have resulted in the identification of five eco-geomorphological classes characterized by a similar range of site conditions:

1) Chuna river flood plain occupied by naturally thinned fragmentary stands of birch, *Pinus sylvestris*, fir, larch, and *P.sibirica* on humic gley and peat bog soils;

2) the first fluvial terrace above flood plain comprised of pure pine stands, with lichens dominating the surface cover, growing on hillock sands;

3) the second fluvial terrace represented by pine forest, with a little of hardwood-green moss forest type, on sandy soil underlaid by quartz sand bedrock;

4) a near-watershed slope covered by pine and polydominant communities supported by sand and loam; and

5) flat interfluves of watersheds with loamy soils creating typical site conditions for dark coniferous forest development.

Sample plots were established, similar in dominant woody species composition and age (within 10 years of age), representing each ecological class. A detailed analysis of local vegetation diversity of each class was done using the method of spatial and temporal analogy (Alexandrova 1964; Vinogradov 1984; Armand and Kushnareva 1989). Also, forest regeneration series were constructed for different site types; boundaries of the sites are determined by meso-relief features, composition of soil-forming constituents, soil-ground layer structure, and soil type.

Biogeocoenotic interpretation of data on ecological conditions and vegetation obtained in the sample site revealed 5 and 3 regeneration series under automorphic and hydromorphic conditions, respectively. For each series, mean values for the major parameters were calculated.

Using ecological scales developed by Ramensky et al. (1956), ecological areals were determined for the series (pine-lichen, pine-green moss, pine-mixed herb, and polydominant series) most representative of the sample site (in the first three series, surface cover is dominated by *Cladonia silvatica*, *C.rangiferina*, and *Vaccinium vitis-idaea*; *Pleurozium schreberi*, *Hylocomium proliferum*, and *Dicranum polyzetum* and *V. vitis-idaea*; *Carex macroura*, *Calamagrostis arundinacea*, *Iris ruthenica*, and *Rubus saxatilis*, respectively. In the latter series, tree layer is represented by *Abies sibirica*, *Picea obovata*, *Pinus sylvestris*, *P.sibirica*, and *Larix sibirica*). These ecological areals were then plotted in eco-phytocenotic coordinates, with the Y-axis axis showing soil richness and the X-axis showing soil moisture capacity. Ordination of the series into a medium of ecological factors (Fig.2) confirms their being ecologically detached and agrees well with the results of our investigation of environmental factors (relief, soil and others). This justifies an ecological approach to the classification of vegetation development stages.

For each regeneration series represented by a certain type of site condition, regeneration dynamics descriptors that include environmental conditions, site-specific wood quality, regeneration similarity between climax and secondary woody vegetation, and site-specific rate of succession were assessed.

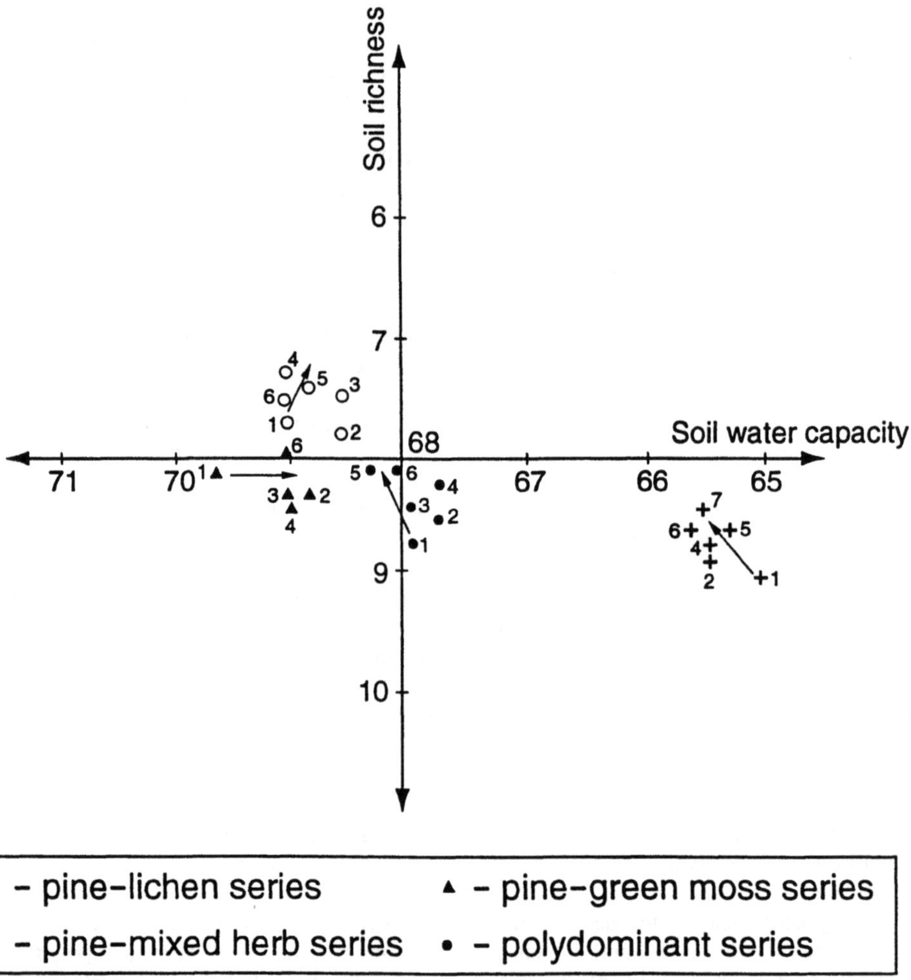

Fig.2. Centers of ecological ranges of successional stages of regeneration series found in the sample sites located in the Central Angara region (ecological coordinate system after Ramensky et al. [1965]). Y-axis: soil richness; X-axis: soil moisture capacity. Numbers indicate regeneration stages within the series (pine-lichen series; pine-green moss series; pine-mixed herb series; pine-polydominant series).

Ecological conditions were assessed using soil richness and soil moisture capacity scales, as well as the Soil-Ecological Index (SEI; Krutko et al. 1982), which is an integral index of the quality of environmental conditions, since it considers the major ecological factors determining particular vegetation quality being estimated by vegetation response manifested in mean stand height variations.

The SEI is calculated by the following equation:

$$SEI = Ch\ f(h) + Cc\ f(c) + Cub\ f(ub) + Cp\text{-}e\ f(p\text{-}e)$$

where

C = factor weight coefficient
h = is humus layer depth
c = clay content
ub = underlying bedrock depth of location
p-e = precipitation-evapotranspiration coefficient
f = function of factor ranging from 0 to 1.

Weight coefficients of the factors were found through the comparison of mean height values for stands supported by soils characterized by relatively similar values for all the factors, except for that considered at the moment. The value for f was calculated using mean stand height plotted against changes in each factor.

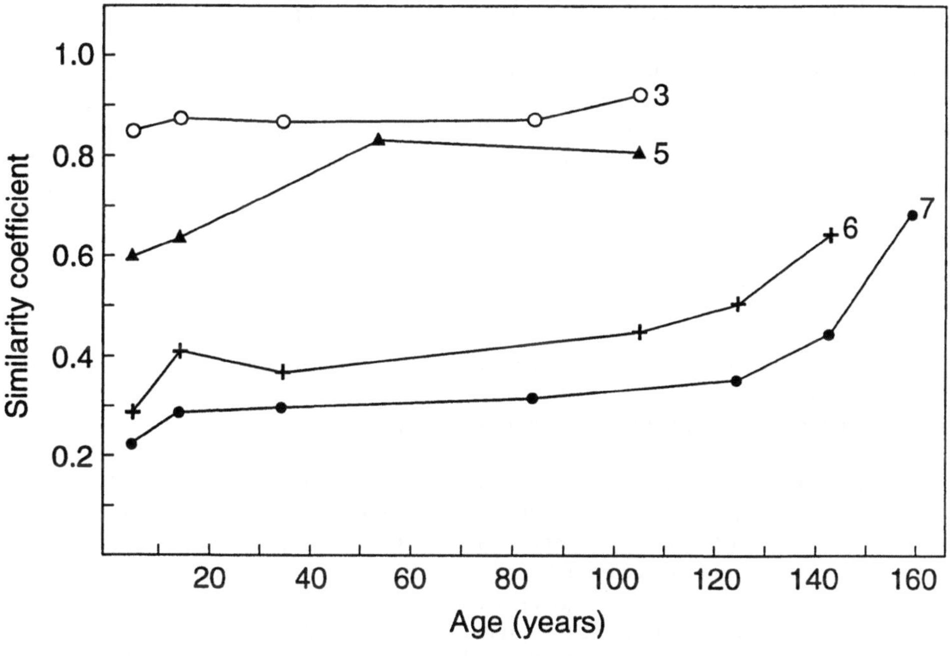

Fig.3. Changing similarity coefficients for various regeneration series (3 - pine-lichen series; 5 - pine-green moss series; 6 - pine-mixed herb series; 7 - pine-polydominant series. X-axis: stand age (yrs); Y-axis: similarity coefficients.

3. Results

To determine rate of succession under various site conditions coefficients were calculated (Vasilevich 1969) showing similarity between regeneration stages and a climax stand of a given series; these coefficients describe a community "level of recover", in per cent (i.e. they show to what extent woody species composition, as well as small brush-grass layer composition of a given community is similar to that of the climax community). Figure 3 gives the similarity coefficient curves for four regeneration stage series. The rate of regeneration varies with site conditions. The two upper curves describing relatively poor site conditions show that by the time trees achieve 100-110 years of age, similarity coefficient is as high as 80-90%, i.e. species regeneration is almost finished. Under relatively rich site conditions (lower curves), the similarity coefficient is about 60%, even for a longer (140-150 years) period of time. This suggests that regeneration is slower under rich site conditions.

Similarity coefficients and methods proposed by Lopatin (1972) were used to calculate succession rate (% per year) indicating the level of changes occurring at each regeneration stage, and relative rate of regeneration showing the percentage of changes for each stage.

As is obvious from Figure 4, the highest rate occurs at initial succession stages. The steepest slope describes relatively poor site conditions (pine-lichen series), while the smoothest one corresponds to relatively rich sites (polydominant series) where changes occur much slower. This is due to succession-induced replacement of overstory and dominant grass layer species.

Estimation of the dynamics of separate forest ecosystem components does not provide the real picture of forest regeneration peculiarities. Soils, tree stands, grass layer, and moss layer disturbed by fire or cutting recover at different rates. Therefore, in order to integrally evaluate the structure of an ecosystem at various regeneration stages, it is reasonable to use energy parameters, which are invariant for all forest components.

The stabilization coefficient (Sc) was taken as a criterium of ecosystem evaluation. This coefficient is derived as:

$$Sc = 1 - \frac{I}{G}$$

where

I = amount of energy lost by a living phytocoenosis to litter and dead trees at various regeneration stages
G = total energy in vegetation and humus layer at the same stages.

Energy characterisics were calculated from phytomass composition data. The analysis of the data in Figure 5 shows that the relationship (ratio) between lost and total energy of ecosystems is not constant along the whole succession. The temporal pattern of community stabilization is of a cyclic character and depends on the periodicity of the processes occuring in community recover back to its original structure.

The curves show three periods characteristic of forest regeneration dynamics. Maximum energy fluxes and minimum stability of the ecological environment are observed at the grass stage-tree stage interface (10 years after disturbance). This period is remarkable for the most profound changes in phytocoenoses composition and structure due to high competition between grasses and trees.

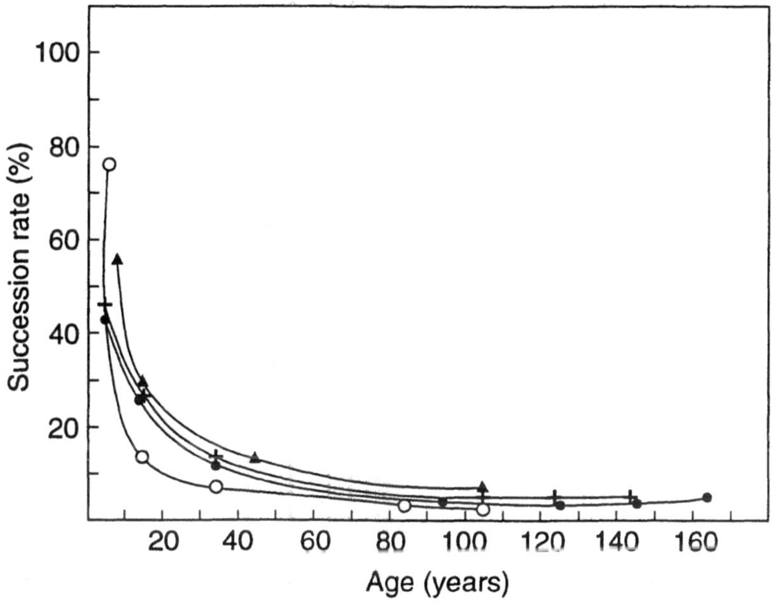

Fig.4. Relative rate of syngenetic succession under various site conditions (% per stage). For series indication: see Figure 2. X-axis: stand age (yrs); Y-axis: rate of succession.

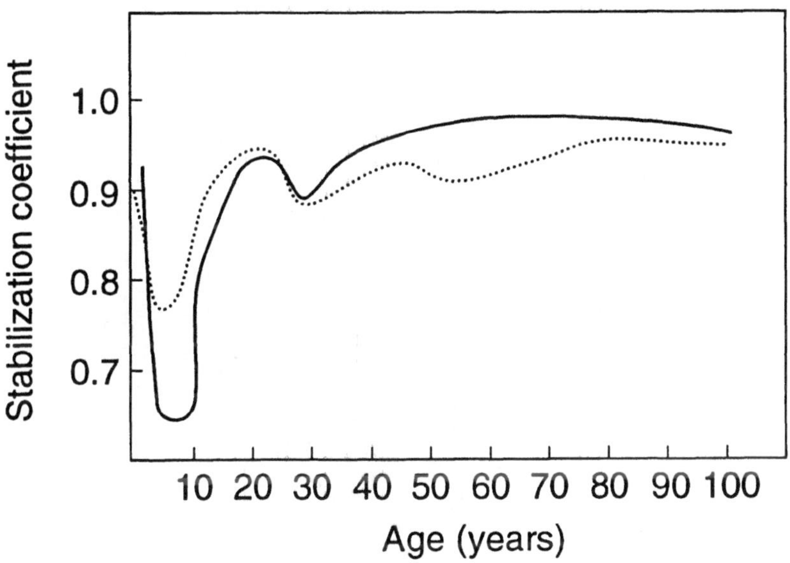

Fig.5. Ecosystem stabilization coefficient changes in a pine-green moss (dashed line) and in a pine-mixed herb (solid line) series. X-axis: stand age (yrs); Y-axis: stabilization coefficient.

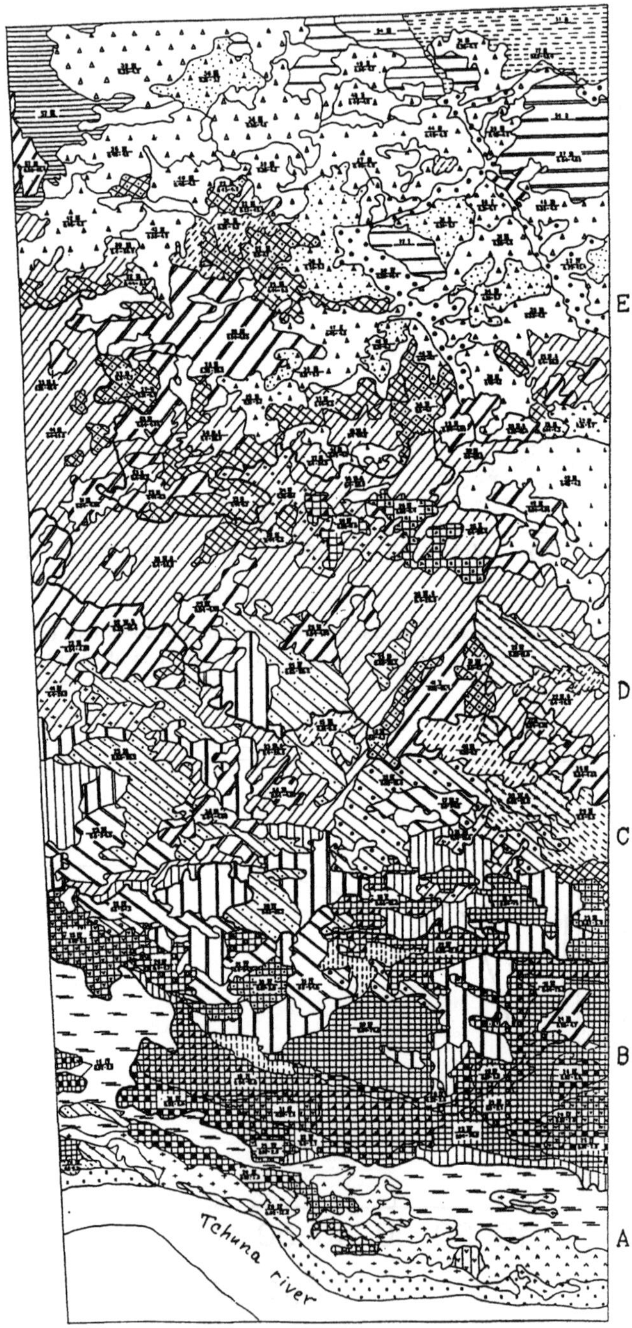

Fig.6. Ecological map of vegetation dynamics observed in the sample sites in the Central Angara region. Legend: Figure 7 (p.401).

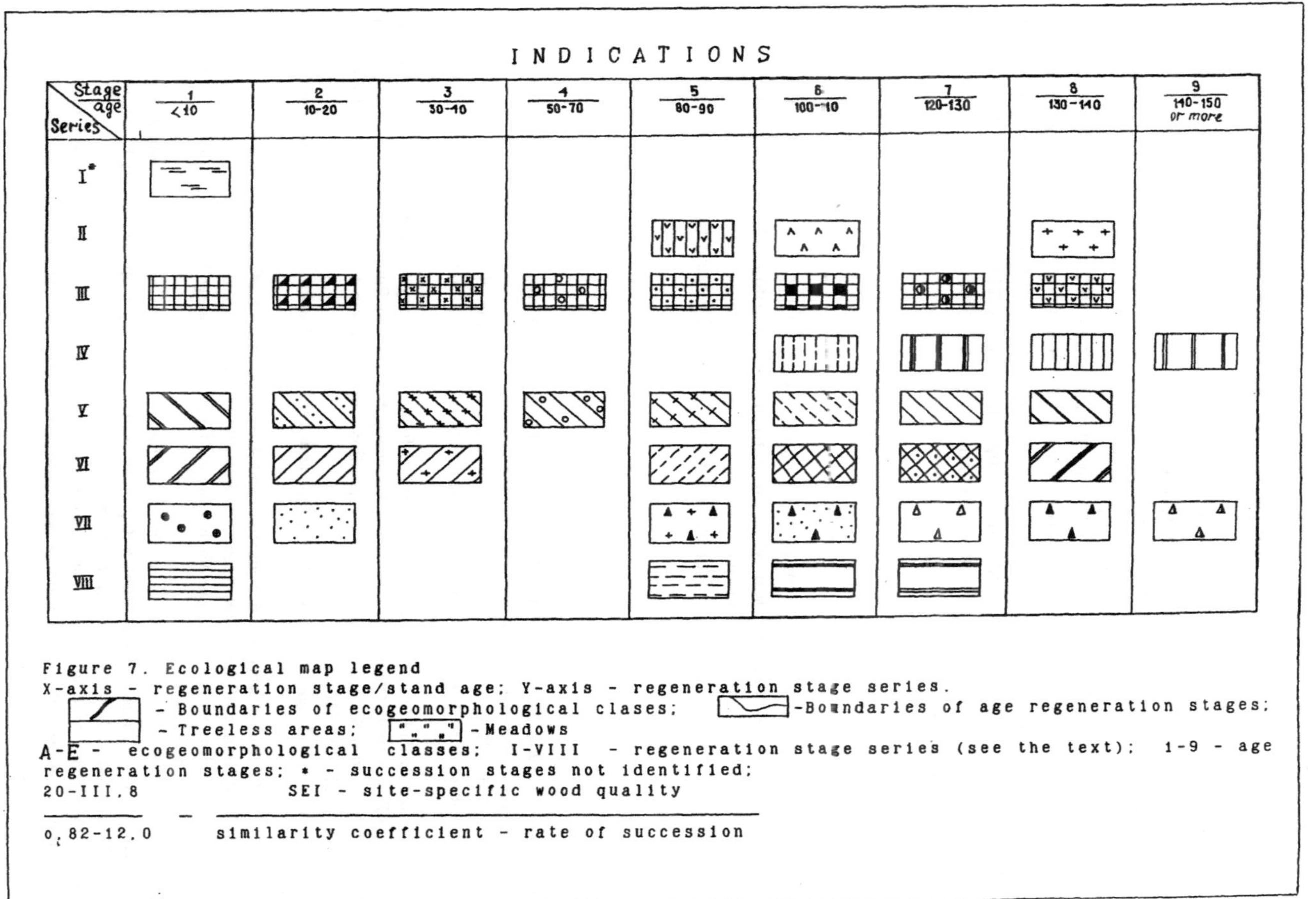

Fig.7. Legend for the ecological map (Fig.6).

One more decrease in ecosystem stability occurs during the period of intensive natural forest thinning, when tree competition is the highest (some 30 years after a disturbing event). Although phytomass losses are high, the stabilization level of ecosystems during this period is higher than that at initial succession stages.

Ecological conditions are the general controls of ecosystem stability and determine individual forest regeneration stage duration. The richer the conditions, the slower the ecosystem stabilization.

The results of the investigation of site-specific vegetation regeneration dynamics and acquired estimates of environmental conditions were used to build a large-scale ecological map reflecting dynamics of forest communities in the study area. The map and its legend are shown in Figures 6 and 7, respectively.

The ecological map shows dynamic vegetation categories, with each contour corresponding to a certain stage of an individual regeneration series being characterized by SEI and site-specific forest productivity (i.e. parameters of environmental conditions and stand productivity); similarity coefficient indicating the regeneration level achieved by a community; and the rate of succession considered as a percentage of changes occurred during a given succession stage.

4. Conclusions

The large-scale dynamics map is, in fact, a spatial model of the general succession pattern in the study area. Parameters presented in the map provide a complete and systematic picture of all variants of forest succession probable in the area in question and allow one to predict both direction and rate of forest succession under various site conditions.

References

Alexandrova, V.D. 1964. Vegetation cover dynamics. Field Geobotanical Investigations (Polevaya geobotanika) Vol.3, 300-450. Moscow-Leningrad <in Russian>.

Armand, A.D., and G.V. Kushnareva 1989. Ecosystems overcoming their critical state in space. In: Ecosystems in Critical State (Ekosistemy v kriticheskom sostoyanii), 75-82. Nauka Publ., Moscow <in Russian>.

Buzykin, A.I., and E.P. Popova. 1978. The influence of fires on forest phytocoenoses and soil properties. In: Pine Forest Productivity (Produktivnost sosnovykh lesov), 5-45. Nauka Publ., Novosibirsk <in Russian>.

Chambless, L.F., and E.S. Nixon. 1975. Woody vegetation-soil relations in a bottomland forest of East Texas. Texas J. Sci. 26, 407-416.

Dyrenkov, S.A. 1984. Taiga spruce forest structure and dynamics. Nauka Publ., Leningrad. 173 pp. <in Russian>.

Filrose, E.M., A.E. Riabchinsky, G.M. Galushko, and A.V. Konashov. 1990. Ecology of Forests of Western Bashkiria. Ural Branch of the USSR Acad. Sci. Sverdlovsk. 180 pp. <in Russian>.

Kopp, D., und W. Schwanecke. 1972. Zur Methodenwahl für großmaßstäbige forstliche Standortskartierungen. Beitr. Forstwirt. 6 (2), 4-10.

Krauklis, A.A. 1975. Specific features of the taiga forest of Angara region. In: Natural Regimes and Topogeosystems of Taiga Forest in Angara Region (Prirodnye rezhimy i topogeosistemy Priangarskoi taigi), 14-27. Nauka Publ., Novosibirsk <in Russian>.

Krauklis, A.A. 1985. Area analysis on a regional scale. In: Ecosystem Dynamics and Forest Resource Exploitation in Angara Region (Dinamika ekosistem i osvoyenie priangarskoi taigi), 6-72. Nauka Publ., Novosibirsk <in Russian>.

Krutko, V.N., S.A. Pegov, D.M. Khomyakov, and P.M. Khomyakov. 1982. Formalization of qualitative estimates of environmental components. All-Union Institute for System Investigation. Moscow <in Russian>.

Layser, E.F. 1974. Vegetative classification. It's application to forestry in the northern Rocky Mountains. J. For. 72, 354-357.

Lopatin, V.D. 1972. Using similarity coefficient equation in the analysis of grass cover development in meadows and in calculating the rate of syngenesis. In: Methods of Quantitative Estimation in the Investigation of Phytocoenoses Structure (Primenenie kolichestvennykh metodov pri izuchenii struktury fitocenozov), 76-83. Nauka Publ., Moscow <in Russian>.

Manko, Yu.I. 1989. Classification of dark coniferous forests of the Far East. In: Dynamic Forest Typology (Dinamicheskaya tipologia lesa), pp.72-80. Nauka Publ., Moscow <in Russian>.

Popov, L.V. 1957. Forests of Chuna-Vikhoreva interfluve. In: Biology Series, Transactions of East-Siberian Branch of the USSR Acad. Sci. Vol. 5, 144 pp. <in Russian>.

Popov L.V. 1982. Southern taiga forests of Central Siberia. Irkutsk State University Publ., Irkutsk, 360 pp. <in Russian>.

Ramensky, L.G, I.A. Tzatzenkin, O.I. Chizhikov, and N.A. Antipin. 1956. Ecological assessment of natural meadowlands using vegetation cover analysis. Moscow. 472 pp. <in Russian>.

Vasilevich, V.I. 1969. Statistical methods in geobotany. Nauka Publ., Leningrad. 232 pp. <in Russian>.

Vinogradov, B.V. 1984. Aero-space ecosystem monitoring. Nauka Publ., Moscow. 321 pp. <in Russian>.

J.G.Goldammer and V.V.Furyaev (eds.), Fire in Ecosystems of Boreal Eurasia, 404-408.

Forest Formation Processes after Fire in the Volga Region

K.K. Kalinin [1]

1. Introduction

The investigations of the forest regeneration processes on the fire areas of the Volga region have been conducted from 1974 to 1990 in the region of Nizhni Novgorod and in the Republic "Mari El" on areas burnt in the large forest fires of 1972.

The region has been the place of several great forest fires during the last 150 years. Great fires took place here in 1815, 1823, 1848-1851, 1891, 1921, and 1972 (Denisov 1979). The causes of these fires were climate, geological and hydrological conditions of the soil and the type of forests. The prevailing vegetation formation here is pine forests, that is stipulated by the geomorphology and soil formation of the region. In respect to vegetation and climatic conditions, this region bears some characteristics of the forest steppe climate. In comparison with other Volga regions (e.g. Kostroma Region), we discover that the probability of droughts during the last 100 years constitutes for the first region -10% and for the second region -5%. The annual atmospheric precipitation constitutes 472 and 558 mm respectively. The annual vaporization according to the level of the water balance is 320 and 295 mm (Kolobov 1968).

The fires of 1921 and 1972 had the most catastrophic effect on the region of investigations. In 1972 an area of 404,400 ha was affected by fire in Nizhni Novgorod region and 184,900 ha in the Republic "Mari El". The fires of 1972 were remarkable for high intensity, e.g. in the Mari Republic 69.6% of burnt areas experienced crown fires (Denisov 1976).

A remarkable feature of the natural regenerated seedling population on the fire areas of 1972 is their even age. It is due to the fact that abundant seeding by conifers happened in the first year after fire. The fires coincided with a good seed year for pine and spruce. In general they burnt at the end of august, when the seeds have practically ripened and even killed trees had preserved their seeds in the cones (Chistyakov and Kreyer 1976). Pine stands which survived on parts of the burnt areas, have produced practically no seeds during the first 5 years after the fire and only later a weak fruiting was observed. This however had no practical effect on the complementary accumulation of undergrowth because of the development of the grass soil cover and soil turfing. Owing to the fact that the isolated trees and groups of trees, left as seed trees, have not fulfilled their role and have not influenced the generation of pine. Broadleaved trees, both of seed and vegetative origin, appeared mainly during the first two years after the fire.

[1] The Mari Polytechnical Institute, 424024 Yoshkar-Ola, Republic Mari El

Tab.1. **Characteristics of naturally regenerated stands 10 years after fire**

Fire type	Average number of individuals by species (x 1000 ha^{-1})		Young stand composition*	Coverage (%)		Pine regrowth (%)	
	Pine	Hard-woods		ine	Hard-woods	Completely shaded	Partially shaded
	Pine-lichen forest type						
Unsteady surface fire of low intensity	1	1.5	4p 4b 2as	1	1	0	0
Moderate-intensity surface fire	1	2	6p 3b 1as	1	1	0	0
Steady surface fire	4.2	0.8	8p 2b	12	1	0	0
	Pine-*Vaccinium vitis idaea* forest type						
Unsteady surface fire of low intensity	3.5	3.3	5p 3b 2as	1	1	0	0
Moderate-intensity surface fire	5	3	6p 2b 2as	6	5	0	0
Steady surface fire	7	7	5p 3b 2as	7	5	1	2
Crown fire	1	6	6b 2as 1p	3	6	0	0
	Pine-*Vaccinium myrtillus* forest type						
Moderate-intensity surface fire	4.5	149	9b 1as +p	3	35	31	9
Steady surface fire	8	108	6b 3as 1p	7	59	45	18
	Pine-*Politrichum commune* forest type						
Steady surface fire	9	61	5b 4as 1p	5	21	6	8
	Pine-*Sphagnum* forest type						
Unsteady surface fire of low intensity	0.4	0	1as	0.1	0	0	0
Steady surface fire	23	5.3	8p	11	2	0	0

* In the Russian terminology the forest composition formula is actually a scale of 1 to 10 units representing the fraction of each woody species in a stand: p = pine; b = birch; as = aspen.

2. Regeneration of Forests After Fire

The character of forest regeneration, formation of young stands and their growth differ considerably depending on the type of forest and on the fire intensity (Tab.1). The investigations show that natural regeneration with a certain quota of coniferous trees is observed only where there were stands of middle-aged, premature and mature pine trees able to bring fruits, or broadleaved stands with a considerable proportion of pines.

Everywhere where there are burnt pine stands, mixed young forests are formed and the quantity of pine trees in the young stands reach on average $N=5,000$ ha^{-1} ten years after fire in the lichen type forest, $N=26,000$ ha^{-1} in the cowberry forests, and $N=75,000$ ha^{-1} in the blackberry forests. The lowest quantity of pine trees observed was $N=100$, 2000 and 1000 ha^{-1} in the respective forest types. The highest quantity of young pine trees is observed in the forests on sandy soils and in the transition zone from dry conditions to moister ones, reaching a maximum in the long moss and sphagnum pine forests. In the forest on sand-loam soils (lime, grass-bog types of forests) pine trees among the young stands are absent or there are only a small quantity of them.

The quantity of pine regeneration and its participation in the young stands is observed to increase with the increase of the intensity of lower surface fires. This is especially apparent in pine forests on swamped soils.This can be explained by the fact that surface fires and fires of average intensity damage only the upper layer of the cover or of the peatbogs, and in some cases, because of the peculiarities of relief, some parts happen to be left undamaged. Therefore seeds, especially of the broadleaved trees, have no favorable conditions for germination and growth as a result of the weak soil mineralization. In the case of crown fires, however, the pine seeds in the cones are under the influence of fire and lose, to a considerable extent, their viability, which negatively influences seedling establishment on the given areas. In the pine forests of cowberry and blackberry types, the pine trees are not considerably shadowed by broadleaved trees 10 years after fire. Under these conditions broadleaved trees and pines were of equal height. In some cases, pines happened to be higher than the broadleaved trees. Here pines are only lower than birch trees but these are present only in a small quantity. In these forest types pre-commercial thinning is demanded only in the places with a high quantity of broadleaved trees. In other cases, the regeneration of pine stands occur naturally.

In *Sphagnum* forests, broadleaved trees do not shadow pine trees due to their unsatisfactory development. In some places, however, pine trees do not prevail, although no change of forest type should be expected because it is hardly possible that birch trees will be able to compete with pine trees, and aspens are to be completely excluded. Data of tree mortality on one of the areas can confirm the above mentioned facts. During the period between 1974 and 1981 mortality was the following: pine 78.5%, birch 86.6%, aspen 99.4%.

In fresh and moist forests two-storied young stands with broadleaved trees (mainly birch in the understory and spruce in the overstory) have been forming in the lime and blackberry types of forests. The quantity of spruce in the moist soil conditions reached 6000 ha^{-1} 16 years after fire. Under less moist soil conditions the quantity of spruce in the young stands was considerably less (1200-2400 ha^{-1}), and was unevenly distributed among the lime trees in the spruce forest - mainly near groups of broadleaved trees. The height of 16-year old spruce was lower than that of broadleaved trees of the upper story. The most severe oppression of spruce 16 years after fire was observed in lime forests. Under more moist soil conditions pure stands of broadleaved trees have been forming.

3. Artificial Regeneration

In the region great experience on artifical forest regeneration has been accumulated. Different technologies and methods of artificial regeneration were tested. Pine has been regenerated according to various technologies generally depending on the age of the perished trees. Areas with marketable timber have been regenerated mainly by planting (mechanically or by hand) pine seedlings on the furrow created by the plough PKL-70. The density of seedlings range within N=4200-7600 ha^{-1}, with the spacing of 2.2-4.0 x 0.6-0.7 m. Under the given conditions the analogous siting of the seedlings mechanized planting, along with the spacing, was sometimes carried out without previous soil scarification.

In young and middle-aged stands with the average diameter up to 14 cm, preliminary clearing was carried out by various methods followed by mechanized planting without previous soil treatment. Here seedlings were planted at a density of 9-11,000 ha^{-1}. The spacing between the rows was 1.2-1.5 m and within the rows ≤0.75 m. More rarely, seedlings were planted into ploughed furrows with the plough PLP-135 at a spacing of 3-4 x 0.6 m. On territories infected with the larvae of cockchafer (*Melolontha vulgaris*), complete ploughing was carried out followed by the application of insecticides and the planting of seedlings. The density of seedlings has been brought up to 12-14,000 ha^{-1}.

Spruce in general was regenerated on pre-fire sites. Usually soil treatment was carried out with the plough PKL-70 or with the cover eliminator PDN-1. Forest planting was carried out with the mechanized or hand method in the planting of 2-yr old spruce seedlings with the plough PKL-70 into bottom or furrow layers, and also into the stripes with the cover eliminator PDN-1. The seedlings were mainly planted at a distance of 4.5-5.0 x 0.5 m with the density 4000-4400 ha^{-1}. Planting of pine and spruce was done within 2-3 years after the fire.

Tab.2. State characteristics of 8-yr old pine stands established by planting

Type of site conditions*	Ploughing method	Site category	Initial density (x 1000 ha^{-1})	Survival (%)	Status of pine		Average height of pine (cm)
					healthy (%)	fire damaged (%)	
A1	complete	burned	11.1	65.5	68.0	34.0	226.5
A1	furrows	cut	5.4	51.6	18.9	72.9	110.6
A1	untreated	cut	6.0	57.7	35.0	55.7	96.4
A2	complete	burned	17.5	83.4	95.3	4.7	257.0
A2	furrows	cut	7.2	75.0	70.1	29.9	246.2
A2	untreated	cut	6.1	62.0	76.2	23.8	206.1
B2	furrow	cut	5.7	63.0	66.2	33.8	151.4
B2-3	furrow	cut	5.7	22.7	63.2	36.8	99.1
B3	furrow	cut	5.7	22.7	84.6	15.4	115.2

* A1 - dry sandy soil; A2 - fresh sandy soil; B2 - fresh loamy sand; B3 - moist loamy sand

Under the conditions of the fresh coniferous (A2) the differences between the 14-16 year seedlings practically disappear (Tab.2). But it should be mentioned that also here the pine seedlings planted into the cultivated soil had higher rates of survival and growth than those planted into cleared out soil. Good survival and growth was observed for seedlings planted in the glades without previous soil treatment. This can be explained by a less disturbed forest environment having a favorable influence on stand growth. Under the conditions of the fresh coniferous forest-sowing also gave good results. In this case both methods of regeneration (planting and sowing) gave equal results.

In fresh and moist (B2, B2-3) forest types there was intensive development of grass and leaf trees. Only by timely carrying out agro-technical and silvicultural works could pine be regenerated successfully.

4. Conclusions

Both planting and sowing of spruce on drained wet land carried out during the first years after the fire gave satisfactory results. In the blackberry forest satisfactory results were achieved only by seeding or planting into the ploughed furrow layers. At present more than 20 years after the fires spruce populations are under the cover of broadleaved trees in considerable areas which now need to be thinned.

References

Chistyakov A.R., and Kreyer V.A. 1976. Natural Regeneration on different types of fire slashes. In: Problems of forest fire (1972). Elimination of its consequences in the Mari Republic, 68-76. Yoshkar-Ola <in Russian>.

Denisov A.K. 1976. Condition and classification of the fire-slashes in the Mari Republic. In: Problems of Forest Fire (1972). Elimination of its consequences in the Mari Republic, 14-43. Yoshkar-Ola <in Russian>.

Denisov A.K. 1979. Forest fires in the middle forest Volga region in 1921 and 1972 and their lessons. In: Combustion and fires in the forest. Krasnojarsk: ILID SO USSR, Part 3, 16-26. <in Russian>.

Kolobov N.B. 1968. Climate of the Middle Volga region. Kazan, 252 pp. <in Russian>.

J.G.Goldammer and V.V.Furyaev (eds.), Fire in Ecosystems of Boreal Eurasia, 409-413.

Response of the Endemic Insect Fauna to Fire Damage in Forest Ecosystems

V.M. Yanovsky and V.V. Kiselev [1]

1. Introduction

Forest fires are one of the most common factors influencing boreal forest ecosystems. The effect of fire on a biogeocenosis varies widely ranging from insignificant disturbance to stand degradation. Correspondingly, the impact of fire is reflected by the rate of changes that occur within the ecosystems affected, including the population dynamics of forest insects.

Long-term observations indicate that site conditions determine specific entomocomplex features (Yanovsky 1979,1987). Within any given altitudinal system of forest types, insect guilds evolve which differ in species composition, population structure, and biogeocenotic relationships. The comparison of the resulting entomocomplexes of similar ecosystems with different levels of disturbance provide information about trends and levels of insect group transformation.

The influence of fire-caused stand injury on insect populations depends on site-specific population features. Entomocomplexes of the sub-taiga/forest-steppe zone are characterized by maximum species diversity, a large number of species dangerous for forest stands, and high periodicity of insect mass reproduction. Low tolerance of these biogeocenoses to any sort and scale of endo- and exogenic influence (including low-intensity surface fires), as well as the high aggressiveness of forest insects, lead to a considerable increase in ecosystem degradation and delay forest regeneration, often resulting in the replacement of forests by non-forest vegetation. However, the extent of forests in the subtaiga/forest-steppe zone is limited due to local and mosaic patterns of forest distribution.

In taiga forests, insect diversity is sufficiently broad, although not as great as in the subtaiga/forest-steppe zone. The number of economically important (pest) species, however, is not large. Moreover, these species are in latency and largely secondary pests. For mass outbreaks to occur, some strong influence unfavorable for a phytocenosis is prerequisite (Kolomietz 1962). Under these conditions, damage by high-intensity fires over a considerable area is necessary for the outbreak of dendrophagous insects. In this case, outbreaks, primarily of defoliators, occur in forests adjacent to burned sites (Rozkov 1965). Large numbers of dead trees promote trunk-infesting pests. Thus, the risk is greatly confined to non-affected, healthy forests adjacent to fire-disturbed areas. Thus, abundant food sources

---

[1] V.N. Sukachev Institute for Forestry and Timber, Siberian Branch, Russian Academy of Sciences, 660036 Krasnoyarsk, Russian Federation

on large areas of fire-disturbed biogeocenoses provide favorable conditions for dendrophagous insects to fully use their population growth potential. In sub-alpine and near-tundra forests, insect diversity is strongly limited by the extreme climatic conditions. Outbreaks are exceptional in this situation. Even heavy disturbance does not influence the producer/consumer interactions in such forests.

Entomocomplex response to various levels of fire damage to biogeocenoses was estimated through the analysis of changes of the entomofauna in forests of Mongolia and Buryatia (Yanovsky 1980, 1983, 1991; Yanovsky et al.1991). This study is different from other works since it addresses transformations in the dendrophagous insect group in general. Most entomological research studies so far stress the influence of fire on the occurrence of outbreaks only.

2. Response of the Insect Fauna to Different Levels of Fire Disturbance

Analyzing the entomofauna, patterns of entomocomplex changes were established using a forest ecosystem disturbance classification developed by Krasnoshekov et al.(1990). Forest fires can be divided into three categories by the rate of influence on forest insect guilds:

Category I

Fires which do not cause any increase in tree mortality but lead to a decrease in tree resistance to other factors. In this case, the ecosystem recovers rapidly (Krasnoshekov et al.1990). Weakening of biogeocenoses by such fires promote primarily quantitative changes of entomocomplexes. A single disturbance event sometimes induces an increase in number of defoliators. For example, light fire damage to larch-grass forests of Eastern Khentey (Mongolia) resulted in a Siberian gipsy moth (*Dendrolimus superans sibiricus* Tschetv.) outbreak. Boldaruyeva (1959) states that, in Buryatia, these outbreaks can be attributed to the fire killing parasites, mainly those feeding on eggs. This results in a much lower level of parasite influence on gipsy moth population. Not ignoring this factor, however, it is assumed that the increase in the pest abundance is primarily due to decreasing physiological resistance (Girs 1979, Girs and Zubareva 1979).

Multiple disturbances by low-intensity fires can lead to fungal diseases as fungi develop around fire scars. In larch stands of Eastern Khentey low-intensity fire damage caused tree infestation by larch fungi. Trunks of fire-injured larches are infested by insects which are highly resistant to tree antibiosis (e.g., *Xylotrechus altaicus* Gebl. and *Scolytus morawitzi* Sem.). The insect activity decreased tree viability and made these trees suitable hosts for less tolerant insect species (e.g., *Tetropium gracilicorne* Rtt. [Cerambycidae] and *Ips cembrae* Heer); their populations began to increase subsequently.

Since ecosystems are only lightly disturbed, no qualitative entomocomplex changes are observed. The composition of insect group species remains virtually unchanged.

Category II

Fires resulting in light and moderate disturbance of the biogeocenosis (up to 50% of climax trees are burned). In such sites, forest regeneration normally occurs without stand replacement and takes 30 years (Krasnoshekov et al.1990). At this level of biogeocenosis disturbance a considerable population growth of phylophagous insects can be expected due to reduced tree vigor and changed micro-climatic conditions. With an increasing number of trees irreversibly damaged by fire, the population density of the trunk-infesting insects also increases.

When tree mortality accounts for up to 20% of the total stand and the forest burned is small in area, xylophagous insects increase slightly in number (local micro-concentrations occur). After the depletion of the food source the insect population decreases back to the pre-fire level. A sudden increase in food supply promotes xylophagous insect outbreaks by a manifold increase in population density. After depletion of the food, the insects move to the vigoros, non-damaged trees adjacent to the disturbed site.

The tree disturbance level is reflected by the group structure of xylophagous insects. In larch stands characterized by lightly disturbed biogeocenoses in which fire-induced reduction of tree vigor is irreversible, the groups are dominated by *Melanophila guttulata* Gebl. (Buprestidae) and *T. gracilicorne* as these are less sensitive to tree antibiosis. The invasion by the latter weakens larch trees and makes them suitable hosts for the Big Larch Bark Beetle (*I. cembrae*), which cannot inhabit relatively resistant trees because of low population density. The Big Larch Bark Beetle produces up to two generations, thus rapidly increasing in number. Once a high population density becomes achieved, the beetle is able to inhabit partly defoliated, weakened host trees and forces *M. guttulata* and *T. gracilicorne* out of this ecological niche since the latter two develop more slowly (one generation per year) (Girs and Yanovsky 1991). In heavily burned biogeocenoses, insect groups steadily dominated by the Big Larch Bark Beetle are immediately formed on larches which cannot restore their viability.

Moderate biogeocenosis disturbance leads to certain qualitative changes of the entomocomplexes. Consumption of insects by fire, along with burning of understory and surface cover, induces a sudden decrease in insect number and elimination of some insect species that develop in brush, surface cover and forest floor layers or feed on flowering plants. The period of time needed for the initial species composition to restore is apparently equal to the biogeocenosis recovery period.

Category III

These are fires resulting in severe ecosystem disturbance (with 50-100% of the total overstory killed). The high disturbance level enhances an ecosystem degradation and dysfunction in all respects. The recovery period ranges from 100 to 300 years (Krasnoshekov et al. 1990).

A high fire damage is primarily reflected in the species composition of the entomocomplex. Insect groups dependant on climax conifer tree species decrease drastically. Regeneration of these insects back to the initial species composition is believed to take more time than the recovery of the overstory. Burned sites are invaded first by insects developing on hardwood tree species and brushes as well as by those common for steppe and meadows. Xylophagous insects present an exception that are trophically related with climax tree species. Abundance of optimal food increases the attractiveness of the site and leads to a rapid increase in population density. Observations conducted in Mongolia and Tuva show that xylophagous insects, once their number is increased, rapidly expand to the non-damaged stands surrounding the burned site, thereby drastically extending the disturbed forest area.

3. Conclusions

According to observations of the relationships between fire, forest ecosystems, populations and the impact of insects, it is concluded that dendrophagous insects are a sufficiently strong synergetic factor contributing to biogeocenosis disturbance. The re-arrangement of the entomocomplex, being determined by the fire disturbance level and ecosystem tolerance to stress conditions, result largely in dominance of insect species highly dangerous for the phytocenoses. In such a situation these insects are very active and promote the degradation of the biogeocenosis.

Furthermore, insect activity may drastically decrease ecosystem resistance to exogenic disturbance factors. Stand mortality caused by fire followed by dendrophagous insect outbreaks does not affect other biogeocenosis components, particularly soil, but profoundly changes water regimes through increasing surface water drainage. The risk of fire in such ecosystems increases greatly and make the occurrence of fire inevitable (Krasnoshekov et al. 1990).

Also, a possibility exists for insects to influence forest regeneration and even changes in the forest community composition. In forests of Mongolia and Buryatia, periodical outbreaks of *Lymantria dispar* L. (a polyphagous insect species) considerably decrease larch stand productivity (including seed production) and lead to low regrowth viability. This makes conifer forest depression possible, but does not have any noticeable effect on more resistant small-leaved trees (e.g., birch, aspen) (Yanovsky 1980, 1991). In this case, secondary ecosystems can exist for a long time, being maintained, among other factors, by periodic fires (Krasnoshekov et al. 1990).

Ecosystem disturbance by fire causes considerable qualitative changes of the entomocomplex, manifested in the transformation of their species composition and structure. These changes can have profound impacts on forest insect guild biogeocenotic relationships and, consequently, on the role of insect guilds on the whole biogeocenose.

References

Boldaruyeva, V.O. 1959. The role of parasitic insects in *Dendrolimus sibiricus* Tschetv. reproduction. Abstract of a report at IV All-Union Ecology Society Congress. Vol.2, pp.149-152. USSR Acad. Sci., Moscow, Leningrad <in Russian>.

Girs, G.I. 1979. Dynamics of green pigments of *Pinus sylvestris* needles following surface fires. In: Conifer Response to Disturbing Factors, pp.22-33. Inst. Forest, Krasnoyarsk <in Russian>.

Girs, G.I., and O.N. Zubareva. 1979. Tolerance of vegetative plant parts to high temperature. In: Conifer Response to Disturbing Factors, pp. 5-14. Inst. Forest, Krasnoyarsk <in Russian>.

Girs, G.I., and V.M. Yanovsky. 1991. Effects of larch defenses on xylophagous insect guilds. In: Forest Insect Guilds: Patterns of Interaction with Host Trees (P.A. Radnor, ed.), 149-152. USDA For. Serv., Northeastern Forest Exp. Station.

Kolomietz, N.G. 1962. The role of *Dendrolimus sibiricus* in coniferous forests of Western Siberia. USSR Acad. Sci. Publ. Novosibirsk <in Russian>.

Krasnoshekov, Yu.N., I.A. Korotkov, Yu.S. Cherednikova, and G. Cedendash. 1990. Methods of present forest ecosystem state estimation and mapping in Mongolia. Joint Russian-Mongolian Expedition. Ulan-Bator. 30pp. <in Russian>.

Rozkov, A.S. 1965. Mass reproduction of *Dendrolimus sibiricus* and measures for its control. Nauka. Moscow. 180 pp. <in Russian>.

Yanovsky, V.M. 1979. Ecofaunistic estimation of dendrophagous insects in Mongolia. In: Wildlife of Baikal Basin, pp.78-92. Nauka Publ., Novosibirsk <in Russian>.
Yanovsky, V.M. 1980. Major forest pests in Mongolia. In: Forests of Mongolia (Management Implications), pp.116-137. Nauka, Moscow <in Russian>.
Yanovsky, V.M. 1983. Pests of near-taiga forests. In: Forests of Mongolia (Larch Forests of Central Khentey), pp.126-142. Nauka, Novosibirsk <in Russian>.
Yanovsky, V.M. 1987. Difference in ecological requirements between dendrophagous insects and plants as an index of forest ecosystem stability. In: Ecological Wildlife Habitat Estimation, pp.5-16. Nauka, Novosibirsk <in Russian>.
Yanovsky, V.M. 1991. Contribution of insects to forest tolerance decrease under antropogenic stress. In: Forest Tolerance to Insect Influence, pp.32-33. Inst. Forest, Krasnoyarsk <in Russian>.
Yanovsky, V.M., A.S. Pleshanov, T.A. Agafonova, and V.I. Epova. 1991. Entomological classification of forests of the Jida river basin. In: Problems of the Baikal Region Forest Ecology, pp.121-148. SIBFIBR, Irkutsk <in Russian>.

J.G.Goldammer and V.V.Furyaev (eds.), Fire in Ecosystems of Boreal Eurasia, 414-430.

Simulation of Forest Insect Outbreaks

A.S. Isaev, V.V. Kiselev, and T.M. Ovchinnikova [1]

1. Introduction

It is not uncommon that phytophagous insects strongly compete with humans for the forest resources of the globe. Evolutionary history has shown that insects are genetically flexible and well adaptable due to large population sizes and high reproduction rates. The management of a particular pest can only be successful if its biology, physiology, behavior, ecology and its role in the forest ecosystem is well understood. An ecological study should include consideration of the population density in space and time.

Study of the population dynamics of the species is important for revealing causes of outbreaks and developing prognosis simulation models as tools of modern forest management. The knowledge about mechanisms of pest dynamics allows one to monitor stands in order to estimate the probability of outbreaks and sizes of damaged areas. The understanding of how the pest interacts with its environment is also important for the development of a strategy to prevent the growth of population density and to control the areas of outbreaks. In this sense, simulation models of the interaction of insects with the forest are a bridge between theoretical and applied ecology. We used this approach to solve the problem of the Siberian forests damage by fir sawer (*Monochamus urussovi* Fisch.) which strongly influences structure of the taiga forest stands (Isaev et al. 1988, 1990).

2. Ecology of the Fir Sawer (*Monochamus urussovi* Fisch.)

The fir sawer is classified as a most important pest in shade-tolerant Siberian forests. All the *Pinaceae* species can be host plants of the fir sawer with *Abies* spp. taking first place on the list. Only in fir forests has the sawer such a relationship with the host plant that allow it to lower the resistance of trees. The ecological plasticity of the fir sawer is very high. It can be found in various forest types. Weakened forest stands provide habitats for the sawer. It colonizes stands infested by defoliating insects and stands weakened by drought or industrial pollution. The fir sawer is concentrated in burned places, wind-fall areas, cuttings, and in dumps of rough timber. In healthy stands it uses the natural die-back, colonizing dead parts of overmature trees.

[1] V.N. Sukachev Institute for Forestry and Timber, Siberian Branch, Russian Academy of Sciences, 660036 Krasnoyarsk, Russian Federation

The majority of individuals in sawer populations have two-year cycles of development. Young beetles leave the larva cradles with their ovaries underdeveloped. To become mature they need to take up additional nutrition in the crown. Both the tree stem and the crown provide the nutrition substrata for the insect at different stages of its development. The sawer colonizes stems only of weakened or dying trees. The development of pre-imago individuals takes place in the stems. The beetles feed on the crown gnawing the bark on branches which subsequently die off. The damage to the green part of the tree may reach such a level that the whole tree become suitable for the colonizing and the development of the sawer.

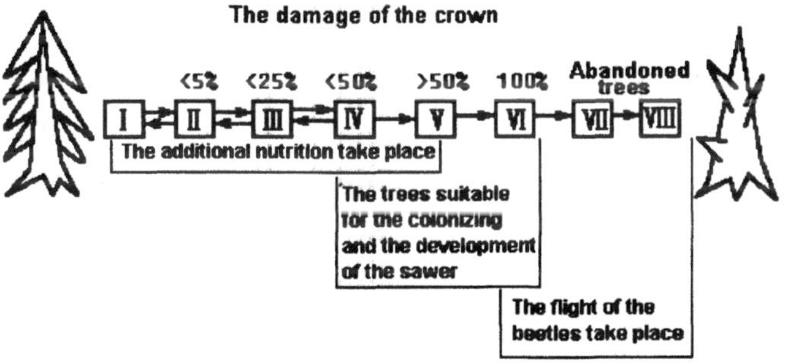

Fig.1. Diagram of the interaction of the fir sawer with the host tree

The degree of damage to trees as a result of additional nutrition determines the sequence of firs colonized by the sawer (Fig.1). The nutrition activity increases as the degree of damage grows to 50%, and then activity decreases. As a measure of the damage degree we use the number of the sawer bites per one linear meter of the first order (largest) branches. The degree of damage of state classes II, III, IV and V corresponds to 1.4, 3.5, 4.0 and 5.7 bites per one linear meter of the first order branches. The parameters of colonization relate closely to the sizes of the trees (Tab.1). A total of 25% of the bole surface area is colonized by the beetle in trees of state class IV, the same parameters for trees of state classes V and VI are 82% and 96% respectively.

Three types of fir sawer population densities (sparse, dense, and extra dense) are distinguished in accordance with the insect activities. The sparse population of the sawer consume weakened and dying trees, thus playing an important cleaning role in the forest stand. A uniform flow of trees suitable for nutrition determines a relative stability of the beetles density in the sparse population; a feedback regulation takes place there. The higher the population density the less the amount of food available.

Tab.1. Characteristics of the colonization zones in fir stems as dependent on the stem diameter

Parameter	Diameter Class (cm)						
	8	12	16	20	24	28	32
Tree height (m)	8.7	13.6	16.0	18.2	21.7	23.1	25.6
Length of the colonization zone (m)	4.0	8.9	11.9	14.2	17.2	19.6	21.6
Diameter limiting the colonization zone (cm)	7.5	7.2	8.0	9.9	10.3	10.0	11.1
Square of the colonization zone (100 cm^2)	100	221	465	706	963	1141	1623
Number of flight holes	2.2	16.7	43.3	55.0	77.3	141.5	146.9
Share of consumed trees in the total dead trees (%)	57	84	100	100	100	100	100
Length of crown (m) of 1st order branches	35	60	75	105	123	137	150

Dense populations of sawer are formed when the beetle colonizes both natural die-back trees and the largest trees of the stand. The populations of the increased density are an important factor in the age dynamics of the fir forests. In fact, the probability of colonization depends on the size of trees. An increased proportion of large trees in a stand favors the propagation of the sawer (Tab.1). Dense populations present a number of colonization spots within mature fir stands. All characteristic features of the outbreak can already be seen in the spots. The sawer actively destroys the upper layer. Clumps of fir undergrowth fill the appearing gaps. The intensive die-back of mature and over-mature trees complicates the age dynamics of fir generations.

An extra dense population of the sawer is the equivalent of a mass outbreak. Complete destruction of forest stands is characteristic of a period when extra dense population takes place. All trees with a diameter greater than 12 cm are attacked by the beetle. The sawer begins with additional nutrition, and starts laying eggs after the trees' resistance is sufficiently lowered (Fig.1). The tree dies in 4-6 years.

The outbreak situation corresponds to other regulation mechanisms in the "resource-consumer" system. When the density of the sawer is increased the beetle is able to influence the flow of food. A positive feedback regulation appears in this case.

3. Criteria for the Resistance of Fir Against the Sawer

Density dynamics of the fir sawer depend considerably on the environment, with the main cause of outbreaks being the lowering resistance of the trees or stands. That is why it is very important to classify fir stands according to their resistance against the pest. The resistance of a stand against the sawer is determined by the possibility for the beetle to colonize tree stems as a result of additional nutrition. This requires a definite relationship between the imago density and size of the crowns. One can say that a given forest stand is resistant against the sawer if the damage to trees due to the additional nutrition does not achieve a critical level. The population of the sawer persists only in a sparse form in such a stand. An outbreak is possible if a flow of food for the larvae can occur due to additional nutrition in the crowns. We can measure the degree of resistance by the intensity of the food flow.

We shall consider an estimation of the potential amount of food induced by the sawer as a criterion of the resistance. Let F stand for the amount of food in the crown of a tree that has the diameter d, the amount of food is measured in linear meters of first order branches. Let N denote a potential number of the born beetles (Tab.1). Suppose, one beetle consumes R linear meters of the first order branches during the period of the additional nutrition. The difference $F\text{-}R \cdot N$ shows the remained food stored in the crown after the nutrition of the imago is finished. It was experimentally found that if the ratio $(F\text{-}R \cdot N)/F$ is lower than a critical value δ ($\delta \in (0,1)$),i.e.

$$(F-R \cdot N)/F < \delta$$

irreversible changes take place in the tree, and the tree can be colonized. Let us transform the inequality in order to separate parameters dependent on tree size and the size-independent parameters. We obtain

$$N/F > (1-\delta)/R$$

The ratio $U=N/F$ can serve as a measure of the resistance of the fir against the sawer.

The value of the R parameter is determined mainly by weather conditions, and δ depends on the initial state of the tree. The ratio $k=(1-\delta)/R$ corresponds to a horizontal line in the plot (U,d) (Fig.2). A monotone increasing function may describe the dependence of the U criterion on the diameter. An ecological interpretation of such a relationship implies that as the diameter of tree grows the resistance of the tree drops. The point of intersection defines the threshold value of the diameter. If the diameter of a tree exceeds the threshold value the tree becomes non-resistant against the sawer (Fig.2). It is the experimental fact that, for the Central Siberia region the critical value of the damage of crown is approximately 50% (δ = 0.5), and the consumption of the first order branches by one beetle is about 1 linear meter (R = 1.0). Thus, the critical value of diameter d_K is 18 cm.

In fact, the values of R and δ may fluctuate due to various environmental factors. This can be interpreted as a transformation of the line $k=(1-\delta)/R$ into a band with the limits:

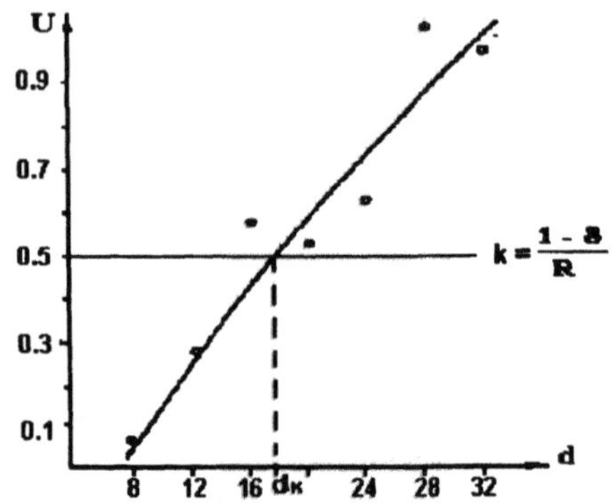

Fig.2. Dependence of the fir resistance criterion on stem size

$$k_{max} = \frac{1-\delta_{min}}{R_{min}} \quad , \quad k_{min} = \frac{1-\delta_{max}}{R_{max}}$$

$$R_{min} \leq R \leq R_{mak} \ , \ \delta_{min} \leq \delta \leq \delta_{max}$$

The graphic presentation reveals the possibility of classifying trees into three types: resistant, non-resistant, and transitional. The main parameter for the classification is the tree diameter. It is diameter which determines the possibility of the flow of food. The trees thinner than d_{MIN} are resistant, trees thicker than d_{MAX} are non-resistant, and the trees whose diameter falls between d_{MIN} and d_{MAX} are classified as transitional.

To formulate a criterion for the resistance of fir stands we shall consider the same phenomenon; a possible flow of food for the sawer. The relation between the number of beetles born in a certain wooden area and the store of food for the imago can be considered as a criterion η of the resistance of the stand.

Therefore,

$$\eta = \frac{\sum_i N_i}{\sum_i F_i}$$

Where N_i is the number of beetles born in the i-th tree, F_i is the sum length of the first order branches in the i-th tree. It is necessary to consider all the trees of a particular plot. The critical value of the criterion (η_C) corresponds to fir stands where there is a positive feedback in the "resource-consumer" system. As the value η grows, the probability of weakening fir stands due to additional nutrition increases, which corresponds to the lowering resistance of the stand against the sawer. The metastable state of the sawer population is essentially impossible in the stands where $\eta < \eta_C$. The value of η_C is not constant; it depends on the ecological condition of the stand and the sawer population. For example, an anthropogenic stress, drought, or other ambient factor would result in a lower value of η_C because of decreased fir resistance. On the other hand, weather conditions which could affect the additional nutrition (rainy short summer) result in a higher value of η_C. A positive difference between η and η_C for a particular fir stand means a certain reserve of stability of the stand, that keeps the whole system from an outbreak. The probability of an outbreak increases with η.

We shall now reveal the relationships of the parameter η to traditional parameters of forest surveying in order to build a classification criterion of the stands' resistance against the sawer. Siberian fir stands are uneven-aged, in the majority of cases, having specific features of age, diameter, and height distributions along the line of development. They can be distinguished into four stages (Falaleev 1964):

Young age: The diameter distribution has two pronounced maxima, in small classes and in average classes.

Middle age: The maximum of the diameter distribution is in the smaller classes.

Under-mature: Most of the trees belong to the interval between thin and middle-sized classes of the diameter distribution.

Mature: The diameter distribution presents a curve of the normal distribution. The values of η calculated for different age stages (Tab.2) gives evidence that the stages defined as "young age" and "mature" are the most favorable for sawer outbreaks.

We shall consider the dependence of η on the average diameter and density of stand for the stages "young age" and "mature". The diameter distribution of the stages have practically the same shape for the diameters more than 8 cm. That is why we shall not distinguish between them from the food supply point of view for the sawer.

With a rise in the average diameter of a stand, the value of η increases within every class of stand density (Fig.3). This implies a drop in the resistance of the stand to a sawer outbreak. A critical value of the average diameter d_C can be found in every class of the stand density, it corresponds to the case when $\eta = \eta_C$. Trees of stands where $d < d_C$ are not at risk of being colonized. The sawer population indefinitely persists in a stable state. Calculations showed that at η_C = 0.6-0.65 the values of d_C are equal to 14 cm, 20 cm, and 25 cm for the high-stocked, middle-stocked, and low-stocked fir stand respectively (Fig.3). From the nutrition point of view the high-stocked stands are most ready for the development of an outbreak. At a constant diameter the increasing density leads to a rise of the value. An analysis of the distribution of outbreak areas gives support to the result obtained. The relative stand density is equal from 0.7 to 0.8 in the areas of heavy damage; at relative density 0.6 stands are slightly damaged; and low-stocked stands (the relative density is 0.4-0.5) are practically not touched by the fir sawer.

Tab.2. Age stages of fir stands

S	$\bar{d}$ (cm)	Distribution over relative diameter classes (η)											Age (years)
		0.4	0.6	0.8	1.0	1.2	1.4	1.6	1.8	2.0	2.2	2.4	
Young age	20	-	18.9	24.2	25.7	17.6	8.8	3.1	0.8	0.2	-	-	0.62
Middle Age	15	-	38.3	26.0	15.7	9.2	4.8	2.7	1.8	0.5	0.4	0.6	0.40
Under-mature	18	-	20.2	24.9	20.4	15.8	11.2	4.5	1.8	0.8	0.2	0.2	0.56
Mature	21	4.4	17.2	24.7	23.9	16.0	8.6	3.6	0.4	0.4	-	-	0.62

Note: η - beetles per linear meter; S - stage of stand development; $\bar{d}$ - average diameter of stand.

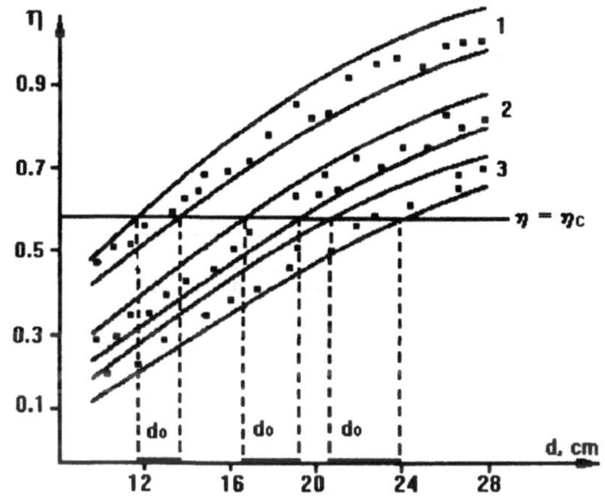

Fig.3. Dependence of the computed parameter on average diameter and stand stock. Bands 1, 2, 3 correspond to high, middle, and low stock respectively.

4. Simulation of Sawer Density Dynamics and the Destruction of the Affected Stands

When making up the simulation system "fir stand - sawer" it was required that the degree of detail of description provide abilities to control the system.

We consider the system as consisting of spatially explicit trees and the consumer (fir sawer). The food resource is represented by two parts: the crown and the stem. The estimation of the population density implies the imago stage. The feeding is considered separately for imago and for the larva because they use different parts of trees. The time of beetle development, overlap of generations in time, the parameters of breeding, and mortality are taken into account. The intensity of the processes is determined by the state of the system and some external conditions.

The diagram of components of the system fir stand-sawer is shown in Figure 4. Two patterns are considered: one feedback loop (Fig.4.1) and two feedback loops (Fig.4.2). The first pattern corresponds to the stable state of the sawer population in an undamaged stand. The second pattern describes the metastable state followed by a destruction of the stand.

Fig.4. A general diagram of interactions in system "stand - sawer". 1 - stable state of population; 2 - metastable state.

The blocks are: N - imago, L - larva, F - food substratum for imago feeding (crown), D - substratum for larva development (stem), M - beetles-emigrants.

The relationships $N \to D$ determine the rate of damage to crown as a result of the additional nutrition of the beetles. The relationship $F \to D$ describes the amount of food prepared by the beetles for succeeding generations. The relationship $F \to N$ determines the movement of the outbreak area borders, i.e. the number of "extra" beetles who migrate to the surrounding stand. In a stable state, a direct relationship exists between the components F and D, a feedback relationship between F and N does not occur (Fig.4.1).It means that migration does not sufficiently influence the density growth of the pest. A direct relationship between F and D and the feedback between F and N appear when the beetles start to prepare food for the imago stages, which corresponds to the transition to the metastabile state. The relationship between N and M results in a mass migration of the beetles so that the area of outbreak widens.

We consider a population of the fir sawer. Let us assume that it is distributed within a wooden area $A=\{a_i\}$, $i=1,2,\ldots$. The elements a_i of the set A will stand for separate trees. The set A may be linear, planar or of some kind which "follows" a local landscape. Let the difference $|i-j|$ for any a_i and a_j have the meaning of distance between the trees a_i and a_j. The model is written for discrete time $t{:}0,1,2,\ \ldots,$ with Δt being equal to one year.

Let functions $F(a_i,t)$, $D(a_i,t)$ and $N(a_i,t)$ be given for every tree a_i in the time moment t. The functions describe the crown, tree stem, and the number of the sawer imago. The function $F(a_i,t)$ shall mean the store of food for the additional nutrition in the crown of the tree a_i at the moment t, i.e. $F(a_i,t)$ is the crown size. As a measure of the size we used the total length of first order branches. The size of crown which is expressed in linear meters allow one to take into account the growth increment of the branches. The negative impact of the sawer on the crown results in shoots with needles spaced from the beetle bites to tips of the branches dying off. Due to this, a change of crown size as a result of grazing can be estimated. $F(a_i,t)>=0$, $F(a_i,0)=F_{i0}$. Let the function $D(a_i,t)$ denote the bole surface area of the tree a_i suitable for colonization by the sawer, i.e. $D(a_i,t)$ is the current amount of food for larva during their growth and feeding under the bark. The value of the function depends on the rate of crown damage and the period when the stem is suitable for colonization. The dependence of $D(a_i,t)$ on crown damage shall be discussed later. $D(a_i,t)>=0$, $D(a_i,0)=D_{i0}$, the unit of measure is squared decimeters. The number of beetles born from the tree a_i at the moment t is denoted by the function $N(a_i,t)$, $N(a_i,t)>=0$ for every t, $N(a_i,0)=N_{i0}$. The value of the function is determined, on the one hand, by the intensity of the colonization by the previous generations, and, on the other, by the state of the tree.

It is known that the choice of trees as the object for feeding depends on the attraction of the trees. It is important to find parameters which reflect the change in the state of the tree in the course of colonization. Research showed that the ratio of the needles remaining at the moment of observation to the total amount of needles before the time when the tree was damaged can serve as such a parameter. Virtually, the parameter is a numerical equivalent of the state class of tree.

Let $F(a_i,t)$ be current size of crown of the tree a_i in case of the beetle attack, with $F_0(a_i,t)$ being the same parameter in case of no damage. Then the state of the tree a_i at the moment t (denoted as $S_i(t)$ or S_i) is defined by the expression:

$$S_i(t)=\frac{F(a_i,t)}{F_0(a_i,t)}$$

The value of the function ranges from 0 up to 1, which corresponds to the state of tree from death to the healthy tree. The function S_i is a continuous equivalent of the state class of tree a_i at the moment t (Fig.1).

In order to estimate the attraction for additional feeding, we use experimental data on the numbers of bites of current year in trees of various state classes (Fig.5a). Let $g(S)$ denote the function that relates the activity of the additional feeding to the state of the tree. The function will be referred to as the function of attraction for additional nutrition. There are data on bite density of current year in trees of various state classes (Fig.5b).

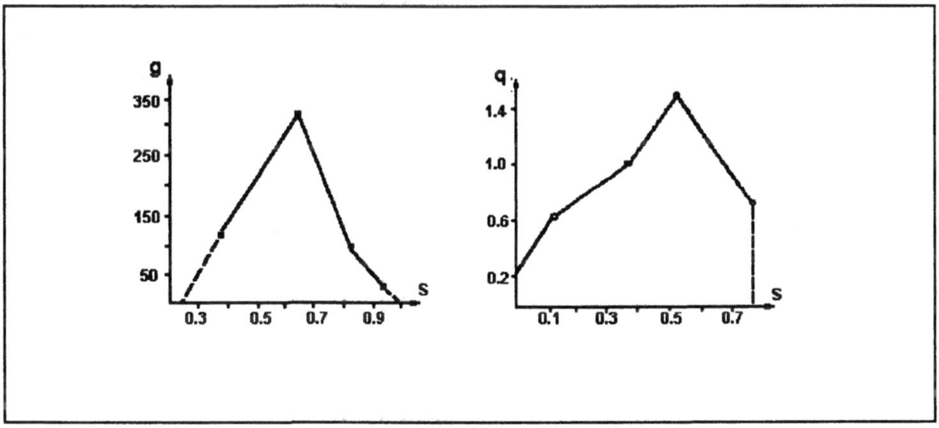

Fig.5. The function of attraction (a) - intensity of the addition nutrition of the sawer, (b) - activity of laying eggs in trees of various state classes.

The relationship will be referred to as the function of attraction for laying eggs and denoted as $q(S)$. The function $g(S)$ and $q(S)$ can be found by piecewise linear approximation of the experimental points.

The behavioral functions $g(S)$ and $q(S)$ allow one to describe the distribution of beetles over the trees (the set A) for the additional nutrition and laying eggs. The algorithm of the distribution is as follows. Suppose that $N(a_i,t)$ imago individuals which were born from the tree a_i first make the additional nutrition and then lay eggs. The additional nutrition takes place in trees a_i such that $|i\text{-}j| <= k_F$, where k_F is considered to be some constant. In other words, the choice of the food resource takes place over some limited forest area A_{ij} which corresponds to the radius of action k_F centered in places of birth $A_{ij}=\{a_j: |i\text{-}j| <= k_F\}$.

The imago which have additional nutrition on tree a_m with the attraction $g(S_m)$ are born from trees placed not farther than k_F from a_m. The number of imago which gather for feeding on tree a_m at the moment t is calculated according to the formula:

$$Y_0(a_m,t) = \sum_{|j-m|\le k_f} N(a_j,t) \frac{F(a_m,t) * g(S_m)}{\sum\limits_{|j-i|\le k_f} F(a_i,t) * g(S_i)} \tag{1}$$

However, a limited number of beetles can have the additional nutrition in a crown. That is why it is worth introducing the notion of crown "capacity", which would correspond to the number of portions of food in a particular situation. If R is the amount of food consumed by a beetle during the season of the additional nutrition, by the crown capacity of tree a_i we shall mean the ratio:

$$\frac{F(a_i,t)}{\gamma \cdot R}$$

The coefficient γ describes the density of the beetles feeding on the food resource. When $\gamma = 1$ there is no over-population of the crown above the number of portions. When $\gamma > 1$ there is a surplus of food due to the lower density of beetles. When $\gamma < 1$ the number of individuals is higher than the number of portions. An inevitable consequence of this is competition of imago in the tree. There are two alternate ways out of the competition: mortality and migration.

Let $Y_C(a_i,t)$ stand for crown capacity, $Y_C(a_i,t)=F/\gamma R$. It is quite possible that the number of beetles "wanting" to have the additional nutrition on tree Y_0 exceeds the capacity of the crown Y_C. Then the number of extra beetles coming to the tree a_i can be estimated as:

$$Y(a_i,t) = \max\{0, Y_0(a_i,t) - Y_c(a_i,t)\}$$

It is natural to assume that some fraction of the extra beetles goes in search of new trees for feeding. When simulating the process, one can require that parts of the imago population die or move beyond the distance k_F.

Let $Y_1(a_i,t)$ stand for the number of beetles on the tree a_i as a result of single distribution. After a reiterated distribution there will be $Y_2(a_i,t)$ beetles on the tree, after the k-th distribution, the number will be equal to $Y_K(a_i,t)$, and after the last distribution there will be $Y(a_i,t)$ beetles on the tree.

The procedure of distributing the imago over a number of stems for laying eggs is the same as for their distribution for the additional nutrition. The beetles which feed on tree a_i and have the total number $Y(a_i,t)$ search for the trees for laying eggs in the vicinity of the tree, i.e. within a subset A_{ij}. The radius of the subset is k_D, $A_{ij}=\{a_j: |i-j| <= k_D\}$. The beetles distribute among the tree a_m according to their attraction $q(S_m)$ and the amount of the bole area available for reproduction $D(a_m,t)$.

Therefore, the number of beetles taking part in breeding on tree a_i at the moment t can be estimated as:

$$X(a_m,t) = \sum_{|i-m| \le k_d} Y(a_j,t) \frac{D(a_m,t) * q(S_m)}{\sum_{|j-i| \le k_d} D(a_i,t) * q(S_i)} \quad (2)$$

The system "forest stand-fir sawer" is described by a number of recurrence equations for the functions $F(a_i,t)$,$D(a_i,t)$ and $N(a_i,t)$. The change of value time results from the activity of the sawer in the stand and the age dynamics of trees. The spatial variations take place due to the migration of beetles searching for food.

There are a number of assumptions used to derive the equation for the current size of the tree crown:

1. The amount of food in crown $F(a_i,t)$ is the result of the natural growth increment of the branches, their deaths, the transformation of branches to an inedible state, and the consumption of the branches by the beetles in the course of taking additional nutrition.

2. The size of crown does not drop till the consumption during one season exceeds the natural mortality plus the transformation of branches into the inedible state.

3. The natural increment of the crown occurs if the level of damage (states) does not drop lower than 0.5.

Therefore, the balance equation has the following form:

$$\begin{aligned} F(a_i,t+1) = F(a_i,t) + \alpha(a_i,t) \cdot H[S_i(t)] - \alpha(a_i,t-\tau_F) - \\ -\max\{0, Y(a_i,t) \cdot R - \alpha(a_i,t-\tau_F)\} \end{aligned} \quad (3)$$

where $\alpha(a_i,t)$ is the function for the natural growth increment of crown, τ is the time. When branches are suitable for the additional nutrition, R is the amount of food one beetle consumes in one season. The function

$$H[S_i(t)] = \begin{matrix} 0, & S_i \le 0.5 \\ 1, & S_i > 0.5 \end{matrix}$$

takes into account the dependence of the natural increment on the level of damage.

The derivation of the equation for $D(a_i,t)$ which describes the part of stem suitable for laying eggs is based on field measurements. It was found that the bole surface area colonized by the sawer relates to the state of the tree. The measure of colonization of the stem is defined as the ratio of the colonized area $D(a_i,t)$ to the total bole surface area $D_0(a_i,t)$. The relationship between this measure and the state of the tree was determined (Fig.6). Let $c(S)$ stand for the relationship.

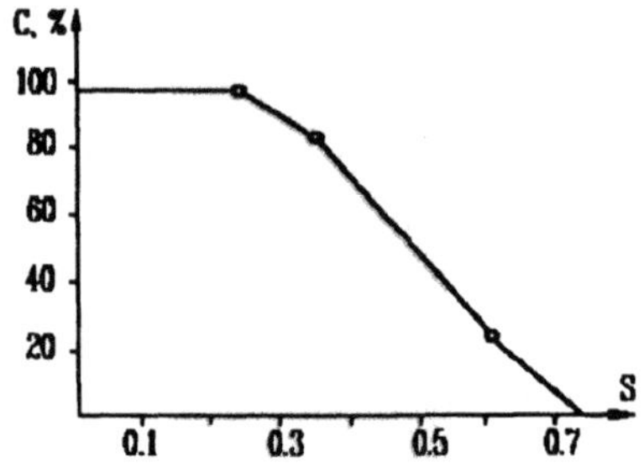

Fig.6. Dependence of colonization area of tree stem on state of the tree

Provided we know the relative colonization $c(S_i)$ and the time span when the stem is suitable for laying eggs τ_D we can write the equation:

$$D(a_i,t) = c[S_i(t)] \cdot D_0(a_i,t) \cdot H[S_i(t-\tau_D)], \tag{4}$$

$$H[S_i(t-\tau_D)] = \begin{cases} 0, S_i(t-\tau_D) < 0.75 \\ 1, S_i(t-\tau_D) \geq 0.75 \end{cases}$$

The function H takes into account the time when the stem is suitable for larva development. The sequence of the colonization of the stem of tree by generations of the sawer is shown in Figure 7. The value of τ_D usually varies from 2 to 4 years. We consider the function $D(a_i,t)$ to be a constant for a given stem diameter because increase of the bole surface area due to tree growth is a much slower process than beetles dynamics. Therefore, the given equation deals with such resource dynamics as they arise due to the sawer activity alone.

We now consider the derivation of the equation for $N(a_i,t)$ that will control the density dynamics in the model. The number of beetles born depends on the number of parent cohort, the sex ratio of the parents, quantity and quality of the food, the density of bites and the density of the developing individuals, the survival of the young generation, and the type of stem colonization. The areas of colonization of a stem by different sawer generations can overlap because the colonization lasts for several years (Fig.7).

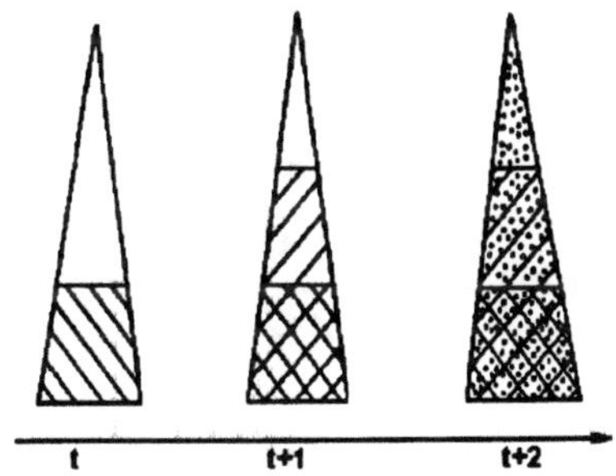

Fig.7. Pattern of stem colonization during three years

We shall take into account the colonization of a stem during three years. The development of the beetle from egg to imago usually takes two years. Thus, the parents of the beetles that are born at time t flew and laid eggs at the time t-2. The bole surface area that can be colonized once at the moment t-2 is as follows:

$$d_1(a_i,t-2) = D(a_i,t-2) - D(a_i,t-3),$$

twice:

$$d_2(a_i,t-2) = D(a_i,t-3) - D(a_i,t-4),$$

three times:

$$d_3(a_i,t-2) = D(a_i,t-4)$$

It is known (cf. Tab.1) that the number of beetles born from a unit of the bole surface area does not exceed some value (the maximal capacity of stem). The value depends on the sizes of trees and conditions of the preimago development. The number of beetles from the second colonization cannot exceed the difference between the maximal capacity and the number of beetles from the first generation. The number of beetles from the third colonization generation cannot be higher than the difference between the maximal value and the total of all beetles born from the tree earlier.

Let U_{id} stand for the maximal number of beetles born from a unit of the bole surface area of the tree a_i with diameter d; W is the percentage of females in the total number of beetles $X(a_i,t)$ on the tree a_i, J the number of eggs per one female, and P the survival probability.

Therefore, from the above assumptions, the number of beetles born from the unit of the bole surface area colonized once at the moment t is equal to:

$$n_1(a_i,t)=\min\{U_{id},\ \frac{X(a_i,t-2)\cdot W\cdot P\cdot J}{D(a_i,t-2)}\}$$

The number of beetles born from the part of the stem colonized twice per unit of area is:

$$n_2(a_i,t)=\min\{\frac{X(a_i,t-2)\cdot W\cdot P\cdot J}{D(a_i,t-2)},U_{id}-n_1(a_i,t-1)\}$$

The number of beetles from the part of the stem suitable for threefold colonization is:

$$n_3(a_i,t)=\min\{\frac{X(a_i,t-2)\cdot W\cdot P\cdot J}{D(a_i,t-2)},U_{id}-n_1(a_i,t-2)-n_2(a_i,t-1)\}$$

Therefore the total number of beetles born from the tree a_i at the moment t is as follows:

$$N(a_i,t)=\sum_{j=1}^{3}n_j(a_i,t)\cdot d_j(a_i,t-2) \quad (5)$$

The equations 3-5 are based on the biology of the sawer and on thorough research of the interactions of the beetle with the host tree. These equations as well as the equations 1-2 are the basis of the simulation prognosis model.

The input data for the model are parameters of the stand and individual trees (size of the plot, number of trees, the diameter distribution, etc.), locations and sizes of the damaged areas, the degree of damage, the initial number of beetles and parameters of their dynamics (survival, fecundity, food consumption), the natural mortality in the stand, and the effect of silvicultural treatments on the sawer.

A user chooses a time interval for which the model computes the parameters of the system. The time dynamics of the sawer density, the spatial distribution of the beetles, and the parameters of damage in the stand are the output information of the model.

5. Verification of the Simulation Model and Analysis of the Solutions

The test of the model was done with the help of experimental data. The data were obtained as a result of research on sparse density populations of the sawer in the mountain forest of western Sayan and high density populations in the region of Enisey Kryag. The model was considered to be fit for a prognosis if the predicted density dynamics of the sawer and the resulting tree mortality correspond to observed values in a sufficiently accurate way. The parameters for the simulation experiment were close to the actual ones (R =1.0; k_F=7; k_D=70; α =2.0; γ =1.0; J =28; W =0.5; P =0.076), with the initial beetle density in the model being equal to the measured one.

The simulation showed a metastable state of the system during a long time span. The annual tree mortality was approximately 2%, and the average density of beetles was 1,700 per ha. The actually observed mortality in a damaged area was 1.5-2.2%, the average density was approximately 1,500 beetles per ha (Vetrova 1986).

We also used data collected over the time period of 20 years in areas of high density populations of the sawer in the region of Enisey Kryag. The data include annual surveying of the trees' diameter, the state of their damage, and the estimations of the beetles' density. In natural conditions, the mortality of trees was 15 stems per ha, the beetle density varied from 1,600 to 2,500 per ha within the time interval of 5 years. The average tree mortality, as computed by the model, was 15.2 trees per ha, the beetles' density varied between 1,600 and 3,000 beetles per ha, with the average density being equal to 2,300 beetles per ha. The dynamics of the metastable state of the system is reproduced by the model in the most exact way. When the outbreak diminishes due to changes in stand structure, the model overestimates the percentage of dead trees by 5-8%.

The numerical investigation of the model showed that the system has two states: one stable and one metastable. The states correspond respectively to sparse and extra dense populations of the sawer. The metastable state has the form of spatial waves of insect density, the amount of food for imago and larvae. The waves correspond to a spreading outbreak area.

The metastable state occurs only under certain conditions. A necessary condition for the wave to appear is a sufficiently high density of beetles in a stand. The high density allows the sawer to weaken the trees as a result of additional feeding. The model allows one to estimate the threshold values which separate the different states of the system.

We explored the conditions for the wave to arise as dependent on the parameters of the model. For example, a bigger radius of flight of the imago results in a wider "active layer" (the border of the outbreak area). On the other hand, the moving wave can exist only if the concentration of beetles in the "active layer" is sufficiently high. If the radius is very big the beetles disperse in the stand and cannot actively influence the state of trees, and, therefore, metastability cannot exist. The finding is very interesting for the development of an effective system of defense of taiga forests from the pest. Isaev (1971) put forward an idea that the forest can be protected from the sawer by a dispersion of the beetles over stands. Our experiments with a higher radius of flight support the idea.

The rate of damage is directly dependent on the food requirements of the sawer during the additional feeding. In natural conditions a decrease of food consumption is observed in cold and rainy periods of summer. The activity of beetles is minimal at these times, sometimes half or less. The model allows one to find a threshold value of R; if the actual value is lower than this an outbreak is not possible. The value of R determines not only the possibility or impossibility of an outbreak but also other features of the process. For example, we found a non-linear relationship between the average number of beetles in the population and the annual consumption of food by the beetles during the additional feeding. As R grows, the density of the population first increases and then drops.

6. Conclusions

In a brief outline, the dynamics of forested area damage by the fir sawer looks as follows. Separated extra dense populations of the sawer live in optimal habitats, with the beetles being concentrated in clumps of dying trees. All characteristic features of the metastable state can be found in the micro breeding grounds but an outbreak does not yet arise. The impact of weather conditions (two succeeding droughts) gives rise to an increase of sawer density. Warm dry weather stimulates the activity of beetles which, in turn, results in higher crown damage. On the other hand a drought affects trees due to insufficient water supply. The two processes go parallel and decrease the resistance of the stands against the sawer. Under these conditions, the micro breeding grounds grow to create a big outbreak area. The borders of outbreak areas spread as running waves which die out in forested areas with increased resistance. The process ends by the destruction of optimal stands, a dispersion of the sawer populations, and the transition to a stable state.

Particular parameters of sawer dynamics can be estimated in the field in the course of pest monitoring. The "readiness" of the stands for an outbreak, with respect to the age structure and the current state, can be assessed by the analysis of surveying data and interpreting space photographs. Not only a composition of the stand adjacent to the damaged area plays an important role but the landscape structure should also be taken into account. Using the parameters mentioned along with weather data allows one to predict the dynamics of sawer density in particular forest ecosystems.

The work was supported in part by the Russian Foundation for Fundamental Investigations under Grant 93-04-20941.

References

Vetrova, V.P. 1986. Influence of sparse and dense populations of fir sawer on fir stands /Lesovedenie, N1, 43-50 <in Russian>.

Isaev, A.S., A.S.Rozhkov, and V.V.Kiselev. 1988. Fir sawyer beetle *Monochamus urussovi* (Fisch.). Novosibirsk, "Nauka" Publishing House, Siberian Branch, 271 pp. <in Russian>.

Isaev, A.S. 1971. Dynamics of density of fir sawer and measures to control it in fir stands of Krasnoyarsk territory.In:Probl.ZaschTaej.Les.Sib., 21-23.Krasnoyarsk,Inst.For.Wood<inRussian>.

Isaev A.S., V.V.Kiselev, and T.M.Ovchinnikova. 1990. Siberian fir stand resistance against the damage by xylophagous insects. Lesovedenie, N 6, 3-10 <in Russian>.

Kiselev V.V., and T.M. Ovchinnikova. 1991. Simulation of *Monochamus urussovi* (Fisch.) mass reproduction area dynamics. Lesovedenie, N 6, 3-9 <in Russian>.

Falaleev E.N. 1964. Fir forests of Siberia and complex use of them. Lesnaya Prim., 166 pp. <in Russian>.

J.G.Goldammer and V.V.Furyaev (eds.), Fire in Ecosystems of Boreal Eurasia, 431-444.

Fire - Vegetation - Wildlife Interactions in the Boreal Forest

H. Gossow [1]

1. Introduction

The distribution and composition of wildlife in a given area is to a great extent the result of its actual vegetation conditions. In the boreal forest biome, arboreal as well as ground vegetation is predominantly a result of local site conditions and corresponding snow and fire regimes. There appears to be rather good agreement among scientists studying boreal forest-fire interactions, that "in the north, most natural forests are either maturing following the last fire or being instantly recycled by the next" (Heinselman 1981) - with emphasis on cycles instead of succession. Only locally, e.g. on the floodplains of rivers, may primary successions occur (Van Cleve and Viereck 1981); whereas secondary successions take place in specific "fire refugia" (cf. Zackrisson 1977).

Forest stand origin mapping and fire scar dating give good evidence for age class mosaics (e.g. Heinselman 1973; Tande 1979), showing that "many northern forests are monotypes in the tree stratum" (Heinselman 1981). Most of the vegetation changes in northern forests, which are dependent on fires, do not fit the traditional concepts of succession (with plant composition changes and progressions toward a steady-state etc.). Most species, and even most individuals become established during the first (few) year(s) after fire from canopy-stored seeds, organic layer seed banks, vegetative regeneration, or windblown invader seeds from nearby. There is mostly no uni-directional development of vegetation, and there are mostly no steady-state communities. The vegetation is a mosaic of patches of even-aged forest, each patch dating from the last fire but varying in species composition with site factors, propagule sources, and fire behavior variables (e.g. Wein 1993). All this may be complemented by fire-independent or non-impacted additional islands as a matter of "ecological fragmentation", and as a basis for increased structural and textural diversity.

What does that mean for local wildlife situations? There are some descriptions on the relationship of fire and wildlife, more generally (e.g. Bendell 1974; Nelson 1974; Wright and Bailey 1982; Peek 1986) or also with special reference to the boreal forest (e.g. Scotter 197O, 1971; Komarek 1971; Viereck 1973; Rowe and Scotter 1973; Fox 1978). There is also, for instance, a Canadian "Forest Fire Prediction System" especially for fuel type characteristics (Alexander et al. 1984; cf. Johnson 1992) which could serve as some basis for wildlife habitat differentiation and wildlife fire-successor types. If the connection between accompanying animal species and plant society or stand characteristics is very strict, a specific description and discussion would not make too much sense in this context. But

[1] Institute of Wildlife Biology and Game Management, Agricultural Univiversity of Vienna, A-1190 Wien

whereas avertebrates often show very strong and close relationships to specific food or host plants or other limiting substrates, vertebrates appear more structure-oriented and stratum-oriented in their habitat demands and use, especially the more mobile groups such as fish, birds, and mammals. Habitat evaluation procedures (HEP) are preferably based on optimal structural parameters.

Snow and fire, floods, storms, strong rains, or droughts and heat more or less alter the structural characteristics of vegetation comparatively quickly, concerning food quality, cover suitability, availability etc. A corresponding and long-lasting co-evolution of floral and faunal elements appears only natural and is to be expected. Also in the hands of aboriginal man, the tool-maker and tool-user, fire has already since long ago been a tool of landscape modifying, and of habitat management (e.g. Pyne 1993). Anthropogenic fires obviously differed from dominating lightning fires, according to seasons and localities. But more recently, also in the more remote northern forests, other anthropogenic eco-techniques and inputs have started to change landscape and vegetation types, partly to a dramatic extent.

The clearing of forests, regulation of rivers, drainage of swamps, clearcutting for timber use - partly as brute "timber mining", partly on a sustainable yielding basis - and plantation forestry, with its opening-up of the landscape for traffic, trade, industry, and settlements, all mean grades of landscape 'development' from near virgin status to pure exploitation or production and housing grounds. But despite a certain 'exoticing' of the original, primeval landscape and flora, there are still existing structural characteristics which appear attractive for some wildlife types, whereas others are more or less vanishing. Prerequisites for over-abundance exist in some instances, whereas ecological "sinks" or "traps" in others have made the long-lasting co-evolutionary adaptedness more or less questionable.

To which extent are such statements or assumptions also relevant for boreal forests, or will become so? Or what specific characteristics of this forest type are of interest especially for higher vertebrate wildlife? These questions shall be discussed here to some extent for 'higher' vertebrate wildlife.

2. Interests in and Demands of Wildlife in Boreal Forests

Man's emphasized interest in wildlife usually concerns remarkable, utilizable, or problematic species. This includes big game because of its special trophy value, rarity, fur or meat quality etc. But quite often such species or trophy animals are available only in small densities and numbers (such as large predators), or decreasing in numbers (such as various tetraonid species), and insofar, already more an object of protection measures. Several wildlife species may also become over-abundant by numbers or distribution densities, and insofar, may become a problem or even a pest in view of specific human land use interests. On the other hand, forest pasturing by cattle and other domestic ungulates, as well as by semi-domestic reindeer, is a pronounced land use interest also in boreal agro-forestry.

With such aspects in mind, one can imagine that quite a few wildlife problem species or populations may get the quality of "strategic key" or "flagship" species, or even as bio-indicators for management objectives such as nature conservation, pest control, habitat improvement, timber production procedures, etc. Regarding higher vertebrate wildlife, cyclic population fluctuations as well as local over-abundance in large mammals appear to be characteristic of natural boreal forest (and tundra) ecosystems; and, consequently, forest damage problems by these animals because of more recent alterations may be inflicted to them by timber exploitation and plantation-like reforestation. Over-abundance in connection with anthropogenic land use impacts and influences in the taiga range can be taken as an (unintended) experimentation with nature (field experiment) which may have interrupted the often postulated co- evolution and "unity" between the woodland and its wildlife.

Here, it may be worthwhile to consider what Laws (1981) stated in a workshop on management problems with over- abundant large mammal populations: "K-selected species of animals are adapted to stable environments and have evolved rather subtle and more delicate mechanisms to respond to ecosystem changes than have the more opportunistic r-selected species. When out of balance they can be expected to influence ecosystems in ways that r-selected species probably do not. Their role in structuring ecosystems may be great, while r-selected species, being opportunists, would be expected to have less influence on the structure of mature (!) ecosystems. Because of their more defined regulatory mechanisms they are less likely to overshoot carrying capacity than r-selected species, but when they do, the effect will be sustained and possibly catastrophic." Besides the life strategy of the respective wildlife species, also that of its main food resource should be considered in this context. Insofar, the concept of r- and K-selection (or -strategy) should be applied also to important food plants or their relevant plant communities, and to necessary (thermal and security) cover types, as long as specific vegetation types are offering suitable structures (Gossow 1987a). If wildlife shall be understood and managed as a sustainable resource, as a vanishing treasure, or even as a sustainable and possibly increasing problem source, it is necessary not only to become informed on the value of these non-timber resources, but also on ecological and socio-economic prerequisites for their preservation, utilization and control facilities. Typically, the same wildife species may belong to the game use category as well as to the preservation or problem species category. In order to differentiate between various causes and triggers of problem situations for/by wildlife as well as to decide about management consequences, sufficient knowledge about habitat demands, population dynamics, interspecific competition/predation and the like, is a necessary prerequisite for the responsible forester, or wildlife manager, or nature conservationist. If necessary, such experts must be integrated into landscape planning/developing teams.

Three aspects shall be discussed here in more detail, as *pars pro toto* of diversity conservation, population cycles, and game damage problems in boreal forests.

3. Wildlife Species Diversity in Boreal Forests

There may be a comparatively high wildlife diversity - in game as well as in non-game bird and mammal species - in pristine boreal forests of the Palaearctic. This refers to fire-recycled forest mosaics, but even more if interspersed with non-fire influenced, or non-forestry-impacted habitats. We can differentiate between more typical "succession", or r-strategic species, and "climax" or K-strategic species, in plants as well as in animals. But the boreal forest, usually an age-class patched mosaic, may reach ages between 50 and 200 years (or even more) years, depending on peak fire intervals, however, only very seldomly on a tree-specific senescence. "Old forests" in the taiga zone are different from what we associate elsewhere with this term. Thus, it may be difficult to detect real "climax" species in boreal wildlife. Perhaps it is only possible to detect those which are able to cope fairly well with longer- lived stands, for instance with a larger amount of dead wood (snags) - e.g. for woodpeckers - resulting from lighter fires which led to some higher structural diversity, or with a sufficient lichen growth after a sufficiently long period without burning - for reindeer or caribou (*Rangifer tarandus*).

On the other hand, quite a few "succession" species should be found in boreal forests. The frequent fire impacts and the fire-caused recycling of whole communities also include arboreal and shrub elements which offer sufficient food and cover resources for a variety of wildlife species such as hare, grouse, deer, and moose - and, insofar, also for their predators.

Small mammals, as a rule, show high mortality losses in areas with surface fires. But burned areas can be recolonized very rapidly - often by genetically different eco-types (e.g. Krebs and Gaines 1971). Arboreal species such as flying squirrels and tree squirrels will suffer from crown fires which eliminate tree canopies. Some species, like deer mice, may also profit as related to an increased seed production of forage species (Peek 1986). However, differentiation must be made between short-term responses (e.g. negative with regard to snowshoe hares) and delayed longer-term responses, when foods and favorable cover proliferate only a few years after the burn.

As pointed out by MacArthur and MacArthur (1961) and Emlen (1970), bird habitat selection should be detectable from structural parameters like leaf densities or other gross visual aspects of vegetative physiognomy (cf. also Hildén 1965; Berndt and Winkel 1974; Gossow 1977). Differences can be found in birds regarding response to fire influences on various foraging types ("guilds"; e.g. Bock and Lynch 1970). No notable response at all to fire is seen in most bird species of a given range (e.g. Bendell 1974; Lyon and Marzluff 1984). Bendell considered this population stability within a rapidly fluctuating environment to indicate that many (especially song) birds are able to control their populations independently of fire disturbances. In some contrast is the situation with raptors, ravens, and tetraonid species. Exposure of potential prey species (like small mammals and ground-nesting birds) may be "considered responsible for the influx of avian predators. Similarly, surface insects become more vulnerable to birds following fire".

Gallinaceous birds, quite generally, but especially many tetraonid species, have »proven to be exceptionally responsive to burning of habitats, with the inevitable exceptions« (Peek 1986). For instance, experience with blue grouse (*Dendragapus obscurus*) appear quite controversial in regards to their density, reproductive, behavioral, and population genetic reactions toward logging and/or fires. Ruffed grouse (*Bonansa umbellus*) react very impressively to fire-managed habitats, especially where prescribed burning of aspen is concerned (e.g. Doerr et al. 1970; Gullion 1970, 1972). The black grouse (*Tetrao tetrix*) and the capercaillie in Fennoscandia (*Tetrao urogallus*) and Russia-Siberia (*Tetrao parvirostris*) are apparently understood especially well for their habitat requirements from their peripheral populations in Central Europe, in mostly man-made landscape types, which are highly influenced by pasturing, clearing, and logging activities (e.g. Gossow et al. 1984, 1987b; cf. also Picozzi et al. 1992). The Fennoscandian situation is discussed by Swenson and Angelstam (1993) for woodland grouse in relation to boreal forest succession. The niche of the black grouse, under natural conditions, appear to be in the birch-rich early successional stages following forest fires, or in bogs (Angelstam and Martinsson 1990) - so, they »conclude that black grouse habitat has been positively affected by modern forestry«, because the dominant fire regime in the main boreal forest was large, intense fires (according to Heinselman 1981) as is the dominant forest management system throughout Fennoscandia based on clear-cuttung areas of 10-60 ha (Swenson and Angelstam 1993). But the short rotation period of only 100 years for managed forests mean that hazel grouse (*Tetrastes bonasia*), and especially more capercaillie, are lacking important long-term habitats. Hazel grouse habitat is offered only in younger age classes when the forest becomes dense, and still contains deciduous trees - eliminating the latter makes these production forests less suitable for hazel grouse. Capercaillie, which use forests more than 70 years old, and select them only when more than 90 years old, find suitable habitats only in too short a period (e.g. for lek tradition). Especially »the vast majority of the fire-maintained old pine forest, to which the capercaillie is adapted, disappears under modern forest management« (Swenson and Angelstam 1993). Similar experiences have been made with roe deer (*Capreolus capreolus*) and moose (*Alces alces*) in Central European and Central Russian habitats, as well as in pronounced boreal habitats. Especially logging or clear-cutting has widely opened natural forest types for invasion-like expansions of these game ungulates to the north.

Both groups, tetraonid game birds and game ruminants, appear to be comparatively broad-niched animals. Although tetraonids or woodland grouse are highly taiga-oriented, they are able to cope with similarly structured production, protection, and pasturing forests in Central Europe. In boreal forests, the tetraonines' high adaptability to these habitats is reflected by the circumstance that locally they may encompass nearly 80 % of the total bird biomass (according to Semenov-Tjan-Shanskij 1960).

That which is characteristic of typical bird and mammal species of the taiga, is considered also characteristic of typical northern trees, many of which also have broad niches. "For example, the natural geographical ranges of quaking aspen, paper birch, jack pine, black spruce, and white spruce are extremely large, they occupy a wide range of soils and land forms; they mingle freely with many different associates in various regional and local contexts. All but white spruce also occur in areas with short or long fire cycles, and in areas with predominant crown fires as well as some surface fires. Such species may be less vulnerable because of their broad niches. Aspen and birch can reproduce both vegetatively and from seed after fire. Black spruce reproduces well from seed after fire, but .... it can (also) reproduce vegetatively by layering. Such multiple strategies allow survival under many circumstances" (Heinselman 1981).

For wildlife, the size and intensity of the respective burns and time intervals are most important. Single fire patches may cover as much as 100.000 - 500.000 ha, and few fires often burn vast amounts of the total area. Insofar, a large number of moors, swamps, creeks, rivers, and lakes diversify or "fragment" the burn patterns. Within sites, structural diversities may be low, and patches showing low diversity may be large because of the often large size of fires. But as Rowe (1979) puts it, »fire continually creates between-site diversity, renewing the spatial mosaic of community types and age classes«. »It is this between- site diversity that accomodates most of the diversity that exists in northern ecosystems« (Heinselman 1981). Mobility of the wildlife birds and mammals in the taiga enables them to cope with respective interspersions of habitat types, offering food, cover, breeding and nesting grounds, etc. If necessary, seasonal migrations may also become part of their life strategies.

Woodland caribou (*Rangifer tarandus caribou*) as well as wild forest reindeer (*Rangifer tarandus fennicus*) are typical, but infrequent inhabitants of the taiga. Most early investigations on their ecology tended to stress the detrimental aspects of fire (destruction of old-growth forest, particularly of its stands of *Cladina* spp. forage lichens, according to Schaefer and Pruitt 1991). More recent investigations questioned this view, or even contended that fire may improve the quality of *Rangifer* taiga ranges (cf. Schaefer and Pruitt [1991] for references).

According to Klein (1982), »a major problem in assessing the relationship between fire and caribou ecology has been the failure to distinguish between short-term versus long-term effects.« Schaefer and Pruitt (1991) tried to consider this in their woodland caribou study on taiga fire effects: "Compared to old-growth forest stands (90 yrs.), most burned upland habitats exhibited enhanced productivity of summer forages but a decline in quality and accessibility of winter forages. This deterioration of winter habitat for caribou resulted from the loss of lichens (*Cladonia* spp.) in the predominant jack pine communities (*Pinus banksiana*), the increase in both thickness and hardness of snow cover, and the accumulation of deadfalls". The taiga does not appear to be very suitable as a winter habitat for *Rangifer* in its recently-burned and intermediate stages (up to 50 years or more following fire). "Yet fire may be necessary to maintain optimal, long-term lichen resources". For instance if the tendency of upland communities is to evolve toward a spruce-fir mix, then under the absence of fire, »by 160 years of age, mixed coniferous habitats exhibited an increased cover of feather mosses at the expense of forage lichens« (Schaefer and Pruitt 1991; cf. also Miller 1989).

The question arises whether northern wildlife will also continue to thrive in a converted boreal forest where intensified forestry is practised !? A production and plantation forest is also an age-class forest, but it usually consists only of the timber-relevant age classes, with an exclusion of most pioneering shrub, non-timber trees and accompanying plant communities. Elder age-classes will be a minimum or non-existent. That means considerable shortcomings in food supplies and cover availability, and of structural diversity for r- as well as K-strategic wildlife species. "The maintenance of (a wildlife species such as the) capercaillie in the managed landscape presents the greatest challenge to forestry, on both the habitat and landscape scales .... forest managers must strive to mimic more closley the natural variation in structure and size of forest stands" (Swenson and Angelstam 1993; cf. also Gjerde 1991; Gossow et al. 1984; Picozzi et al. 1992).

4. The Case Example of the Moose in a Forest-Converted Taiga

Especially the r-strategic game ungulates such as roe deer, white-tailed deer (Odocoileus virginianus), and moose can profit from modern and intensified forestry practices like clear-cutting and plantation forestry. Roe deer have 'conquered' the Fennoscandian North - and in their wake, their main predator, the European lynx (*Lynx lynx*), did so too. The moose populations in Russia have expanded their ranges considerably towards the north, the south, and even to the west (Austria and Czechia host meanwhile moose too).

Moose densities in the Baltic republics of the former USSR were estimated at up to 5 moose per 1000 ha in the 80s, with a SE/E gradient of decreasing densities. But the moose is, according to his feeding type, not only a selective browser, he is also great in stripping bark, thrashing and breaking stems. Insofar, the increased and expanded moose range in Fennoscandia and in Russia-Siberia does not only offer the possibility of more game use and trophy antler hunting but also brings increasing problems for the conifer plantation forestry (Peek et al. 1976, 1978; Bergström and Hjeljord 1987; Kuznetsov 1987; cf. also Oldemeyer and Regelin 1987).

According to our experiences with roe deer, also a selective browser in Central Europe, age-class conifer forests may act as 'ecological traps', especially in connection with some hunting pressure, but also because of sharp structural edge effects from clear-cutting (e.g. Gossow and Reimoser 1985; Reimoser 1985, 1990; Gossow 1987a; Donaubauer et al. 1990). Roe deer, and probably also moose, are quite susceptible to edge effects. And security, or even thermal cover demands, are well offered within thicket and pole stands of pine and spruce plantations. However, these appear to be in obvious imbalance with browsable and available food supplies - with the exception of young conifer twigs, buds, and bark - after all deciduous shrubs and trees and herbs etc. have been browsed out, or even systematically eliminated. Buffer foods are lacking, or minimally due to overbrowsing, or being overgrown too quickly, especially by spruce or even grasses and perennials.

Insofar, the "moose problem" also appears to be mostly "home- made" by forestry or the timber industry, itself. "Game damage" cannot be measured simply in browsing or bark-stripping percentages of planted or growing tree densities. In order to evaluate such quite natural feeding practices causing game damage, obligatory "SOLLWERT" prescriptions (debit quota guidelines) for fores ry production or protection goals are a necessary prerequisite. And of greater importance than the impacted young trees are those which survive without game impacts and are sufficient to build up the future forest (Reimoser 1985).

Regarding moose habitat management, there are a lot of recommendations to be found in the literature, including the use or rejection of prescribed burning (e.g. Oldemeyer and Regelin 1987; Peek et al. 1978). Bergström and Hjeljord (1987) "conclude that modern forestry fits well into the habitat and forage requirements of moose .... (and emphasize) that the forest cycles were formerly in the same order as in the cycles now created by man.« However, as shown by Van Wagner (1978), age class distribution can be completely different in forests affected by fire compared to the human-managed forests of today. Another difference may be the size (up to 50 ha) and distribution of clear-cuts, which mean that cover is always near, with 15-30 % of a land area should as good human-made winter habitats.

Clearcuts in Fennoscandia, Eastern Europe, Russia and Siberia may be better accessible winter feeding grounds because of a low to moderate snow cover as compared to many places in boreal North America. Insofar the comparatively small-sized clear-cuts may more easily be overbrowsed (cf. Kuznetsov 1987; Oldemeyer and Regelin 1987), or a bit later become like pole stands, susceptible to bark stripping.

5. Wildlife Population Cycles in the Boreal Forest

A most remarkable characteristic of many wildlife populations in northern tundra and taiga ranges are their cyclic oscillations (cf. Elton 1942; Elton and Nicholson 1942; Keith 1963, 1973). Best documented, in respect to a longer time-scale, and rather well-known are the interconnected cycles of snowshoe hare (*Lepus americanus*) and Canada lynx populations (*Lynx canadensis*) which average at 9-10 year peak intervals.

For smaller mammals such as lemmings and voles, a 3 - 4 years periodicity is known. But also bird species, such as tetraonids or raptors may show such cycles in abundance. Especially the cycles of predatory species appear to be only 'secondary', and more so for more secondary or buffer prey species in times of need. The cycle orginates primarily from herbivorous animals like the forest-dwelling snowshoe hare or the tundra-inhabiting lemming. In a comprehensive report and analysis of evidence documented in literature, Fox (1978) has discussed the remarkable coincidence between the Canada lynx cycle (as the better and for longer periods documented data base) and the occurrence of forest and brush fires in the taiga of North America. In this, he follows an already earlier published hypothesis by Grange (1949, 1965) that fire and plant succession are the driving mechanisms for the snowshoe hare cycle. A simple predator-prey model is not compatible with the systems dynamics (e.g. Schaffer 1984).

In order to exclude other environmental factors - like summer precipitation or snow depth, and moon phases etc. - as "causes of the hare-lynx cycle, one has to prevent secondary succession (i.e. fire) over so large an area that immigration is negligible. The U.S. Forest Service appears to have done the experiment .... In Wisconsin, where fires have been most radically suppressed, fluctuations appear to have been stopped .... Damping of the hare cycle in the Lake States could also be attributed to land development" (Fox 1978; cf. Keith 1963; Dolbeer and Clark 1975). But a disappearance of the cycle in more southern areas is also typical in unmanaged areas (Breitenmoser and Breitenmoser-Würsten 1993).

6. Discussion

The several findings and statements, given by Fox (1978) in his paper, are in part somewhat controversial, for instance, concerning conclusions for theoretical ecology, like on stable vs. instable limit cycles and how to discriminate between instrinsic limit cycles and an external driving cause of the hare-lynx cycle. "If coincidence in periodicity could be relied upon, lynx cycles should be explained by sunspots, which exhibit a very neat 10-11 years peak in their variance spectrum. Sunspots and lynx time series marched in phase and amplitude for a string of years, about as long as our fire time series: now the two are completely opposite in phase" (Fox 1978, according to Moran 1949).

A more recent paper by Sinclair et al. (1993) brings together convincingly sun spot periodicity and hare densities as well as snow accumulation trends. An earlier paper, by Arditi (1979), had already presented evidence for a 1-2 year metachrony of the long-lasting 10-year rhythm of precipitation and temperatures, and the hare-lynx cycles, respectively. Obviously, the solar cycle provides a phase-locking mechanism for the hare cycle, and insofar »hare cycles in Siberia should be in phase with those (studied) in Canada« (Sinclair et al. 1993). Solar activity affects also weather systems, and insofar also fire regimes which show such good correlation with hare-lynx cycles. But that boreal forest fires should drive the snowshoe hare cycle primarily and sufficiently (as according to Fox 1978) appears a bit oversophisticated. It also appears interesting that more recent discussions of the hare- or vole-cycles again emphasize a driving role in the included predators (cf. Lindstrom et al. 1991; Krebs et al. 1992; Breitenmoser et al. 1993).

More important in applied respects appears to be that »the biological and historical evidence that increases in land areas, undergoing secondary succession, can directly cause increased growth, survival, and reproduction of snowshoe hares or other wildlife species possessing similar food and habitat requirements« (Fox 1978). There is a lot of literature documenting the importance of successional food plants among 'successional' wildlife species of the boreal forest, such as hare, various grouse species, deer, and moose. Post-fire successional habitats contain enough of both summer and winter food to allow increased population densities on such sites (regarding 'functional' as well as 'numerical' responses).

On the other hand, decreases of such populations may also take place after long-term increases, when post-fire or post-logging secondary succession is completed and its carrying capacity reduced to elder growth-standards. Such examples in North America or elsewhere date mostly from the past 40 - 100 years, and they are not related to any 'cycles' but reflect more the history of logging and fire protection in these areas (cf. also Fox 1978 or Pyne 1993). Locally, of course, also the beaver (*Castor canadensis* and *C. fiber*) can become very efficiently operative as a landscape manager, and interact, for instance, with moose or other wildlife (e.g. Johnston et al. 1993).

Modern forestry tries to operate in a sustainable manner, while increasing the predictability of forests as wildlife habitats. This may be a good prerequisite for sustainable game management (such as with supplemental feeding for deer too). But the question arises as to how essential the boreal cycles in ecosystem dynamics are, and if they are needed by somehow the floral and faunal wildlife? And finally, what is the ecosystemic value of boreal forest fire influences and of their cyclic characteristics? (This has to be answered primarily by the vegetation ecologists, and much less decisively by wildlife ecologists.)

More problematic and acute is probably the more recent developmental trend towards taiga conversion and devastation in several parts of the world, especially obvious in extended parts of Siberia (e.g. Petrov 1992). This causes not only survial and existence problems for taiga wildlife, but also, increasingly, for many indigenous peoples there (Bartels and Bartels 1988; Anderson 1991). In Fennoscandia and NE Siberia, reindeer husbandry is highly dependent on forest pasturing, and insofar, on ground and arboreal lichen forage supplies; in other words: On old-growth forest types which are highly susceptible to ground and crown fires. Fire protection efforts especially like those in NE Siberia (cf. Valendik, this volume) appear necessary. But for potential long-term benefits of boreal forest fires one should take into consideration also the findings by Miller (1980) or Schaeffer and Pruitt (1991).

Whether nature or biosphere reserves and reservations may help in this devastative situation, and to what extent, has to be investigated and the results awaited. Obviously, only species and landscape protection through legislative means is not enough. A corresponding management of biotopes (or habitats) as well as of people (or land use, tourism etc.) is also imperative too.

However, additionally, today one must also ask as to how far these type of conservation efforts may be sufficient on a long-term scale in times of a progressive global climate change. The corridor (and step stone) strategies have meanwhile become a »biological conservation principle« (Harris and Scheck 1991). But corridors are usually conceived for remnant patches of formerly connected habitats of specific floral or faunal species (designed often as "flagship species") - hoping that in their wake other species or species-associations should benefit too. Inherent in the recent on-site conservation through the establishment of national parks and nature reserves, is the »idea that the global environment is stable or near stable« (Hobbs and Hopkins 1991). Major climatic changes in the past were obviously developed very slowly and were considered mostly in evolutionary respects. Changes associated with the greenhouse effect could develop much faster.

"Rapid changes in global climate are likely to have dramatic effects on the world's natural biota, necessitating large geographical shifts in distribution and migration at unprecedented rates. Studies of vegetation (and fauna) responses to past climate changes suggest that species will migrate in an individualistic manner, rather than as intact communities" - meaning (like Hobbs and Hopkins, for instance, but also various other authors), that there is a high degree of uncertainty associated with predicted biota response to rapid climate change. The role of corridors in assisting such migrations is correspondingly uncertain too. But in boreal forests these very cycles of whole communities appear typical and perhaps they are also somehow predisposed - insofar one may wonder if such community 'shifts' might be more easily possible in the north than elswhere?

If through such reserves also some kind of gene-banking should be facilitated and secured, gene flow must remain possible to some extent by means of appropriate corridor facilities (such as rivers and riverine forests or mountain crests and ridges; cf. Gossow 1992). Corridor conceptualization and design is a common, or at least obvious, strategy in face of landscape and habitat fragmentation. But fragmented landscapes "no longer allow the biota (and its dwellers, respectively) to migrate freely" in response to progressive climatic changes, making predictions still more complex and uncertain. "However, if we assume that climatic changes are likely to follow existing climatic gradients, corridors should ideally run parallel to these gradients« - to quote Hobbs and Hopkins once more, and which may be especially valid for temperature gradients and rainfall gradients, and may well correspond with the large northward-running streams in Russia-Siberia and North America as well as with the north-south oriented mountains, also in Scandinavia (e.g., the Urals, the Scandinavian Fjeller ridge, the Cascades, and the Rocky and Coast Mountains).

Another aspect of climatic change toward warmer and drier conditions would be corresponding implications for large, free-ranging ungulates, especially as far as winter food sources are concerned (cf. Romme and Turner 1991). But it could also affect the summer season-distribution patterns of caribou and reindeer because of increased insect activities - or even a decrease in this respect because of a reduction of permafrost and wetlands (!?), including then dramatic losses of waterfowl breeding habitats as well, especially in tundra ranges. Taiga forests will come "under increasing stress and wildfires will increase in frequency and intensity and reinforce the greenhouse effect" (Wein 1990).

A characteristic of contemporary nature conservation in Central Europe is the establishment of national parks and the like in more peripheral localities, being often especially climate- impacted, and mostly not large enough to act sufficiently as genetic resources. But even northern coniferous forests are very likely not able to adapt quickly enough to climate changes in the coming century (e.g. Grassl 1992). As stated already, "biotic response to future climate change cannot be expected to result in a simple geographic displacement of existing communities, but rather in individual species responding to changes in particular climatic parameters which affect their growth or reproduction and migrating at different rates" (Hobbs and Hopkins 1991).

The above means that contemporary and future forestry (or the management of other so-called renewable natural resources correspondingly), to an "adaptive forestry" (e.g. sensu Stoszek 1992; cf. Schmidt-Vogt 1985, or Thompson and Welsh 1993) becomes imperative. And not only old-growth forest phases should again be preserved and sustained, but forest and vegetational succession at all should be more thoroughly considered and respected in sustainable and adaptive forest management - including the use and control of fire. For instance, prescribed burning strategies should also consider wildlife habitat demands, sustainability, and diversity. Then, wake effects, as well as space demands of the more fastidious wildlife species should be fulfilled, as for instance, in Scotland for red deer (*Cervus elaphus*) and red grouse (*Lagopus scoticus*) through sophisticated heather burning. In North America, minimum and optimum size demands for burning and clear-cuts respectively are given, for instance, for moose (Oldemeyer and Regelin 1987), and ruffed grouse (e.g. Bendell 1974; Gullion 1972, 1984; Dolgaard et al. 1976). For Fennoscandia and/or Russian Eurasia, similar recommendations should be elaborated - and checked in the field for suitability, feasibility, and acceptability.

References

Alexander, ME., B.D. Lawson, B.J. Stocks, and C.E. Van Wagner. 1984. User guide to the Canadian forest fire behavior prediction system: rate of spread relationships. Environment Canada, Canadian Forestry Service Fire Danger Group, interim edition.

Anderson, D.G. 1991. Turning hunters into herders: A critical examination of Soviet development policy among the Evenki of SE Siberia. Arctic 44, 12-22.

Angelstam, P., and B. Martinsson. 1990. The importance of appropriate spatial and temporal scales in population studies - conservation lessons based on the population dynamics of black grouse in boreal forest in Sweden. In: Symp. Proc. De toekomst van de wilde hoenderachtigen in Nederland (J.T. Lumeij and Y.R. Hoogeveen, eds.), 82-96. Amersfoort, NL.

Arditi, R. 1979. Relation of the Canadian lynx cycle to a combination of weather variables. Oecologia 41, 219-233.

Bartels, D., and A. Bartels. 1988. Are Siberian native peoples part of a "Fourth World"? Dialectical Anthropology 12, 245-252. (cit. after Anderson 1991).

Bendell, J.F. 1974. Effects of fire on birds and mammals. In: Fire and Ecosystems (T.T. Kozlowski and C.E. Ahlgren, eds.), 73-137. Academic Press, New York.

Bendell, J.F., D.G. King, and D.H. Mossop. 1972. Removal and repopulation of blue grouse in a declining population. J. Wildl. Manage. 36, 1153-1165.

Bergström, R., and O. Hjeljord. 1987. Moose and vegetation interactions in NW Europe and Poland. Swed. Wildl. Res. Suppl. 1(1), 213-226.

Berndt, R., and W. Winkel. 1974. Ökoschema, Rivalität und Dismigration als öko-ethologische Dispersionsfaktoren. J. Ornith. 155, 398-417.

Bock, C.E., and J.F. Lynch. 1970. Breeding bird populations of burned and unburned forests in the Sierra Nevada. Condor 72, 182-189.

Breitenmoser, U., and C. Breitenmoser-Würsten. 1993. Der Schneeschuhhasen-Luchs-Zyklus. Die Suche nach dem Taktgeber. Wildbiol. Intern. 5/8. Zurich (CH)

Breitenmoser, U., B.G. Slough, and C. Breitenmoser-Würsten. 1993. Predators of cyclic prey: Is the Canada lynx victim or profiteer of the snowshoe hare cycle? Oikos 66, 551-554.

Doerr, P.D., L.B. Keith, and D.H. Rush. 1970. Effects of fire on a ruffed grouse population. Proc. Ann. Tall Timbers Fire Ecol. Conf. 11, 25-46. Tall Timbers Research Station, Tallahassee, Florida.

Dolbeer, R.A., and W.R. Clark. 1975. Population ecology of snowshoe hares in the central Rocky Mountains. J. Wildl. Manage. 39, 535-549.

Dolgaard, S.J., G.W. Gullion, and J.C. Haas. 1976. Mechanized timber harvesting to improve ruffed grouse habitat. Techn. Bull. 308, Forestry Ser. 23, 1-11. Univ. Minnesota.

Donaubauer, E., H. Gossow und F. Reimoser. 1990. "Natürliche" Wilddichten oder forstliche Unverträglichkeitsprüfung für Wildschäden. Österr. Forstz. 101 (6), 6-9.

Elton, C. 1942. Voles, mice and lemmings. Clarendon Press, Oxford, 496 pp.

Elton, C., and M. Nicholson. 1942. The ten-year cycle of lynx in Canada. J. Anim. Ecol. 11, 215-244.

Emlen, J.T. 1970. Habitat selection by birds following a forest fire. Ecology 51, 343-345.

Fox, J.F. 1978. Forest fires and the snowshoe hare-Canada lynx cycle. Oecologia 31, 349-374.

Gjerde, I. 1991. Cues in winter habitat selection by capercaillie. I. Habitat characteristics. II. Experimental evidence. Ornis Scand. 22, 197-204; 205-212.

Gossow, H., 1977: Waldstrukturen und Wildstandsentwicklung. Tagungsber. Wald und Wild, Wien, 1977. 1-27.

Gossow, H. 1986. Ungulate game overabundance in central Europe and management integration with forestry demands. 18th IUFRO World Congr. Proc. Div.I, pp.625-636.

Gossow, H. 1987a. Alpine Rotwildvorkommen im Konflikt mit verschiedenen Landnutzungs-Interessen. Cbl. Ges. Forstw. 104, 82-95.

Gossow, H. 1987b. Human land use activities and grouse. In: Proc. 4th Intern. Grouse Symp. (Lam) (T. Lovel and P. Hudson, eds.), 85-93.

Gossow, H. 1992. Political borders as specific ecotones in Central Europe: Problems and options in nature conservation and forestry-wildlife management. IUFRO Centennial Congr. (Berlin - Eberswalde), Abstracts, p.186.

Gossow, H, K. Pseiner, H.G. Jeschke, and B. Pokorny. 1984. On the suitability of some forest parameters in capercaillie habitat evaluation in the Eastern Alps. In: Proc. 3rd Intern. Grouse Symp. (York) (T. Lovel and P. Hudson, eds.), 363-375.

Gossow, H., und F. Reimoser. 1985. Anmerkungen zum Zielkonflikt Wald - Wild - Weide - Tourismus. Schweiz. Z. Forstw. 136, 913-929.

Grange, W.B. 1949. The way to game abundance. Scribner's, New York, 365 pp.

Grange, W.B. 1965. Fire and tree growth relationships to snowshoe rabbits. Proc. Ann. Tall Timbers Fire Ecol. Conf. 4, 110-125. Tall Timbers Research Station, Tallahassee, Florida.

Grassl, H. 1992. Klimaänderung: Folgen für naturnahe Ökosysteme. NNA (Schneverdingen) Ber. 5/1, 50-52.

Gullion, G.W. 1970. Factors influencing ruffed grouse populations. Trans. N.A. Wildl. and Nat. Res. Conf. 35, 93-105.

Gullion, G.W. 1972. Improving your forested lands for ruffed grouse. Minn. Agr. Expt. Sta. Misc. J. Ser. Publ. No. 1439.

Gullion, G. 1984. Managing northern forest. Minnesota Agric. Exper. Station/St. Paul, Miscell. J. Ser. 13/442, 1-72.

Harris, L.D., and J. Scheck. 1991. From implications to applications: the dispersal corridor principle applied to the conservation of biological diversity. In: Nature Conservation 2. The Role of Corridors (D.A. Saunders and R.J. Hobbs, eds.), 189-220. Surrey Beatty & Sons, Chipping Norton.

Heinselman, M.L. 1973. Fire in the virgin forests of the Boundary Waters Canoe Area, Minnesota. Quat. Res. 3 (3), 329-382.

Heinselman, M.L. 1981. Fire and succession in the conifer forests of northern North America. In Forst Succession (D.C. West, H.H. Shugart, and D.B. Botkin, eds.), 374-405. Springer-Verlag, New York.

Hildén, O. 1965. Habitat selection in birds: a review. Ann. Zool. Fenn. 2, 53-75.

Hobbs, R.J., and A.J.M. Hopkins. 1991. The role of conservation corridors in a changing climate. In: Nature Conservation 2. The Role of Corridors (D.A. Saunders and R.J. Hobbs, eds.), 281-290. Surrey Beatty & Sons, Chipping Norton.

Johnson, E.A. 1992. Fire and Vegetation Dynamics: Studies from the North American Boreal Forest. Cambridge Univ. Press.

Johnston, C.A., J. Pastor, and R.J. Naiman. 1993. Effects of beaver and moose on boreal forest landscapes. In: Landscape Ecology and GIS (R.Haines-Young et al., eds.), 237-254. Taylor & Francis Ltd., London.

Keith, L.B. 1963. Wildife's ten-year cycle. Univ. of Wisconsin Press, Madison.

Keith, L.B. 1973. Some features of population dynamics in mammals. Proc. Int. Congr. Game Biol. (Stockholm) 11, 17-58.

Klaus, S., A.V. Andreev, H.H. Bergmann, F. Müller, J. Porkert und J. Wiesner. 1986. Die Auerhühner *Tetrao urogallus* und *T. urugalloides*. Ziemsen, Wittenberg.

Klein, D.R. 1982. Fire, lichens, and caribou. J. Range Manage. 35, 390-395.

Knight, D.H., and L.L. Wallace. 1989. The Yellowstone fires: Issues in landscape ecology. BioSience 39, 700-706.

Komarek, E.V. 1971. Principles of fire ecology and fire management in relation to the Alaskan environment. Proc. Fire in the Northern Environment (Fairbanks), 3-22.

Krebs, C.J., R. Boonstra, S. Boutin, M. Dale, S. Hannon, K. Martin, A.R.E. Sinclair, J.N.M. Smith, and R. Turkington. 1992. What drives the snowhoe hare cycle in Canada's Yukon? In: Wildlife 2001: Populations (D.McCullough and R. Barrett, eds.), 886-896. Elsevier, London.

Krebs, D.J. and M.S. Gaines. 1971. Genetic changes in fluctuating vole populations. Evol. 25, 702-723.

Kuznetsov, G.V. 1987. Habitats, movements, and interactions of moose with forest vegetation in USSR. Swed. Wildl. Res. Suppl. 1 (1), 201-211.

Laws, R.M. 1981. Large mammals feeding strategies and related overabundance problems. In: Problems in Management of Locally Abundant Wild Mammals (P.A. Jewell and S. Holt, eds.), 217-232. Academic Press, New York.

Lindstrom, E., H. Andrén, P. Angelstam, and P. Widén. 1991. Synchronous population cycles among northern mammals and birds - where and why? Trans. 18th IUGB Congr. (Krakow 1987), pp. 597-603.

Lyon, L.J., and J.M. Marzluff. 1984. Fire's effects on a small bird population. In: Symp. Proc. Fire's Effects on Wildlife Habitat (Missoula) (J.E. Lotan and J.K. Brown, eds.), pp. 16 - 22. Ogden (Utah).

MacArthur, R.H., and J.W. MacArthur. 1961. On bird species diversity. Ecology 42, 594-599.

Miller, D.R. 1989. Wildfire effects on barren-ground caribou wintering on the taiga of northcentral Canada: a reassessment. Proc. Int. Reindeer and Caribou Symp. (Rõros) 2, 84-98.

Moran, P.A.P. 1949. The statistical analysis of the sunspot and lynx cycles. J. Anim. Ecol. 18, 115-116.

Nelson, J.R. 1974. Forest fire and big game in the Pacific Northwest. Proc. Ann. Tall Timbers Fire Ecol. Conf. 15, 85-102. Tall Timbers Research Station, Tallahassee, Florida.

Oldemeyer, J.L., and W.L. Regelin. 1987. Forest succession, habitat management, and moose on the Kenai National Wildlife Refuge. Swed. Wildl. Res. Suppl. 1 (1), 163-179.

Peek, J.M. 1986. A Review of Wildlife Management. Prentice-Hall, New Jersey.

Peek, J.M., D.L. Urich, and R.J. Mackie. 1976. Moose habitat selection and relationships to forest management. Wildl. Monogr. 48.

Peek, J.M., D.J. Pierce, D.C. Graham, and D.L. Davis. 1978. Moose habitat use and implications for forest management in Northcentral Idaho. Swed. Wildl. Res. Suppl. 1 (1), 195-199.

Petrov, D. 1992. Kahlschlag in der sibirischen Taiga. In: Klassenfeind Natur. Die Umweltkatastrophe in Osteuropa (V. Thurn and B. Clausen, eds.), pp. 170-178.

Picozzi, N., D.C. Catt, and R. Moss. 1992. Evaluation of capercaillie habitat. J. Appl. Ecol. 29, 751-762.

Pyne, S.J. 1993. Keeper of the flame: a survey of anthropogenic fire. In: Fire in the Environment: The Ecological, Atmospheric, and Climatic Importance of Vegetation Fire (P.J. Crutzen and J.G. Goldammer, eds.), 245 266. Dahlem Workshop Reports. Environmental Sciences Research Report 13. John Wiley & Sons, Chichester.

Reimoser, F. 1985. Wechselwirkungen zwischen Waldstruktur, Rehwild und Rehwildbejagbarkeit in Abhängigkeit von der waldbaulichen Betriebsform. Fallstudie aus einem montanen Bergwaldrevier. Diss. Agricult. Univ. Vienna. 318 pp.

Reimoser, F. 1990. Über die Problematik der objektiven Kontrolle von Wildschäden im Zusammenhang mit forstlichen Verbiß-Gutachten als Grundlage für die Abschußplanung bei Schalenwild. Holzwirtsch. 46 (8/9), 26-30.

Romme, W.H., and M.G. Turner. 1991. Implications of global climate change for biogeographic patterns in the Greater Yellowstone Ecosystem. Conserv. Biol. 5, 373-386.

Rowe, J.S. 1979. Large fires in the large landscapes of the north. Symp. Proc. Fire Management in the Northern Environment, pp. 8 - 32. U.S. Dpt. Interior.

Rowe, J.S., and G.W. Scotter. 1973. Fire in the boreal forest. Quat. Res. 3 (3), 444-464.

Schaffer, W.M. 1984. Stretching and folding in lynx fur returns: evidence for a strange attractor in nature? Am. Nat. 124, 798-820.

Schmidt-Vogt, H. 1985. Struktur und Dynamik natürlicher Fichtenwälder in der borealen Nadelwaldzone. Z. Schweiz. Forstwes. 136, 977-994.

Schaefer, J.A., and W.O. Pruitt Jr. 1991. Fire and woodland caribou in SE Manitoba. Wildl. Monogr. 116, 1-39.

Scotter, G.W. 1970. Wildfires in relation to the habitat of barren-ground caribou in the taiga of northern Canada. Proc. Ann. Tall Timbers Ecol. Conf. 10, 85-105.

Scotter, G.W. 1971. Fire, vegetation, soil, and barren-ground caribou relationship in northern Canada. Proc. Fire in the Northern Environment (Fairbanks), pp. 209-230.

Semenov-Tjan-Shanskij, O.I. 1960. Ecology of tetraoids. Trudy Laplands. Zapov. 5, 1-318 <in Russian, cited after Klaus et al. 1986>.

Sinclair, A.R.E., J.M. Gosline, G. Holdsworth, C.J. Krebs, S. Boutin, J.N.M. Smith, R. Boonstra, and M. Dale. 1993. Can solar cycle and climate synchronize the snowshoe hare cycle in Canada? Evidence from tree rings and ice cores. Amer. Nat. 141, 173-198.

Stoszek, K. 1992. Adaptive Waldbewirtschaftung. Österr. Forstz. 4/92, 14-16.

Swenson, J.E., and P. Angelstam. 1993. Habitat separation by sympatric forest grouse in Fennoscandia in relation to boreal forest succession. Can. J. Zool. 71, 1303-1310.

Tande, G.F. 1979. Fire history and vegetation pattern of coniferous forests in Jasper National Park, Alberta. Can. J. Bot. 57, 1912-1931.

Thompson, I.D., and D.A. Welsh. 1993. Integrated resource management in boreal forest ecosystems - impediments and solutions. For. Chron. 69, 32-39.

Van Cleve, K., and L.A. Viereck. 1981. Forest succession in relation to nutrient cycling in the boreal forest of Alaska. In: Forest succession (D.C. West, H.H. Shugart, adn D.B. Botkin, eds.), pp.185-211. Springer, New York.

Van Wagner, C.E. 1978. Age-class distribution and the forest fire cycle. Can. J. For. Res. 8, 220-227.

Viereck, L.A. 1973. Wildfire in the taiga of Alaska. Quatern. Res. 3 (3), 465-495.

Wein, R.W. 1990. The importance of wildfire to climate change - hypotheses for the taiga. In: Fire in ecosystem dynamics. Mediterranean and northern perspectives (J.G. Goldammer and M.J. Jenkins, eds.), 185-190. SPB Academic Publishing, The Hague, 199p.

Wein, R.W. 1993. Historical biogeography of fire: circumpolar taiga. In: Fire in the Environment: The Ecological, Atmospheric, and Climatic Importance of Vegetation Fire (P.J. Crutzen and J.G. Goldammer, eds.), 267-276. Dahlem Workshop Reports. Environmental Sciences Research Report 13. John Wiley & Sons, Chichester.

Wright, H.A., and A.W. Bailey. 1982. Fire Ecology. Wiley & Sons, New York.

Zackrisson, O. 1977. Influence of forest fires on the North Swedish boreal forests. Oikos 29, 22-32.

J.G.Goldammer and V.V.Furyaev (eds.), Fire in Ecosystems of Boreal Eurasia, 445-452.

Fire Ecology in Sweden and Future Use of Fire for Maintaining Biodiversity

A. Granström [1]

1. Introduction

Except for the southernmost part and the high mountains in the northwest, Sweden lies in the boreal and hemiboreal (Ahti et al. 1968) zones. Although the country spans nearly 1500 km from North to South, mature forest vegetation is surprisingly similar, especially on poor to medium rich soils: ericaceous dwarf shrubs often dominate in the field layer, above a mat of pleurocarpous mosses. There are two dominant conifers, and a handful of boreal broadleaved tree species. In the hemiboreal zone several temperate broadleaved tree species occur too, but they usually constitute only a small fraction of the stands.

From the composition of the vegetation alone one would suspect fire to be important. However, due to suppression of wild-fires, becoming quite effective even in remote locations already in the mid 1800s, and a decline in the cultural use of fire, the area burnt has decreased steadily and is now on average only a few hundred hectares per year, out of a total forest area of 28 million hectares. Therefore, the study of fire ecology in Sweden is largely an archaeological exercise. The near absence of recent large fire fields makes it difficult to assess the role of fire in the landscape, in particular, questions regarding natural tree regeneration patterns, and typical fire behaviour. However, today, answers to these questions are needed by foresters in search of management methods that will maintain the diversity created by disturbance regimes of the past.

2. Fire History

The fire history of earlier times has been reasonably well worked out for the northern half of Sweden. The first attempt at this was already made in the 1920s by Wretlind (1932), who dated fires back to the 1400s in an area in southern Lappland. One of his observations was a differentiation in fire return intervals between dry lichen dominated sites and mesic, moss dominated sites. In an extensive analysis of fire history along a 290 km section of the Vindelälven river valley, Zackrisson (1977) found average fire return intervals to vary with vegetation type, topography and exposition. The average fire return interval was 52 years for lichen types, 91 years for the *Vaccinium myrtillus* type and over 150 years for moist forest types. Roughly the same picture was found by Engelmark (1984) in Muddus National Park,

[1] Department of Forest Vegetation Ecology, Swedish University of Agricultural Sciences, S-901 83 Umeå

250 km further North. Such differentiation between forest types should be expected if the fires often burn during conditions when only the more xeric types will burn. During extreme drought very few locations are fire proof. Studies from the southern part of the boreal region proper have reported fire return intervals of around 60 years for *Vaccinium*-type forests (Kohh 1975, Linder 1988), perhaps indicating somewhat higher frequencies in this region than further north.

The time period covered in these studies is 500-600 years, but until the the mid 1800s there is no marked trend in mean area burned, i.e. nothing that could indicate an influence of climate fluctuation (Zackrisson 1977, Engelmark et al. 1994). The same holds in one remarkably long chronology, from the lower part of the Luleälven river. There 18 fires were dated between A.D. 1081 and 1888, with a mean fire interval of 47 years (Zackrisson and Östlund 1991).

From the early part of the 18th century agriculture expanded into the interior part of the northernmost part of Sweden, but this did not result in any major change in fire frequencies, which could be taken as an indication that the fire regime in northern Sweden was essentially natural until the mid 19th century. However, it is known that settling involved the burning of forest land, primarily for improved grazing (Tirén 1937). For control reasons, and also in order to produce a desired impact on the vegetation, these burns seem to have been conducted under moderate burning conditions, and were for the most part relatively small (Wretlind 1935; Tirén 1937). Therefore it may be assumed that with agricultural expansion there was a gradual shift from a situation with few large burns, mainly lightning started, to a situation with a large number of small burns, of which many were accidentally or willfully set by people. It is also clear that there were also efforts to control catastrophic fires before forestry expanded into the area, such as during the notable fire year 1831: "Last summer people east and west of here were occupied with fighting fires. It was then an unusually severe drought and the fires were spreading unbelievably fast, and fear and anxiety preceded, surrounded and followed them everywhere" (Zetterstedt 1833). Incidentally, Zetterstedt argued for lifting the restrictions on burning: "One who has the intention of burning forests in these desolate areas for his own purpose, is not likely to reveal this, and what use is then a legislation which strictly forbids forest burning, when no one can be tied to the act" (Zetterstedt 1833). But then, around 1850, as timber cutting began on a massive scale in the interior of northern Sweden, intentional burning ceased (Wretlind 1935), and fire fighting gradually became more aggressive. This is seen in a steadily reduced fire load in the decades after 1850 (Zackrisson 1977). The legislation had in fact been very restrictive from the mid 1600s regarding the use of fire both for slash and burn agriculture and for improving grazing. This legislation seemed to have been initiated to conserve wood for the mining and smelting industry in central Sweden, but was applied throughout the country, with some modifications (Wieslander 1936). It is unclear if the legislation did indeed alter peoples old habits, but it is likely that it forced them to operate clandestinely.

Although most forests have a history of past fires there are areas that have no readily observed evidence of past fires. Usually they are wet forest habitats, or mesic habitats that are protected from fire for topographic reasons (Zackrisson 1977; Engelmark 1984), but charcoal analyes of peat stratigraphies in such forests usually reveal evidence of at least a few fires in the past (Hörnberg et al. 1995). In the forested areas near to the western mountains in the north, a couple of studies have likewise documented a much lower influence of fire than what appears to be the case over most of the interior of northern Sweden (Steijlen and Zackrisson 1986; Hofgaard 1993).

Further south in Sweden much less is known of the former influence of fires, and it is generally more difficult to obtain fire chronologies due to a much longer history of timber exploitation. There are, however, a few observations indicating very frequent fires on productive forest sites. One example is from a site in the northern part of the province Värmland, where 22 fires were dated between 1395 and 1815 within a 100 hectare area (Per Linder, pers. comm.). Several of the fires did not cover the whole investigation area which suggests that they were mainly management fires, set to improve grazing. The cessation of burning here also coincides with the first commercial cutting operations in the area.

There are only a couple of paleoecological studies that traces the influence of fire further back in Sweden. Bradshaw and Hannon (1992) provide circumstantial evidence for frequent fires up until the late 1700s at a site in eastern central Sweden. Here the fire-sensitive *Picea abies* did not start to expand on dryer ground until 200 years ago, although it has been present in the area since about 2000 years BP. A number of charcoal peaks in the stratigraphy support the interpretation of fire as a causal agent, although quantitative estimates of fire frequencies may not be possible to extract.

Unfortunately anthropogenic fire cannot be separated from natural fire in these records either. The lack of cereal pollen through most of the stratigraphy rules out slash and burn agriculture, but a lot of these fires may well have been set to improve grazing.

There are reasons to believe that there have been marked gradients in the country in the frequency of natural fire. Ignition density from lightning show a large-scale variation (Granström 1993). Ignition densities are highest in the dry southeastern provinces of Östergötland and Kalmar (Fig.1). From there densities decline both to the north and to the west. Ignition density is around 4-5 times higher in the southeast than in the boreal region proper. The province with the lowest ignition density is Västerbotten, where incidentally the most detailed fire history work has been done. The steep east-west gradient in ignition density in southern Sweden correspond to a gradient in summer precipitation (Granström 1993). Although natural fire frequencies cannot be directly inferred from ignition densities it is likely that the marked gradient in summer precipitation would enhance any difference caused by ignition density alone. The eastern distribution in southern Sweden of two annual fire-adapted plant species (Granström and Schimmel 1993; Granström 1993) give some support to the hypothesis of a particularly high influence of fires on the dry eastern side of the peninsula. Land management with the use of fire may for partly the same reason have had its greatest impact on the forests on the western side. Renberg et al. (1993) have given proof of increased lake-water pH in a number of small lakes in southwestern Sweden 1500-2000 years ago, probably a result of anthropogenic burning of forest land with little influence of natural fire. In many instances these areas were then maintained as *Calluna* heathlands, kept open by burning on a rotation of less than 10 years (Atlestam 1942). When these areas were planted with *Picea abies* in the late 1800s, lakewater pH started to decrease, later reinforced by acid rain (Renberg et al. 1993).

3. Fire Behaviour and Fire Effects

There is very little direct information regarding the spectrum of behaviour of past fires in the Swedish landscape, but circumstantial evidence suggests that crown fires have not been the norm. First, the topography is generally broken and the landscape is intersected with numerous lakes and open mires. Secondly, broadleaved tree species, primarily *Betula* spp. and *Populus tremula*, usually constitute a considerable proportion of the canopy, also in advanced stages of succession. This contrasts with much of the interior boreal part of North America, and should discourage the type of violent fire behaviour, with extensive crown fires, which seem to be the norm there (Johnson 1992). In many places the forest structure itself speaks of the relatively low intensity of former fires: on dry and mesic sites there have often been stands of old and large pines that have survived repeated surface fires (Wretlind 1935; Zackrisson 1977). This timber resource was in fact the base of the lumber industry in the 19th century (Zackrisson and Östlund 1991; Östlund and Lindersson 1995).

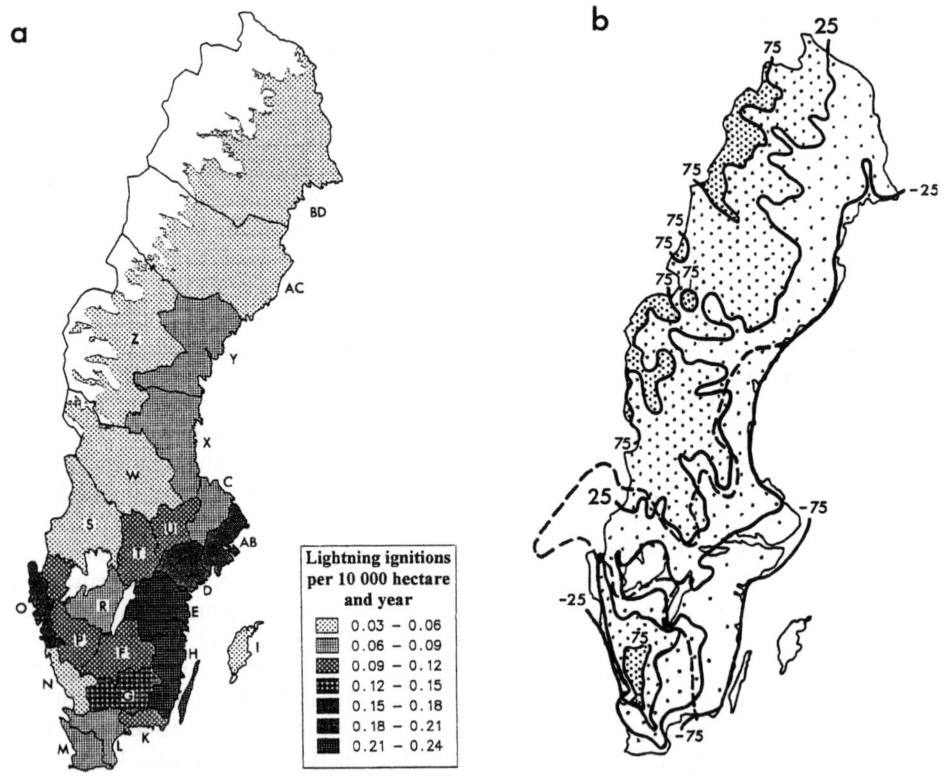

Fig.1. (a) Density of lightning ignitions (1953-1975) in the 24 provinces of Sweden. The white area in the northwest is alpine area and land above the limit of commercial forestry. (b) Isolines for average steps of 50 mm in precipitation minus evapotranspiration (from Eriksson 1986) for the period June to August. The broken line gives the distribution limit for the fire annual *Geranium bohemicum* (present east of the broken line, from Hultén 1971). The distribution of *G.bohemicum* extends into the dry eastern part of Norway (Granström 1993).

Also the behaviour of wildfires during this century show that Sweden has a relatively benign fire climate. The largest fires recorded have been in the order of 2000 to 3000 hectares and have been possible to stop within two or three days. Maximal spread rates have been around

20 m min^{-1}, associated with extensive spotting, but the actual crown fire areas have covered only a minor proportion of the burns (Granström, unpubl.). Another factor apparently limiting the potential for both fire spread and fire intensities in the past is the slow fuel accumulation. The only study of fuel succession conducted so far in Sweden (Schimmel 1993), suggest that in the north, fine fuels on the ground reach a steady state some time between 50 and 100 years after fire. The bulk of the fuel on the ground then consist of pleurocarpous mosses and lichens. In the first couple of decades after fire, fires are unlikely to spread. This corresponds to the shortest fire intervals observed in fire history studies (Linder 1988, Niklasson and Granström, unpubl.). One may assume the relative fire intensity to increase with age since last fire up to the age of 50 years or more (Schimmel 1993). Detailed mapping of past fires with the use of crossdating have also revealed several instances where young burns (up to the age of ca. 15 years) have stopped the advance of new fires (Niklasson and Granström, unpubl.).

Survival in the tree stand is of course directly related to fire intensity. Of the common tree species *Betula pubescens* and *P. abies* are the least resistant to fire damages, although it is not uncommon to find individuals of *P. abies* which have survived fire of low intensity. However, *Pinus sylvestris*, and to a lesser degree *Betula pendula*, are the only species with a capacity to survive repeated fires of moderate intensity. Therefore it is likely that there have been extensive areas of productive forest land where *P. sylvestris* has been able to dominate due to repeated fires (Zackrisson 1977). Without fire *P. abies* is the superior competitor, except on very poor soils.

When it comes to fire effects on ground flora, the depth of burn seems to be the most important variable. In an experimental study in northern Sweden where the depth of burn varied (Schimmel and Granström 1996), there were dramatic differences in the early development of vegetation: fire of low severity allowed *Vaccinium* spp. and the grass *Deschampsia flexuosa* to recover rapidly. Fire of higher severity depressed these species and allowed more colonization of *Luzula pilosa* from a buried seed bank. Fire of still higher severity eliminated the rhizomatous species as well as much of the seed bank and opened for rapid moss colonization of pioneer species as well as for the colonization from species dispersing their seeds into the site.

There is still no information available on the typical depth of burn associated with wildfires in Sweden. The presence of seeds of pioneer species well into the mineral soil (Granström 1982) suggests that the humus, at least sometimes, has been reduced considerably. It is of note, however, that the dominant rhizomatous species in northern Sweden have their rhizomes relatively superficial in the mor layer (Schimmel and Granström 1996). The ubiquity of these species may be taken as an indication that complete mor consumption is rare. In the southern half of the country there are some very common forest species with deep lying rhizomes lying at a few cm depth in the mineral soil: *Calamagrostis arundinacea* and *Pteridium aquilinum*. They usually survive even severe fires and expand vigorously thereafter (Kujala 1926).

Tree colonization varies tremendously both in the length of the colonization process, in tree densities and in which species become dominant. The governing factors have not been analysed more than on a very general level. On the poorer soils *Pinus sylvestris* is usually completely dominant in regeneration (Sarvas 1937). On richer soils broadleaved tree species do better, often to the exclusion of the relatively light-demanding *P. sylvestris* (Wretlind 1935). If broadleaved trees are able to form a closed canopy their dominance can sometimes prevail for over 100 years, until they are finally overtopped by invading *Picea abies* (Wretlind 1935).

4. Conservation Problems Connected with Fire

Although we still have only a rough picture of the role of fire in the structuring of the Swedish forest landscapes, it is clear that the present land use creates severe conservation problems. The most obvious problem concerns the former fire refugia, where clear-felling can be supposed to extripate a number of sensitive species. However, there is still only limited data in support of the hypothesis that these areas have indeed functioned as refugia for fire-sensitive plants. Certain lichens and fungi (Esseen et al. 1981) may be the best examples, but little is known of their capacity to recolonize after disturbance. Another conservation problem is that ordinary forest management does not create the extensive areas with broadleaved tree dominance that were common on more productive sites after fire, and which harbour a particular fauna. Nor does ordinary forest management provide habitat for fire-adapted plants and animals.

In fact, except for the fire refugias, all these problems are evident also inside the forest reserve areas and national parks, which have been created to maintain the properties of the natural landscape. They are in most cases far too small for lightning ignitions alone to create an acceptable fire regime, and at present all fires are also attacked in reserve areas. Even in the southeast, where lightning ignition density is highest, a reserve area of 100 hectares (large for this part of the country) would receives, on average, only one ignition in 430 years! Clearly a system of planned ignitions is necessary in order to maintain fire structured forests within reserve areas.

Within the framework of commercial forestry a lot can be done. Fire refugia can be treated with selective cutting systems or completely exempted from cutting. Areas with dominance by broadleaved trees can be created with or without the use of fire. Also mechanical soil scarification results in good colonization of most broad-leaved tree species on highly productive sites. For maintaining fire-adapted flora and fauna however, fire has to be used. It should then be remembered that fauna is mobile and seem to be able to also locate rather small burns (Wikars 1992). Fire-adapted plants on the other hand are virtually sessile since they are mainly seed bank species. For these plants, burning will have to be repeated over time on the same piece of land. At present no one knows the depletion rate of their seeds in the soil. It may well be that their survival in the soil is in the order of several hundred years. In that case we still have a lot of time left to discuss their proper management.

5. Conclusions

Yet we have only an outline of the natural and cultural history of fire in Sweden, and nearly all information concerns the last few hundred years when human impact has been considerable. In order to better understand the natural role of fire, more paleoecological work is needed. From a conservation standpoint it would be useful to understand more of the factors governing the fire regime. For example there are many unsolved questions as to the extent of fire refugias in different parts of the country, their stability over time and their biological function.

References

Ahti, T., L. Hämet-Ahti, and J. Jalas. 1968. Vegetation zones and their sections in northwestern Europe. Ann. Bot. Fenn. 5, 169-211.

Atlestam, P-O. 1942. Bohusläns ljunghedar. En geografisk studie (Dissertation). Meddelanden från Göteborgs Högskola, Geografiska Institutionen 30.

Bradshaw, R., and G. Hannon. 1992. Climatic change, human influence and disturbance regime in the control of vegetation dynamics within Fiby forest, Sweden. J. Ecology 80, 625-632.

Engelmark, O. 1984. Forest fires in the Muddus National Park (Northern Sweden) during the past 600 years. Can. J. Bot. 62, 893-898.

Engelmark, O., L. Kullman, and Y. Bergeron. 1994. Fire and age structure of Scots pine and Norway spruce in northern Sweden during the past 700 years. New Phytologist 126, 163-168.

Eriksson, B. 1986. The precipitation and humidity climate of Sweden during the vegetation period. Reports in meteorology and climatology from the Swedish Meteorological and Hydrological Institute 46.

Esseen, P-A., L. Ericson, H. Lindström, and O. Zackrisson. 1981. Occurrence and ecology of *Usnea longissima* in central Sweden. Lichenologist 13, 177-190.

Granström, A. 1982. Seed banks in five boreal forest stands originating between 1810 and 1963. Can. J. Botany 60, 1815-1821.

Granström, A. 1993. Spatial and temporal variation in lightning ignitions in Sweden. J. Vegetation Sci. 4, 737-744.

Granström, A., and J. Schimmel. 1993. Heat effects on seeds and rhizomes of a selection of boreal forest plants and potential reaction to fire. Oecologia 94, 307-313.

Hofgaard, A. 1993. Structure and regeneration patterns of a virgin Picea abies forest in northern Sweden. J. Vegetation Sci. 4, 601-608.

Hultén, E. 1971. Atlas of the distribution of vascular plants in northwestern Europe. Generalstabens Litografiska Anstalt, Stockholm.

Johnson, E. 1992. Fire and vegetation dynamics. Studies from the North American boreal forest. Cambridge University Press, New York.

Kohh, E. 1975. A study of fires and hard pan in the forests of Älvdalen. Sveriges Skogsvårdsförbunds Tidskrift 73, 299-336.

Kujala, V. 1926. Untersuchungen über den Einfluss von Waldbränden auf die Waldvegetation in Nord-Finnland. Comm. Inst. Quaest. For. Fin. 10, 1-37.

Linder, P. 1988. Jämtgaveln, En studie av brandhistorik, kulturpåverkan och urskogsvärden i ett mellannorrländskt skogsområde. Report 3, Provincial Government of Västernorrland, Härnösand.

Östlund, L., and H. Linderson. 1995. A dendrochronological study of the transformation of a boreal forest stand. Scandinavian J. For. Res. 10, 56-64.

Renberg, I., T. Korsman, and H.J.B. Birks. 1993. Prehistoric increases in the pH of acid-sensitive Swedish lakes caused by land-use changes. Nature 362, 824-827.

Sarvas, R. 1937. Uber die natürliche Bewaldung der Waldbrandflächen (German summary). Acta For. Fennica 46, 76-145.

Schimmel, J. 1993. On fire. Fire behavior, fuel succession and vegetation response to fire in the Swedish boreal forest. Dissertation, Swedish University of Agricultural Sciences, Umeå.

Schimmel, J., and A. Granström. 1996. Fire severity and vegetation response in the boreal Swedish forest. Ecology (in press).

Steijlen, I., and O. Zackrisson. 1986. Long term regeneration dynamics and successional trends in a northern Swedish coniferous forest stand. Can. J. Bot. 65, 839-848.

Tirén, L. 1937. Skogshistoriska studier i trakten av Degerfors i Västerbotten. Meddelanden från Statens Skogsförsöksanstalt 30, 67-322.

Wieslander, G. 1936. Skogsbristen i Sverige under 1600- och 1700-talen. Sveriges Skogsvårdsförbunds Tidskrift 1936, 593-663.

Wikars, L. 1992. Skogsbränder och insekter. Entomologisk Tidskrift 113.4, 1-11.

Wretlind, J. 1932. Om hyggesbränningarna inom Malå revir. Norrlands Skogsvårdsförbunds Tidskrift 1932, 243-331.

Wretlind, J. 1935. Naturbetingelserna för de nordsvenska järnpodsolerade moränmarkernas tallhedar och mossrika skogssamhällen. Sveriges Skogsvårdsförbunds Tidskrift 32, 329-396.

Zackrisson, O. 1977. Influence of forest fires on the north Swedish boreal forest. Oikos 29, 22-32.
Zackrisson, O. and L. Östlund. 1991. Branden formade skogslandskapets dynamik. Skog och Forskning 1991 (4), 13-21.
Zetterstedt, J.W. 1833. Resa genom Umeå Lappmarker i Vesterbottens län förrättad år 1832. Örebro.

J.G.Goldammer and V.V.Furyaev (eds.), Fire in Ecosystems of Boreal Eurasia, 453-464.

Impacts of Prescribed Burning on Soil Fertility and Regeneration of Scots Pine (*Pinus sylvestris* L.)

E. Mälkönen [1] and T. Levula

1. Introduction

The decomposition of organic matter is a slow process in boreal coniferous forests. Consequently, poorly decomposed organic matter accumulates on the soil surface in the course of stand development. In mature coniferous stands the thick raw humus layer deteriorates the ecological conditions of soil by lowering soil temperature, increasing water retention capacity, retarding decomposition and acidifying soil, and hinders the natural regeneration of stands (e.g. Sirén 1955). In virgin forests regeneration can usually take place only after fire or other natural catastrophe.

At the end of the last century, as shifting cultivation gradually declined in Finland, methods based on burning began to be applied in forest regeneration (Heikinheimo 1915). One recommendation was that the logging residues and nonmerchantable standing trees should be burned to promote the establishment of new growth. Broadcast seeding of Scots pine (*Pinus sylvestris*) on snow, preceded by prescribed burning during the previous year, became a common practice in the 1920s. Especially in the 1950s and 1960s, prescribed burning was widely used as a silvicultural measure (Fig.1). In this way, controlled fire, in place of wildfires, has been employed in forestry as a means of maintaining soil conditions suitable for regeneration.

The share of artificial regeneration in reforestation increased rapidly in the 1960's, but during the same period prescribed burning has been almost totally abandoned in Finland. The main reasons for the decline of prescribed burning are the following: 1) the reliance on weather conditions and difficulties in organising the work, 2) risk of fire getting out of control, 3) widespread use of mechanised site preparation, and 4) uncertainty as to loss of nutrients. In addition, the high costs involved in the management of small-size regeneration areas used especially in private forestry dropped the prescribed burning from site preparation practice. The recent growing interest in nature-oriented management practices and in maintaining biodiversity favor the revival of prescribed burning.

Prescribed burning has been practised on ca 0.5 million hectares of forested land in Finland. This corresponds to 12% of the areas seeded and planted, and less than 3% of the total area of forested land. Wildfires are almost totally eliminated in Finland (Aarne 1993).

[1] Finnish Forest Research Institute, Department of Forest Ecology, FIN-01301 Vantaa
[2] Finnish Forest Research Institute, Parkano Research Station, FIN-39700 Parkano

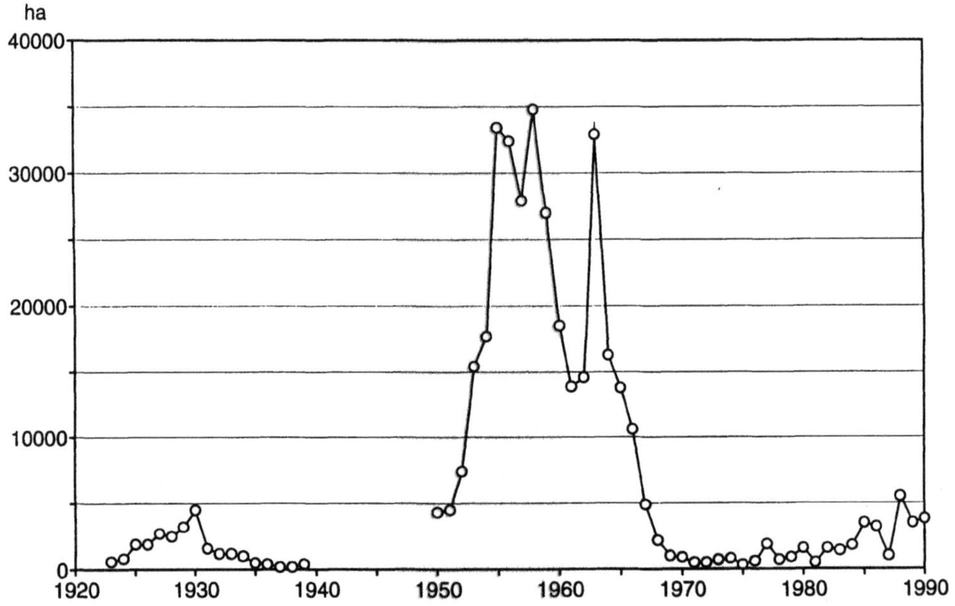

Fig.1. Annual area treated with prescribed fire in Finland (Aarne 1993).

Although forest regeneration sites in the Scandinavian countries have been subjected to prescribed burning for about a century, our understanding of the appropriateness of prescribed burning as a silvicultural measure is still partly based on experiences of wildfires (e.g. Sarvas 1937a, 1937b). Some detailed studies have been carried out on the impact of prescribed burning, for example, on soil fertility (Viro 1969), soil microbial activity (Fritze et al. 1993), and the development of ground vegetation (Lindholm and Vasander 1987), but the experimental basis of most studies is rather limited. Owing to the complexity of fire effects on nutrient cycling our knowledge on the productivity of different site types after prescribed burning is very inadequate.

The purpose of this paper is to present some results on the impacts of prescribed burning on soil properties and on the initial development of Scots pine plantations in southern Finland.

2. Material and Methods

2.1 Study Sites

The material for this study was collected from prescribed burning experiments located in southern Finland (Fig.2). The general characteristics of the experimental sites are presented in Table 1. Experiments 651-658 were established in accordance with the split-plot design with unburned control and prescribed burning as the main treatments, and three different soil tillage measures as the sub-treatments. This paper deals only with the impacts of prescribed burning without mechanical soil preparation. In these regeneration experiments, the usual plot size was 50 x 50 m. The experiments usually consisted of three replications, but experiment 655 has exceptionally only two replications. Owing to three sub-treatments and the mantle around the plots, the size of continuously burned ground in every replication was more than one hectare. The plots were treated with prescribed burning during the summer following final cutting, at a time when the site was thought to be sufficiently dry for burning. During very dry periods prescribed burning is, however, prohibited due to fire-hazard warnings. Owing to the different site and weather conditions fire intensity varied quite a lot between the experiments.

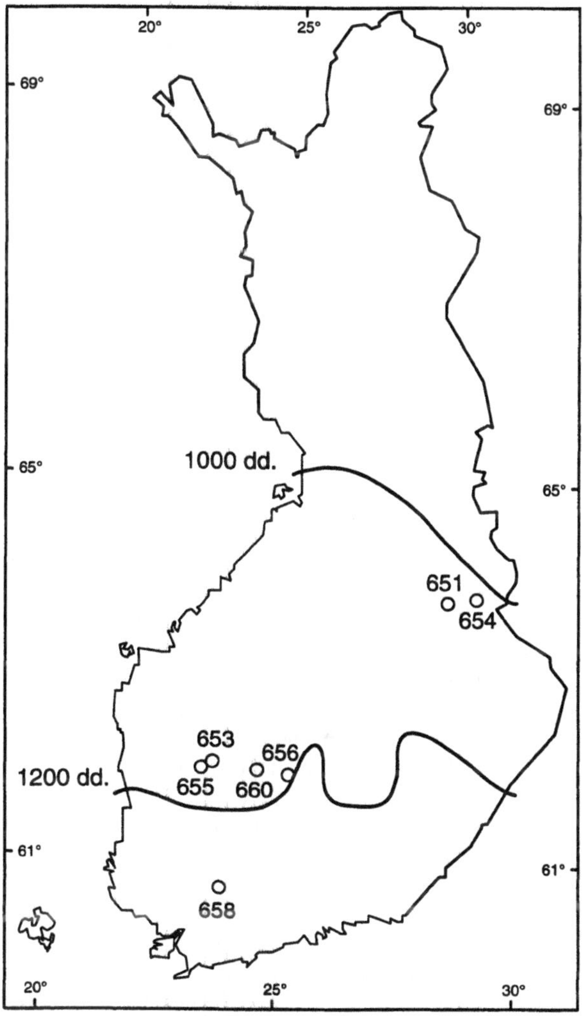

Fig.2. Location of the prescribed burning experiments and mean annual effective temperature sum (threshold value 5°C).

Bare-rooted seedlings of Scots pine were planted the following spring applying a spacing of 2000 seedlings per hectare. When planting on unburned plots, the ground vegetation was manually removed with the planting hoe from the planting spot.

Experiment 660 differs from the other experiments in having been established in a stand of 100 year-old Scots pine spaced out at 330 trees per hectare and with a stem volume of 181 m^3 ha^{-1}. The size of the plots marked out on the site prior to prescribed burning was 30 x 30 m. The experiment has four replications. Logging residues (ca 19 t ha^{-1} dry matter) brought over from a nearby cutting area were spread evenly on the plots.

Tab.1. General characteristics of prescribed burning experiments

Experiments Number and Location	Soil Texture Class	Humus Layer (cm)	Compositions of earlier Stands (%)			Year of Burning
			Pine	Spruce	Broadl.	
651 Valtimo	Fs till	2.7	10	73	16	1976
653 Kuru	Fs till	2.6	12	77	11	1982
654 Nurmes	Fs till	2.3	12	58	30	1981
655 Parkano	Fs till	2.8	48	37	15	1982
656 Jämsänkoski	Fs till	2.8	8	92	0	1983
658 Tamala	Cs till	3.8	15	78	7	1988
660 Kuorevesi	Cs	2.5				1990

Abbreviations: Fs = fine sand, Cs = coarse sand

Sampling and Chemical Analyses

The humus layer was sampled systematically with a soil auger (ø 58 mm) from twenty sampling points per plot at the establishment of experiments. The humus samples were combined to one bulk sample per plot. Experiment 660 was an exception in this respect as it had only fifteen sampling points owing to the smaller plot size. The thickness of the humus layer was measured at the same time as the humus sample was taken. Sampling of the humus layer was repeated in some experiments soon after burning.

After desiccation at room temperature, the humus samples were weighed and milled to pass through a 1-mm sieve. The moisture content of the air-dry samples was determined by oven-drying at 105°C. The pH was measured in water. Total C and total N were determined using an automatic CHN analyser (LECO). The organic matter content was arrived at by multiplying the carbon content by 1.72. Total P, K, Ca and Mg were determined by dry-ashing and by using methods described by Halonen et al. (1983). The concentrations of P, K, Ca and Mg were determined by using inductively coupled plasma atomic emission spectrophotometry (ICP/AES) (ARL, model 3580 OES).

Needle samples were collected from four experiments (651, 653, 655 and 656). Fifteen sample trees per plot were selected randomly and the current-year's needles from the second branch whorl were collected during the period of dormancy. The needle samples were then combined and analysed plot-by-plot in accordance with the aforementioned methods

(Halonen et al. 1983). Needle analysis depicts the nutrient status of trees at an average of six years after stand establishment.

Soil water was collected from experiment 660 at the depth of 20 cm using four percolation lysimeters per treatment. The lysimeters (ø 20 cm) were emptied 4-8 times per year during the period 1990-1992. The volumes of the water samples were determined immediately while the pH measurements were done on arrival at the laboratory. The filtered samples were stored in a deep freezer to await analysis. Total N and NH_4-N concentrations were determined by flow injection analysis (FIA); NO_3-N by ion chromatography, and Ca, Mg and K by means of inductively coupled plasma atomic emission spectrometry (ICP/AES) (Jarva and Tervahauta 1993).

The seedling establishment inventory was conducted with a 25% systematic sampling in each plot. The parameters for survival, total height, annual height increment and identifiable cause of seedling damage were determined for ca 200 seedlings per plot.

3. Results and Discussion

3.1 Organic Matter and Soil Physical Properties

Loss of organic matter is one of the most important effects of fire on forest soil. Prescribed burning resulted in the burning up of most of the logging residues and ground vegetation, as well as varying degrees of the uppermost part of the humus layer (Fig.3). Depending on the quantity and quality of the logging residues, the moisture of the humus layer, and the weather conditions at the time of burning, there may be considerable differences in fire intensity. Viro (1969) reported a reduction of organic matter in the humus layer by 25 % from 33 to 25 t ha^{-1} in prescribed burnings in Finland. In some cases the organic matter content of the humus layer even increases following prescribed burning through incompletely combusted matter from logging residues and ground vegetation on the soil surface.

The maximum surface temperatures recorded by Vasander and Lindholm (1985) during prescribed burning were more than 800°C. Due to moisture condensation, the humus layer is an excellent thermal insulator, which prevents the fire from penetrating deep into the humus layer.

A thinner humus layer after burning means higher soil temperatures and a reduced water retention capacity. Under a thick cover of moss and humus, soil temperature is far below its optimum value for microbial activity or root growth. This being the case, the effect of prescribed burning on soil temperature conditions is almost entirely a favorable one (Tab.2).

On morainic soils there is usually a satisfactory supply of water; coarse-textured sites are the exception as burning there may reduce the water retention capacity of the soil (Uggla 1967). According to Viro (1969), prescribed burning should be performed on a degree intense enough to consume the majority of the humus layer. This recommendation is not, however, valid on dry sites (sandy sediments with a thin humus layer) on which burning often results in long-lasting soil deterioration (e.g. Uggla 1958; Huss and Sinko 1969). Loss of soil organic matter in burning can lead to the creation of heatlands on coarse-textured poor soils.

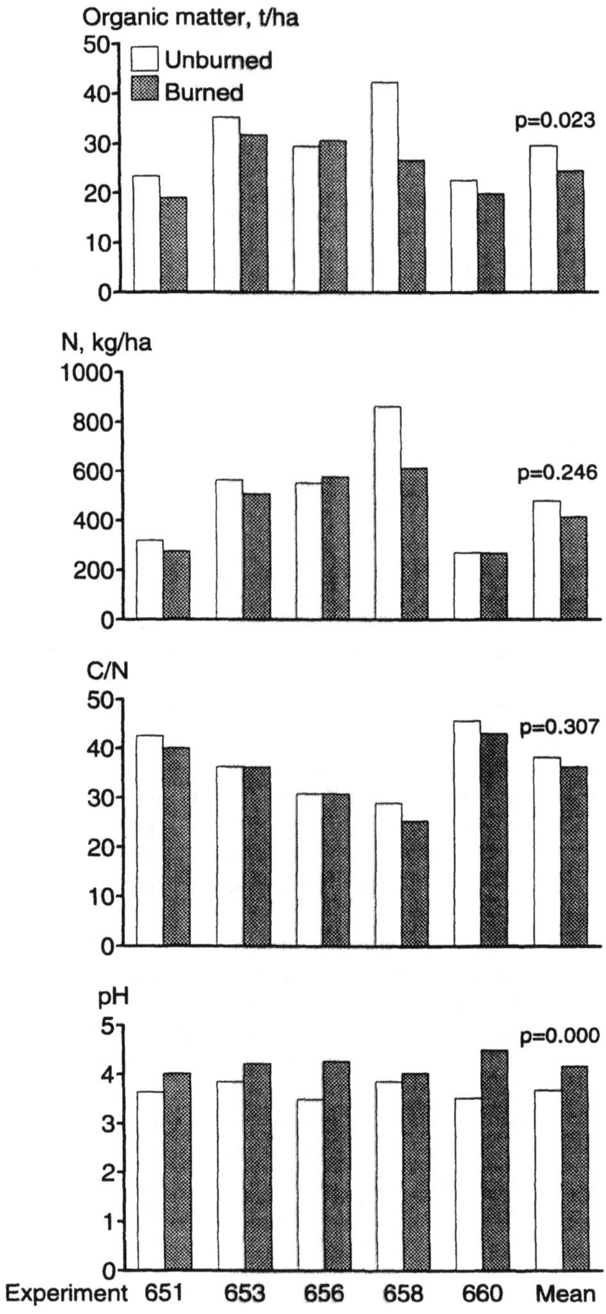

Fig.3. Effect of prescribed burning on the amount of organic matter and total nitrogen, the C/N ratio, and the pH of the humus layer. Burned means the situation immediately after prescribed burning.

Tab.2. Soil temperatures in a mature Norway spruce stand, and clearcut area without and with prescribed burning (Viro 1974)

Monitoring Depth	Soil Temperatures (°C)		
	Forest	Clearcut Area	Burned Clearcut
Surface Humus Layer	18.0	24.4	31.3
Mineral Soil Surface	9.1	12.8	16.0
10 cm	8.7	10.6	12.5
20 cm	8.3	10.2	12.0

3.2 Acidity and Nutrient Conditions

From the point of view of soil fertility, the most important impact of prescribed burning is in the transfer of nutrients contained in the combusted material or in the intensely heated surface soil via volatilisation, particulate movement, or leaching through the rooting layer (Feller 1988).

The nitrogen contained in the burned organic material is volatilized entirely and lost into the atmosphere (Little and Ohman 1988). Consequently, the nitrogen resources of a forest site are reduced by prescribed burning (Fig.3). Given the circumstances prevailing in Finland, merely the logging residues left on the site can contain 150-200 kg of N per hectare. The loss of nitrogen is, however, of minor importance and does not constitute a severe problem with respect to the development of the new tree generation. Of the total nitrogen resources of the rooting layer in forest soils (2-4 t ha^{-1}), only a very small fraction is in a mineralised form suitable for uptake by plants; prescribed burning promotes the mineralisation of nitrogen (Viro 1969).

In addition to nitrogen, sulphur, phosphorus and boron are also subject to gasification losses at high temperatures caused by fire (Raison et al. 1985). Some amounts of nutrients may be removed from the site during the burning in the form of fly ash.

The mineral nutrients bound into organic matter are released mainly as oxides and carbonates, and these raise the pH value of the soil (Fig.3). Both the rise in pH of the humus layer caused by prescribed burning, as well as the duration of its influence (from a few years to tens of years) vary greatly according to results obtained (e.g. Uggla 1957, Viro 1969). This variation is probably explained by differences in soil properties and in the amount and quality of ash. The rise in pH in the top layers of the mineral soil has been modest. The total amounts of mineral nutrients in the humus layer may be increased considerably in burning (Tab.3).

According to the results obtained from the lysimeter experiment, the acidity of the percolation water was significantly reduced and the potassium concentration rose immediately after prescribed burning (Fig.4). Potassium is known to be readily leached following prescribed burning (Viro 1969). However, the calcium and magnesium concentrations of the percolation water on the non-burned plots during the first two years after prescribed burning were, on average, higher than on the burned plots. These results are contradictory to the changes of potassium and calcium concentrations of soil solutions monitored by Starr (1985) on experiment 653. Soil analyses results of previous studies (e.g.

Viro 1969) have led researchers to conclude that leaching of calcium and magnesium is not a serious problem in prescribed burning. The results obtained also indicate that nitrification is promoted by prescribed burning (Uggla 1958; Viro 1969). Lysimeter measurements are characterised by their wide deviation, which complicates conclusions.

Tab.3. Amount of nutrients in humus layer before and after burning

Nutrient	Treatment	Amount of Nutrients (kg ha^{-1})							
		Experiment							
		651		656		658		660	
		x	s_x	x	s_x	x	s_x	x	s_x
P	Unburned	26.3	3.7	41.1	2.9	54.0	6.4	23.0	3.0
	Burned	25.6	4.1	52.9	1.2	44.0	5.0	29.5	1.6
K	Unburned	27.2	5.0	40.5	4.5	51.0	10.6	23.4	4.4
	Burned	29.5	7.1	84.2	4.2	44.5	6.7	29.4	2.3
Ca	Unburned	88.0	17.0	148.0	17.0	221.0	44.0	86.0	12.0
	Burned	107.0	33.0	261.0	18.0	180.0	25.0	180.0	23.0
Mg	Unburned	15.5	2.7	15.6	1.9	29.1	3.5	10.0	1.2
	Burned	18.6	5.2	28.8	2.6	27.3	6.3	16.8	2.2

3.3 Initial Stand Development

The development of ground vegetation has an important role in maintaining soil fertility. On dry sites the recovery of vegetation takes place slowly and a lot of nutrients released in burning may be leached away. On moist sites, very favorable nutrient conditions after burning stimulate the fast development of ground vegetation which maintains nutrient cycling (Viro 1969; Lindholm and Vasander 1987).

Prescribed burning decreases microbial biomass and changes the microbe population. The decrease of soil acidity and development of vegetation gradually promote microbial activity, which probably increases up to crown closure (Fritze et al. 1993).

The effects of prescribed burning on the initial development of Scots pine were strongly dependent on site conditions (Fig.5). Burning has not had an effect on stand development on experiments 654 and 655 which are on rather wet sites with some indication of paludification. For the other experiments, burning promoted survival and/or height growth of pine. Needle analyses indicated (Tab.4) that the nitrogen concentration of needles of seedlings growing on burned plots was slightly lower, while their boron concentration was slightly higher than the corresponding values obtained for seedlings of untreated plots. From the point of view of seedling nutrient status, however, these differences were quite small.

The percentage of survival of pine seedlings has been generally higher on well burned rather than unburned areas (Kolehmainen 1955). However, on insufficiently burned areas the survival of pine is often extremely low (Yli-Vakkuri 1961). It is also important for stand development that burning favors deciduous trees, whose litter decomposes rapidly. According to Sirén (1974), prescribed burning has a long-lasting positive effect only on the development of pine stands on moist site types. Reforestation with Norway spruce (*Picea abies*) on burned sites has not been considered meaningful (Uggla 1967).

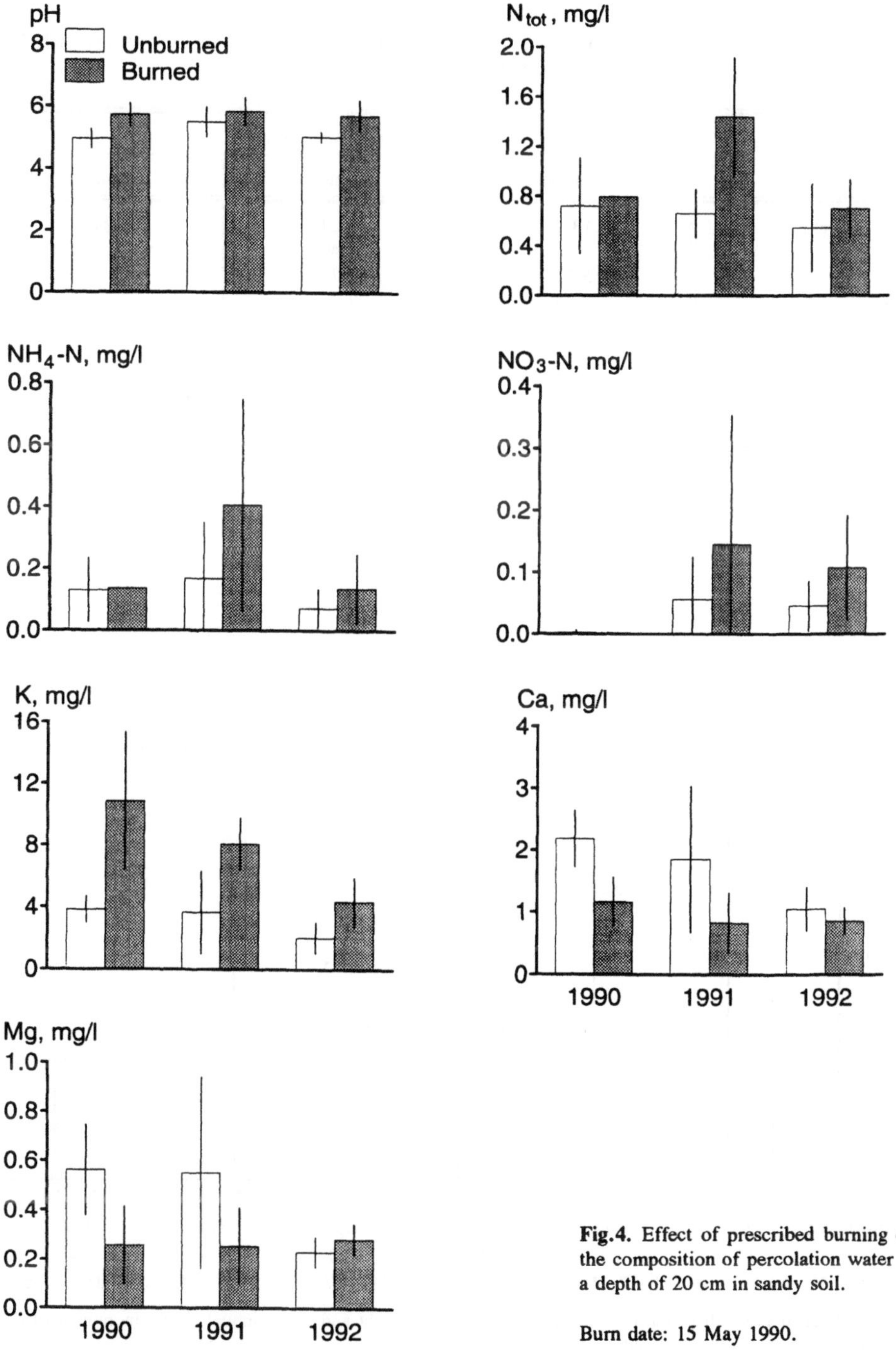

Fig.4. Effect of prescribed burning on the composition of percolation water at a depth of 20 cm in sandy soil.

Burn date: 15 May 1990.

Tab.4. Mean nutrient concentrations of Scots pine needles on the prescribed burning experiments (651, 653, 655 and 656) about 6 years after stand establishment

Nutrient	Mean Nutrient Concentrations		
	Unburned	Burned	F-Value
N (g kg^{-1})	13.7	12.4	4.85*
P "	1.46	1.40	
K "	5.21	5.72	
Ca "	2.14	2.31	
M (g)	0.92	0.89	
Mn (mg kg^{-1})	535.0	493.0	
Zn "	44.1	42.2	
Cu "	4.17	3.89	
B "	12.6	14.5	8.15***

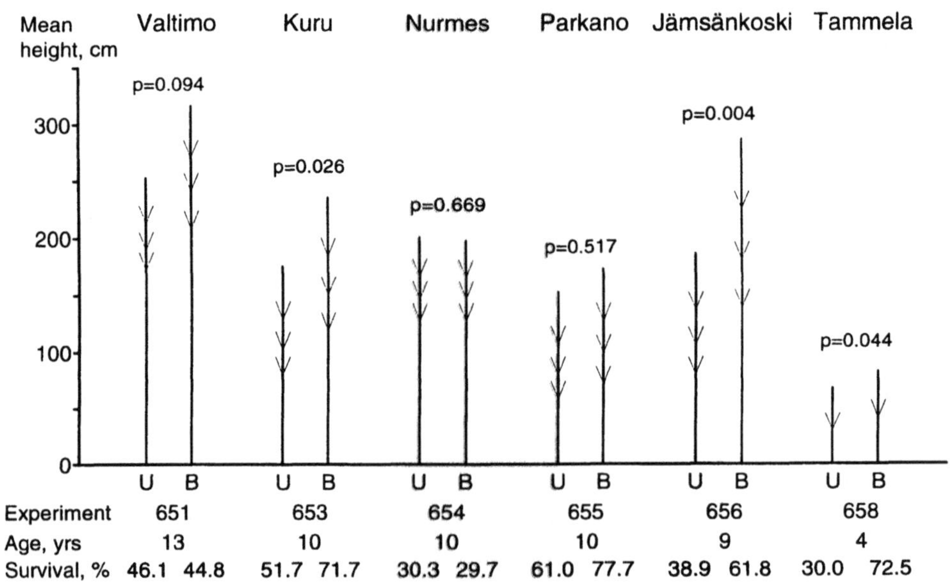

Fig.5. Effect of prescribed burning on the survival, mean height and mean annual height increment of Scots pine (U = unburned, B = burned).

4. Conclusions

Prescribed burning can possess several beneficial impacts with respect to the physical, chemical and biological properties of the soil. Through these impacts, prescribed burning also influences succession in the forest ecosystem. The final outcome depends especially on the fertility of the site and the intensity of the fire.

Prescribed burning converts unavailable mineral nutrients of organic matter to easily soluble and available nutrients for plants. From the point of view of soil fertility, prescribed burning may be recommended to be applied mainly when dealing with moist and dryish moraine soils where the supply of water is generally adequate. On such sites prescribed burning does not lead to a deterioration of soil water conditions. On the other hand, prescribed burning serves no purpose if the site is in need of drainage. The direct impact of prescribed burning on soil fertility should be seen as being fairly short-lived. However, as a factor guiding succession in the forest ecosystem and a maintainer of biodiversity, prescribed burning is of great importance.

5. References

Aarne, M. (ed.) 1993. Yearbook of forest statistics 1992. The Finnish Forest Research Institute. Helsinki. 317 pp.

Feller, M.C. 1988. Relationships between fuel properties and slashburning-induced nutrient losses. For. Sci. 34, 998-1015.

Fritze, H., T. Pennanen, and J. Pietikäinen. 1993. Recovery of soil microbial biomass and activity from prescribed burning. Can. J. For. Res. 23, 1286-1290.

Halonen, O., H. Tulkki, and J. Derome. 1983. Nutrient analysis methods. Metsäntutkimuslaitoksen tiedonantoja 121. 28 pp.

Heikinheimo, O. 1915. Kaskiviljelyksen vaikutus Suomen metsiin. Referat: Der Einfluss der Brandwirtschaft auf die Wälder Finnlands. Acta Forestalia Fennica 4. 149 pp.

Huss, E. and M. Sinko. 1969. Effekt av hyggesbränning. Sveriges Skogsvårdsförbunds Tidskrift 4.

Jarva, M. and A. Tervahauta. 1993. Vesinäytteiden analyysiohjeet. Methods of water analysis. Metsäntutkimuslaitoksen tiedonantoja 477. 171 pp.

Kolehmainen, V.A. 1955. Havaintoja kulotuksen merkityksestä metsiemme uudistamisessa. Referat: Beobachtungen über die Bedeutung des Abschwendens bei Verjüngung von finnischen Wäldern. Silva Fennica 85, 1-32.

Lindholm, T. and H. Vasander. 1987. Vegetation and stand development of mesic forest after prescribed burning. Silva Fennica 21 (3), 259-278.

Little, S.N. and J.L. Ohmann. 1988. Estimating nitrogen lost from forest floor during prescribed fires in Douglas-fir / western hemlock clearcuts. For. Sci. 34, 152-164.

Raison, R.J., P.K. Khanna, and P.V. Woods. 1985. Mechanisms of element transfer to the atmosphere during vegetation fires. Can. J. For. Res. 15, 132-140.

Sarvas, R. 1937a. Kuloalojen luontaisesta metsittymisestä. Referat: Über die natürliche Bewaldung der Waldbrandflächen. Acta Forestalia Fennica 46.1. 146 pp.

Sarvas, R. 1937b. Havaintoja kasvillisuuden kehityksestä Pohjois-Suomen kuloaloilla. Referat: Beobachtungen über die Entwicklung der Vegetation auf den Waldbrandflächen Nord-Finnlands. Silva Fennica 44. 64 pp.

Sirén, G. 1955. The development of spruce forest on raw humus sites in northern Finland and its ecology. Acta Forestalia Fennica 62.4. 363 pp.

Sirén, G. 1974. Some remarks on fire ecology in Finnish forestry. Proceedings Ann. Tall Timbers Fire Ecology Conferences 13, 191-209. Tall Timbers Research Station, Tallahassee, Florida.

Starr, M.R. 1985. Prescribed burning and soil acidification. In: Symposium on the effects of air pollution on forest and water ecosystems, Helsinki April 23-24, 1985. The Foundation for Research of Natural Resources in Finland, p.101-106.

Uggla, E. 1957. Mark- och lufttemperaturer vid hyggesbränning samt eldens inverkan på vegetation och humus. Summary: Temperatures during controlled burning. The effect of the fire on the vegetation and the humus cover. Norrlands skogsvårdsförbunds tidskrift 4, 443-500.

Uggla, E. 1958. Ecological effects of fire on north Swedish forests. Almqvist & Wiksell AB, Uppsala. 18 pp.

Uggla, E. 1967. En studie över bränningseffekten på ett tunt råhumustäcke. Summary: Effects of the fire on a thin layer of raw humus. Sveriges skogsvårdsförbunds tidskrift 2, 155-170.

Vasander, H. and T. Lindholm 1985. Fire intensities and surface temperatures during prescribed burning. Silva Fennica 19 (1), 1-15.

Viro, P.J. 1969. Prescribed burning in forestry. Communicationes Instituti Forestalis Fenniae 67.7, 49 pp.

Viro, P.J. 1974. Effects of forest fire on soil. In: Fire and ecosystems (T.T.Kozlowski and C.E. Ahlgren, eds.), 7-45. Academic Press, New York.

Yli-Vakkuri, P. 1961. Emergence and initial development of tree seedling on burnt-over forest land. Acta Forestalia Fennica 74.1. 51 pp.

J.G.Goldammer and V.V.Furyaev (eds.), Fire in Ecosystems of Boreal Eurasia, 465-475.

Composition of Smoke from North American Boreal Forest Fires

W.R. Cofer III [1], E.L. Winstead [2], B.J. Stocks [3],
D.R. Cahoon [1], J.G. Goldammer [4], and J.S. Levine[1]

1. Introduction

The release of gaseous and aerosol products into the atmosphere during controlled and naturally occurring vegetation fires is exerting a global influence on both atmospheric chemistry and climate (Goldammer and Crutzen 1993). In many cases, assessments of the impacts from biomass burning are not sufficiently quantitative to support serious environmental policy decisions or legislation that might lead to immediate economic and social consequences. Although large amounts of burning occur naturally, most burning in the last century has been the result of human activities (Andreae 1991). Burning for agricultural purposes is extensively practiced worldwide. Fire is used commonly in deforestation and in land management. Substantial amounts of environmentally important gases and aerosols are released into the atmosphere during combustion, and biomass fires may be modifiers of longer-term biogenic processes involved in the interchange of gases between the atmosphere and biosphere (Levine et al. 1990; Zepp 1988). Fires are also shaping the present and future population and structure of vegetation (Richter et al. 1982), and may be among the key proximate factors determining how terrestrial vegetation responds to future climate change (Walker 1991; Clark and Reid 1993).

Since Crutzen et al. (1979) first suggested that biomass fires could be a sig- nificant contributor to the budgets of several important atmospheric trace gases (CO_2, CO, H_2, CH_4, N_2O, NO_x, COS, hydrocarbons, and CH_3Cl), substantial amounts of research have been conducted toward understanding trace gas production from biomass fires. Many combustion products of significant climatic and photochemical importance are released into the atmosphere during biomass burning (Hegg et al. 1987; Andreae et al. 1988; Cofer et al. 1990). Three of the major trace gases produced during biomass burning are greenhouse gases -- CO_2, CH_4, and N_2O -- and are increasing in the modern atmosphere, although perhaps not at the rates of the last decade (Khalil and Rasmussen 1992, 1993; Keeling et al. 1984; Sarmiento 1993). While the causes of these increases are not explicitly known, global biomass burning is certainly a significant contributor. In addition to the gaseous emissions from biomass fires, smoke aerosols and particulates are also likely to influence global

[1] Atmospheric Sciences Division, NASA-Langley Research Center, Hampton, VA 23681-0001, U.S.A.
[2] Science Applications International Corporation, Hampton, VA 23666, U.S.A.
[3] Canadian Forestry Service, Ontario Region, Sault Ste Marie, Ontario P6A 5M7, Canada
[4] Max Planck Institute for Chemistry, Fire Ecology and Biomass Burning Research Group, c/o University of Freiburg, D-79085 Freiburg, Germany

climate and atmospheric chemistry (Radke et al. 1991; Ward et al. 1991). Smoke aerosols can interact directly with solar radiation (Lenoble 1991; Kaufman 1987), serve as a source of condensation nuclei (Penner et al. 1992; Rogers et al. 1991), or form substrates for a variety of heterogeneous atmospheric chemical reactions (Andreae 1991). There are also complex biogeochemical interactions between biomass burning and ecosystems, often involving the intricate exchange of aerosol dispersed nutrients over large distances (Lacaux et al. 1991; Andreae 1991).

Most current attention to the environmental aspects of biomass burning has been directed toward the tropics, specifically to the tropical rain forests of Brazil and to the savannas of Africa and South America, where most of the world's biomass burning occurs (Andreae 1991). In terms of biomass burned and emissions released into the atmosphere, tropical savanna fires clearly dominate (~65%) other biomass fires (Goldammer 1991). This can be expected to continue since savanna burning is used for agricultural and land management purposes. In fact, the very persistance of savanna systems depends largely upon recurring fires. Savannas, once burned, quickly re-incorporate atmospheric carbon dioxide via rapid new growth. Thus, the time scales for re-establishment of savanna vegetation, or "CO_2-recycling," is relatively fast. Mature forests, however, present a much different picture. While not intended to diminish the importance of tropical burning, the following discussion is intended to emphasize the potential importance to atmospheric chemistry and climate of burning in northern boreal forests.

The boreal forest, or taiga, contains extensive tracts of coniferous woodlands composed essentially of species of pine (*Pinus*), larch (*Larix*), fir (*Abies*), and spruce (*Picea*). About 12 million square kilometers of boreal forest, generally between 45° and 70°N latitude, is found in North America and Eurasia (Stocks 1991). As much as 25% of the carbon stored in terrestrial ecosystems is contained in boreal forests (Dixon and Krankina 1993; Kauppi et al. 1992). A significant part (~33 %) of the fuel carbon stored in the boreal biome consists of a partially decomposed and compacted organic surface layer (duff layer) formed from years of accumulation of forest litter. The boreal forest is a fire-dependent ecosystem with a low frequency of fire. About 5-8x10^6 ha of boreal forest burn annually (Stocks 1991). While fire emissions from the boreal and temperate forests may presently account for no more than 3-5% of the average annual global biomass fire emissions budget (Andreae 1991), there are several unique factors in the boreal system that may make boreal fire emissions environmentally more significant than the 3-5% would suggest.

Boreal forests are located at northern latitudes, an environmentally sensitive region of the atmosphere. The lower amounts of water vapor in the colder northern atmosphere make absorption by other greenhouse gases (e.g., CO_2) more significant. Since average fuel consumptions during burning are about an order of magnitude more per unit area for boreal and temperate zone forest fires than for most other ecosystems, fires are usually much more intense. As a result, the altitudes to which the atmospheric emissions are initially injected are typically higher, often producing smoke columns that extend upward more than 8 km. Thus, the need to employ mechanisms other than the initial smoke cloud bouyancy to loft emissions into the free troposphere is often not necessary. Emissions, therefore, have a better chance of escaping normal atmospheric removal processes, extending their atmosperic lifetime, and becoming part of the global circulation. Gravitational settling of fire-produced particles also requires more time for higher initial injections. The tropopause is lower at more northern latitudes, and many large boreal fires generate enough energy to penetrate the tropospause. Major boreal fires tend to be episodic, resulting in few, but massive, injections into the atmosphere. The proportions and quantities of "reactable" carbon combustion products (e.g. CO, hydrocarbons, etc.) resulting from boreal fires are usually higher due to the nature of the combustion-emission chemistry, which will be discussed later in this chapter. Boreal forests, once burned, cannot quickly reincorporate equivalent amounts of carbon via CO_2 into new growth or into a new rich organic surface layer.

Lastly, there are legitimate concerns that environmental effects from biomass burning in the boreal system may be coupled to global warming in a manner that is self-feeding. That is, global warming may cause increases in the fire frequency of the boreal, which in turn causes more surface albedo changes and further greenhouse emissions, thus more warming, forming a self-feeding esclating cycle (Clark and Reid 1993; Dixon and Turner 1991; Kurz et al. 1994).

2. Biomass Combustion

The principal elemental components of vegetation (cellulose, $C_6H_{10}O_5$) are carbon, oxygen, and hydrogen. Carbon constitutes ~45 % of the weight of cellulose. "Ideal combustion" of biomass leads exclusively to water and carbon dioxide as reaction products according to the following scheme:

$$(C_6H_{10}O_5)_n + (6O_2)_n = (6CO_2)_n + (5H_2O)_n$$

However, real fires produce varying amounts of CO, CH_4, H_2, hydrocarbons, particulates, etc. This departure from the ideal reveals the importance of the nature of the combustion on fire emissions. Factors such as wind, diffusion, moisture, and temperature all play important roles in determining the products of biomass combustion. Some of these factors are determined (or at least strongly influenced) by basic ecosystem characteristics. For example, the partially decomposed and compacted organic surface layer typical of the boreal forest serves as surface fuel during biomass fires. Combustion in this layer would not be expected to proceed as in a well-arrayed fuel. Thus, the ecosystem property of a duff layer dictates a certain type of combustion and resulting emissions chemistry.

Deviations from ideal or complete combustion result in combustion products other than CO_2 and H_2O. The combustion efficiency (CE) for biomass fires can be expressed as the ratio of the CO_2 produced by the fire to the total carbon product in terms of the mass of carbon, expressed in the equation below.

$$CE = \frac{CO_2}{(CO_2 + CO + CH_4 + \text{carbonaceous particles} + \text{hydrocarbons})}$$

Since quantitative measurements of all the carbon products produced during biomass burning can be very difficult, if not impossible, to obtain, the measured ratio of CO/CO_2 in smoke plumes is often used as a measure of combustion efficiency (Hegg et al. 1990; Ward et al. 1991). The larger the CO/CO_2 ratio, the less efficient the combustion.

Biomass fires will almost always follow a qualitatively predictable combustion pattern. Easily ignitable, well-aerated, small-size surface fuels will ignite and burn early in the development of a fire. These fuels may be grasses, lichen, litter, brush, etc. If the ecosystem consists largely of small, well-arrayed, essentially dry fuels, combustion will produce large flames. When flames are visually observed to be the dominant characteristic of the combustion, a biomass fire (or region of a fire) is considered to be in the flaming stage of combustion. The most efficient combustion (low CO/CO_2) usually occurs during the flaming stages of a biomass fire. A biomass fire then makes transitions, as the fuels become harder to ignite and burn, into smoldering phases of combustion. During smoldering, combustion efficiency drops and a proportionately higher fraction of non-CO_2 carbon is produced. This can be explained by the nature of smoldering combustion. Smoldering combustion primarily occurs with the larger fuel sizes (high volume-to-surface area), the less aerated fuels (litter/duff), and the green or wet fuels. Temperatures are significantly lower (~ 300-600°C) with no notable flaming (Crutzen and Andreae 1990; Susott et al. 1991).

There is no quantitative way to define the combustion phase in a large biomass fire. Large biomass fires consist of mixtures of simultaneously occurring flaming, transitional, and smoldering combustion. The utility of these terms exists only in their ability to descriptively identify the dominant form of ongoing combustion.

While smoldering combustion typically produces the largest proportion of incompletely oxidized carbon products, substantial amounts of these can also be produced during especially intense flaming combustion. An advancing intense fire radiatively vaporizes fuels in its path--these hot, volatilized, primarily hydrocarbon fuel fragments subsequently combine with oxygen during combustion. The combination of fuels with oxygen can be limited by several physical/chemical processes. If the vaporized hydrocarbon fuel fragments can cool before reaction with oxygen, many incompletely oxidized or unoxidized compounds will result. Temperature drops of several hundred degrees over relatively short distances within a fire often occur. During very intense combustion, the potential for rapid thermal quenching and/or transient pockets of depleted oxygen exists (Cofer et al. 1989; Susott et al. 1991). These regions do not have to persist long and can be visualized as being forced through regions of radiatively volatilized fuel fragments by intense inward and upward fire-generated wind flows. If combustion can be restricted in these pockets for brief periods during cooling, many incompletely reacted products may result. Measurements of oxygen concentration within a fire would not necessarily indicate this. In addition, extremely violent wind flows feeding into the fires can be generated that actually thermally quench (cool) vaporized fuels before combination with oxygen.

3. Biomass Fire Emissions

Many important trace gases are released into the environment during biomass fires. Most of these, as would be expected, are carbon-based. CO_2, CO, and CH_4 are produced in greatest quantity. This can be seen in Table 1. Gaseous non-methane hydrocarbons (NMHC) produced by burning are too numerous to list. A substantial fraction of these hydrocarbons are alkenes (ethylene, propylene, etc.) and are thus highly reactive in the atmosphere (Greenberg et al. 1984; Bonsang et al. 1991). Hydrogen is produced during biomass fires (Crutzen et al. 1979) in quantities that may account for as much as 25% of its global source (Laursen et al. 1992), and may play an important role in stratospheric photochemistry (Crutzen and Andreae 1990). Depending upon the biomass fuel nitrogen, sulfur, and chlorine content, gaseous and particulate forms of these will also be released in small, but often very important amounts. Nitrogen oxides (e.g., NO_x) and hydrocarbons emitted from biomass fires may be very important in tropospheric chemistry, serving as precursors for ozone production in the troposphere (Delany et al. 1985; Crutzen et al. 1985). Satellite data have suggested that enhanced tropospheric ozone concentrations in the southern hemisphere may result from biomass burning in South America and Africa (Fishman et al. 1990). Large biomass fires appear to be an important source of nitrous oxide (Cofer et al. 1991), although there is some disagreement with laboratory work (Hao et al. 1991), and may be the main terrestrial source of atmospheric methyl chloride (Rasmussen et al. 1980), and methyl bromide (Khalil et al. 1993; Manö and Andreae 1994), and likely contribute to ozone depletion in the stratosphere (Andreae 1991).

Tab.1. Biomass combustion produced emissions of atmospheric significance

Species	Emissions (Tg yr^{-1})	Atmospheric Impact
CO_2	9.0 - 15 x 10^3	greenhouse gas
C O	4.0 - 9.0 x 10^2	photochemistry
CH_4	1.7 - 3.8 x 10^1	greenhouse gas
NMHC[a]	1.5 - 3.5 x 10^1	photochemistry
Tot. Part.	4.0 - 17 x 10^1	radiation
O C	3.0 - 9.0 x 10^1	
E C	0.5 - 2.0 x 10^1	
NO_x	0.5 - 3.8 x 10^1	photochemistry
H_2	3.0 - 9.0	photochemistry
N_2O	0.4 - 2.0	stratosphere
NH_3	0.5 - 6.0	greenhouse gas
RCN	0.5 - 2.0	stratosphere
SO_x	1.0 - 7.0	tropospheric chemistry
CH_3Cl	0.2 - 0.9	photochemistry
COS	0.7 - 3.1 x 10^{-1}	tropospheric chemistry
		stratosphere
		stratosphere

Biomass burning is also recognized as a major source of particulate carbon in the atmosphere (Seiler and Crutzen 1980; Andreae 1991). Many particulates are generated during biomass fires. Particle size distributions determined from biomass fire smoke appear to show little variation from fire to fire, although particle concentrations vary substantially (Radke et al. 1988). A nucleation mode (<0.1 μm), an accumulation mode (0.1-2.0 μm), and a coarse mode (>2.0 μm) are all observed (Radke et al. 1991). Generally, the largest number of smoke particles are found in the accumulation mode, centered around 0.3 μm. Radke et al. (1991) found particles about this size to consist mostly of tarry condensed hydrocarbons of spherical shape. Andreae et al. (1988) have reported results from the Amazon that conclude about 35% by mass of biomass fire smoke particles are soluble in water as ionic species, and have used potassium as a tracer for smoke aerosols (Andreae 1983). It is clear that a very significant portion of smoke particles consist of inorganic material (Cofer et al. 1988a). This may help explain why smoke particles can act as very efficient cloud condensation nuclei (Penner et al. 1992; Radke et al. 1991; Rogers et al. 1991). What fraction of the inorganic particulates are produced chemically during combustion, much later in the aging smoke plume via gas-to-particle reactions, or simply advected soil particles suspended in the smoke during vigorous updrafts, is unclear. Nevertheless, it is quite clear that carbonaceous aerosols are the dominant aerosol product of biomass fires (Andreae et al. 1988; Cachier et al. 1991). Carbonaceous aerosols are usually divided into organic carbon (OC) and elemental or black carbon (EC) aerosols (Mazurek et al. 1991; Ward et al. 1991). The OC aerosol fraction predominates the EC fraction in smoke plumes. Results from biomass fires in the Canadian boreal system indicate ratios of OC/EC of about 20:1 (Mazurek et al. 1991). It has been determined that EC, or black carbon, is primarily produced during flaming combustion. This is qualitatively apparent by the "whiter" smokes visible during low intensity or smoldering combustion. This is significant since the OC particles impact radiation primarily through scattering, while black carbon has an absorption component as well (Lenoble 1991; Penner et al. 1991).

4. Emission Ratios

Gaseous and particulate smoke plume measurements are frequently reported as emission ratios (Crutzen et al. 1979). Emission ratios (ERs) typically provide a method of comparing the behavior of an individual emission(s) relative to one of the primary combustion products, most often CO_2 (Crutzen et al. 1979; Greenberg et al. 1984; Cofer et al. 1988b). A fire-emitted species measured in a smoke plume is first adjusted by subtraction of its ambient background concentration to yield an excess. This excess is then assumed proportional to its production in the fire. As long as all excesses among the smoke plume emissions chosen for intercomparisons (including CO_2) are derived at the same time and in the same fashion, the proportionality assumption should be valid. CO_2-normalization yielding an ER for species X is shown in equation form below.

$$ER_x = \frac{(plume\ X - background\ X)}{(plume\ CO_2 - background\ CO_2)}$$

ER's are ratios based on a mole/mole (v/v) ratio, or on a mass/mass basis, and are often expressed as percentages. Since CO_2 accounts for $\sim 90\%$ of the carbon combustion product, ER's based on CO_2-normalization tend to inherently place the relative magnitude of fire emissions in their proper perspective. CO is favored as the normalizing agent in some cases (Andreae et al. 1988). The use of CO as a normalizing parameter can be advantageous when smoke plume CO_2 concentrations are not sufficiently elevated above background concentrations to provide an accurate determination of excess CO_2. This typically occurs in aged or well-diluted smoke plumes for the following reason. Background levels of CO_2 are about 350 ppmv in the troposphere. CO_2 is seldom encountered at levels of more than a factor of 2-3 above background, even in the most concentrated of atmospheric smoke plumes. CO concentrations measured under the same conditions, however, are often enriched by factors of several hundred above their atmospheric background (~0.1 ppmv). Thus, as an aging smoke plume mixes and dilutes with ambient air, in-plume CO_2 concentrations converge with their background concentrations more rapidly than corresponding CO concentrations. Consequently, the determination of excess CO_2 becomes much more difficult than for CO (small difference between two large numbers) in well-diluted smoke plumes.

5. Emission Factors

Emission factors (EFs) relate the mass of a particular species released into the atmosphere during burning to the mass of fuel combusted (Ward et al. 1979). EF's are usually expressed as the ratio of grams of product to kg of fuel burned. EFs are developed from experimental fires where fuel types and fuel consumptions can be measured before and after a fire. While the development of EFs requires many emission measurements correlated with fuel consumption for particular types of vegetations and/or fire conditions, once developed, they can have broad application. The establishment of EFs provides a pathway for the development of global biomass burning emission budgets. For example, a set of EFs developed for boreal fires has been used to estimate emissions from remotely sensed fires in China and Siberia (see Cahoon et al. 1994). While EF development lends itself well to applications on experimental or prescribed fires, where measurements before/after a fire can be made, it is not so suitable for application to wildfires. Ward et al. (1979) and Nelson (1982) have developed a technique referred to as the carbon mass balance (CMB) technique for calculating EFs, which can be used to develop EFs for wildfires and fires where no prefire fuel characterizations have been done. The following synopsizes the CMB approach:

$$EF(X) = \frac{EX's(X)}{(2)TEX'sC}$$

Total excess carbon (TEXC) is calculated from all measured excesses of carbon products, i.e., CO_2 + CO + CH_4 + NMHC + Particles. This calculation is done on a mass of carbon/volume basis. The TEXC is related to the original mass of fuel through multiplication by 2 since C content in woody fuels is about 50% of the dry mass. By convention, it is expressed in g/kg. Thus the EF for species X can be determined by, where EX(X) is excess of species X. Since more than 97% of the carbon released during biomass combustion is in the form of CO_2, CO, CH_4, and particulates, EFs determined by the CMB technique would be expected to be very reasonable. This technique has been applied to aircraft obtained smoke plume measurements to determine EFs by Radke et al. (1988).

Emission factors determined for North American boreal and temperate forest fires by fixed-wing, helicopter, and ground-based measurement techniques are presented in Table 2. The EFs in Table 2 represent the averaged emissions from flaming, transitional, and smoldering combustion. The agreement among the EFs determined by different techniques and by the different research groups can be seen to be exceptionally good. The higher variabilities shown in the ground measurements should be expected due to less atmospheric mixing.

Tab.2. Average emission factors (g kg^{-1}) for major combustion products determined from fires in North American boreal and temperate forests (S = sources: 1,2 = fixed wing aircraft (Radke et al. 1991; Laursen et al. 1992), 3 = helicopter (Cofer et al. 1990), 4,5 = tower measurements (Susott et al. 1991; Ward et al. 1992).

CO_2	CO	CH_4	TNMHC's	Particles	N_2O	NO_x	H_2
1650±35	93±16	3.8±1.0	1.8±0.4	20±15	0.23±0.05	4±6	1.8±0.7
1595±45	105±20	3.5±1.1	3.7±1.1	15±10	0.15±0.05	NM	1.4±0.6
1625±85	107±45	4.1±1.7	3.2±2.1	20±15	NM	2±2*	1.9±0.7

NM = not measured
* = NO not NO_x

Tab.3. A comparison of mean emission ratios (in %) obtained from helicopter measurements of boreal forest fire smoke.

	CO	H_2	CH_4	TNMHC
F (28)*	6.7 ± 1.2	2.0 ± 0.5	0.61 ± 0.21	0.61 ± 0.19
S (22)*	12.3 ± 1.9	3.1 ± 0.7	1.22 ± 0.29	1.15 ± 0,27

* = combustion phase (F = flaming; S = smouldering) and number of samples (from Cofer et al. [1990])

Caution must be used in developing source budgets from such emission factors. Since there are no quantitative ways to determine the fraction of the total emissions resulting from purely flaming, transitional, or smoldering combustion in a large fire, they are most often weighted about equally (as were the EF's presented in Tab.2). For most large fires consisting of significantly different vegetations, fuel sizes, etc., there is no strong basis for this assumption. If the composition of the emissions did not change substantially with the combustion phase, the assumption would present no serious problem. A comparison of CO_2-normalized emission ratios for CO, H_2, CH_4, and TNMHC determined for flaming and smoldering combustion in Canadian boreal forest (shown in Tab.3) illustrates that emissions do change substantially as a function of combustion stage. Proportions of CO, H_2, CH_4, and TNMHC produced relative to CO_2 can be seen to double between the flaming and smoldering stages of combustion. While the impact of assuming a qualitatively inaccurate partition for the amount of flaming and smoldering combustion emissions on the CO_2 source budget is arguably small (~5% in the worst case), the significance of a similarly inaccurate partition on the calculated emissions budget for CO, H_2, CH_4, and TNMHC would be substantially larger (~50% worst case). This illustrates the difficulty of estimating the emissions from boreal fires. Until further research is conducted to develop a quantitative method of partitioning the amount of emissions resulting from flaming, transitional, and smoldering combustion, global emissions budget development for boreal and temperate forest fires will remain too speculative.

6. Conclusions

Additional work is justified due to the high variability of the composition of fire emissions; the environmental sensitivity to greenhouse effects at northern latitudes; the typically higher intensities of boreal fires (lofting emissions to near stratospheric altitudes); and the potential self-feeding loop between global warming and increasing numbers of fires in the boreal forest.

References

Andreae, M.O. 1983. Soot carbon and excess fine potassium: Long-term transport of combustion derived aerosols. Science 220, 1148-1151.

Andreae, M.O. 1991. Biomass burning: Its history, use, and distribution and impact on ennvironmental quality and global climate. In: Global Biomass Burning: Atmospheric, Climatic, and Biospheric Implications (J.S. Levine, ed.), 3-21. MIT Press, Cambridge, MA.

Andreae, M.O., E. Browell, M. Garstang, G.L. Gregory, R.C. Harris, G.F. Hill, D.J. Jacob, M.C. Pereira, G. Sachse, A. Setzer, P.L. Silva Dias, R.T. Talbot, A.L. Torres, and S.C. Wofsy. 1988. Biomass-burning emissions and associated haze layers over Amozonia. J. Geophys. Res. 93, 1509-1527.

Bonsang B., G. Lambert, and C. Boissard. 1991. Light hydrocarbon emissions from African savanna burnings. In: Global Biomass Burning: Atmospheric, Climatic, and Biospheric Implications (J.S. Levine, ed.), 155-161. MIT Press, Cambridge, MA.

Cachier, H., J. Ducret, M. Bremond, V. Yoboue, J. Lacaux, A. Gaudichet, and J. Baudet. 1991. Biomass burning aerosols in a savanna region of the Ivory Coast. In: Global Biomass Burning: Atmospheric, Climatic, and Biospheric Implications (J.S. Levine, ed.), 174-180. MIT Press, Cambridge, MA.

Cahoon, D.R., B.J. Stocks, J.S. Levine, W.R. Cofer III, and J.M. Pierson. 1994a. Satellite analysis of the severe 1987 forest fires in northern China and southeastern Siberia. J. Geophys. Res. 99(D9), 18627-18638.

Clark, J. S., and C.D. Reid. 1993. Sensitivity of unmanaged ecosystems to global change. In: Non-linear Responses to Global Change Resources for the future (J.Darmstader and M.A.Toman eds.), pp. 53-89.Washington,D.C.

Cofer, W.R. III, J.S. Levine, D.I. Sebacher, E.L. Winstead, P.J. Riggan, J.A. Brass, and V.G. Ambrosia 1988a. Particulate emissions from a mid-latitude prescribed chaparral fire. J. Geophys. Res. 93, 5207-5212.

Cofer, W.R. III, J.S. Levine, P.J. Riggan, D.I. Sebacher, E.L. Winstead, E.F. Shaw, J.A. Brass, and V.G. Ambrosia. 1988b. Trace gas emissions from a mid-latitude prescribed chaparral fire. J. Geophys. Res. 93, 1653-1658.

Cofer, W.R. III, J.S. Levine, D.I. Sebacher, E.L. Winstead, P.J. Riggan, B.J. Stocks, J.A. Brass, V.G. Ambrosia, and P.J. Boston. 1989. Trace gas emissions from chaparral and boreal forest fires. J.Geophys. Res. 94, 2255-2259.

Cofer, W.R. III, J.S. Levine, E.L. Winstead, and B.J. Stocks. 1990. Gaseous emissions from Canadian boreal forest fires. Atmos. Environ. 24A, 1653-1659.

Cofer, W.R. III, J.S. Levine, E.L. Winstead, and B.J. Stocks. 1991. Nitrous oxide emissions from biomass burning. Nature 349, 689-691.

Crutzen, P.J., and M.O. Andreae. 1990. Biomass burning in the tropics: Impact on atmospheric chemistry and biogeochemical cycles. Science 250, 1669-1678.

Crutzen, P.J., L.E. Heidt, J.P. Krasnec, W.H. Pollock, and W. Seiler. 1979. Biomass burning as a source of atmospheric gases CO, H_2, N_2O, NO, CH_3Cl, and COS. Nature 282, 253-256.

Crutzen, P.J., A.C. Delany, J. Greenberg, P. Haagenson, L. Heidt, R. Lueb, W. Pollock, W. Seiler, A. Wartburg, and P. Zimmerman. 1985. Tropospheric chemical composition measurements in Brazil during the dry season. J. Atmos. Chem. 2, 233-256.

Delany, A.C., P. Haagensen, S. Walters, A.F. Wartburg, and P.J. Crutzen. 1985. Photochemically produced ozone in the emissions from large-scale tropical vegetation fires. J. Geophys. Res. 90, 425-429.

Dixon, R.K., and O.N. Krankina. 1993. Forest fires in Russia: Carbon dioxide emissions to the atmosphere. Can. J. For. Res. 23, 700-705.

Dixon, R.K., and D.P. Turner. 1991. The global carbon cycle and climate change: Responses and feedbacks from below-ground systems. Environ. Pollut. 73, 245-262.

Fishman, J., C.E. Watson, J.C. Larsen, and J.A. Logan. 1990. Distribution of tropospheric ozone determined from satellite data. J. Geophys. Res. 95, 599-617.

Goldammer, J.G. 1991. Tropical wildland fires and global changes: Prehistoric evidence, present fire regimes, and future trends. In: Global Biomass Burning: Atmospheric, Climatic, and Biospheric Implications (J.S. Levine, ed.), 83-91. MIT Press, Cambridge, MA.

Goldammer, J.G., and P.J. Crutzen. 1993. Fire in the environment: Scientiific rationale and summary of results of the Dahlem Workshop. In: Fire in the Environment: The Ecological, Atmospheric, and Climatic Importance of Vegetation Fires (P.J. Crutzen and J.G. Goldammer, eds.), 1-14. John Wiley & Sons, London.

Greenberg, J.P., P.R. Zimmerman, L. Heidt, and W. Pollock. 1984. Hydrocarbon and carbon monoxide emissions from biomass burning in Brazil. J. Geophys. Res. 89, 1350-1354.

Hao, W.M., D. Scharffe, J.M. Lobert, and P.J. Crutzen. 1991. Emissions of N_2O from the burning of biomass in an experimental system. Geophys. Res. Lett. 18, 999-1002.

Hegg, D.A., L.F. Radke, P.V. Hobbs, C.A. Brock, and P.J. Riggan. 1987. Nitrogen and sulfur emissions from the burning of forest products near large urban areas. J. Geophys. Res. 92, 701-709.

Hegg, D.A., L.F. Radke, P.V. Hobbs, R.A. Rasmussen, and P.J. Riggan. 1990. Emissions of some trace gases from biomass fires. J. Geophys. Res. 95, 669-675.

Kaufman, Y.I. 1987. Satellite sensing of aerosol absorption. J. Geophys. Res. 92, 4307-4317.

Kauppi, P., K. Mielikainen, and K. Kuusela. 1992. Biomass and the carbon budget of European forests: 1971-1990. Science 256, 70-74.

Keeling, C.D., A.F. Carter, and W.G. Mook. 1984. Seasonal, latitudinal, and secular variations in abundance and isotopic ratios of atmospheric CO_2. J. Geophys. Res. 89, 4615-4628.

Khalil, M.A.K., and R.A. Rasmussen. 1992. The global sources of nitrous oxide. J. Geophys. Res. 97, 651-660.

Khalil, M.A.K., and R.A. Rasmussen. 1993. Decreasing trend of methane: Unpredictability of future concentrations. Chemosphere 26, 803-814.

Khalil, M.A.K., R.A. Rasmussen, and R. Gunawardena. 1993. Atmospheric methyl bromide: trends and global mass balance. J. Geophys. Res. 98, 2887-2896.

Kurz, W.A., M.J. Apps, B.J. Stocks, and J.A. Volney. 1994. Global climate change: Disturbance regimes and biospheric feedbacks of temperature and boreal forests. In: Biotic Feedbacks in the Global Climate System: Will the Warming Speed the Warming? (G.Woodwell, ed.), 119-133. Oxford University Press, Oxford, UK.

Lacaux, J.P., R.A. Delmas, B. Cros, B. Lefeivre, and M.O. Andreae. 1991. Influence of biomass burning emissions on precipitation chemistry in the equatorial forests of Africa. In: Global Biomass Burning: Atmospheric, Climatic, and Biospheric Implications (J.S. Levine, ed.), 167-173. MIT Press, Cambridge, MA.

Laursen, K.K., P.V. Hobbs, L.F. Radke, and R.A. Rasmussen. 1992. Some trace gas emissions from North American biomass fires with an assessment of regional and global fluxes from biomass burning. J. Geophys. Res. 97, 20687-20701.

Lenoble, J. 1991. The particulate matter from biomass burning: A tutorial and critical review of its radiative impact. In: Global Biomass Burning: Atmospheric, Climatic, and Biospheric Implications (J. S. Levine, ed.), 381-386. MIT Press, Cambridge, MA.

Levine, J.S., W.R. Cofer III, D.I. Sebacher, R.P. Rhinehart, E.L. Winstead, S.S. Sebacher, C.R. Hinkle, P.A. Schmalzer, and A.M. Koller. 1990. The effects of fire on biogenic emissions of methane and nitric oxide from wetlands. J. Geophys. Res. 95, 1853-1864.

Manõ, S. and M.O. Andreae. 1994. Emission of methyl bromide from biomass burning, Science 263, 1255-1258.

Mazurek, M.A., W.R. Cofer III, and J.S. Levine. 1991. Carbonaceous aerosols from prescribed burning of a boreal forest ecosystem. In: Global Biomass Burning: Atmospheric,Climatic, and Biospheric Implications (J.S. Levine, ed.), 258-263. MIT Press, Cambridge, MA.

Nelson, R.M., Jr. 1982. An evaluation of the carbon mass balance technique for estimating emission factors and fuel consumption in forest fires. U.S.D.A. Forest Service, Southeastern Forest Experimental Station, Res. Paper SE-231. Asheville, NC.

Penner, J.E., S.J. Ghan, and J.J. Walton. 1991. The role of biomass burning in the budget and cycle of carbonaceous soot aerosols and their climate impact. In: Global Biomass Burning: Atmospheric,Climatic, and Biospheric Implications (J.S. Levine, ed.), 387-393. MIT Press, Cambridge, MA.

Penner, J.E., R.E. Dickinson, and C.A. O'Neil. 1992. Effects of aerosol from biomass burning on the global radiation budget. Science 256, 1432-1433.

Radke, L.F., D.A. Hegg, J.H. Lyons, C.A. Brock, P.V. Hobbs, R. Weiss, and R. Rasmussen. 1988. Airborne measurements on smokes from biomass burning. In: Aerosols and Climate (P.V. Hobbs and M.P. McCormick, eds.), 411-422. Deepak Publishing, Hampton, VA.

Radke, L.F., D.A. Hegg, P.V. Hobbs, J.D. Nance, J.H. Lyons, K.K. Laursen, R.E. Weiss, P.J. Riggan, and D.E. Ward. 1991. Particulate and trace gas emissions from large biomass fires in North America. In: Global Biomass Burning: Atmospheric, Climatic, and Biospheric Implications (J.S. Levine, ed.), 209-224. MIT press, Cambridge, MA.

Rasmussen, R.A., L.E. Rasmussen, M.A.K. Khalil, and R.W. Dalluge. 1980. Concentration distribution of methyl chloride in the atmosphere. J. Geophys. Res. 85, 7350-7356.

Richter, D.D., C.W. Raalston, and W.R. Harms. 1992. Prescribed fire: Effects on water quality and forest nutrient cycling. Science, 215, 661-663.

Rogers, F.R., J.G. Hudson, B. Zielinska, R.L. Tanner, J. Hallett, and J.G. Watson. 1991. Cloud condensation nuclei from biomass burning. In: Global Biomass Burning: Atmospheric, Climatic, and Biospheric Implications (J.S. Levine, ed.), 431-438. MIT Press, Cambridge, MA.

Sarmiento, J.L. 1993. Atmospheric CO_2 stalled. Nature 365, 697-698.

Seiler, W., and P.J. Crutzen. 1980. Estimates of gross and net fluxes of carbon between the biosphere and the atmosphere from biomass burning, Climate Change 2, 207-247.

Stocks, B.J. 1991. The extent and impact of forest fires in northern circumpolar countries. In: Global Biomass Burning: Atmospheric, Climatic, and Biospheric Implications, J.S. Levine, ed.), 197-202. MIT Press, Cambridge, MA.

Susott, R.A., D.E. Ward, R.E. Babbitt, and D.J. Latham. 1991. The measurement of trace emissions and combustion characteristics for a mass fire. In: Global Biomass Burning: Atmospheric, Climatic, and Biospheric Implications (J.S. Levine, ed.), 245-257. MIT Press, Cambridge, MA.

Walker, B.H. 1991. Ecological consequences of atmospheric and climate change. Climatic Change 18, 301-316.

Ward, D.E., R.M. Nelson, and D.F. Adams. 1979. Forest fire smoke plume documentation. Paper presented at the 77th Annual Meeting, Air Pollution Assoc., Air and Waste Management Assoc., Pittsburg, PA.

Ward, D.E., A.W. Setzer, Y.J. Kaufman, and R.A. Rasmussen. 1991. Characteristics of smoke emissions from biomass fires of the Amazon region-Base-A experiment. In: Global Biomass Burning: Atmospheric, Climatic, and Biospheric Implications (J.S. Levine, ed.), 394-402. MIT Press, Cambridge, MA.

Ward, D.E., R.A. Susott, J.B. Kauffman, R.E. Babbitt, D.L. Cummings, B. Dias, B.N. Holben, Y. Kaufman, R.A. Rasmussen, and A.W. Setzer. 1992. Smoke and fire characteristics for cerrado and deforestation burns in Brazil: Base B experiment. J. Geophys. Res. 97, 14,601-14,619.

Zepp, R.G. 1988. Environmental photoprocesses involving natural organic matter. In: Humic Substances and their Role in the Environment (F.H. Frimmel and R.F. Christman, eds.), 193-214. John Wiley and Sons, N.Y.

J.G.Goldammer and V.V.Furyaev (eds.), Fire in Ecosystems of Boreal Eurasia, 476-480.

The Effects of Forest Fires on the Concentration and Transport of Radionuclides

S.I. Dusha-Gudym [1]

1. Introduction

Studies on the radionuclear contamination of forests caused by the Chernobyl disaster of 1986 show that as of the year 1992, in the Bryansk Region, 84% of the radionuclides (averaged over all the forest types examined) have been concentrated in the forest fuels of soil cover and forest litter. The specific radioactivity of the debris, litter, moss and grass is much higher than that of the tree crown fuels. As for the crown fuels, it is their bark that accounts for the highest content of radionuclides.

2. Radioactive Contamination of Forest Fuels and Their Combustion Products

The level of ^{137}Cs radionuclides is 3 to 50 times higher in the ash and partially burned fuels compared to the aerial (crown) fuels. During forest fires, stem wood usually does not burn entirely, while forest debris, litter, needles and leaves, bark, etc. are consumed by fire more completely. Therefore, burning of these fuel types produces much more ash than that of wood, and their ash output may change from 1.5% to more than 10-20% of the original weight (Tab.1).

The ash, partially burned organic matter and smoke aerosols, generated during forest fires, are actually open sources of ionizing radiation on the areas where the density of radiocesium contamination exceeds 10 $Ci \cdot km^{-2}$. The concentration of radionuclides in the ash and partially burned matter amounts to up to millions of bequerels per kg.

The fires, generating hundreds and thousands of kilograms per one hectare of forest land of open ionizing radiation sources, may be called accordingly "radioactive forest fires" (Dusha-Gudym 1993).

In June 1992, smoke samples of a ground fire and control samples were taken from an area contaminated with radionuclides of ^{137}Cs; the soil contamination density was 15-40 $Ci \cdot km^{-2}$. The smoke samples were taken at a height of about 1.5m from the ground. Thus, the data obtained characterize a typical environment to which fire fighters are exposed during the combat of ground fires. The analyses of the filters for background air sampling revealed that aerosols contained only radionuclides ^{137}Cs of permissible amounts. The gamma

[1] Forest Pyrology Laboratory, All-Russia Research Institute of Forestry Chemization, VNIIKhleskhoz, 141250 Ivanteevka, Moscow Region, Russian Federation

spectrometric analysis of the smoke samples showed that the specific contamination of smoke aerosols of ^{137}Cs ranged from 0.022 to 1.73 $Bq \cdot m^{-3}$, and in the most contaminated samples, these values were 62-228 times higher compared to that of the background. The analysis of radionuclides in the four filters most affected by ^{137}Cs contamination revealed that smoke aerosols contained the beta- and alfa-emitting ^{90}Sr (strontium-90) and ^{239}Pu (plutonium-239). The concentration of ^{90}Sr in the smoke aerosols varied from 8.1×10^{-16} to 1.62×10^{-14} $Ci \cdot l^{-1}$, whereas that of ^{239}Pu ranged from 2.295×10^{-17} to 4.59×10^{-16} $Ci \cdot l^{-1}$.

Tab.1. Specific radioactivity of forest fuels and their combustion products

Fuel Types and their combustion products	Soil Contamination by ^{137}Cs	
	10 Ci kg^{-1}	30 Ci kg^{-1}
Forest Debris	2.3×10^{-7}	3.6×10^{-7}
Forest Debris Ash	1.0×10^{-5}	2.1×10^{-5}
Litter (upper Layer)	6.5×10^{-7}	1.1×10^{-6}
Litter Ash	1.2×10^{-7}	1.2×10^{-5}
Moss	2.9×10^{-6}	4.5×10^{-5}
Moss Ash	3.9×10^{-5}	6.8×10^{-5}
Pine Crowns (Needled Branches)	3.2×10^{-8}	5.2×10^{-8}
Pine Crowns (Ash)	4.4×10^{-7}	6.8×10^{-7}

1 Curie (Ci) = 3.7×10^{10} Bequerel (Bq)

3. Forest Fire Occurrence and the Transport of Radionuclides

Out of the three regions of Russia subjected to the highest levels of soil contamination with radionuclides (Bryansk, Kaluga and Ryazan Regions), forest fires are a most acute problem in the Bryansk and Kaluga Regions; in the Ryazan Region only few forest areas are contaminated. In 1985, i.e. before the Chernobyl accident, there had been 97 forest fires in these three regions, whereas in the course of 6 years following the accident (1986-91) a total of 1440 fires occurred there, or on the average, 240 fires a year. In 1992, 1173 forest fires occurred in these regions. In the national forest enterprise districts of the Bryansk Region, 590 fires were recorded; this is the highest value for the last 40 years. A similar situation was observed in both the Kaluga and Ryazan Regions. In 1992, in the Bryansk Region, 341 fires were recorded on radioactively contaminated forest areas belonging to different agencies; the area of each fire being less than one hectare.

However, the height of the convection columns of this category of forest fires often exceeded 300-400 m. Depending on the direction and speed of prevailing winds, the smoke plumes drifted to adjacent forests, fields, and populated areas. Nine of the registered fires covered a forest area of 2-10 ha, two of the fires with an area above 10 ha, and the area of one of them being 25 ha. Therefore, conditions could allow the formation of more powerful convection columns to transport the radionuclides to comparatively greater distances.

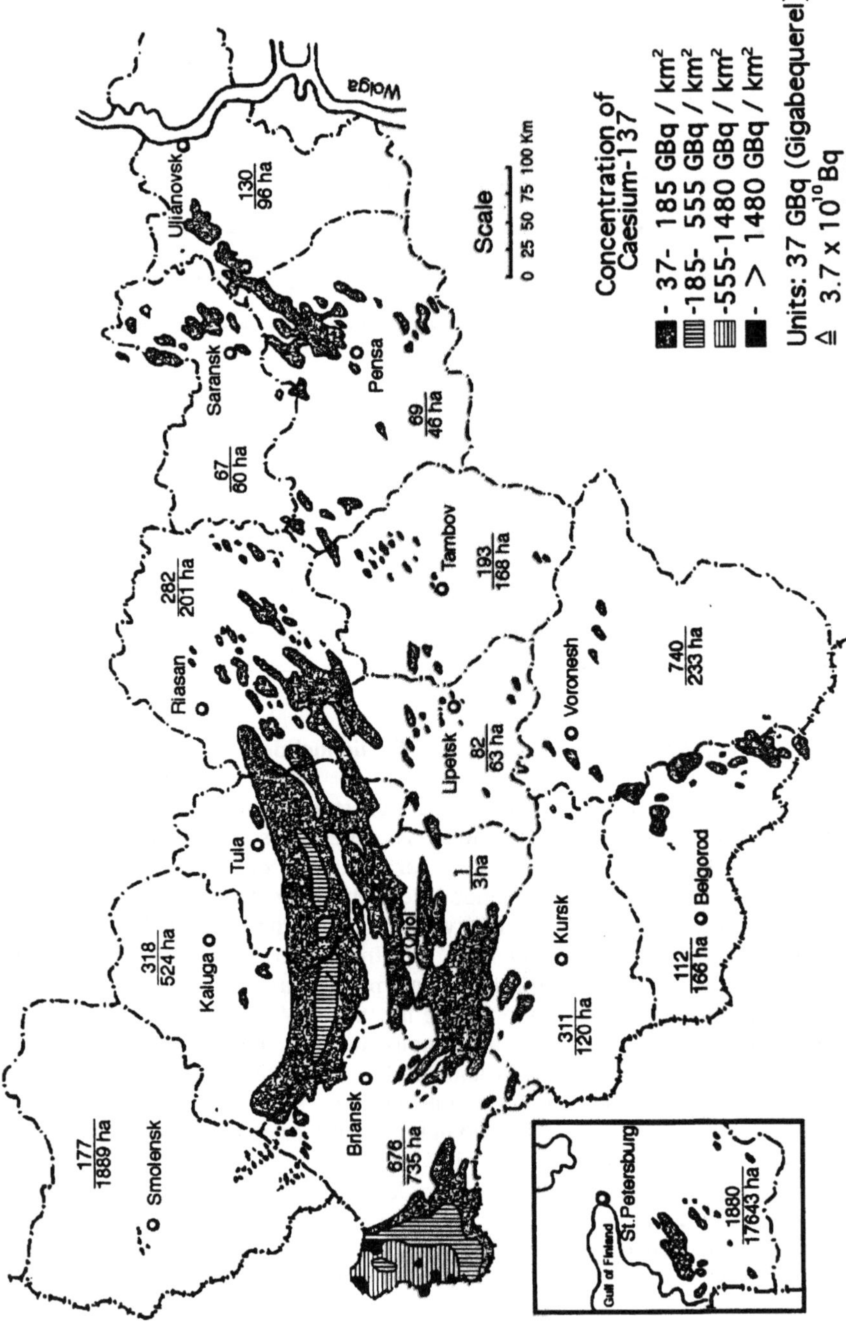

Fig. 1. Contamination of the land area in the vicinity of the Chernobyl nuclear power plant and in St.Petersburg region by Caesium-137. The numbers indicate forest fires and area burned in 1992 in the most contaminated districts.

In 1992 relatively few big forest fires occurred in the Bryansk Region and no significant changes of the radioactive situation had been expected. However, in the summer of 1992, the experts of the Experimental Meteorology Institute recorded both an increased content of radioactive aerosols in the atmosphere and radioactive depositions in the vicinity of Novozybkovo. In 1992 the concentration of ^{137}Cs in the atmosphere over the contaminated territory of Russia, was on the average, $0.884x10^{-5}$ Bq·m^{-3}, varying between 0.352 and $1.750x10^{-5}$ Bq·m^{-3} in different months. Near the town of Novozybkov radiocaesium concentrations in the atmosphere ranged from $30x10^{-5}$ to $70x10^{-5}$ Bq·m^{-3} during the period from May to October. In order to identify the sources of these radioactive emissions, fires occuring on adjacent forest lands had to be analyzed.

In the summer of 1992 our neighbours in the Polesye Regions of the Ukraine and Belorussia faced a very dangerous fire situation. The press reported that in May 1992 there were six sites exposed to most severe wildfires, burning for several days, within the 3-km zone of the Chernobyl nuclear power plant (Abduradimov and Odnoljko. 1993). The area of each of them ranged from 170 to 900 ha, and the total burned area was about 2000 ha. Forest fires covering 150-200 ha and more usually produce convective activity and may transport radioactive materials to significant heights and great distances from fire sites. In the Chernobyl regions, fires burned large tracts of dry grass, and the bulk of its ash was transported as smoke aerosol. The output of ash from bush grass (*Calamagrostis epigeios*), residues of grass crops and moss averaged to 7% of the total fuel burned of 0.5 kg·m^{2}. Thus, the ash yield is 350 kg·ha^{-1}. Fires on 2000 ha have potentially generated ca. 700 tons of radioactive ash.

The first mathematical models of the radioactive transfer affected by forest fires, demonstrated that the smoke of big forest fires may transport radioactive contamination of the magnitude $nx10^{1}$ Ci·km^{-2} to a distance of 1-1.5 km if these fires occur in areas with high levels of radioactive contamination and with a wind velocity of 7-10 m·sec^{-1}. Figure 1 shows the number and areas of forest fires which occurred in 1992 in those regions of the Russian Federation where there are sites of radionuclide contamination. Most of these regions have rather low levels of radioactive contamination and minimal areas of forest fires. But in the highly contaminated parts of the Bryansk Region, forest fires are a grave challenge.

4. A Radioactive Forest Fire Model

The prediction of the behavior, spread and effects of forest fires under various conditions involves the use of physico-mathematical models of all the processes involved. Such models should be based on the distinction of various fuel types. Based on previous research and fuel classification and while regarding the forest as a multi-storey structure, it is reasonable to distinguish 10 layers within the forest phytocoenoses, out of which eight layers are the fuels proper, and the other two layers are the terrestrial and the planetary boundary layer of the atmosphere (Grishin 1992).

I consider it necessary to construct radioactive-pyrological models of forest stands, unforested areas and non-forested areas of the forest lands. A radioactive-pyrological model would serve as a basis for the estimation of the total heat flow of forest fuel combustion products, i.e. ash, partially burned matter, smoke aerosols, as well as the amount of radionuclides concentrated in solid and gaseous products of forest fuel combustion. Radioactive pyrological models will provide a basis for the development of various types of forest fire models. Figure 2 depicts a principal radioactive-pyrological model of a forest where the height of the layers and the amount of forest fuels (except for the the soil layer of 0-5 cm) are assumed to correspond to the Maximum Model of Grishin (1992); the radionuclide content in forest fuels and their combustion products are based on studies carried out on the areas contaminated by radionuclides.

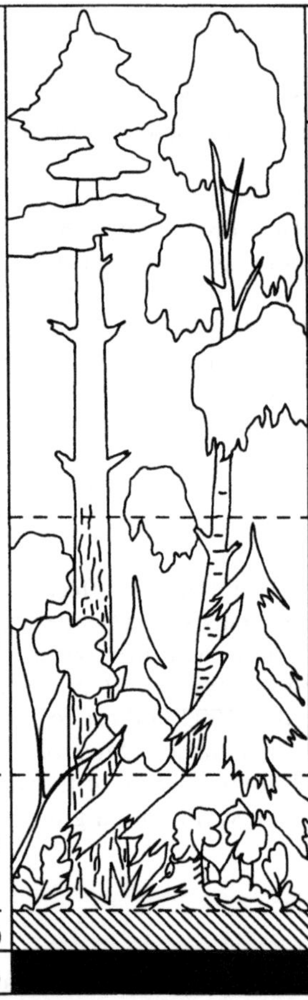

	Height of Fuel Layer	Fuel Load	Calorific Value	Specific Contami-nation of Fuels	Specific Contami-nation of Ash
H_3	22m	5.1 $\frac{kg}{m^2}$	21,949 $\frac{kJ}{kg}$	$n \cdot 10^{-7}$ $\frac{Ci}{kg}$	$n \cdot 10^{-6}$ $\frac{Ci}{kg}$
H_2	6m	2 $\frac{kg}{m^2}$	21,949 $\frac{kJ}{kg}$	$n \cdot 10^{-7}$ $\frac{Ci}{kg}$	$n \cdot 10^{-6}$ $\frac{Ci}{kg}$
H_1	2m	1.6 $\frac{kg}{m^2}$	21,949 $\frac{kJ}{kg}$	$n \cdot 10^{-7}$ $\frac{Ci}{kg}$	$n \cdot 10^{-6}$ $\frac{Ci}{kg}$
H_0	15cm	3	19,446	$n \cdot 10^{-6}$	$n \cdot 10^{-5}$
F_0	5cm	50		$n \cdot 10^{-6}$	

A radioactive fuel model of a forest in Briansk Region contaminated by Caesium-137
Units: 1 Curie (Ci) = 3.7 x 10^{10} Bequerel (Bq).

Fig.2. A radioactive-pyrological forest model

References

Abduradimov I., and A. Odnoljko. 1993. Forest fire danger. Nauka y Zhiznj No.2, 42-43 <in Russian>.

Dusha-Gudym, S.I. 1993. Forest fires on the territories contaminated by radionucleids. Vnitslesresurs, Moscow, 52 pp. <in Russian>.

Grishin A.M. 1992. Mathematic model of forest fires and new methods of their control. Nauka, Novosibirsk, 408 pp. <in Russian>.

Mahonjko, K.P. (ed.). 1993. The radiation situation on the Russian territory and the border states in 1992. NPO Tayfun, Obninsk, 290 pp. <in Russian>.

J.G.Goldammer and V.V.Furyaev (eds.), Fire in Ecosystems of Boreal Eurasia, 481-487.

Fire Weather Climatology in Canada and Russia

B.J. Stocks and T.J. Lynham [1]

1. Introduction

The boreal forest, or taiga, predominates as a vegetation type in northern circumpolar countries, covering in excess of 12 million square kilometres, primarily in Russia and Canada, with lesser amounts in Scandinavia, China, and the United States (Alaska). Lying generally between 45° and 70° N latitude, and accounting for approximately 25% of the world's forested area, the boreal zone contains extensive tracts of coniferous forest, primarily pine, spruce and larch, which provide a vital natural and economic resource in northern countries. Forest fire is the major disturbance in the boreal zone, boreal species having adapted over millennia to large-scale natural fires, to the point where fire is generally required for adequate species regeneration. When viewed on a large scale, the boreal forest is an irregular patchwork mosaic of even-aged, fire-origin stands, and is a classic example of a fire-dependent ecosystem, capable, during periods of extreme fire weather, of sustaining the large, high-intensity, stand-replacement wildfires which are responsible for its existence.

Human settlement of large sectors of the boreal zone has been accomplished largely through the development of highly efficient forest fire management systems designed to detect and suppress fires at an early stage. This approach has been largely successful in certain regions of Canada and Russia, but forest fires continue, under periodically extreme fire weather conditions, to exert a tremendous influence on the forest resource in both countries. Ecological and economic concerns over the attempt to suppress all fires in Canada has led to a multiple-use approach to fire management, where consideration is given to the natural ecological role of fire, the economics of fire suppression, and the priority of high-value resources. As a result, large areas of the boreal zone in northern Canada receive a "modified"or "selective" form of fire management, in which fires are prioritized on the basis of values-at-risk, resulting in numerous fires in low-priority areas receiving no suppression action and being permitted to perform their natural function in cycling the boreal forest. This practice, in conjunction with several extreme fire weather events, contributed to an annual area burned of 2.5 million hectares in Canada during the 1980s (Stocks 1991). In Russia, fires in a large part of the heavily populated regions of the country are actively suppressed, primarily through the use of airborne fire brigades, although the annual area burned in these regions ranged between 0.3 and 2.0 million hectares during the 1980-89 period (Dixon and Krankina 1993; Korovin, this volume). However, in many areas of central and eastern Siberia fires are not suppressed or even well-monitored. Recently, satellite imagery has been

---

[1] Forest Fire Research, Canadian Forest Service, Sault Ste. Marie, Ontario P6A 5M7, Canada

used to partially document large fires in this region of Siberia, revealing that at least 10 million hectares burned in 1987 (Stocks 1991, Cahoon et al. 1994), while additional statistics provided by Dixon and Krankina (1993) indicate that an annual average of 6.8 million hectares burned during the 1988-1991 period.

Clearly, forest fires are the major natural disturbance in the world's boreal forests. Economic realities and potential climate change impacts are likely to combine to ensure that the future rate of burning in boreal forests will increase. It is therefore imperative that fire scientists and managers work collectively to ensure that the role of fire in the boreal zone continues to evolve in a manner beneficial to both nature and society. This will require the combined and cooperative efforts of Russian and North American fire scientists, which, after decades of independent progress, the world's political climate will now permit. Towards this end, this study investigates fire weather climatology during the 1980s in Canada and Russia, the two countries sharing the vast majority of the world's boreal forest.

2. General Climatology of Canada and Russia

The climate of Canada is under the influence of three great air masses that originate in the Arctic, the Gulf of Mexico, and the North Pacific Ocean. Cold and usually dry air moves southward, usually by way of the Mackenzie River Valley and Hudson Bay. These outbreaks are usually more frequent during winter, and warm, moist air from the Gulf of Mexico and the Atlantic Ocean moves north to cover much of southeastern Canada in the summer. The third great air mass is the mild, moist air from the North Pacific Ocean that almost continually affects the coast of British Columbia, particularly bringing heavy winter precipitation. Sometimes a fourth air mass, originating in the southwestern United States, invades southern Canada in midsummer, causing extensive hot and dry periods. The Canadian Arctic, north of the tree line, has long days with very cold winters and cool summers, and receives only 125-250 mm of precipitation annually. The boreal zone has a growing season of approximately 180 days, with the western section receiving 300-350 mm of precipitation, mostly in summer. The eastern boreal zone receives two to three times as much precipitation, primarily through much heavier winter snowfall. As a result, snowfall disappears much earlier throughout the prairie provinces than it does at corresponding latitudes in eastern Canada. In general, annual precipitation decreases with movement westward and northward through the boreal zone from Ontario, and increases with movement eastward through Quebec and Newfoundland. The Canadian prairies receive moderate precipitation (300-400 mm annually), most of it falling as summer rain. Atlantic Canada receives very heavy winter snowfall, and an average of 1000-1400 mm of annual precipitation. Southern and central Ontario and Quebec experience the same winter conditions as the Atlantic provinces, but spring comes earlier and the summer is longer and warmer, while precipitation is seasonally uniform, averaging 750-1000 mm annually. In British Columbia coastal areas are moist most of the year, but interior regions, particularly in the south, experience hot and dry summer conditions.

Two fundamental factors mold the climate of Russia: the disposition of mountain ranges, and the existence of two vast pressure systems. With major mountain ranges located along its southern and eastern periphery, almost all of the country is exposed to the movement of Arctic air masses from the north and Atlantic air masses from the west, resulting in no clearly defined climatic regions, only a series of very gradual transitions. The two large pressure systems, the winter high pressure which is centered near Lake Baikal, and the summer low pressure with its focus over the south of central Asia, are both greatly intensified as they build up over the world's largest land mass. In particular, the winter high-pressure system is very stable and prolonged, beginning to build in the early fall and lasting through April, and deflecting depressions of Atlantic origin to the north and south. The result is very little winter precipitation and extremely cold temperatures. The summer low, conversely, tends to draw in depressions, with the result that almost all of Russia has a summer precipitation maximum. The effect of these depressions diminishes with movement eastward and northward. Whereas annual precipitation in European Russia averages 500-600 mm, this steadily decreases eastward and northward, to only 200-250 mm in the Lena River Basin, and only 100 mm at the delta of the Lena River. A cold monsoon regime along the Pacific coast of Russia results in very high precipitation amounts (1000-1500 mm annually). Summer temperatures across Russia show relatively little variation, except in the extreme south where temperatures are significantly higher. In winter, however, the temperature falls steadily northeastward into Siberia as continentality increases, reaching extremely low temperatures in the mountain basins of northeastern Siberia.

Landsberg et al. (1965), in a study of world climatology, described the climates of the boreal zones of Canada and Russia in terms of degree of continentality. A highly continental boreal climate exists throughout virtually all of the Siberian taiga east of the Yenisey River, and across the Canadian boreal zone in the northern regions of Manitoba, Saskatchewan, and Alberta, and throughout the Northwest and Yukon Territories. Highly continental boreal climates have very long and extremely cold, dry winters with widespread permafrost, and a short summer, with a growing season of < 100 days. The remainder of the boreal zone in eastern Canada, and in Russia west of the Yenisey River, experiences a continental boreal climate, defined as having long and very cold winters with prolific snowfall, and relatively warm summers with a growing period of 100-150 days.

3. Measuring Fire Danger in Canada and Russia

Over the past sixty years a number of fire danger rating systems have been developed by federal fire research scientists in Canada. The current system, under development since 1968, is termed the Canadian Forest Fire Danger Rating System (CFFDRS) (Stocks et al. 1989). The CFFDRS is used throughout Canada and has recently been adopted by other countries. At the present time the CFFDRS has two modules or subsystems: the Canadian Forest Fire Weather Index (FWI) System (Van Wagner 1987), introduced in 1970, which is strictly weather-based, and the Canadian Forest Fire Behavior Prediction (FBP) System (Forestry Canada Fire Danger Group 1992), introduced in 1989, which predicts quantitative fire behavior in major Canadian fuel types. The FWI System depends solely on daily measurements of temperature, wind speed, relative humidity, and precipitation, and has six components that individually and collectively account for the effects of fuel moisture and wind on fire behavior. The final component in the FWI System, the Fire Weather Index (FWI) itself, represents a relative measure of the potential intensity of a single spreading fire in a standard fuel complex such as a mature jack pine stand. Numerous studies undertaken in Canada have shown strong correlations between various measures of forest fire business and the component codes and indices of the FWI System (Turner 1973, Stocks 1974, Kiil et al. 1977).

The FWI can also be converted to a measure of daily or seasonal fire severity, using a technique developed by Williams (1959), and modified by Van Wagner (1970), which permits the integration of fire weather over periods of various lengths, and therefore the comparison of fire weather between different periods or different regions. Using rate of spread and resistance to control data from many test fires, Williams developed severity factors to represent the relative severity of fire danger classes, leaving moderate fire danger values untouched, reducing the relative severity of low danger values, and increasing the relative severity of high and extreme danger values. The simple averaging of fire danger values over a period of time could be misleading, as a season composed primarily of low and extreme days is much more severe than a season of primarily moderate values, although the average fire danger would be the same in both cases. For this reason the severity rating scheme devised by Williams, with its relative weighting of danger classes, provides a much more realistic measure of the fire problems likely to be encountered over any given period of time. In 1970 Van Wagner converted the Williams' severity rating scheme for use with the newly-developed FWI, formulating an equation, {$DSR = 0.0272\ (FWI)^{1.77}$ that calculated a Daily Severity Rating (DSR) for each FWI value, rather than weighing factors for each danger class only. DSR values could then be summed to produce a severity rating over any period of time. Stocks (1971) used DSR values summed over 10-day periods to track the pattern of fire danger in Ontario throughout the fire season. For the purposes of this study Monthly Severity Rating (MSR) values are used.

Numerous fire danger systems have been developed over the past 50 years in Russia. To date the most widely used is a relatively simple ignition index developed by Nesterov (1949). The Nesterov Index (NI) requires daily observations of dry-bulb temperature, dewpoint temperature, and precipitation. The difference between daily temperature is multiplied by temperature and cumulatively summed over the number of days since 3 mm of precipitation to provide a general index of ignition potential. In general, NI values below 300 are considered low, values between 300 and 1000 are moderate, with NI values from 1000-4000 considered high, and extreme conditions prevail when the NI is above 4000. Various Russian fire researchers have calibrated the NI for regional conditions (Valendik 1970, Snytkin 1971, L'vov and Orlov 1972), and Telitsyn (1970) produced a logarithmic version of the NI in order to decrease disparity between regional NI scales.

4. Methodology

Data from 150 Canadian and 150 Russian weather stations were used in this study. Typical of both countries, weather station locations are weighted towards populated or well-travelled regions, with the result that weather station network density decreases from south to north. Weather stations selected in this study were chosen to create a balanced geographical distribution as much as possible. Emphasis was placed on selecting as many stations as possible in the boreal zone, although stations both north and south of this zone were also used to aid in contouring. Weather data for Canada was supplied by the Atmospheric Environment Service of the Canadian Department of the Environment, while weather data for Russia was obtained from the National Climate Data Center of the National Oceanic and Atmospheric Administration. Ten years of weather data was used for Canada (1980-89), and nine years being used for Russia (1982-90), since 1980 and 1981 weather data was unavailable.

In general the fire season in both countries runs from May through September, and these five months were used in this analysis. Daily local noon standard time weather data were used for Canada, while 1400 h local weather data (the nearest to noon available) were used for Russia. DSR and NI values were computed daily for each station and averaged monthly for the period of record for each country. These Monthly Severity Rating (MSR) and Monthly Nesterov Index (MNI) values were then contoured on an Intera-Tydac Spans Geographic Information System (GIS). Class breakdowns for contouring were determined after evaluating the MSR and MNI frequency distributions for each country. Although fire danger classes cannot be determined when averaging monthly fire danger indices, it can be assumed that the highest 10% of values (90th percentile) represent monthly conditions where effective fire control is next to impossible (Fosberg et al. 1996), and that the 70-90th percentile values represent conditions where fire control options are severely limited. From Table 1, which shows monthly values averaged for all Canadian and Russian weather stations over the complete fire season, including those stations in the far north where the fire season is extremely short, it can be generally assumed that MSR values above 2.5 and MNI values above 3000 represent high fire danger conditions while MSR and MNI values above 4.5 and 4500 represent extreme conditions.

Tab.1. MSR and MNI frequency distributions for Canada and Russia.

Monthly Severity Rating (% occurrence)			Monthly Nesterov Index (% occurrence)		
Scale	Canada	Russia	Scale	Canada	Russia
0.00 - 0.25	13.11	12.90	0 - 500	9.31	9.84
0.25 - 0.50	13.54	6.45	500 - 1000	17.46	10.16
0.50 - 0.75	8.35	7.29	1000 - 1500	22.85	12.06
0.75 - 1.00	12.26	8.03	1500 - 2000	13.23	14.80
1.00 - 1.25	7.29	7.82	2000 - 2500	11.32	11.75
1.25 - 1.50	7.82	7.40	2500 - 3000	10.05	11.01
1.50 - 1.75	4.55	6.77	3000 - 3500	5.19	7.20
1.75 - 2.00	6.66	6.66	3500 - 4000	2.86	4.55
2.00 - 2.50	7.08	9.62	4000 - 4500	2.43	5.93
2.50 - 3.00	3.91	6.34	4500 - 5000	1.38	2.86
3.00 - 4.00	5.81	8.14	>5000	3.92	9.84
4.00 - 6.00	6.13	7.61	-	-	-
6.00 - 8.00	1.48	2.75	-	-	-
>8.00	2.01	2.22	-	-	-

5. Results and Discussion

Contoured MSR and MNI values are presented using a polar projection in Plates 2 to 6 *. From these maps a definite seasonal chronology of fire danger development (current) emerges for both Canada and Russia:

* Plate numbers refer to the "current indices" in the color plate section between this and the following chapter

* In May extreme fire danger in Canada is restricted to the southern portions of the prairie provinces, while virtually all of southern Siberia is experiencing extreme fire danger conditions. Northern regions in both countries remain at very low to low fire danger.

* During June fire danger rises quickly over large portions of both countries, moving northward dramatically. In Canada fire danger levels are high to extreme over all western Canada, rising to relatively moderate levels in the eastern region of the country. At the same time, extreme fire danger conditions now exist over virtually all of Siberia, with particularly extreme conditions throughout southern and northeastern Siberia.

* By July fire danger conditions in southern Siberia have decreased slightly while rising in the northern parts of the region. Extreme conditions prevail throughout central and northeastern Siberia. In Canada the fire danger situation has modified only slightly from June, with the same regional pattern.

* August shows a continuing decline in fire danger levels throughout southern Siberia, and diminishing fire danger in the north. However, east-central Siberia continues to experience high to extreme fire danger levels. In Canada, fire danger conditions are slightly lower than July, particularly in the northern region of western Canada.

* By September low fire danger conditions prevail over much of Canada and Russia. The exceptions are in the southern region of Canada's prairie provinces, and in extreme southwestern Siberia, both areas which are virtually unforested. A large portion of east-central Siberia, however, continues to experience moderate fire danger levels.

Three immediate observations can be drawn from a study of Plates 2 to 6:

* The striking similarity of the MSR and MNI monthly maps for both Canada and Russia. Both approaches track the development of the fire season in each country, generally moving northward in late spring and southward in late summer, in a very consistent manner.

* A second major observation is the immense area of Siberia that experiences high to extreme fire danger conditions during the summer months. This area is approximately three times the size of the similarly-affected region of western Canada, and this fact alone probably explains why so many large fires occur in Siberia each year.

* There is a strong disparity between eastern and western Canada in terms of monthly fire danger conditions. A high degree of continentality in interior western Canada, and the fact that eastern Canada's weather is strongly influenced by the Great Lakes and the Atlantic Ocean, are the primary reasons for higher fire danger conditions in western Canada.

This study was undertaken primarily to provide a better understanding of relative fire danger conditions in both Canada and Russia. Although the spatial and temporal scales used were necessarily coarse, the results are quite useful in visualizing the seasonal development of fire danger on a national scale. More detailed comparisons, at a regional scale and over shorter timeframes, are the next logical step, and this work is now beginning. Under the auspices of the International Boreal Forest Research Association, fire research scientists from Canada and Russia will, among many research initiatives, cooperate in evaluating the performance of the CFFDRS in various regions of Russia, with particular emphasis on Siberia.

Acknowledgements

The authors gratefully acknowledge the assistance of John Mason and Gary Hartley of the Canadian Forest Service (Ontario Region) for data base manipulation and graphical display development.

References

Cahoon, D.R., B.J. Stocks, J.S. Levine, W.R. Cofer, and J.M. Pierson. 1994. Satellite analysis of the severe 1987 forest fires in northern China and southeastern Siberia. J. Geophys. Res. 99 (D9), 18627-18638.

Dixon, R.K., and O.N. Krankina. 1993. Forest fires in Russia: carbon dioxide emissions to the atmosphere. Can. J. For. Res. 23, 700-705.

Forestry Canada Fire Danger Group. 1992. Development and structure of the Canadian Forest Fire Behavior Prediction System. For. Can., Ottawa, Ont., Inf. Rep. ST-X-3, 49 pp + Appendices.

Fosberg, M.A., B.J. Stocks, and T.J. Lynham. 1996. Risk analysis in strategic planning: fire and climate change in the boreal forest (this volume).

Kiil, D.A., R.S. Miyagawa, and D. Quintilio. 1977. Calibration and performance of the Canadian Fire Weather Index in Alberta. Can. For. Serv., North. For. Res. Cent. Inf. Rep. NOR-X-173. 45 pp.

Landsberg, H.E., H. Lippmann, K. Paffen, and C. Troll. 1965. World maps of climatology. New York: Springer-Verlag.

L'vov, P.N., and A.I. Orlov. 1972. Determination of fire hazard in forests of Arkhangelsk Oblast. Lesnoye Khozyaystvo 6, 70-72 <in Russian>.

Nesterov, V.G. 1949. Combustibility of the forest and methods for its determination. USSR State Industry Press <in Russian>.

Snytkin, G.V. 1971. Determining forest fire danger in forests of the extreme northeast. Lesnoye Khozyaystvo 4, 67-69 <in Russian>.

Stocks, B.J. 1971. Fire severity index distribution in Ontario: 1963 to 1968. Can. For. Serv., Great Lakes For. Res. Cent. Inf. Rep. O-X-151. 11 pp + Appendix.

Stocks, B.J. 1974. Wildfires and the Fire Weather Index System in Ontario. Can. For. Serv., Great Lakes For. Res. Cent. Inf. Rep. O-X-213. 17 pp.

Stocks, B.J. 1991. The extent and impact of forest fires in northern circumpolar countries. In: Global Biomass Burning: Atmospheric, Climatic, and Biospheric Implications (J.S. Levine, ed.), 197-202. MIT Press, Cambridge, Mass.

Stocks, B.J., B.D. Lawson, M.E. Alexander, C.E. Van Wagner, R.S. McAlpine, T.J. Lynham, and D.E. Dube. 1989. The Canadian Forest Fire Danger Rating System: an overview. For. Chron. 65 (6), 450-457.

Telitsyn, G.P. 1970. Logarithmic fire danger index. Lesnoye Khozyaystvo 11, 6-10 <in Russian>.

Turner, J.A. 1973. A fire load index for British Columbia. Can. For. Serv., Pac. For. Cent. Inf. Rep. BC-X-80. 15 pp.

Valendik, E.N. 1970. Degrees of fire danger in the Krasnoyarsk Region. Lesnoye Khozyaystvo 7, 4-8 <in Russian>.

Van Wagner, C.E. 1970. Conversion of William's Severity Rating for use with the Fire Weather Index. Can. For. Serv., Petawawa For. Exp. Sta. Inf. Rep. PS-X-21. 5 pp.

Van Wagner, C.E. 1987. Development and structure of the Canadian Forest Fire Weather Index System. Can. For. Serv., Ottawa, Ont., For. Tech. Rep. 35. 37 pp.

Williams, D.E. 1959. Fire season severity rating. Can. Dep. Nor. Aff. Nat. Res., For. Res. Div. Tech. Note No. 73.

J.G.Goldammer and V.V.Furyaev (eds.), Fire in Ecosystems of Boreal Eurasia, 488-494.

Colour Plate Section to Chapters "Fire weather climatology in Canada and Russia (B.J.Stocks and T.J.Lynham) and "Risk analysis and strategic planning: Fire and climate change in the boreal forest" (M.A.Fosberg et al.).

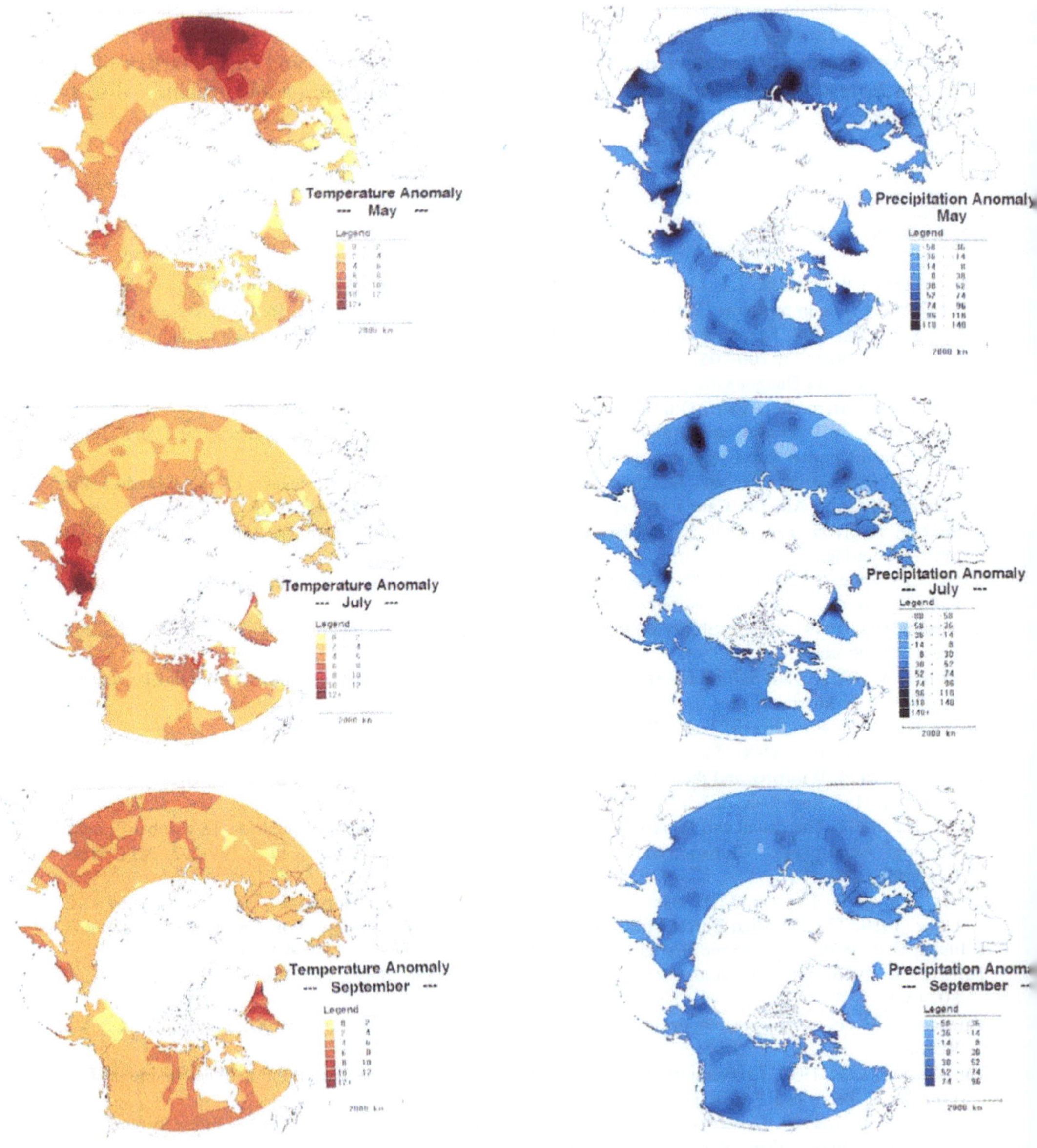

Plate 1. Fire season changes in temperature and precipitation. 1a change in May temperature, 1b July temperature, 1c September temperature, 1d percentage change in May precipitation, 1e, July precipitation, 1f September precipitation.

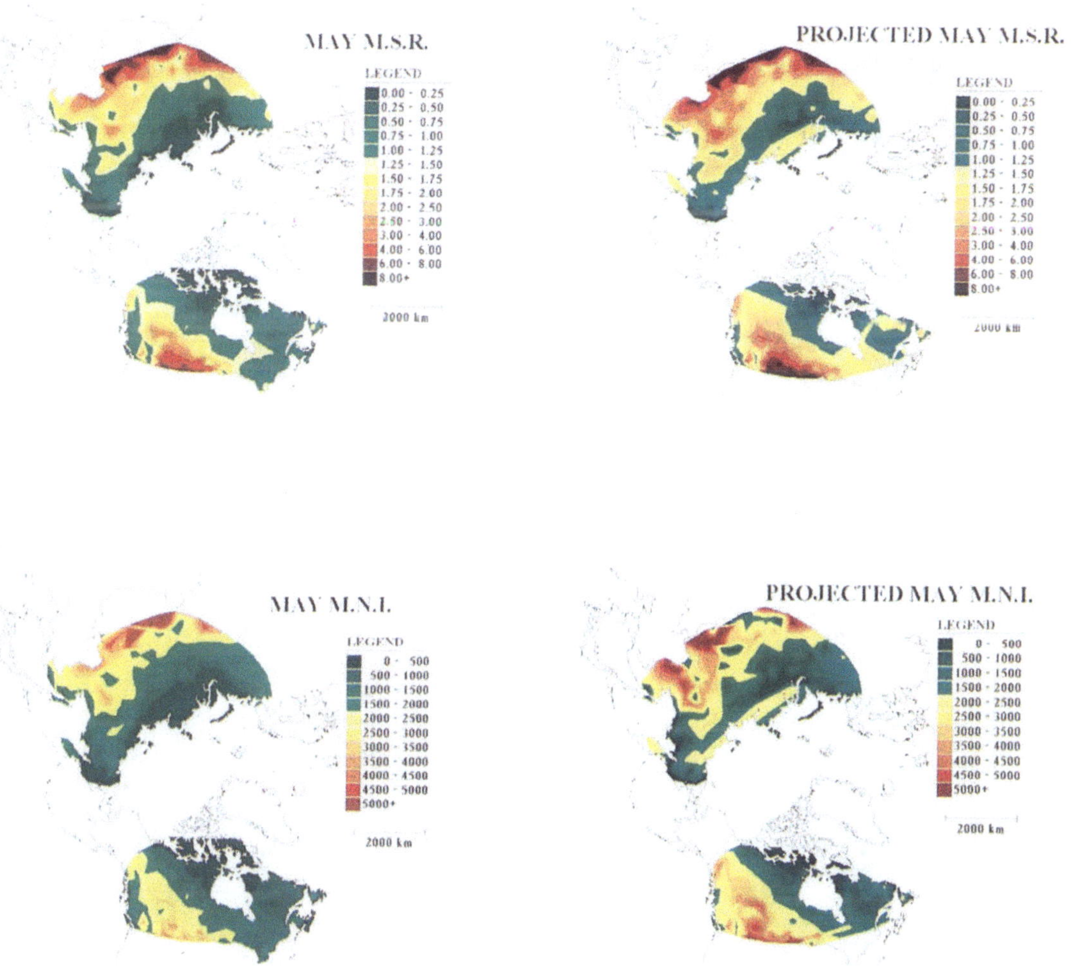

Plate 2. Current (1980s) and projected mean fire indices for Canada and Russia for May. 2a current severity (MSR) index, 2b projected severity (MSR) index, 2c current ignition (MNI) index, 2d projected ignition (MNI) index.

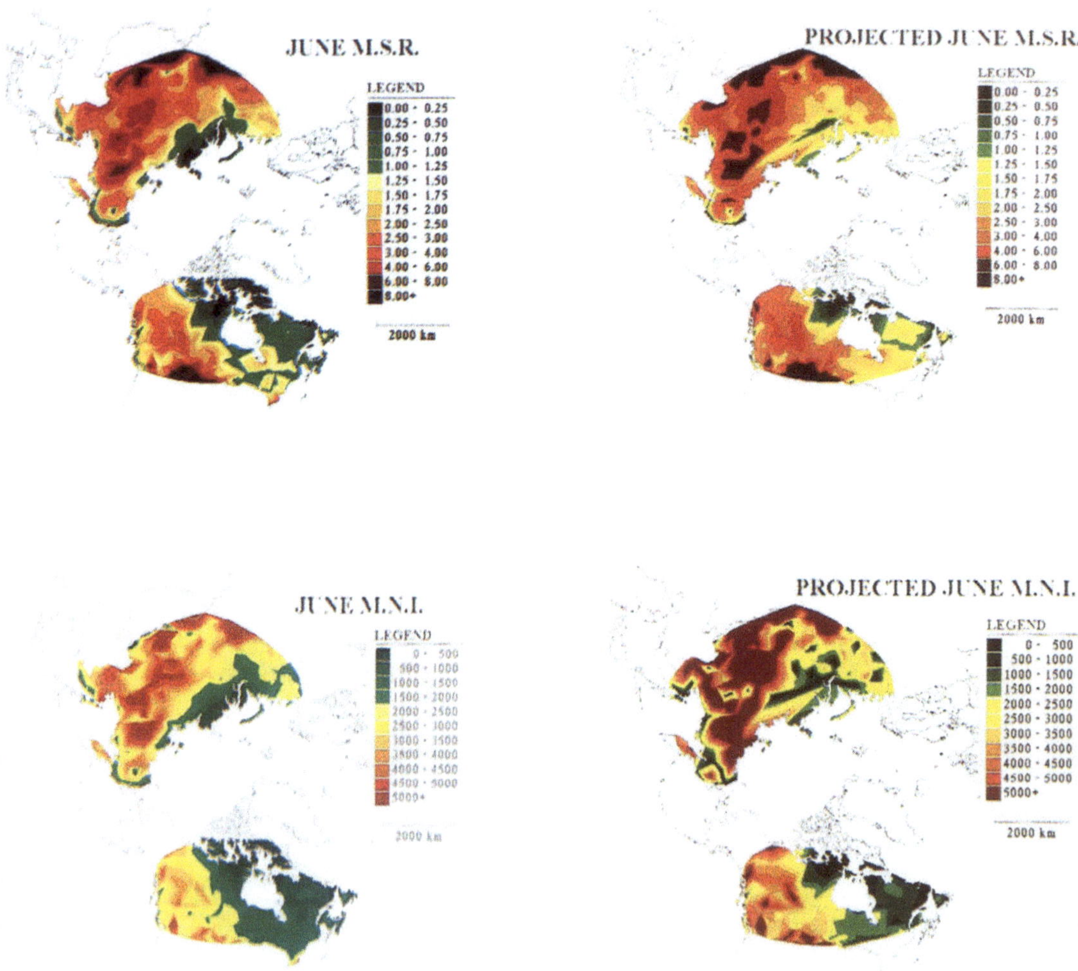

Plate 3. Current (1980s) and projected mean fire indices for Canada and Russia for June. 3a current severity (MSR) index, 3b projected severity (MSR) index, 3c current ignition (MNI) index, 3d projected ignition (MNI) index.

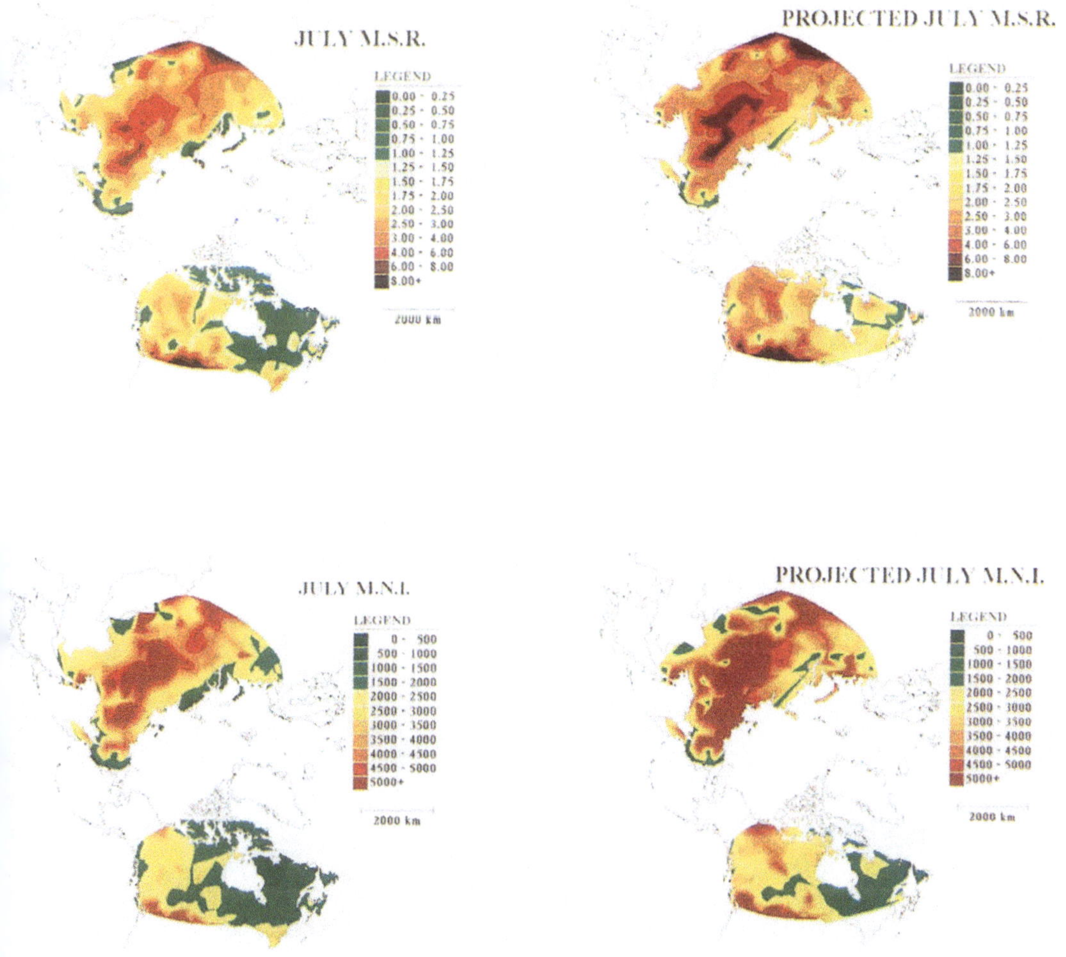

Plate 4. Current (1980s) and projected mean fire indices for Canada and Russia for July. 4a current severity (MSR) index, 4b projected severity (MSR) index, 4c current ignition (MNI) index, 4d projected ignition (MNI) index.

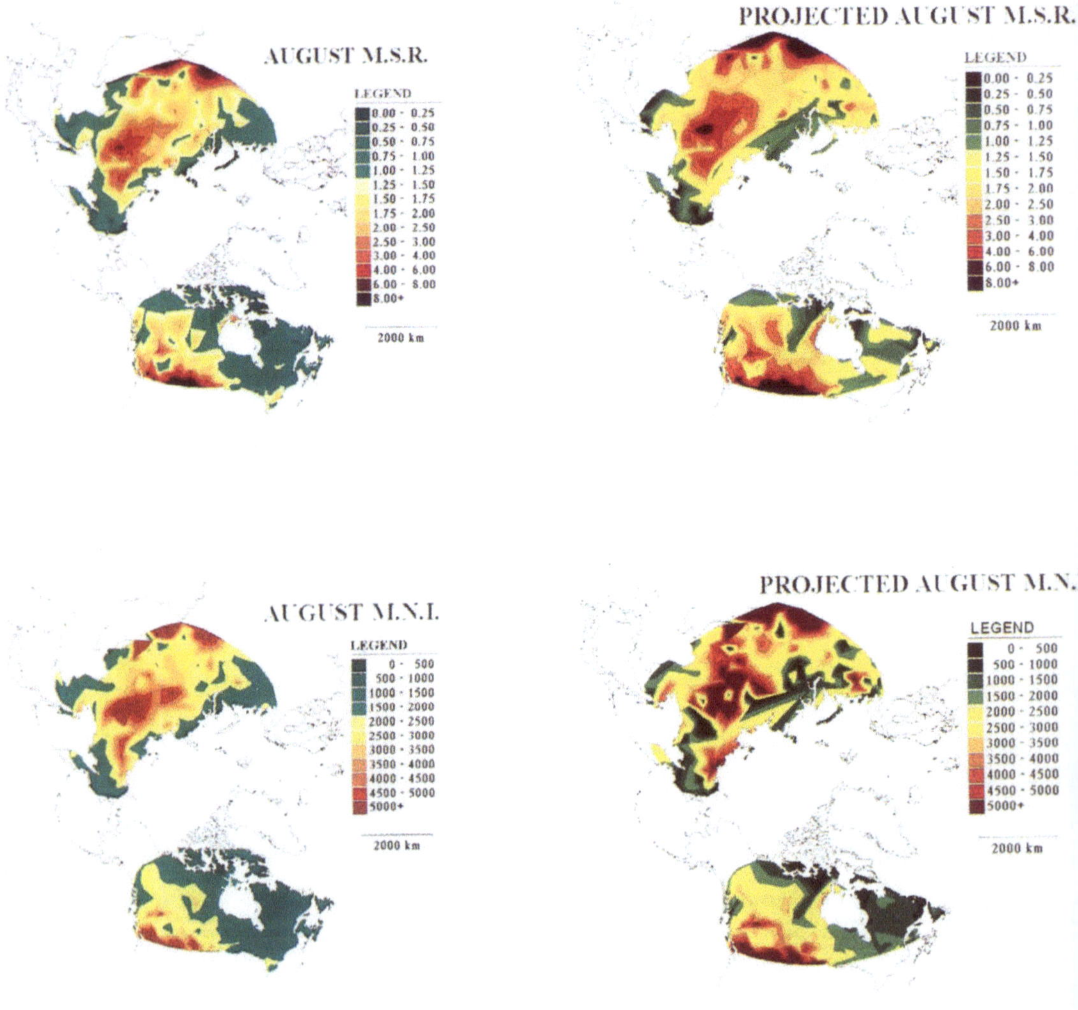

Plate 5. Current (1980s) and projected mean fire indices for Canada and Russia for August. 5a current severity (MSR) index, 5b projected severity (MSR) index, 5c current ignition (MNI) index, 5d projected ignition (MNI) index.

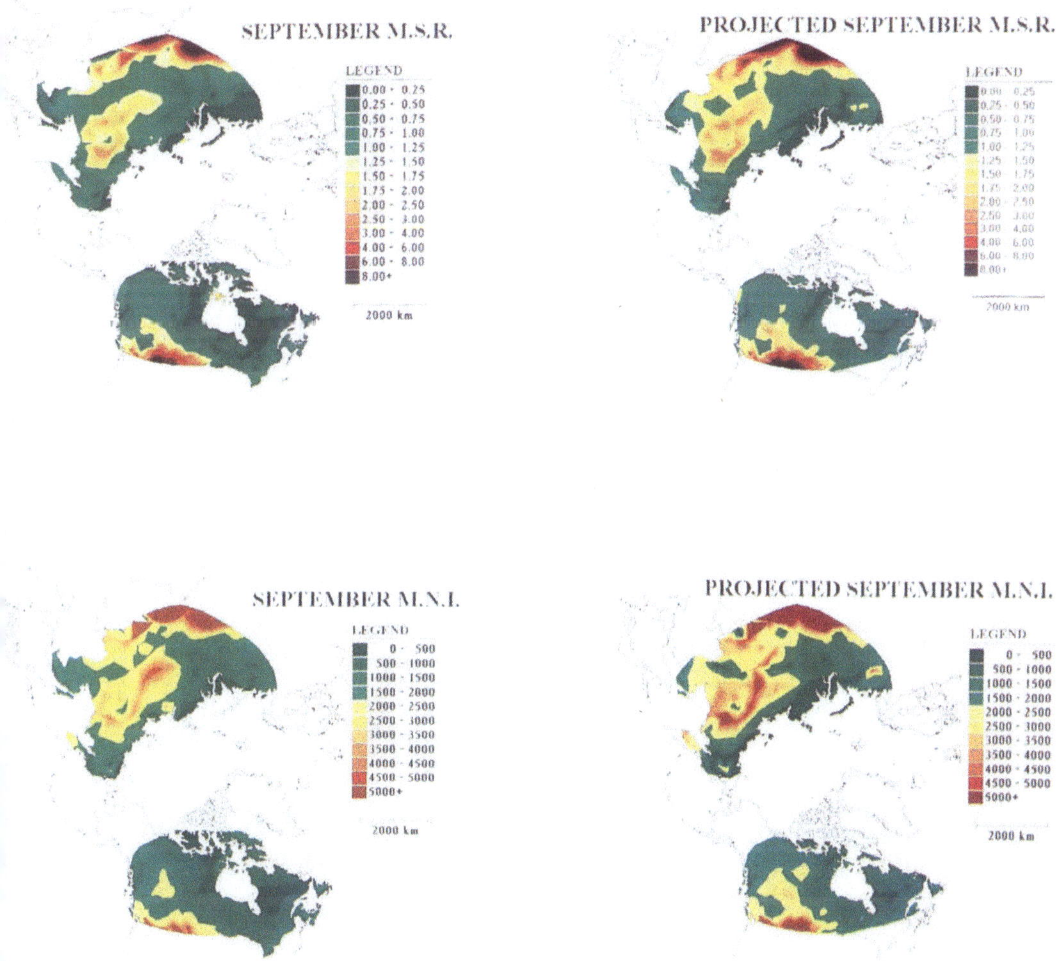

Plate 6. Current (1980s) and projected mean fire indices for Canada and Russia for September. 6a current severity (MSR) index, 6b projected severity (MSR) index, 6c current ignition (MNI) index, 6d projected ignition (MNI) index.

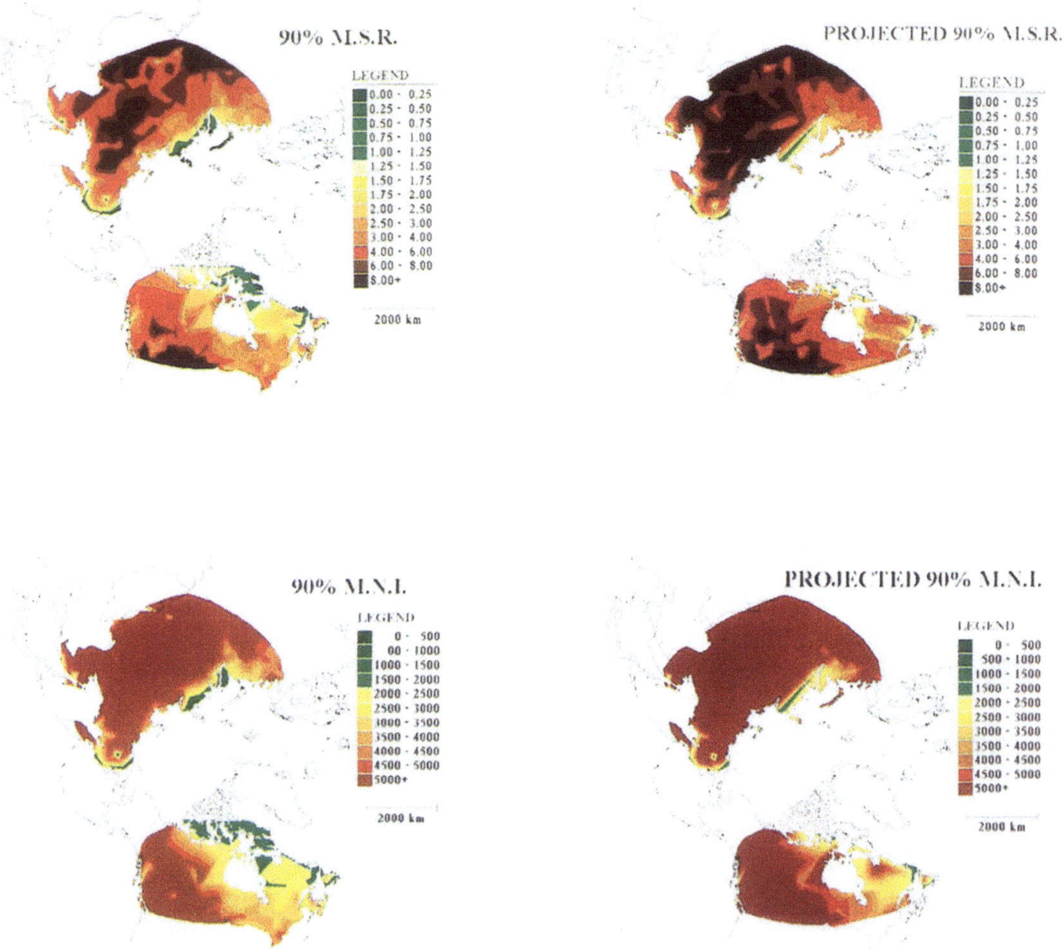

Plate 7. Current (1980s) and projected worst case fire danger. 7a current fire severity (MSR), 7b projected fire severity (MSR), 7c current ignition (MNI) index, 7d projected ignition (MNI) index.

J.G.Goldammer and V.V.Furyaev (eds.), Fire in Ecosystems of Boreal Eurasia, 495-505.

Risk Analysis in Strategic Planning: Fire and Climate Change in the Boreal Forest

M.A. Fosberg [1], B.J. Stocks [2], and T.J. Lynham [2]

1. Introduction

Boreal ecosystems contain approximately one fourth of the world's forests. These forests are in Russia, Canada, the United States, China, Norway, Sweden, and Finland (Fig.1). Main tree species of boreal ecosystems are pine (*Pinus* spp.), larch (*Larix* spp.), spruce (*Picea* spp.), fir (*Abies* spp.) for conifers and birch (*Betula* spp.), poplar (*Populus* spp.), alder (*Alnus* spp.) for broadleaf trees. While there are distinct differences in varieties of these species between North American and Northern Eurasian forests, form and function are remarkably similar.

A convenient method of classifying climate of the boreal zone (Kuusela 1992) is by maritime, continental and high continental sub-zones. The maritime sub-zone has summer temperatures of 10 to 15°C, winter temperatures of 2 to 3°C, and annual precipitation of 400 to 800 mm. The continental sub-zone has long cold winters with mean temperatures of -20 to -40°C and summer mean temperatures of 10 to 20°C. The growing season is between 100 and 150 days. Annual precipitation is between 400 and 600 mm per year. The high continental sub-zone is by far the largest portion of the boreal region. This zone is characterized by more extreme winters and milder summers. Northern portions of the boreal zone experience discontinuous permafrost with strong microclimate influence on the mosaic of plant communities.

The current boreal ecosystem developed during the last 11,000 years following the retreat of the glacial period icecap. On the southern edge of the boreal ecosystem, the mosaic of conifer and broadleaf species is determined by available soil moisture. On the northern edge of the boreal zone, temperature and nutrient availability determine the ecotone between tundra and the taiga. Flora migration following the retreat of the glaciers has resulted in ecosystem disassociation (Davis 1981), and in expansion of refugia of species. Spruce, fir, birch, popular, and fir have successfully followed the glacial retreat. North American boreal pine, specifically lodgepole pine (*Pinus contorta*), has migrated northward at a much slower rate than temperate climate species of pine (*Pinus ponderosa* and *Pinus banksiana*). The Northern Eurasian equivalents to these species are for ponderosa pine, scots pine (*Pinus*

---

[1] USDA Forest Service, Forest Fire and Atmospheric Sciences Research, Washington, D.C. 20250, U.S.A.

[2] Forest Fire Research, Great Lakes Forestry Center, Canadian Forest Service, Sault Ste. Marie, Ontario P6A 5M7, Canada

sylvestris) and for lodgepole pine and jack pine (*Pinus banksiana*), siberian pine (*Pinus cembra* L. var. *sibirica*). Both lodgepole pine and siberian pine have serotinous cones which require heat to release seeds. Survival strategies for each differ, siberian pine moved from a refugia both north and south from the glacial polar icecap and the highlands glaciers of the Himalayas, while lodgepole pine is still migrating north from its refugia in the south. Both lodgepole and siberian pine are fire dependent species. Unlike ponderosa or scots pine which thrive on frequent surface (understory) fire, lodgepole and siberian pine require stand replacement fire for survival.

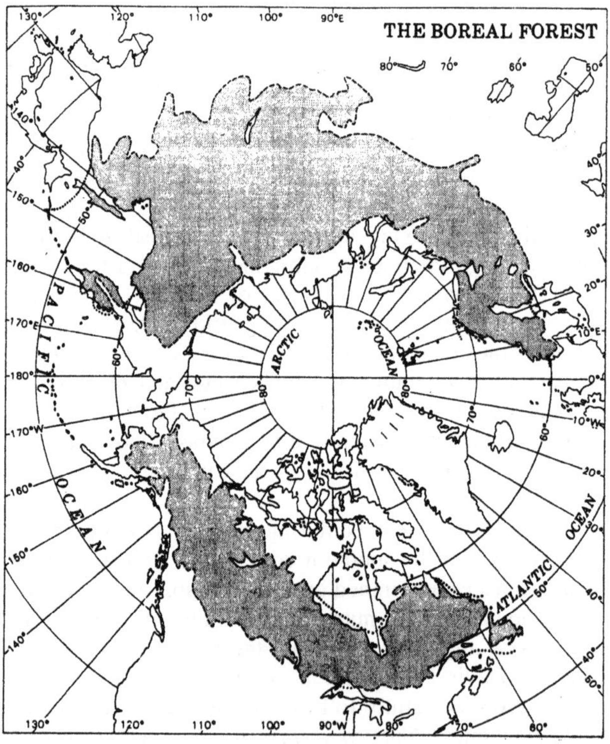

Fig.1. Global distribution of boreal forests.

Because of cold temperatures, decomposition of litter is slow, leading to deep organic soils and slow turn-over of nutrients. Fire is a major factor in nutrient turn-over in that mineral nutrients are released during combustion. Boreal forests are born of fire and die of fire. Fire and nutrient cycling are crucial to the health, diversity and productivity of the taiga.

Human activity in the taiga has been limited. Population centers north of 50°N in North America are south of the boreal zone. Europe and western Eurasia have a large number of population centers between 50 and 60°N, but Siberia and the Far East of Russia are mainly unpopulated. The taiga is the largest expanse of undeveloped forest in the world. Current use of the taiga is predominantly low impact gathering of forest products such as nuts, fruits, mushrooms, or low level extraction of game and fish by people living in or near the taiga, not employed in forest development. This low-level use in a sparsely populated region is now faced with change. We are beginning to see a trend toward intensive use of renewable forest products such as timber, or the extraction of non-renewable mineral resources. Change to the taiga from increased human demand for products, along with the specter of global climate warming which could alter the future environment, requires that we begin to think about the long-term future of the boreal region. Strategic plans for sustainable development of the taiga require that we know the environmental constraints of the boreal zone as well as the potential for development.

The Brundtland Commission (World Commission on Environment and Development, 1987) defined sustainable development as to "...insure that [humanity] meets the needs of the present without compromising the ability of future generations to meet their own needs. The concept of sustainable development does imply limits - not absolute limits, but limitations imposed by the present state of technology and social organization on environmental resources and by the ability of the biosphere to absorb the effects of human activities."

Strategic planning and risk management involve five steps. These are, in generic form:

- Establish goals (sustainable development);
- Assess current state of the environment;
- Predict change and assess risk (both natural and human caused);
- Develop and evaluate options, and
- Implement policies.

We will focus our attention in this paper on steps 2 and 3 within this broad framework on strategic planning and risk management. Specifically, we will focus on future risk for forest fire management as a result of global climate change.

2. Global Climate Change

Atmospheric concentrations of greenhouse gases have been increasing since the beginning of the industrial revolution (IPCC 1990). These increases are expected to reach a level twice that of the pre-industrial background concentrations of the 18th century within the next 100 years. This predicted build-up is based on population growth and per capita consumption of energy. Impact of this build-up of greenhouse gases, is generally determined through use of atmospheric general circulation models. These models are normally run with a fixed concentration level of greenhouse gases twice that of the pre-industrial background. The model results then represent the equilibrium future of a warmer climate, and cannot provide information on either the timing of change or on the rate of change. In general, the models project a mean global warming of 2 to 4°C, with greatest warming taking place at high latitudes. The models also project this warming to be realized through increased nighttime and winter temperatures rather than through daytime or summer temperatures. For the most part, the hydrologic cycle is poorly depicted in the models, and there is greatest uncertainty in regional precipitation patterns. Efforts to improve the regional depiction of climate change have relied on embedding meso-scale models in atmospheric general circulation models (Giorgi and Mearns 1991).

Inter-annual and inter-decadal climate variabilities are not addressed systematically in the atmospheric general circulation models, nor is the variability associated with the transition from one equilibrium climate to a new equilibrium climate. Large variations in climate during the transition from the Pleistocene to the Holocene, specifically the Younger Dryas period, show rapid flickering between the two equilibria (Alley et al. 1993; Taylor et al. 1993). Meehl and Washington (1993) have shown an increase in inter-annual and inter-decadal variability for a doubled carbon dioxide atmosphere for South-East Asia, but such analysis has not been carried out on a systematic global basis.

Major shortcomings of projections for a future climate are the lack of systematic regional projections as opposed to the coarse resolution of general circulation models, a large uncertainty in the projections associated with the hydrologic cycle, and a lack of systematic evaluation of both the inter-annual/inter-decadal variability and the transient response of the atmosphere to climate change.

Even with these shortcomings, there is useful information and some basis for accepting the projections. Analysis of early Holocene climate change and forest response to that change provides a limited degree of confidence in projections of the future. Guetter and Kutzback (1990) used a climate change analog to map the ecoregions of the earth. Their analysis of climate change correlates with observed species migrations of woody plants (Bernabo and Webb 1977; MacDonald 1992). While much concern has been raised about disassociation and fragmentation of ecosystems because of differing migration rates of individual species (Davis 1981), use of climate analogs of ecoregions such as the Köppen system (Bailey 1989) or the Holdrige life zone (Holdrige 1947) appear to provide a first approximation of forest response to climate change. Detailed process-based models of the ecosystem (Agren et al. 1991) have been applied to climate change impacts on forests (Botkin et al. 1988; Solomon and West 1987) along the southern ecotone of the North American boreal forest. These model results differ in fine detail, but are in general agreement with the Köppen climate analog projections of Shugart and Smith (1992) which show a transition from boreal forest to cool temperate mixed forest.

Paleoclimate records of fire (Clark 1988) show that maximum fire occurrence was during warm, dry periods and that during cool, wet periods fire occurrence was minimal during the last 750 years on the southern edge of the North American boreal forest. Fire frequency in the boreal forest ranges from approximately 20 year intervals in the south with pine forests to 100 to 500 year intervals in the north with spruce and in maritime areas with fir forests (Wein 1993). The correlation between climate, forest cover and fire frequency is clear, but it is not clear what relationship exists between forest cover and fire frequency. Boreal forest species such as lodgepole pine are strongly dependent on fire. Lodgepole pine has migrated northward at a much slower rate than spruce or fir following the glacial retreat and brings a fire environment to the southern boreal forest.

There are three facets of fire and climate change. First is the changed vegetation, or fuels facet. Second is the changed potential for fire occurence, either from lightning or human cause. Third is the change in fire severity associated with changes in fire weather.

Climate analog methods to project the future forest of the boreal zone indicate that the traditional boreal species of spruce, fir, pine, larch, poplar, and birch will be replaced with broadleaf species on the southern edge. Shugart and Smith (1992) used several atmospheric general circulation models and the Koppen climate analog for ecoregions to depict this change. Using the correlation model of potential fire intensity and fire spread and the Koppen climate analog, Fosberg et al. (1993) suggested that the fire intensity would double in the southern portion of the current boreal zone because of changed forest fuel type. This correlation model also indicated that potential fire intensity would remain unchanged in the central and northern portions of the boreal zone because the forest fuel type would be largely unchanged. This correlation model also indicated that potential fire spread associated with fuel type would not change markedly throughout the boreal zone. That is, the overall availability of fine fuels to carry fire would remain nearly constant even though the species mix changed.

The second facet of climate change impacts on fire in the boreal zone is from changed probability of ignition. Probability of ignition is based on both natural ignition (lightning) and human caused ignition. Increase in lightning frequency from 50 to 70°N is expected to be between 30 to 40 percent (Fosberg et. al. 1990; Price and Rind 1994a,b). This change in lightning frequency is based on total lightning discharges in the boreal zone, and does not distinguish between cloud to cloud and cloud to ground lightning, nor is there a regional projection of lightning activity. Human induced fire starts will undoubtedly increase as use of the region increases, but it is beyond the scope of this paper to address that component.

The third facet of climate change impacts on fire in the boreal zone is based on changes in fire weather. Fire weather is a catch all phrase for hot, dry, windy weather which favors high intensity and fast spreading forest fires. In the taiga, extreme fire weather means large area stand replacement fires. Fire weather also refers to the length of the fire season, which may change as a result of climate change. Wotton and Flannigan (1993) used the Canadian Climate Center general circulation model to project that the period in which fire control organizations would need to be prepared for fires would increase by 22 percent for Canada as a result of a doubled carbon dioxide concentration in the atmosphere. Strategic planning for fire management implies financial needs as well as achieving resource protection goals. Wotton and Flannigan's analysis requires consideration of when to begin seasonally supplemental fire suppression crew employment and when to end that seasonal employment. A second consideration is size (personnel) and resources (equipment) of that seasonal fire suppression force. Personnel and equipment resources for fire suppression are based on both rapid deployment to small fires in order to limit damage, and to have adequate resources to contain and suppress major forest fires. The latter is based on some estimate of probable worst case fire management situation. This worst case scenario is typically based on the worst 10 percent or worst 5 percent or other low probability occurrence.

Planning decisions in fire management associated with climate change can be structured around three levels of effects. Primary effects of climate change are associated with changes in ecosystem structure and composition. Fire management planning and decisions are keyed to the future fuel load. Secondary effects of climate change are those associated with inter-annual and inter-decadal variability of climate, and the nature in which climate will transit from the current equilibrium to a changed equilibrium in the future. Fire management planning for these secondary effects is mainly concerned with changed requirements for suppression forces, length of fire season and other strategic issues. Tertiary aspects of fire and climate change are related to day-to-day decisions of a tactical nature and are driven by the climate change realization of chance of exceeding thresholds of fire danger. Nearly all fire management decisions associated with climate change will be made to address the secondary and tertiary effects. Our analysis in this paper addresses the changed risk from secondary and tertiary effects of climate change.

3. Fire Severity and Frequency for a Changed Climate

We base our climate change projections on the output files from the Canadian Climate Center atmospheric general circulation model (McFarlane et al. 1992; Boer et al. 1992). The Canadian Climate Center model is one of the most advanced climate models. The model contains 10 layers vertically, is based on a nominal 3.75° lat/long grid, and incorporates interactive cloud parameterization, improved solar and terrestrial heating calculations, a sophisticated treatment of land surface processes (including vegetation and soil type), and a simple ocean mixed layer with a sea ice component. The Canadian Climate Center model projects a global mean temperature increase of 3.5°C for a doubled carbon dioxide equivalent greenhouse gas forcing. The model also projects a 4 percent increase in global precipitation and evaporation and a 2 percent decrease in global cloud cover.

This model, like all other atmospheric general circulation models, contains a climate bias in which depiction of the current climate differs from the observed climate. Details for this bias are well documented and are beyond the scope of this paper (Mitchell 1989; Schlesinger 1989; Fosberg et al. 1993). We have used a standard method of reducing this model bias by expressing temperature for the doubled carbon dioxide atmosphere as a temperature difference between the control run for the current climate and the doubled carbon dioxide equilibrium run. Similarly, we reduce the bias in precipitation projections by expressing precipitation change as a percent change between the control and the doubled carbon dioxide run. This approach implicitly assumes that the temporal variance in climate remains unchanged and that only the mean of the statistical distribution has changed. This approach does allow enhancing the spatial distribution of climate change by applying the model gridded differences between the control (current climate) and the doubled carbon dioxide equilibrium climate run of temperature and precipitation to the rich climatological base, and therefore recapturing the mesoclimate variability lost because of the coarse spatial resolution of the general circulation model. We have applied the model grided change to 224 climate stations in Russia and to 191 stations in Canada. The base climatology for this analysis is given by Stocks and Lynham (1994). Monthly data from the climate model and the climate data were aggregated for December, January, and February (winter); March, April and May (spring), and September, October and November (fall). The fire season, May through September were evaluated separately on a month-by-month basis.

We use two measures of fire danger and severity in this analysis. The *Nesterov* index of ignition (Nesterov 1949) is based on the difference between temperature and dewpoint, and is weighted by temperature. The index is calculated on a daily basis and is cumulative since a last rainfall greater than 3 mm. Values of the Nesterov index greater than 300 indicate moderate ignition potential; 1000 to 4000, high ignition potential; and above 4000, extreme potential. The Nesterov index is the most widely used measure of fire danger in Russia. For this analysis, we averaged the daily values of the Nesterov index for individual months. The other measure is based on the Canadian fire weather index system (Van Wagner 1987), a subsystem of the Canadian fire danger rating system (Stocks et al. 1989). The fire weather index system (Fig.2) is comprised of a number of component indices that individually and collectively account for the effects of fuel moisture and wind on fire behavior. The final component, the fire weather index, is a relative measure of frontal fire intensity, and can be converted into a measure of daily or seasonal fire severity (Williams 1959, Van Wagner 1970) which weight fire danger classes relative to fire severity. In this analysis, the daily severity rating (DSR) is averaged over a month to produce a monthly severity rating (MSR). DSR values above 7 represent extreme fire severity, values between 1.9 and 7 high to very high severity, while values between 0.3 and 1.9 represent moderate severity, and values below 1.3 constitute low fire severity. The two measures of fire problems, the Nesterov Index for ignition potential, and the Canadian Fire Weather Index for fire severity, were calculated for both the current climate and the doubled carbon dioxide forced climate.

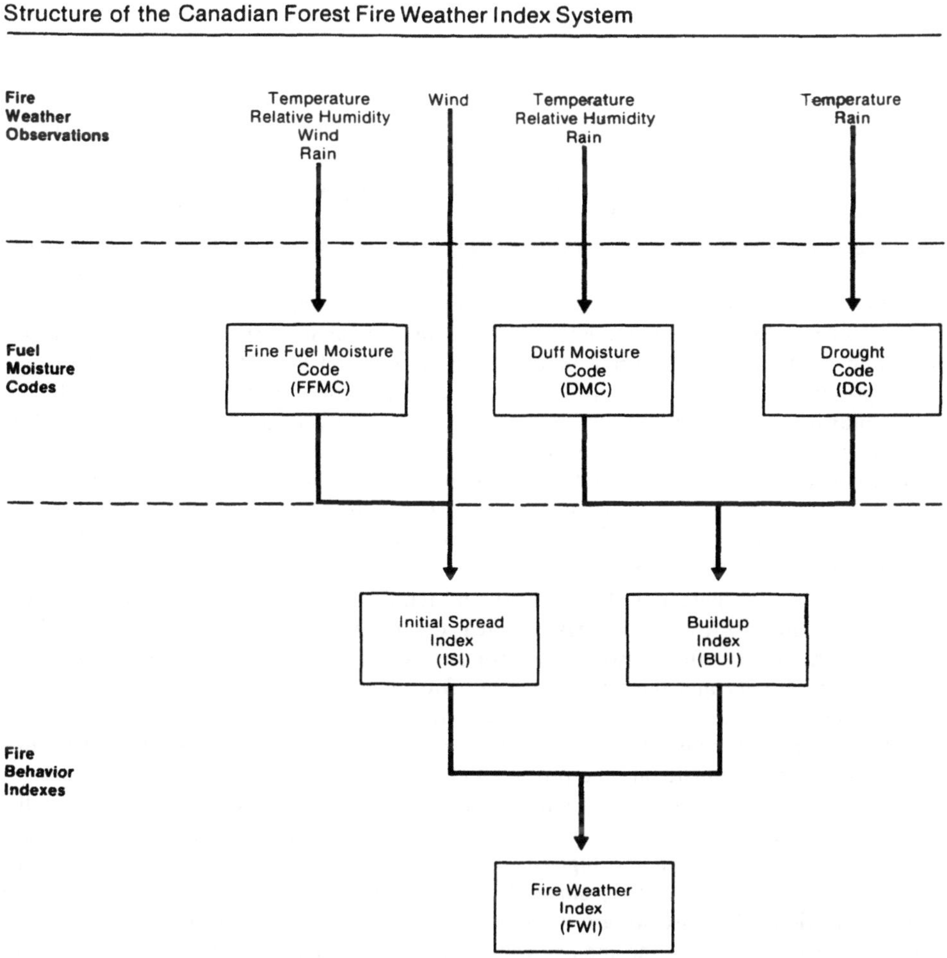

Fig.2. Schematic structure of the Canadian fire weather index system.

Risk management is based on managing the probable worst case situation. For fire management, the probable worst case is set at the 10, 5 or 3 percent or similar level of an historic record. While these events occur rarely, they create a significant portion of the fire protection requirements, in fact, 10 percent of the weather related fires result in 90 percent of environmental and social impact, 5 percent result in 95 percent of the impact, and 3 percent result in 97 percent of the impact. In this analysis, we will define the worst case at the 10 percent level, not for policy reasons, but for the statistical integrity of our data base. Specifically, we have constructed cumulative probability distributions of both the Nesterov Index and the Monthly Severity Rating for each of the climatological stations. We have used the 90th percentile level (corresponding to the worst 10 percent) of the indices at each climatological station as a basis to assess risk in a future warmed climate. The projected 90th percentile level of the fire danger indices were compared to the current 90th percentile levels to project the climate change scenario impacts on fire suppression requirements. Data and projections for the entire fire season were used in this portion of the analysis.

4. Results

Fire management planning and decisions are based on the evaluation of a number of indicators or indices. Temporal trends and spatial distribution of these composite indicators are more important than individual indicators. For this reason, we describe our results as seasonal and spatial analysis of the indices.

The Canadian Climate Center model projects warming of 6 to 8°C in the continental regions of Siberia and Canada, warming of 2 to 4°C for Alaska, and little change for Scandinavia during winter. Projected winter precipitation shows little change from the current climate. Spring temperatures are projected to be uniformly 2 to 6°C warmer throughout the boreal zone. Spring precipitation is projected to be 8 to 30 percent greater than at present. This warmer winter, and warmer and wetter spring, describe the climate leading into the fire season. Early fire season temperature changes (May) show 2 to 4°C warming except for western Siberia where temperatures are projected to be 4 to 6°C warmer (Plate 1a). Mid fire season (July) and late fire season (September) temperatures are projected to be nearly the same as present (Plate 1b,c). Early fire season precipitation is projected to be somewhat greater than present (Plate 1d), while mid and late fire season precipitation is projected to decrease (Plate 1e,f).

The mean, or 50th percentile fire severity and ignition potential for the early part of the fire season (Plate 2) shows an increase of area low and moderate fire severity and of high ignition potential. Mid fire season values for fire severity show little change between the current mean severity and the projected mean fire severity (Plate 3a,b). Ignition potential shows a substantial increase in extreme potential between the current climate and the projected warmer climate (Plate 3c,d). Mean values of both indices show little change between the current climate and the projected climate for late fire season (Plate 4).

A more meaningful way to evaluate climate change on fire potential is through the changes in the statistical distribution of the indices for the entire fire season. In particular, evaluation of the change in the highest 10 percent of the indices give a more accurate depiction of the change in risk, since this is the range in which fire control becomes extremely difficult. Because we are working with the tail of the statistical distribution, it is expected that the values of the indices would fall into the high or extreme class. For this reason, changes in the spatial distribution of the areas in these classes are more meaningful than the absolute value change. By this we mean that areas in which the worst 10 percent are now classed as moderate or high, may in the future be classed as extreme fire severity or ignition potential. The current worst 10 percent which are classified as high or worse in monthly severity (Plate 5a) show substantial increase in geographic expanse in Eurasia and close to doubling of area in North America (Plate 5b). The Nesterov index is at the extreme value for nearly all of Eurasia at present and is expected to virtually saturate Eurasia in the future. For North America, highest values of the Nesterov index are in western Canada. Projection of the future shows that the ignition problems will continue in the west, and will expand in area (Plate 5c,d).

5. Summary and Conclusions

This paper describes one scenario of the impacts of global climate change from a single atmospheric general circulation model. This scenario is based on an equilibrium climate rather than a transient solution in which the climate is allowed to change from the current equilibrium to the future equilibrium. Other atmospheric general circulation models would undoubtedly produce different results than those presented here. Thus, these results cannot be considered as the definitive projection of the future.

Perhaps the most significant aspect of this analysis was the development of the formal methodology in which risk associated with climate change could be expressed for strategic planning for natural resources.

Acknowledgments

We thank John Mason and Gary Hartley of the Department of Natural Resources Canada for managing the complex data base used in the analysis and for developing the graphics displays.

References

Agren, G.I., R.E. McMurtie, W.J. Parton, J. Pastor, and H.H. Shugart. 1991. State of the art of models of production-decomposition linkages in conifer and grassland ecosystems. Ecolog. Appl. 1, 118-138.

Alley, R.B., D.A. Meese, C.A. Shuman, A.J. Gow, K.C. Taylor, P.M. Grootes, J.W.C. White, M. Ram, E.W. Waddington, P.A. Mayewski, and G.A. Zielinski. 1993. Abrupt increase in Greenland snow accumulation at the end of the Younger Dryas event. Nature 362, 527-529.

Bailey, R.G. 1989. Explanatory supplement to ecoregions map of the continents. Envir. Conserv. 16, 307-309.

Bernabo, J.C., and T. Webb III. 1977. Changing patterns in the Holocene pollen record of northeastern North America: A mapped summary. Quat. Res. 8, 64-96.

Boer, G.J., N.A. McFarlane, and M. Lazare. 1992. Greenhouse gas-induces climate change simulated with the CCC second-generation general circulation model. J. Climate 5, 1045-1077.

Botkin, D.B., R.A. Nisbet, and T.E. Keynales. 1988. Effects of climate change on forests of the Great Lakes. Final report to EPA. University of California, Santa Barbara. 40pp.

Clark, J.S. 1988. Effects of climate change on fire regimes in Northwestern Minnesota. Science 334, 233-235.

Davis, M.B. 1981. Quatenary history and the stability of forest communities. In: Forest succession: concepts and applications (D.C. West, H.H. Shugart, D.B. Botkin, eds.), 132-153. Springer-Verlag, New York.

Fosberg, M.A., J.G. Goldammer, D. Rind, and C. Price. 1990. Global change: effects on forest ecosystems and wildfire severity. In: Fire in the tropical biota. Ecosystem processes and global challenges. (J.G. Goldammer, ed.), 463-486. Ecological Studies 84, Springer-Verlag. Berlin.

Fosberg, M.A., L.O. Mearns, and C. Price. 1993. Climate change - fire interactions at the global scale: predictions and limitations of methods. In: Fire in the environment (P.J. Crutzen and J.G. Goldammer, eds.), 123-137. John Wiley and Sons, Chichester.

Giorgi, F. and L.O. Mearns. 1991. Approaches to the simulation of regional climate change: a review. Rev. Geophysics 29, 191-216.

Guetter, P.J. and J.E. Kutzback. 1990. A modified Köppen classification applied to model simulations of glacial and interglacial climates. Climatic Change 16, 193-215.

Holdrige, L.R. 1947. Determination of world formulations from simple climatic data. Science 105, 367-368.

IPCC. 1990. Climate Change: The IPCC scientific assessment (J. Houghton, G.J. Jenkins, and J.J. Ephraums, eds.). Cambridge Univ Press. Cambridge.

Kuusela, K. 1992. The boreal forest: an overview. Unasylva 43 (170), 2-13.

McFarlane, N.A., G.J. Boer, J.-P. Blancet, and M. Lazare. 1992. The Canadian Climate Centre second-generation general circulation model and its equilibrium climate. J. Climate 5, 1013-1044.

MacDonald, G. M. 1992. Long-term monitoring of boreal forests using fossil pollen records. Unasylva 43 (170), 14-21.

Meehl, G.A. and W.M. Washington. 1993. South Asian monsoon variability in a model with doubled atmospheric carbon dioxide concentration. Science 260, 1101-1104.

Mitchell, J.F.B. 1989. Climate sensitivity: model dependence of results. In: Climate and geosciences: a challange for science and society in the 21st century (A. Berger, S. Schneider, and J.-C. Duplessy, eds.), 417-433. Kluwer Academic Publishers, Dordrecht.

Nesterov, V.G. 1949. Combustibilty of the forest and methods for its determination. USSR State Industry Press.

Price, C., and D. Rind. 1994a. Modeling global lightning distributions in a general circulation model. Mon. Wea. Rev. 122, 1930.

Price, C., and D. Rind. 1994b, Possible implications of global climate change on global lightning distributions and frequencies. J. Geophys. Res. 99, 10823.

Schlesinger, M.E. 1989. Model projections of the climatic changes induced by increased atmospheric CO_2. In: Climate and geosciences: a challange for science and society in the 21st century (A.Berger, S. Schneider, and J.-C. Duplessy, eds.), 375-415. Kluwer Academic Publishers, Dordrecht.

Shugart, H.H. and T.M. Smith. 1992. Modelling boreal forest dynamics in response to environmental change. Unasylva 43 (170), 30-38.

Solomon, A.M. and D.C. West. 1987. Simulating forest ecosystem responses to expected climate change in eastern North America: applications to decision making in the forest industry. In: The greenhouse effect, climate change and U S forests (W.E. Shands and J.S. Hoffman, eds.), 189-218. The Conservation Foundation. Washington DC.

Stocks, B.J., B.D. Lawson, M.E. Alexander, C.E. Van Wagner, R.S. McAlpine, T.J. Lynham, and D.E. Dubé. 1989. The Candian forest fire danger rating system: an overview. For. Chron. 65, 450-457.

Stocks, B.J. and T.J. Lynham. 1994. Fire weather climatology in Canada and Russia (this volume).

Taylor, K.C., G.W. Lamorey, G.A. Doyle, R.B. Alley, P.M. Grootes, P.A. Mayewski, J.W.C. White, and L.K. Barlow. 1993. The 'flickering switch" of late Pleistocene climate change. Nature 361, 432-436.

Van Wagner, C.E. 1970. Conversion of Williams' severity rating for use with the fire weather index. Can. For. Serv. Petawawa Exp. Sta. Inf. Rep. 35. 37pp.

Van Wagner, C.E. 1987. Development and structure of the Canadian forest fire weather index system. Canadian Forestry Service Tech. Rep. 35. Ottawa. 37pp.

Wein, R.W. 1993. Historic biogeography of fire: circumpolar taiga. In: Fire in the environment (P.J. Crutzen and J.G. Goldammer, eds.), 267-277. John Wiley and Sons, Chichester.

Willams, D.E. 1959. Fire season severity rating. Can. Dept. Nor. Aff. Nat. Res. For. Res. Div. Tech. Note No. 73.

Wotton, B.M. and M.D. Flannigan. 1993. Length of the fire season in a changing climate. Forestry Chronicle 69, 187-192.

World Commission on Environment and Development. 1987. Our common future. Oxford Univ. Press, Oxford.

J.G.Goldammer and V.V.Furyaev (eds.), Fire in Ecosystems of Boreal Eurasia, 505-512.

Fire - Climate Change Hypotheses for the Taiga

R.W. Wein and W.J. de Groot [1]

1. Introduction

Residents of Canada and other northern circumpolar countries are concerned with the scenarios of climate change since Global Circulation Models predict that global warming over the next 30-50 years will be most evident in the northern regions (Bolin et al. 1986; Roots 1989; Maxwell 1992). Climates in the north would be supportive of much more southern types of biotic systems (Emanuel et al. 1985).

We recognize that fire may be the most important (widespread) driving force in changing the taiga under climatic warming conditions. At the time of burning, CO_2 is released to the atmosphere where this greenhouse gas will act as a feedback loop to global warming. In addition, CO_2 release continues for one or more decades after the fire because of higher decomposition rates of organic matter, particularly in northern soils. As for climate change stresses on the biota of the ecosystem, it is our hypothesis that these energy and nutrient conservative ecosystems change very slowly even if the climate changes; however, fire can be a triggering event to remove species that are poorly adapted to the new climate regime. More importantly, fire modifies the physical environment and disrupts the population dynamics to such an extent that there can be strong changes in species abundance and new species may invade the burned area.

2. Wildfires in the Circumpolar Taiga

More than ever before Canada is exploiting the boreal forest so Canadians, and most recently Albertans, are recognizing the extent and importance of this biological zone. The taiga in Canada and Alaska is about 373×10^6 ha of the circumpolar total of 1214×10^6 ha. Russia has almost triple the amount of commercial forested land when compared to North America.

Fire is the most widespread agent of unplanned change in the boreal zone (Wein and MacLean 1983, and more recently, Johnson 1992); fire rapidly modifies the carbon budget not only of the frequently dry upland forests but also of the forests with deep organic soils, particularly after periodic severe droughts. This is not to negate other disturbances

---

[1] Department of Forest Science and Canadian Circumpolar Institute, University of Alberta, Edmonton, Alberta T6G 2E9, Canada

(Overpeck et al. 1990) of disease or insects (there is an increased probability that fire may follow these disturbances as well) and especially harvesting, which covers millions of hectares annually, but periodically fire is the overriding force.

In the past decade there has been an unprecedented increase in the annual area burned by wildfires in Canada (Van Wagner 1988). The 10 year running mean of annual area burned ranged from 1.5×10^6 ha in the 1930s and 1940s to 1.0×10^6 to the late 1970s. The years 1981 and 1982 were spectacular years with values of about 5.0 and 5.5×10^6 ha, respectively. These years were overshadowed by the burned area in 1989 of 7.39×10^6 ha, 3.28×10^6 ha of which was in northern Manitoba (Hirsch 1991).

To add to the temporal pattern above it is instructive to understand the spatial dimension of past fires. During most years, fires are scattered across the boreal zone of the circumpolar north, for example in the year 1980, fires were ignited across much of the western Canadian boreal forest. In other years there are concentrations of fires at the regional level.

Tab.1. The extent of the boreal (taiga) zone in the circumpolar North (in 10^6 ha and % of circumpolar total). Extracted from Kuusela (1992).

	Russia	Canada/ Alaska	Norway/ Sweden/ Finland	Total
Partly forested	790 (65)	373 (31)	51 (4)	1214
Closed forest	673 (73)	202 (22)	44 (5)	920
Commercial forest	450 (71)	149 (23)	40 (6)	639

Tab.2. Selected years with major regional fires in circumpolar taiga

Year	Location	Total Area (10^6 ha)	Area Burned		Comments	Reference
			Total (10^6 ha)	%		
1981	Northern Alberta	32 (11)*	1.36	4 (12)*	29 fires over 200 ha	Delisle and Hall (1987)
1987	Daxinganling, NE China	8	1.33	17	at least 4 major fires joined	Yun-Qian and Ji-Zhong (1989) Di and Ende (1990)
1989	Northern Manitoba	36 (19)*	3.28	9 (17)*	60* fires over 5,000 ha	Hirsch (1991)

* numbers in brackets represent greatest concentrations of multiple fires as estimated from maps

3. Wildfire Scenarios under Global Warming

A few Canadian studies have explored possible wildfire scenarios. Street (1989) provided the first Canadian study using a monthly average approach to expected weather changes in Ontario. He used a 2 x current atmospheric carbon dioxide concentration (CO_2) scenario (typical of climate change models) to estimate temperature and precipitation changes, and used these values to calculate the subsequent change in average monthly burning conditions, or the monthly Fire Severity Rating (FSR) (Williams 1959). He determined that the fire season would increase in length, and the fire regime would increase in severity as a result of doubling atmospheric CO_2 levels. The most severe portion of the fire season would also shift to late summer.

Although averages usually provide a good indication of trends by removing the statistical noise of variability, in this situation, the greater daily variability in burning conditions is more critical than the average change. For instance, the average increase in fire intensity may only be slight, but it can result in a large number of moderate intensity days becoming high or extreme intensity days. Over 90% of all fire losses occur in the high to extreme range (Pyne 1984), so even a doubling of these days (i.e., from 5% to 10% of all days) can have disastrous results. Using averages can mask this effect by indicating only a minor average increase within the moderate range. Stocks (1990) states that GCMs indicate an increase in the occurrence of extreme weather events. This only supports the fact that monthly averages are too coarse for measuring climate change effects on wildfires.

This point was mentioned by Flannigan and Van Wagner (1991) who opted for using a seasonal severity rating (SSR) (Van Wagner 1970). The SSR is a direct function of the daily Fire Weather Index value (Van Wagner 1987) and was originally designed to indicate the difficulty in suppressing a fire. It can be used for spatial and temporal comparison of fire danger (e.g., Harvey et al. 1986) which makes this ability a useful tool for modelling climate change effects on wildfires.

Flannigan and Van Wagner (1991) used a similar approach as Street (1989) by developing a $2xCO_2$ scenario. They selected six stations across Canada to compare changes in wildfire activity using the SSR and GCM estimates. Although their study did not take into account the effect of a longer fire season or dry early springs, their results showed a conservative increase of about 50% in SSR and a commensurate increase in area burned. They also found that the greatest increase occurred for the lowest SSR values. In other words, the net effect was not just an average increase in all daily SSR values, but a change that increased with decreasing SSR.

The same study noted that 85% of the historical area burned in Canada occurs west of Lake Nipigon. Therefore, it was suggested that the greatest proportion of any increase in area burned due to climate change will likely occur in the west. Although a 40% increase in area burned appears possible, Flannigan and Van Wagner (1991) also point out that there has already been a 50% increase since 1970 (Van Wagner 1988).

Surprisingly, the Flannigan and Van Wagner (1991) study showed that changes due to precipitation anomalies had little effect on SSR values. All stations received increased precipitation amounts from the GCM models (a typical global warming scenario), which were distributed according to historical observed sequences. This resulted in little effect on SSR because the cumulative precipitation amount is much less important than the sequence of precipitation events. If there is a greater occurrence of extreme weather events associated with climate change as stated by Stocks (1990), it is possible that extended droughts will interfere with regular precipitation patterns and cause increased wildfire problems.

4. Changes to Fuels on the Landscape

A secondary effect of climate change on wildfire is the role it plays in vegetation distribution, or more aptly, fuels distribution. The Canadian Forest Fire Behaviour Prediction System (Forestry Canada Fire Danger Group 1992) illustrates the wide range in fire characteristics that result from different fuel types. Fuels can also affect fire occurrence, rate of spread, fuel consumption and fire intensity.

Obviously, climate change will result in different temperature and moisture regimes which will adjust the present population boundaries of plant species. Several studies have presented estimates of vegetation shifts, including Emanuel et al. (1985) who predict that the boreal forest will shift northwards into the tundra and shrink in size by about 37%. Kauppi and Posch (1988) estimate the circumpolar boreal forest to shift north about 500-1000 km. Wheaton et al. (1987) used two GCMs to estimate a southern shift of the boreal forest by 250-920 km north, and a northern boundary shift of 80-730 km north. Sargent (1988) estimates a loss of about 100×10^6 ha. in the boreal forest as it shifts north.

Some scientists studying the vegetation of more humid climates, suggest the shift of vegetation regions is expected to be gradual. Solomon (1986) explored forest succession in eastern North America using a stand-level model. He predicted that temperature and precipitation changes resulted in distinctive dieback of trees that were near the boundaries of their natural range. The southern coniferous-deciduous boundary had a less distinct dieback with a rapid recovery of deciduous species. The northern transition areas showed vegetation changes that were sharp as boundary species were replaced with ingressing species. The new species were also sharply replaced by other species shifting northwards as climate change progressed. These responses continued for up to 300 years following the simulated climatic changes.

The above study and others focus on vegetation changes that occur from differences in growth regimes. However, these same climatic changes will also create different fire regimes. In the continental areas of the taiga, climatically stressed and killed vegetation will provide quantitative and qualitative changes in fuel. On most landscapes of the taiga of North America most of the branches and all of the boles of fire killed stands fall to the front floor within one or two decades and decompose slowly over many more decades. Under increased drought more of this fuel becomes available for burning during a second fire. In addition, under increased drought stress more soil organic matter becomes available from not only forest floors but also from wetland areas that support trees (eg. treed bogs) and other non-treed bog types.

5. CO_2 Feedback Loop to Global Warming

We have emphasized the importance of fire in the northern circumpolar forest but it has been estimated that global vegetation burning and forest conversion to agriculture releases 4 to 6×10^{12} kg of carbon to the atmosphere annually; of this, the circumpolar countries contribute only 1.2-1.8% and Canada contributes 0.50-0.75% (Crutzen and Andreae 1990). Stocks (1991) provided more detailed estimates of carbon release for the circumpolar boreal forests over the past decade. He estimated the burned area at 5.6×10^6 ha with Canada contributing 3.0×10^6 ha. If 25×10^3 kg ha^{-1} is an average level of fuel consumption and if 50% by weight of this fuel is carbon, then the carbon release was 7.0×10^{10} kg for the circumpolar countries and 3.8×10^{10} kg for Canada. The reader is referred to Cofer et al. (1990, this volume), Houghton (1991) and Levine (1991) for further details.

These average values are valuable but it must be emphasized that burns far exceed these averages in some years and there are many other factors not included in these calculations. There is much variability of fire intensity from hour to hour and there may be large unburned or lightly burned patches within burned areas (Eberhart and Woodard 1987). Deep burning fires in exceptionally dry years in forests with thick soil surface organic matter layers and in peatlands are not included. Multiple burning of areas is also not considered; during the first fire, trees are killed but not consumed, but in a second fire, the dry trees would be oxidized to release considerable carbon dioxide. In addition, the accelerated post-fire rate of carbon release by decomposition can contribute as much carbon to the atmosphere as during the fire. A.N.D. Auclair (personal communication) conducted a simulation experiment and concluded that emissions during and after biomass burning may have been a significant feedback loop to global warming in the past 10-15 years. Release from tropical deforestation is very rapid and has been emphasized in the literature but northern ecosystems continue to release CO_2 for one to two decades after the fire because of accelerated decomposition under warmer soil conditions.

In calculating global carbon budgets the scaling problems become even greater. Fire is a major force in the Eurasian boreal forest (Stocks 1991) yet, until recently, few data have been available on carbon release rates from those vegetation types which tend to burn with surface fires rather than high intensity crown fires as in North America; even the spatial and temporal nature of burned areas were not readily available.

6. Forest Fires and the Carbon Cycle

Although determination of carbon release levels by fire is of major interest in modelling the effects of fire on climate change, it must be remembered that forest fires are a natural event. Therefore, carbon release should be viewed as part of a larger ecological process, in this case it is carbon cycling.

Kurz et al. (1991, 1992) provide the first carbon budget model that deals with carbon cycling on a Canada-wide scale. This model has incorporated the effect of fire as a disturbance in terms of releasing carbon, as well as altering the rate of carbon uptake by vegetation growth following fire.

Using this modelling approach, disturbances such as fire are viewed as altering the rate of change in the net carbon balance rather than as just a source of release. The model uses three major biomass pools: forest biomass, forest soils, and forest products. There are five disturbance types including wildfire, stand mortality by insect, clearcut logging without slash burning, clearcut logging with slash burning, and partial cutting (Kurz et al. 1991, 1992).

The model uses a 10-year average of area burned during the 1980s to spatially allocate disturbance by fire. In this manner, the most recent fire statistics are utilized, and the wide fluctuations in yearly statistics is dampened to accommodate long-term averaging calculations. The level of fire impact is contained in disturbance matrices which describe the reallocation of carbon at the time of disturbance.

The disturbance matrices contain 12 carbon sources and 16 carbon sinks. Matrix coefficients quantify the proportion of carbon that is moved from the source to the sink. There is a separate disturbance matrix for each of 41 spatial units and disturbance types.

Operating the model for a one year period (1986) indicated that wildfire in Canada caused a release of 18.7 Tg of carbon into the atmosphere from forest biomass, and 14.3 Tg of carbon released into the atmosphere from forest soils (Kurz et al. 1992). Wildfire also caused a transfer of 20.9 Tg carbon from forest biomass to forest soils. To put these amounts into perspective, the net growth of forest biomass caused a carbon accumulation of 92.0 Tg, while peatlands acquired an additional 26.2 Tg of carbon. The Canadian forest sector had a calculated net carbon accumulation of 78.6 Tg for 1986.

Although the carbon released by fire into the atmosphere is small compared to the amount sequestered by biomass growth, it is still the major contributor of atmospheric carbon in the forest sector. However, it is not known what the net effect of wildfire is on the carbon budget. For instance, if wildfire was reduced by 50%, what difference would this have on the overall rate of carbon accumulation in forest biomass? Similarly, how would carbon accumulation rates change in forest biomass if the area burned by wildfire was doubled? Also, what effect would this have on the important role fire plays in shifting forest biomass carbon to the soil?

7. Vegetation Responses

Many predictions of biotic changes under climate change have and will be raised for discussion and testing but we will address only vegetation responses. Vegetation types, especially in nutrient-limited and energy-limited northern ecosystems, are conservative and show little change over time. Even under gradually increasing air temperatures the responses could be subtle, but increased growth and reproductive capacity might be recognized as early changes. The response of vegetation types could be much more dramatic following fire. Fire not only removes part of the vegetation but also changes the energy and nutrient budgets dramatically. Changes in species abundances and even species numbers will be highly probable. Our research group has explored several case studies that illustrate these points.

We have examined an analogue of climate change in northern marshes by modifying moisture regimes (and indirectly temperature) of the vegetation. We moved soil-vegetation blocks upslope to higher elevations where flooding was less frequent (Hogenbirk and Wein 1991). Little change in species abundance occurred over a two year period even though drought stress occurred. By adding the stress of fire there was a strong shift to opportunistic Eurasian weedy species (*Tanacetum vulgare* L., *Sonchus arvensis* L., *Cirsium arvense* [L.] Scop., *Chenopodium album* L.) especially under warmer soil conditions (Hogenbirk and Wein 1992).

A somewhat similar example of the impact of severe fire at the arctic forest-tundra boundary has been studied recently by our team. It is well established that this ecotone is strongly controlled by climate; tree advance and retreat has occurred periodically over centuries and millennia as climate changed. Over the short term of two to five decades tree establishment in the tundra is minor; however, after a 1968 deep-burning fire near Inuvik, Northwest Territories and a subsequent climatic warming trend, *Populus balsamifera*, and *Betula papyrifera* moved well beyond their pre-fire locations (Landhäusser and Wein 1993). In addition, Olsen (1993) has determined experimentally that the post-fire tree establishment success was not strongly influenced by changed moisture regimes (+30%, normal, -30%) and/or changed temperature regimes (+3°C, normal); the main overriding force in increasing tree seedling establishment success and growth was increased fire severity (deeper burning).

The above studies consider only one fire but under climate change scenarios an increased probability of fire may lead to several fires in less time than normal. Multiple fires will lead to rapid and stepwise rates of change, replacing species that are poorly adapted with new combinations of species that flourish under the new environmental conditions. Plant species that disseminate propagules over wide areas have high light requirements, rapid growth rates, reach maturity quickly, and will be most successful in moving northward.

Acknowledgements

Funding for the climate change program of the senior author includes operating grants from the Natural Sciences and Engineering Research Council, the University of Alberta Central Academic Fund and the Atmospheric Environment Service "Mackenzie Basin Impact Study". Logistic support and accommodation by the Polar Continental Shelf Project, the Inuvik Scientific Research Centre of the Science Institute of the Northwest Territories and Wood Buffalo National Park was critically important to the study.

References

Bolin, G., B.R. Döös, J. Jäger, and R.A. Warrick. 1986. The greenhouse effect, climate change and ecosystems. SCOPE 29. J. Wiley and Sons, Chichester, England.

Cofer, W.R. III, J.S. Levine, D.I. Sebacher, E.L. Winstead, P.J. Riggan, B.J. Stocks. 1990. Gaseous emissions from Canadian boreal forest fires. Atmos. Environ. 24A, 1635-1659.

Crutzen, P.J., and M.O. Andreae. 1990. Biomass burning in the tropics: impact on atmospheric chemistry and biogeochemical cycles. Science 250, 1669-1678.

Delisle, G.P., and R. Hall. 1987. Forest fire history maps of Alberta, 1931 to 1983. Canadian Forest Service, Northern Forest Centre, Edmonton, Alberta.

Di, Xueying, and Ende, Ju. 1990. The forest conflagration of May 1987 in Northeastern China. In: Fire in ecosystem dynamics (J.G. Goldammer and M.J. Jenkins, eds.), 169-174. SPB Academic Publ., The Hague.

Eberhart, K.E., and P.M. Woodard. 1987. Distribution of residual vegetation associated with large fires in Alberta. Can. J. For. Res. 17, 1207-1212.

Emanuel, W.R., H.H. Shugart, and M.P. Stevenson. 1985. Climatic change and the broad-scale distribution of terrestrial ecosystem complexes. Climatic Change 7, 29-43.

Flannigan, M.D., and C.E. Van Wagner. 1991. Climate change and wildfire in Canada. Can. J. For. Res. 21, 66-72.

Forestry Canada Fire Danger Group. 1992. Development and structure of the Canadian Forest Fire Behaviour Prediction System. Forestry Canada, Ottawa. Inf. Rep. ST-X-3.

Harvey, D.A., M.E. Alexander, and B. Janz. 1986. A comparison of fire-weather severity in northern Alberta during the 1980 and 1981 fire seasons. For. Chron. 62, 507-513.

Hirsch, K. 1991. A chronological overview of the 1989 fire season in Manitoba. For. Chron. 67, 358-365.

Hogenbirk, J.C., and R.W. Wein. 1991. Fire and drought experiments in northern wetlands: a climate change analogue. Can J. Bot. 69, 1991-1997.

Hogenbirk, J.C., and R.W. Wein. 1992. Temperature effects on seedling emergence from boreal wetland soils: implications for climate change. Aquatic Bot. 42, 361-373.

Houghton, R.A. 1991. Biomass burning from the perspective of the global carbon cycle. In: Global biomass burning: Atmospheric, climatic, and biospheric implications (J.S. Levine, ed.), 321-325. The MIT Press, Cambridge, Massachusetts.

Johnson, E.A. 1992. Fire and vegetation dynamics: studies from the north American boreal forest. Cambridge University Press, Cambridge. 129 p.

Kauppi, P., and M. Posch. 1988. A case study of the effects of CO_2-induced climatic warming on forest growth and the forest sector: A. Productivity reactions of northern boreal forests. In: The impact of climatic variations on agriculture. Vol 1: Assessments in cool temperature and cold regions, 183-195. Reidel, Dordrecht, The Netherlands. pp.

Kurz, W.A., M.J. Apps, T.M. Webb, and P.J. McNamee. 1991. The contribution of biomass burning to the carbon budget of the Canadian forest sector: a conceptual model. In: Global biomass burning: Atmospheric, climatic, and biospheric implications (J.S. Levine (ed.), 339-344. The MIT Press, Cambridge, Massachusetts.

Kurz, W.A., M.J. Apps, T.M. Webb, and P.J. McNamee. 1992. The carbon budget of the Canadian forest sector: phase I. Forestry Canada, North. For. Cent., Edmonton, Alberta. Inf. Rep. NOR-X-326.

Kuusela, K. 1992. The boreal forests: an overview. Unasylva. 170: 3-13.

Landhäusser, S.-M., and R.W. Wein. 1993. Post-fire vegetation recovery and tree establishment at the arctic treeline: climate change-vegetation response hypotheses. J. Ecol. 81, 665-672.

Levine, J.S. (ed.) 1991. Global biomass burning: Atmospheric, climatic, and biospheric implications. The MIT Press, Cambridge, Massachusetts.

Maxwell, B. 1992. Arctic climate: potential for change under global warming. In: Arctic ecosystems in a changing climate (F.S. Chapin, R.L. Jefferies, J.F. Reynolds, G.R. Shaver, and J. Svoboda, eds.), 11-34. Academic Press, New York.

Olsen, S.K. 1993. Fire severity and seedling growth at treeline: a climate change analogue. M.Sc. Thesis, University of Alberta.

Overpeck, J.T., D. Rind, and R. Goldberg. 1990. Climate-induced changes in forest disturbances and vegetation. Nature (London) 343, 51-53.

Pyne, S.J. 1984. Introduction to wildland fire management in the United States. John Wiley and Sons, New York, 455 pp.

Roots, E.F. 1989. Climate change: high latitude regions. Climatic Change 15, 223-253.

Sargent, N.E. 1988. Redistribution of the Canadian boreal forest under a warmed climate. Climatological Bulletin 22 (3), 23-34.

Solomon, A.M. 1986. Transient response of forests to CO_2-induced climate change: simulation modelling experiments in eastern North America. Oecologia 68, 567-579.

Stocks, B.J. 1990. Global warming and the forest fire business in Canada. In: Proc. Canada/US Symp. on the Impacts of Climatic Change and Variability on the Great Plains (11-13 Sept. 1990, Calgary, Alberta), pp. 223-229.

Stocks, B.J. 1991. The extent and impact of forest fires in northern circumpolar countries. In: Global biomass burning: Atmospheric, climatic, and biospheric implications (J.S. Levine, ed.), 197-202. The MIT Press, Cambridge, Massachusetts.

Street, R.B. 1989. Climate change and forest fires in Ontario. In: Proceedings of the 10*th* Conference on Fire and Forest Meteorology, 17-21 Apr. 1989 (D.C. MacIver, H. Auld, and R. Whitewood, eds.), 177-182. Forestry Canada, Ottawa.

Van Wagner, C.E. 1970. Conversion of Williams's Severity Rating for use with the Fire Weather Index. Can. For. Serv. Petawawa Natl. For. Inst. Inf. Rep. PS-X-21.

Van Wagner, C.E. 1987. Development and structure of the Canadian Forest Fire Weather Index System. Can. For. Serv. For Tech. Rep. 35.

Van Wagner, C.E. 1988. The historical pattern of annual area burned in Canada. For. Chron. 64, 182-185.

Wein, R.W. and D.A. MacLean (eds.). 1983. The role of fire in northern circumpolar ecosystems. John Wiley & Sons, Toronto.

Wheaton, E.E., T. Singh, R. Dempster, K.O. Higginbotham, J.P. Thorpe, G.C. Van Kooten and J.S. Taylor. 1987. An exploration and assessment of the implications of climatic change for the boreal forest and forestry economies of the prairie provinces and Northwest Territories: phase one. SRC Tech. Rep. No. 211, Publ. No. E-906-36-B-87, Saskatchewan Res. Counc., Saskatoon, Saskatchewan.

Williams, D.E. 1959. Fire Season Severity Rating. Dep. Nor. Affairs and Nat. Res. For. Branch, For. Res. Div. Tech. Note No. 73. 13 p.

Yun-Qian, Y., and J. Ji-Zhong. 1989. The climatic background and weather conditions of 1987 Daxinganling extreme forest fire in China. In: Proceedings of the 10*th* Conference on Fire and Forest Meteorology, 17-21 Apr. 1989 (D.C. MacIver, H. Auld, and R. Whitewood, eds.), 364-376. Forestry Canada, Ottawa.

J.G.Goldammer and V.V.Furyaev (eds.), Fire in Ecosystems of Boreal Eurasia, 513-524.

ANNEX I

Understanding Boreal Ecosystems
Opening Remarks at the International Conference "Fire in Ecosystems of Boreal Eurasia", Krasnoyarsk, 29 June - 2 July 1993

Eldon W. Ross [1]

Introduction

Communities have always recognized certain lands, such as parks, as belonging to the community as a whole. These common lands were shared and maintained by all. Ehrlich (1988) has proposed a global commons shared and maintained by all of humanity as consisting of the atmosphere, the oceans and diversity of nature. Ehrlich maintains that since we all share these global commons, that we all have the responsibility to maintain them in order to provide security and sustainable development. I add to Ehrlich's list of global commons, the forests of the world and global economics. Recognition and acceptance of these global commons by the world's leaders at the United Nations Conference on Environment and Development in June 1992 at Rio de Janeiro (the Rio Earth Summit) was marked by the signing of the Climate Convention, the Biodiversity Convention, the Forestry Principles; and our framework for action, Agenda 21. Concern for the world's forests can be found in each of the agreements reached at Rio. As examples; the Preamble of the Forestry Principles states that forests are essential to economic development and the maintenance of all forms of life, while Principle 4 identifies the vital role of all types of forests in maintaining the ecological processes and balance at the local, national, regional and global levels through ... their role in protecting fragile ecosystems, watersheds and freshwater resources and as rich storehouses of biodiversity... Principle 12 states ... capacities in education, training, science, technologies, economics, anthropology and social aspects of forests and forest management are essential to the conservation and sustainable development of forests.

Clearly, the Rio Earth Summit focused our attention on critical environmental issues for the 1990s. Global climate change and the carbon budget, sustainable development, biodiversity, and the relationship between people and the environment were on all participant's agendas in Rio. Associated with these issues are growing demands for water resources, both water quality and quantity, and concerns for pollution impacts. The Rio Earth Summit also recognized the inter-connectedness of the environment, and society through the focus on sustainable development. This inter-connectedness is more clearly

---

[1] Associate Deputy Chief for Research, USDA Forest Service (retired), and President, International Boreal Forest Research Association (IBFRA) (June 1991-February 1994)

defined as taking an ecosystem perspective of natural resources. This implies that an integrated multidisciplinary science approach is needed. That is, the social and natural sciences are structured to work towards common goals and objectives.

The Rio Earth Summit and the issues of the 1990s bear strongly on boreal forests. Most food and fiber (except sea food) come from terrestrial ecosystems. One third of the earth is forest covered, and of those forests, one fourth are boreal forests. Boreal forests may be more important than tropical forests in addressing the issues of the 1990's and present more opportunities for science based management. Climate change is expected to be most pronounced at high latitudes. Carbon in boreal forests is as great or greater than tropical forests if soil carbon is included. For the most part, boreal forests are largely undeveloped, yet are found in developed nations. Opportunities to achieve the Rio agenda with integrated multidisciplinary science are great in boreal forests. The science infrastructure in the nations of the boreal forest are highly advanced and can be focused on both understanding ecosystem processes and use of that science to develop and manage boreal forests. Understanding the role that disturbance from insects, disease and fire play in ecosystem processes, and in the future management of the boreal forest is crucial. This workshop, which summarizes our understanding of fire in the circumpolar boreal forest, and the research campaign to improve our understanding of fire from an integrated multidisciplinary science perspective is an important step in addressing the Rio agenda and the issues of the 1990's. It is particularly heartening to see the human dimension integrated into this program.

The International Boreal Forest Research Association

The International Boreal Forest Research Association (IBFRA) provides a vehicle for an integrated multidisciplinary science effort to focus on the boreal forest and to bring that science to bear on management of those forests. IBFRA was established as a result of recommendations from a 1990 science conference in Arkhangelsk. In 1991, representatives from then USSR, Canada, Finland, the United States and the United Nations Environmental Program met in the Ukraine to establish the Association. Current membership in IBFRA consists of Canada, Russia, Finland, Norway, Sweden, and the United States of America. Participation in IBFRA is not limited to governmental agencies in the member states, and in fact includes non-governmental organizations. The Working Group level is open to scientists from non-member countries.

Goals and objectives of IBFRA are to promote research, cooperation, communication, and transfer of knowledge to problems of managing boreal forests. Three functional activities were established to meet these goals and objectives: (1) inventory, monitoring and classification; (2) forest management and biodiversity; and (3) global change and ecosystem function. Within each of these functional areas, working groups have been established to carry out the specific activities and programs. The Stand Replacement Fire Working Group, one of the sponsors of this workshop and experimental campaign, is one of these working groups. The Stand Replacement Fire Working Group is doing exactly what IBFRA was established for through this workshop and experiment: research, cooperation, communication, and transfer of knowledge. This Stand Replacement Fire Working Group is an example to other working groups within IBFRA.

Closing Thoughts

There are three thoughts that I want to leave with you:

- Forests are part of the global commons, along with the atmosphere, the oceans, biodiversity, and the global economy.

- The Rio Earth Summit provided the environmental agenda for the 1990s: sustainable development, global change, protection of our renewable natural resources, and the necessity to consider the human dimensions in natural resources.

- Environmental research will, of necessity, be integrated, multidisciplinary, and include the social sciences as well as the natural sciences.

Reference

Ehrlich, P.R. 1988. The global commons and national security. In: A challenge for science and society in the 21st century (A.Berger, S.Schneider, and J.Cl. Duplessy, eds.), pp. 553-562. NATO ASI Series C Mathematical and Physical Sciences, Climate and Geosciences, Vol. 285. Kluwer Academic Publ., Dordrecht.

ANNEX II

International Boreal Forest Research Association (IBFRA) Stand Replacement Fire Working Group [2]

The **International Boreal Forest Research Association** (IBFRA) established a working group on stand replacement ecology with an emphasis on fire in ecosystems at an Association meeting in Krasnoyarsk Russia on 18-22 May 1992. The working group will foster cooperative research on stand replacement ecoloy with the goal of better understanding boreal ecosystems and assuring sustainable development and exploitation of resources in the taiga. The International Boreal Forest Research Association is composed of forestry research organizations in the United States, Canada, Russian Federation, Norway, Sweden and Finland.

Two documents were developed and signed at this meeting. The first, known as the Protocol, establishes the long term frame of reference for the working group and details eight hypotheses that will guide cooperative research among the participants. The eight hypotheses are related to quantitatively understanding boreal ecosystems, the role of fire in boreal ecosystems, and modeling and predicting boreal forest dynamics. The Protocol will be implemented through mutually agreed annual work plans. A work plan for 1992-1993 was developed and signed in Krasnoyarsk. The eight hypotheses specified in the Protocol are:

1. Large forest fire behavior can be quantitatively related to weather conditions and forest fuel types in boreal forests of North America and Siberia.

2. Spatial and temporal cycles of large fires in boreal forests of Siberia and North America are similar.

3. The influence of stand replacement fires on structure and function of Siberian and North American boreal forests are similar.

4. Ecosystem processes, such as nutrient cycling, soil respiration, and carbon accretion quickly recover to pre-fire rates in fire cycled boreal forests.

5. Ecosystem responses (nutrient cycling, species composition, biological diversity, vegetative cover, etc.) following stand replacement fire is unchanged by salvage logging.

6. Many small fire events will substitute for single large fire events in terms of species and ecosystem response effects.

---

[2] This summary is based on the report by Fosberg (1992)

7. Predict and develop a strategy for an enlarged role for fire management in response to increased fire activity under global climate change, including an increased opportunity to use prescribed fire to advance natural ecosystem response to mitigate the effects of climate change.

8. Models for forest resources and forest protection can be constructed and applied for conditions in North America and Siberia.

The annual work plan for 1992-1993 contains seven tasks. Five tasks were agreed upon for hypothesis 1. These include an exchange of data to evaluate seasonal patterns of fire danger and the behavior of large fires, an intercomparison of fire danger rating systems, and an experimental program of fire behavior.

Under hypothesis 3, it was agreed to conduct a field study of post fire recovery in the United States with both Russian and American scientists. Also, the United States and Canada will pursue obtaining a satellite downlink (NOAA AVHRR) image analysis system in Krasnoyarsk for monitoring and understanding forest fires from a global perspective. Resulting data will be shared and available to the scientific community.

The International Boreal Forest Research Association was first proposed at a meeting of the International Panalon Boreal Forests in Arkhangelsk, Russia. In June 1991, an organizational meeting was held in the Ukraine between the United States, Canada, and Russia. Member countries are represented by coordinators for priority research areas. Two priority areas were identified at this organizational meeting: Inventory and Monitoring, including classification; and Global Climate Change and Ecosystem Function. Norway, Sweden and Finland have joined the Association. The Stand Replacement Fire Working Group is the first Working Group established by the Association.

Reference

Fosberg, M. 1992. International Boreal Forest Research Association, Stand Replacement Fire Working Group. Int. Forest Fire News (ECE/FAO) No.7 (August 1992), 6-8.

ANNEX III

Fire Research in the Boreal Forests of Eurasia: A Component of a Global Fire Research Program

1. Introduction

The International Geosphere-Biosphere Program (IGBP) provides a scientific-organizational framework for interdisciplinary research programs devoted to explore the fundamental relationships between ecosystems (terrestrial and marine), atmosphere and climate. Within the International Global Atmospheric Chemistry (IGAC) Project, one of several IGBP "Core Projects", one of the activities of its "Focus 2: Natural Variability and Anthropogenic Perturbations of the Tropical Atmospheric Chemistry" is devoted to investigate the impact of biomass burning on the atmosphere and biosphere ("Biomass Burnig Experiment" [BIBEX]). The goals of the BIBEX research activities are summarized as follows (IGAC 1992; Goldammer 1994; Goldammer and Furyaev 1995):

- To characterize the production of chemically and radiatively important gases and aerosols species from biomass burning to the global atmosphere.

- To assess the consequences of biomass burning on regional and global atmospheric chemistry and climate.

- To determine the short- and long-term effects of fire on post-fire exchanges of trace gases between terrestrial ecosystems and the atmosphere.

- To understand the biogeochemical consequences of atmospheric deposition of products of biomass burning.

2. Examples of Past, Ongoing and Planned International Fire Research Campaigns

2.1 The First Intercontinental Fire Experiment: The Southern Tropical Atlantic Regional Experiment (STARE)

The first BIBEX research activity was the **Southern Tropical Atlantic Regional Experiment (STARE)**, an aircraft- and ground-based measurement program initiated in 1990. Its objective was to investigate the sources of trace gases, their atmospheric transport, and the chemical processes in the atmosphere which lead to elevated levels of ozone, CO, and other trace gases, as well as pyrogenic aerosols over the southern tropical Atlantic Ocean. Experience collected by fire research campaigns in Brazil (Artaxo et al. 1993) and in West Africa (Lacaux et al. 1993; Lacaux 1995) were important precursor events which facilitated this first intercontinental fire experiment.

The STARE research activities of 1992 consisted of two major components, TRACE-A and SAFARI. The **Transport and Atmospheric Chemistry Near the Equator - Atlantic (TRACE-A)** program covers the western area of STARE-92. The main objective of TRACE-A was to investigate the chemical composition, transport and chemistry of the atmosphere over the southern tropical Atlantic Ocean and the adjacent South American continent. The research focused on understanding the seasonal enhancement in tropospheric ozone that has been observed over the tropical Atlantic Ocean and that is caused most likely by widespread vegetation fires (forest conversion, savanna fires) in South America and Southern Africa. It also investigated the potential contribution of land-use changes in both continents on the observed levels of other trace gases such as CH_4 and N_2O.

The **Southern African Fire-Atmosphere Research Initiative (SAFARI)** was primarily the eastern component of STARE-92. The research objectives of SAFARI were similar to TRACE-A. SAFARI-92 involved ground and airborne chemical and meteorological measurements in the source or near-source regions of South Africa, which were carried out primarily as a cooperative international campaign. In addition it included the investigation of the ecological role of fire in African savannas and the study of trace gas emissions from soils on fire-affected sites.

The results of STARE have verified the hypotheses described above. The results are in the stage of publication at present (Special issue of the Journal Geophysical Research (ca. fall 1996; van Wilgen et al. 1996).

2.2 The Fire Research Campaign Asia-North (FIRESCAN)

The concept of the **Fire Research Campaign Asia-North (FIRESCAN)** was first designed in 1991 as a more or less bilateral initiative between Germany and the Soviet Union. It gradually developed into an "in tandem" exercise of a symposium with field campaign financed by the Volkswagen Foundation and co-sponsored by IGAC/BIBEX and the International Boreal Forest Research Association (IBFRA). In the first part of FIRESCAN the updated state of knowledge on the fire ecology of the Taiga was compiled and discussed in the international conference on "Fire in Ecosystems of Boreal Eurasia" (this volume).

The objective of the first phase of the field campaign is the investigation of a high-intensity stand-replacement forest fire by using methods developed in the East and the West. The fire experiment (**The Bor Forest Island Fire**) took place in the Northern Krasnoyarsk Region (6 July 1993). Besides the comparison of descriptive methods (e.g. vegetation classification, fuel quantification, fire behaviour, etc.) attempts were made to recover historical fire and climate data (tree ring and lake sediment analyses). In addition trace gases and aerosols emitted from the experimental fire were measured by NASA, Max Planck Institute for Chemistry and the Novosibirsk Institute, using sampling devices mounted on Russian airborne platforms. The first phase of FIRESCAN will be followed by long-term research (1994 onward) and provide the fire research component of the upcoming IGBP Northern Eurasia Study (Steffen and Shvidenko 1996).

The first results of the Bor Forest Island Fire Experiment are published (FIRESCAN Science Team 1994, 1996).

2.3 The South East Asia Fire Experiment (SEAFIRE)

In the late 1990s it is envisaged to set up a new BIBEX research focus in Southeast Asia. Large-scale forest fire activities in tropical SE Asia first became visible during the extreme drought period of 1982-83 which was a consequence of the El Niño-Southern Oscillation (ENSO) event (synthesis cf. Goldammer et al. 1996; cf Goldammer 1990, 1993). The interactions of interannual climate variability and the escalating human pressure on South East Asia's tropical forest resources led to an unprecedented amount of wildfires which severely damaged several millions ha of primary and secondary tropical forest in the whole region. Again in 1987, 1991 and 1994 smoke emitted from forest conversion fires in Kalimantan (Borneo island) raised international concern about the negative impacts of wildland fire emissions on human health and safety and on atmosphere and climate.

The regularly occuring smog layers over insular and mainland South East Asia are originating from a variety of vegetation types:

- Slash-and-burn agriculture: Traditional but expanding small-scale clearing of primary forest and secondary vegetation in the perhumid rain forest zone

- Other forest conversion fires: Large-scale clearcuts of primary and secondary rain forest vegetation and subsequent slash burning for conversion into other land-use types (e.g. exploitation of mineral, coal and oil resources; conversion of natural vegetation into agricultural plantations and human-made forests)

- Regularly occuring fires in seasonally dry deciduous and semideciduous forests (monsoon forests, "savanna" forests) in mainland South Asia

- Regularly occuring fires in the submontane and montane coniferous (pine) forests of insular and mainland South Asia

- Agricultural residue burning, mainly rice straw, throughout the whole region

The South East Asia Fire Experiment (SEAFIRE) will be designed to investigate this highly complex and diverse fire theater. The aim of a set of experiments will be to identify the magnitude, patterns, quality and impacts of fire on the local terrestrial and regional atmospheric ecology. Unlike the conceptual design of STARE, the SEAFIRE campaign will not entirely be conducted as a set of short-term and directly linked experiments. Infrastructural constraints of observations to be carried out in a variety of countries (Myanmar, Thailand, Kampuchea, Lao, Viet Nam, Malaysia, People's Republic of China, The Philippines, Indonesia) will require a multi-year approach. Hoewever, it is aimed to concentrate research on a large-scale smog situation over the region. This will require a set of well prepared stand-by instrumentation and personnel.

One of the main objectives of SEAFIRE will be to investigate how the zonal circulation along the equator, the Walker Circulation, which set in motion by the tropical convective clouds, is influencing the regional and global distribution of pyrogenic trace gases and aerosols. Th so-called "hot towers" act as cylinders in the global heat engine to pump energy from the tropical sea and land surfaces to great heights before that energy can be exported poleward.

In the "High Phase" deep clouds and rain are centered over the hot spots. In the "Low Phase" the warm waters of the Warm Pool are transported to the eastern Pacific and the circulations are reversed. The sequence of events described above is the evolution of the Southern Oscillation in which the strong trades of the High Phase are replaced by the weak trades of the Low Phase and the El Niño is born. The repercussions of these events in the Pacific are global in extent, triggering droughts and fires in the Ocean Continent, Australia, South America and Africa.

SEAFIRE conducted under High Phase conditions will carry biomass burnig products to the high troposphere and export these products to the global system. Direct and indirect effects to the global system will follow in terms of trace gas chemistry, Greenhouse gases and the effects of aerosols. SEAFIRE conducted under Low Phase conditions will exercise a substantially different effect upon the global syystem. Products of biomass burning are likely to be trapped in the lower atmosphere, have different effects in terms of Greenhouse gas and aerosols than in the High Phase and be connected to differing parts of the global system.

Plans are prepared at present to conduct SEAFIRE as a sub-component of the planned IGBP Southeast Asia Land-Use Change Study.

References

Artaxo, P., M.A. Yamasoe, J.V. Martins, S. Kocinas, S. Carvalho, and W. Maenhaut. 1993. Case study of atmospheric measurements in Brazil: Aerosol emissions from Amazon Basin fires. In: Fire in the environment: The ecological, atmospheric, and climatic importance of vegetation fires (P.J.Crutzen and J.G.Goldammer, eds.), pp.139-158. John Wiley & Sons, Chichester-New York, 400 pp.

FIRESCAN Science Team. 1994. Fire in Boreal Ecosystems of Eurasia: First results of the Bor Forest Island Fire Experiment, Fire Research Campaign Asia-North (FIRESCAN). World Resource Review 6, 499-523.

FIRESCAN Science Team. 1996. Fire in ecosystems of boreal Eurasia: The Bor Forest Island Fire Experiment, Fire Research Campaign Asia-North (FIRESCAN). In: Biomass burning and global change (J.S.Levine, ed.). The MIT Press, Cambridge, MA (in press).

Goldammer, J.G. (ed.) 1990. Fire in the tropical biota. Ecosystem processes and global challenges. Ecological Studies 84, Springer-Verlag, Berlin-Heidelberg-New York, 497 pp.

Goldammer, J.G. 1993. Feuer in Waldökosystemen der Tropen und Subtropen. Birkhäuser-Verlag, Basel-Boston, 251 pp.

Goldammer, J.G. 1994. Interdisciplinary research projects for developing a global fire science. A paper presented at the 12th Confer. Forest Meteorology, October 26-28, 1993, Jekyll Island, Georgia. Society of American Foresters Publ. 94-02, 6-22.

Goldammer, J.G., and V.V. Furyaev. 1995. Global change, the boreal forest, and fire: Search for new strategies in science policies and research mechanisms. Science Policy: New Mechanisms for Scientific Collaboration between East and West (V.A.Koptyug and J.Klerkx, eds.), 45-61. NATO ASI Series 4, Science and Technology Policy Vol.1. Kluwer Academic Publishers, Dordrecht-Boston-London, 256 p.

Goldammer, J.G., B.Seibert, and W.Schindele. 1996. Fire in dipterocarp forests. In: Dipterocarp forest ecosystems: Towards sustainable management (A.Schulte, ed.). World Scientific Publ., Singapore (in press).

IGAC (International Global Atmospheric Chemistry Project). 1992. Biomass Burning Experiment: Impact on the Atmosphere and Biosphere. An Activity of the International Global Atmospheric Chemistry (IGAC) Project, IGAC Core Project Office, The MIT, Cambridge, MA, 19 pp. + App.

Lacaux, J.P., H. Cachier, and R. Delmas. 1993. Biomass burning in Africa: An overview of its impact on atmopcheric chemistry. In: Fire in the environment: The ecological, atmospheric, and climatic importance of vegetation fires (P.J.Crutzen and J.G.Goldammer, eds.), pp.159-191. John Wiley & Sons, Chichester-New York, 400 pp.

Lacaux, J.P. (ed.). FOS/DECAFE '91 Experiment. Special issue J. Atmospheric Chemistry 22 (1&2, October 1995), 242 pp.
Steffen, W.L., and A.Z. Shvidenko (eds.). 1996. The IGBP Northern Eurasia Study. Prospectus for integrated global change research. The International Geosphere-Biosphere Programme: A Study of Global Change (IGBP) of the International Council of Scientific Unions (ICSU), IGBP Stockholm (in press).
van Wilgen, B., M.O.Andreae, J.G. Goldammer, and J. Lindesay (eds.) 1996. Fire in Southern African savannas. Ecological and atmospheric perspectives. The University of Witwatersrand Press (in press).

Tab.1. Acronyms of regional and global fire research activities (Fig.1 and 2)

List of Acronyms of Regional and Global Fire Research Activities

ABLE	Amazon Boundary Layer Experiment (→ GTE)
BIBEX	Biomass Burning Experiment (→ IGAC)
EXPRESSO	Experiment for Regional Sources and Sinks of Oxidants
FIRE	Fire in Global Resource and Environmental Monitoring (CEC-JRC)
FIRESCAN	Fire Research Campaign Asia - North (→ IGAC)
FIRESCHEME	Fire Information Systems Research in the Ecology, Socio-Culture and History of the Mediterranean Environment
FOS/DECAFE	Fire of Savannas/Dynamique et Chimie Atmosphérique en Forêt Équatoriale
GEIA	Global Emissions Inventory Activity (→ IGAC)
GIS	Geographic Information System
GTE	Global Tropospheric Experiment
IBFRA	International Boreal Forest Research Association
ICSU	International Council of Scientific Unions
IGAC	International Global Atmospheric Chemistry Project (→ IGBP)
IGBP	International Geosphere-Biosphere Programme (→ ICSU)
SAFARI	Southern African Fire-Atmosphere Research Initiative (→ IGAC)
SCAR	Smoke, Clouds, Aerosols, and Radiation Experiment
SEAFIRE	South East Asian Fire Experiment (→ IGAC)
STARE	Southern Tropical Atlantic Regional Experiment (→ IGAC)
SRFWG	Stand Replacement Fire Working Group (→ IBFRA)
TRACE-A	Transport and Atmospheric Chemistry Near the Equator - Atlantic
VFIS	Vegetation Fire Information System (→ Dahlem Konferenz Model)

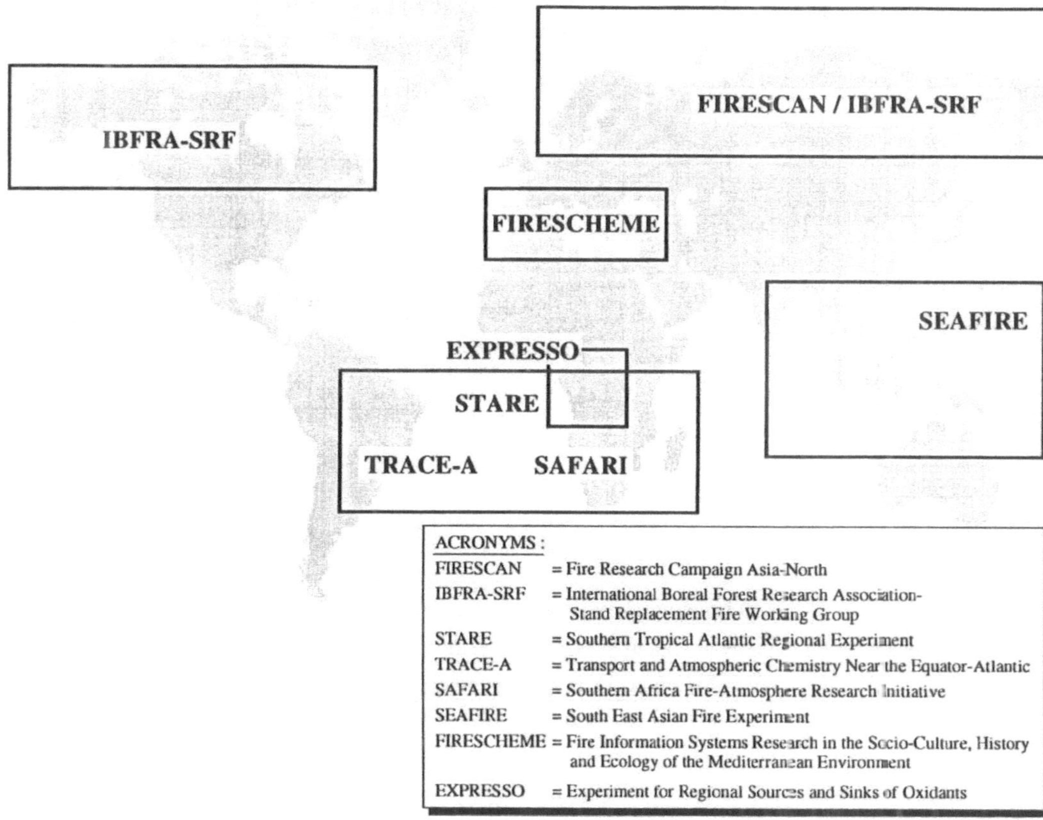

Fig. 1. World map showing the target areas of regional fire research campaigns which are in the stage of implemementation or planning (after Goldammer 1994).

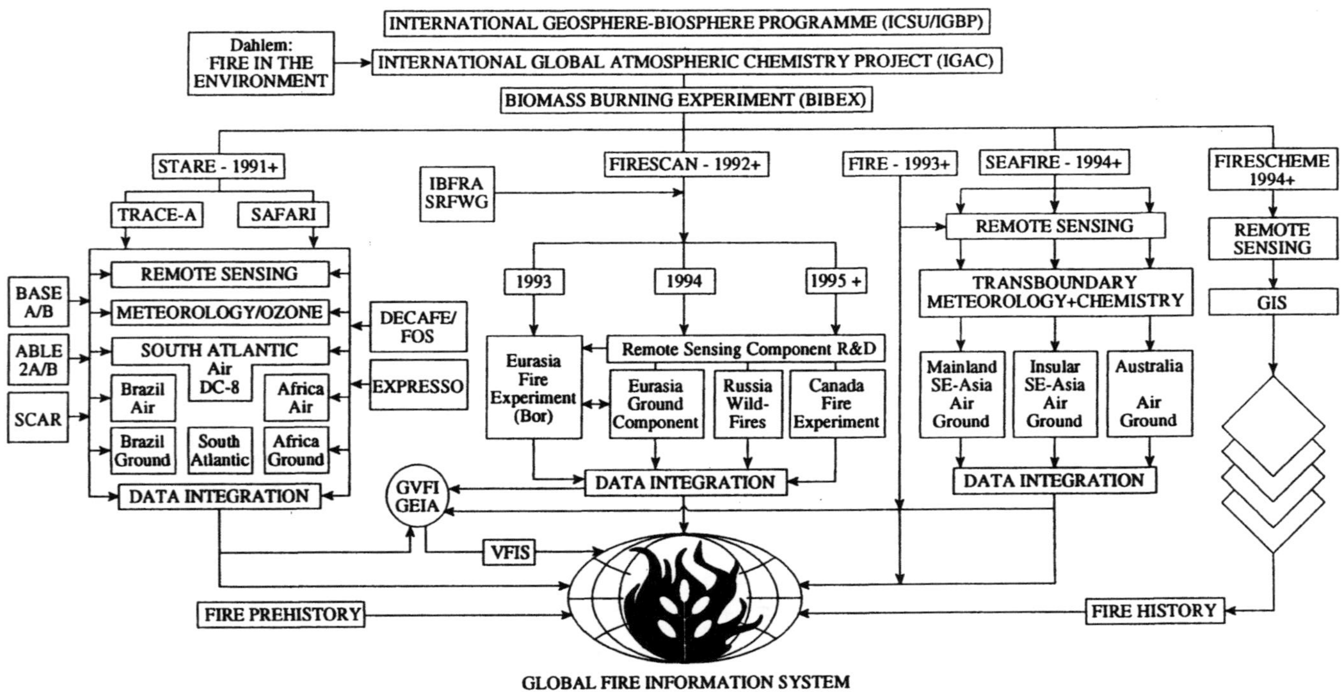

Fig.2. Organizational diagram of fire research activities which will be integrated into the development of a global fire model. Acronyms are explained in Table 1 (after Goldammer and Furyaev 1995).

Index

FORESTRY SCIENCES

1. P. Baas (ed.): *New Perspectives in Wood Anatomy.* Published on the Occasion of the 50th Anniversary of the International Association of Wood Anatomists. 1982
ISBN 90-247-2526-7
2. C.F.L. Prins (ed.): *Production, Marketing and Use of Finger-Jointed Sawnwood.* Proceedings of an International Seminar Organized by the Timber Committee of the UNECE (Halmar, Norway, 1980). 1982 ISBN 90-247-2569-0
3. R.A.A. Oldeman (ed.): *Tropical Hardwood Utilization.* Practice and Prospects. 1982
ISBN 90-247-2581-X
4. P. den Ouden (in collaboration with B.K. Boom): *Manual of Cultivated Conifers.* Hardy in the Cold- and Warm-Temperate Zone. 3rd ed., 1982
ISBN Hb 90-247-2148-2; Pb 90-247-2644-1
5. J.M. Bonga and D.J. Durzan (eds.): *Tissue Culture in Forestry.* 1982
ISBN 90-247-2660-3
6. T. Satoo: *Forest Biomass.* Rev. ed. by H.A.I. Madgwick. 1982 ISBN 90-247-2710-3
7. Tran Van Nao (ed.): *Forest Fire Prevention and Control.* Proceedings of an International Seminar Organized by the Timber Committee of the UNECE (Warsaw, Poland, 1981). 1982 ISBN 90-247-3050-3
8. J.J. Douglas: *A Re-Appraisal of Forestry Development in Developing Countries.* 1983
ISBN 90-247-2830-4
9. J.C. Gordon and C.T. Wheeler (eds.): *Biological Nitrogen Fixation in Forest Ecosystems.* Foundations and Applications. 1983 ISBN 90-247-2849-5
10. M. Németh: *Virus, Mycoplasma and Rickettsia Diseases of Fruit Trees.* Rev. (English) ed., 1986 ISBN 90-247-2868-1
11. M.L. Duryea and T.D. Landis (eds.): *Forest Nursery Manual.* Production of Bareroot Seedlings. 1984; 2nd printing 1987 ISBN Hb 90-247-2913-0; Pb 90-247-2914-9
12. F.C. Hummel: *Forest Policy.* A Contribution to Resource Development. 1984
ISBN 90-247-2883-5
13. P.D. Manion (ed.): *Scleroderris Canker of Conifers.* Proceedings of an International Symposium on Scleroderris Canker of Conifers (Syracuse, USA, 1983). 1984
ISBN 90-247-2912-2
14. M.L. Duryea and G.N. Brown (eds.): *Seedling Physiology and Reforestation Success.* Proceedings of the Physiology Working Group, Technical Session, Society of American Foresters National Convention (Portland, Oregon, USA, 1983). 1984
ISBN 90-247-2949-1
15. K.A.G. Staaf and N.A. Wiksten (eds.): *Tree Harvesting Techniques.* 1984
ISBN 90-247-2994-7
16. J.D. Boyd: *Biophysical Control of Microfibril Orientation in Plant Cell Walls.* Aquatic and Terrestrial Plants Including Trees. 1985 ISBN 90-247-3101-1
17. W.P.K. Findlay (ed.): *Preservation of Timber in the Tropics.* 1985
ISBN 90-247-3112-7
18. I. Samset: *Winch and Cable Systems.* 1985 ISBN 90-247-3205-0

FORESTRY SCIENCES

19. R.A. Leary: *Interaction Theory in Forest Ecology and Management.* 1985 ISBN 90-247-3220-4
20. S.P. Gessel (ed.): *Forest Site and Productivity.* 1986 ISBN 90-247-3284-0
21. T.C. Hennessey, P.M. Dougherty, S.V. Kossuth and J.D. Johnson (eds.): *Stress Physiology and Forest Productivity.* Proceedings of the Physiology Working Group, Technical Session, Society of American Foresters National Convention (Fort Collins, Colorado, USA, 1985). 1986 ISBN 90-247-3359-6
22. K.R. Shepherd: *Plantation Silviculture.* 1986 ISBN 90-247-3379-0
23. S. Sohlberg and V.E. Sokolov (eds.): *Practical Application of Remote Sensing in Forestry.* Proceedings of a Seminar on the Practical Application of Remote Sensing in Forestry (Jönköping, Sweden, 1985). 1986 ISBN 90-247-3392-8
24. J.M. Bonga and D.J. Durzan (eds.): *Cell and Tissue Culure in Forestry.* Volume 1: General Principles and Biotechnology. 1987 ISBN 90-247-3430-4
25. J.M. Bonga and D.J. Durzan (eds.): *Cell and Tissue Culure in Forestry.* Volume 2: Specific Principles and Methods: Growth and Development. 1987 ISBN 90-247-3431-2
26. J.M. Bonga and D.J. Durzan (eds.): *Cell and Tissue Culure in Forestry.* Volume 3: Case Histories: Gymnosperms, Angiosperms and Palms. 1987 ISBN 90-247-3432-0
    Set ISBN (Volumes 24-26) 90-247-3433-9
27. E.G. Richards (ed.): *Forestry and the Forest Industries: Past and Future.* Major Developments in the Forest and Forest Industries Sector Since 1947 in Europe, the USSR and North America. In Commemoration of the 40th Anniversary of the Timber Committee of the UNECE. 1987 ISBN 90-247-3592-0
28. S.V. Kossuth and S.D. Ross (eds.): *Hormonal Control of Tree Growth.* Proceedings of the Physiology Working Group, Technical Session, Society of American Foresters National Convention (Birmingham, Alabama, USA, 1986). 1987 ISBN 90-247-3621-8
29. U. Sundberg and C.R. Silversides: *Operational Efficiency in Forestry.* Vol. 1: Analysis. 1988 ISBN 90-247-3683-8
30. M.R. Ahuja (ed.): *Somatic Cell Genetics of Woody Plants.* Proceedings of the IUFRO Working Party S2.04-07 Somatic Cell Genetics (Grosshansdorf, Germany, 1987). 1988 . ISBN 90-247-3728-1
31. P.K.R. Nair (ed.): *Agroforestry Systems in the Tropics.* 1989 ISBN 90-247-3790-7
32. C.R. Silversides and U. Sundberg: *Operational Efficiency in Forestry.* Vol. 2: Practice. 1989 ISBN 0-7923-0063-7
    Set ISBN (Volumes 29 and 32) 90-247-3684-6
33. T.L. White and G.R. Hodge (eds.): *Predicting Breeding Values with Applications in Forest Tree Improvement.* 1989 ISBN 0-7923-0460-8
34. H.J. Welch: *The Conifer Manual.* Volume 1. 1991 ISBN 0-7923-0616-3
35. P.K.R. Nair, H.L. Gholz, M.L. Duryea (eds.): *Agroforestry Education and Training.* Present and Future. 1990 ISBN 0-7923-0864-6
36. M.L. Duryea and P.M. Dougherty (eds.): *Forest Regeneration Manual.* 1991 ISBN 0-7923-0960-X

FORESTRY SCIENCES

37. J.J.A. Janssen: *Mechanical Properties of Bamboo.* 1991 ISBN 0-7923-1260-0
38. J.M. Bonga and P. Von Aderkas: *In Vitro Culture of Trees.* 1992 ISBN 0-7923-1540-5
39. L. Fins, S.T. Friedman and J.V. Brotschol (eds.): *Handbook of Quantitative Forest Genetics.* 1992 ISBN 0-7923-1568-5
40. M.J. Kelty, B.C. Larson and C.D. Oliver (eds.): *The Ecology and Silviculture of Mixed-Species Forests.* A Festschrift for David M. Smith. 1992 ISBN 0-7923-1643-6
41. M.R. Ahuja (ed.): *Micropropagation of Woody Plants.* 1992 ISBN 0-7923-1807-2
42. W.T. Adams, S.H. Strauss, D.L. Copes and A.R. Griffin (eds.): *Population Genetics of Forest Trees.* Proceedings of an International Symposium (Corvallis, Oregon, USA, 1990). 1992 ISBN 0-7923-1857-9
43. R.T. Prinsley (ed.): *The Role of Trees in Sustainable Agriculture.* 1993 ISBN 0-7923-2030-1
44. S.M. Jain, P.K. Gupta and R.J. Newton (eds.): *Somatic Embryogenesis in Woody Plants,* Vol. 3: Gymnosperms. 1995 ISBN 0-7923-2938-4
45. S.M. Jain, P.K. Gupta and R.J. Newton (eds.): *Somatic Embryogenesis in Woody Plants,* Vol. 1: History, Molecular and Biochemical Aspects, and Applications. 1995 ISBN 0-7923-3035-8
46. S.M. Jain, P.K Gupta and R.J. Newton (eds.): *Somatic Embryogenesis in Woody Plants,* Vol. 2: Angiosperms. 1995 ISBN 0-7923-3070-6
    Set ISBN (Volumes 44-46) 0-7923-2939-2
47. F.L. Sinclair (ed.): *Agroforestry: Science, Policy and Practice.* Selected Papers from the Agroforestry Sessions of the IUFRO 20th World Congress (Tampere, Finland, 6–12 August 1995). 1995 ISBN 0-7923-3696-8
48. J.H. Goldammer and V.V. Furyaev (eds.): *Fire in Ecosystems of Boreal Eurasia.* 1996 ISBN 0-7923-4137-6

KLUWER ACADEMIC PUBLISHERS – DORDRECHT / BOSTON / LONDON

FORESTRY SCIENCES